AF767708

Modern Spectrum Analysis of TIME SERIES

Modern Spectrum Analysis of TIME SERIES

Prabhakar S. Naidu

CRC Press
Boca Raton New York London Tokyo

Library of Congress Cataloging-in-Publication Data

Naidu, Prabhakar S.
Modern spectrum analysis of time series / Prabhakar S. Naidu.
p. cm.
Includes bibliographical references (p. -) and index.
ISBN 0-8493-2464-5
1. Time-series analysis. 2. Spectral theory (Mathematics)
I. Title.
QA280.N34 1995
519.5'5—dc20

95-32335
CIP

This book contains information obtained from authentic and highly regarded sources. Reprinted material is quoted with permission, and sources are indicated. A wide variety of references are listed. Reasonable efforts have been made to publish reliable data and information, but the author and the publisher cannot assume responsibility for the validity of all materials or for the consequences of their use.

Neither this book nor any part may be reproduced or transmitted in any form or by any means, electronic or mechanical, including photocopying, microfilming, and recording, or by any information storage or retrieval system, without prior permission in writing from the publisher.

CRC Press, Inc.'s consent does not extend to copying for general distribution, for promotion, for creating new works, or for resale. Specific permission must be obtained in writing from CRC Press for such copying.

Direct all inquiries to CRC Press, Inc., 2000 Corporate Blvd., N.W., Boca Raton, Florida 33431.

© 1996 by CRC Press, Inc.

No claim to original U.S. Government works
International Standard Book Number 0-8493-2464-5
Library of Congress Card Number 95-32335
Printed in the United States of America 1 2 3 4 5 6 7 8 9 0
Printed on acid-free paper

Prologue

The power spectrum (or simply spectrum) analysis of time series has a long and varied history. It was A. Schuster[1] who first applied in 1898 the periodogram analysis to geomagnetic observations. The evolution of spectrum analysis techniques from the days of Schuster to modern times will be outlined later, in Chapter One. Spectrum analysis can be considered as a topic in statistics as well as a topic in digital signal processing (DSP). As a topic in statistics it may appear as a collection of theorems and lemmas (typical work of this type is the book by Brillinger[2]). On the other hand, spectrum analysis as a topic in DSP tends to be technique oriented with emphasis on estimation (typical work of this type is the recent book by Kay[3]). The present book takes a middle course. It emphasizes the time series models and their impact on the spectrum analysis. In particular, it is argued that by increasing the sophistication of the model it is possible to achieve an improved spectrum. The modern developments in the domain of spectrum analysis are explained as a natural consequence of the assumed models of the time series. We may broadly divide the time series models into two groups, namely, rational filter model and sum of sinusoids in the presence of white noise. The rational filter model appears to be the most appropriate in some physical problems, for example, speech waveform, seismic wave propagation in layered earth, random vibrations, biomedical signals, etc. On the other hand, the signals used in communication are generally sinusoids of discrete frequencies, and also in array signal processing a plane wavefront incident on a linear array leads to a spatial sinusoidal signal. The spectrum of the time series from the first group is generally smooth, possibly with broad peaks; in contrast, the spectrum of the time series from the second group consists of sharp peaks rising above the smooth spectrum of the background. The subject of spectrum analysis of *stationary* time series is well developed with many practical applications. However, there are many real-life situations where the assumption of stationarity is questionable, at least over a long duration of time. There exists a group of methods, often known as time frequency distributions (TFD), which are useful for analyzing the slowly varying stochastic processes. Wigner-Ville (WV) distribution, short time Fourier transform (STFT), and wavelet transform are some examples from this group. There is also a special class of nonstationary processes known as cyclostationary processes having potential applications in the modeling of modern communication signals for which the spectrum analysis approach has been used.

Though the spectrum is adequate to characterize a Gaussian process, it is not appropriate when the process is non-Gaussian. While it is true that most real-life time series tend to be Gaussian or nearly Gaussian, there are important situations where the time series is essentially non-Gaussian; for example, an output of a nonlinear system is non-Gaussian. Also, it is found that a linear system driven by non-Gaussian input is capable of being completely identified from the higher order spectrum of its output even if it is not minimum phase. A study of higher order spectra, in particular bispectrum and trispectrum, forms the unique feature of the present work.

A time series has traditionally been modeled as a stochastic process, that is, an output of a linear system driven by a pure white noise process. An alternate viewpoint has emerged in recent times. A time series is considered as an output of a nonlinear dynamic system. It has an appearance of a stochastic process, yet it is an output of a completely deterministic system. Deterministic chaos as a model for time series is gaining acceptance. Correlation dimension, fractal dimension, and Lyapunov spectrum are often used to characterize a chaotic time series. This important current viewpoint is briefly discussed in the second chapter.

The book is divided into five chapters. The first chapter begins with elements of probability theory and goes on to introduce the theory of stationary stochastic processes. The depth of coverage is moderately extensive but not too intensive. Many topics of concern to spectral characterization of Gaussian and non-Gaussian time series, scalar and vector time series, are covered. A section is devoted to the emerging areas of nonstationary and cyclostationary time series. Chapter Two is devoted to time series models. Starting from a general filter model of the time series we go on to the more specific time series models, e.g., MA, AR, ARMA, and sum of random sinusoids. The emphasis is more on showing how these models specify the spectrum as well as the bispectrum of a time series. Finally, a brief description of deterministic chaos as an alternate model of the time series is included. The first two chapters, it is hoped, will provide ample theoretical background for a good understanding of the current developments in spectrum analysis.

The next two chapters are devoted to spectrum estimation methods which may be broadly divided into two groups, namely, the low-resolution and the high-resolution methods. In the first group we have the classical methods like Blackman-Tukey, Welch's direct method, Thomson's multiple window method, etc. All these methods share a common property that their resolution is limited, that is, inversely proportional to the length of data. But, they are linear in the sense that the spectrum of the sum of two uncorrelated time series is equal to the sum of the spectra. We also look into the estimation of a time varying spectrum within the general framework of low-resolution methods. In the second group of methods a knowledge of the time series model is assumed. Such prior knowledge is responsible for high spectral resolution. These methods are nonlinear in the sense that the spectrum of the sum of two time series is not equal to the sum of the spectra. Not only that, the high-resolution methods are more difficult to characterize; in particular, the error analysis gets more involved. The recent developments in spectral analysis, in particular, under the high-resolution category have been phenomenal. To keep the scope of the book well demarcated and at the same time to keep the coverage up-to-date, an attempt has been made to select some of the more promising new concepts and illustrate them through simple examples. These include data-adaptive windows, maximum entropy and minimum cross entropy, rational spectrum, subspace approach to frequency estimation and extrapolation of band limited time series. Finally, the last chapter (Chapter Five) looks at some of the practical issues in spectrum estimation. An attempt is made to show how one can handle such practical pitfalls as missing data, uneven sampling, presence of outliers, etc. The artifacts introduced by a spectral window (which are covered in Chapter Three) are absent if the time series is white. Hence, whitening of the time series is one possible way to minimize the effects of spectral window. The central theme is to make the spectrum estimation data adaptive.

The present work is organized more as a text book rather than a reference book. In each chapter we have included many examples to illustrate the concepts described therein. In addition, at the end of each chapter several exercises are included. The level of presentation is appropriate for graduate and research students. Indeed, the material included here is drawn from the lecture notes on modern spectrum analysis taught by the author to the graduate and research students for over a decade. About two thirds of the material could be covered in one semester of four months duration (two hours per week). The first two chapters covering the basics of the stochastic process and models of time series were compulsorily covered every time. The choice was between Chapter Three and Chapter Four, depending upon the students' interests. In practice it turned out that these two chapters were covered on alternate years. Selected parts of Chapter Five were covered within the remaining time, depending upon which of the chapters, three or four, was covered previously. The students taking this course had previously taken or were concurrently taking a course in probability theory, stochastic process, linear algebra, and a first course in digital signal processing. At the end of each chapter there is a list of problems, the purpose of which is twofold. The first purpose is to reinforce some of the subtle concepts through simple numerical problems. The second purpose is to introduce some new topics as extensions to the already covered main line of thought through problems, whose solution may be found in the reference cited in the problem. It is expected that the reader consults the reference while attempting the problem. In this way the reader is encouraged to look around beyond the boundaries of the book. The classroom lectures were supplemented by computer exercises followed by a miniproject.

April 1995

Prabhakar S. Naidu
Indian Institute of Science
Bangalore 560012
India.

References

1. A. Schuster, On the investigation of hidden periodicities with application to a supposed 26 day period of meteorological phenomenon, *Terra Magnet*, 3, 13-41, 1898.

2. D. R. Brillinger, *Time Series: Data Analysis and Theory*, Holt, Rinehart and Winston, New York, 1975.

3. S. M. Kay, *Modern Spectrum Analysis*, Prentice-Hall, Englewood Cliffs, 1989.

Acknowledgment

The thought of formalizing the lecture notes into a text crossed my mind in one of the quiet moments I was fortunate to enjoy at the Naval Postgraduate School, Monterey, CA, where I was spending my sabbatical leave in 1988-89 as an NRC Research Associate. Much of the ground work was done during this period. I am grateful to NRC as well as NPS, in particular Prof. J. P. Powers, who was my host. Many of my students, who credited the course on Modern Spectrum Analysis, have contributed by way of working out the examples cited in the text. I am particularly grateful to the following: T. Ganesan, B. Madhukar, Anil Prasad, L. Anand, A. Gupta, A. S. Bopadikar, K. Maheswara Reddy, M. Ramakumar, G. Viswanath, R. Subramaniyan, H. S. Udayshankar, B. V. Ramakrishnan, A. Vasuki, A. Bhuvaneswari, C. Usha Padmini, George Matthew, and B. Prabhakar Rao. Dr. K. V. S. Hari read the manuscript at an early stage and made many constructive suggestions. I wish to thank the CRC Press Inc., in particular Ms. Nora Konopka and Ms. Debbie Didier for their promptness and patience.

Finally I owe deep gratitude to my family — my wife, Madhumati, and my sons, Srikanth, Sridhar, and Srinath for their forebearance. I must specially thank my son, Srinath, who carefully scrutinized the manuscript.

Modern Spectrum Analysis of Time Series

Contents

Chapter Four

Spectrum Estimation: *High-Resolution Methods* 241

Chapter Five

Spectrum Estimation: *Data Adaptive Approach.* 337

Dedication

This work is dedicated to the memory of the great visionary, J. R. D Tata, who shaped the Indian Institute of Science for many decades.

Chapter One

Stochastic Characterization of Time Series

An observed signal, presumed to be sampled or intrinsically discrete (e.g., rainfall time series), is treated as a stochastic process which, if Gaussian, may be characterized within the framework of second order statistics, namely, covariance function or spectrum density function. Higher order spectra such as bispectrum, trispectrum, etc., are required to bring out the non-Gaussian character of a stochastic time series. Stationarity is often assumed mostly for the ease of analysis but rarely encountered in practice, at least not a perfect stationarity. A slowly varying statistics is a useful model which is amenable to analysis. These topics form a backbone of the material covered in this chapter.

1.1 Time Series as a Stochastic Process

A time series is a set of sequential observations made at discrete and equal time intervals. The discrete samples are presumed to have been obtained by sampling a continuous process, for example, a temperature record of a reactor. Let x(t), t=0, 1, 2, ... T − 1 denote a time series of length T. The sampling interval, Δt, is often suppressed or set to unit time interval. The time index, t, may range over $\pm\infty$ (infinite prerecorded observations), or over 0 to ∞ (infinite past observations), or over 0 to T − 1 (finite past observations). A time series is often modeled as a sum of a systematic component and a random component,

$$x(t) = s(t) + \varepsilon(t),$$

where s(t) is the systematic component and $\varepsilon(t)$ the random component, which may be pure measurement error. The desired signal component is s(t), which may be represented by a low order polynomial. In such a situation the object of time series analysis is to estimate the coefficients of the polynomial. A least squares approach is often used to estimate the polynomial coefficients. An example of this kind of time series analysis is the fitting of a straight line to the observed data. When s(t) is a highly complex waveform requiring very high order polynomial representation, the approach of fitting a polynomial becomes impractical. One must look for an alternate method of modeling and analyzing a stochastic time series. One such alternative is the stochastic approach where we assume that s(t) is a stochastic function but different from $\varepsilon(t)$. Consequently, x(t) is modeled as a stochastic time series. Another approach is to treat a time series as deterministic chaos, a topic of growing interest. We shall only briefly deal with deterministic chaos in Section 2.7. A stochastic process has traditionally been modeled in two different ways.

A. Filter Model

A stochastic time series is considered as an output of a filter driven by white noise (see Figure 1.1). The transfer function of the filter plays a crucial role in modeling a stochastic time series. In this context, it is convenient to use the spectral representation of a stationary stochastic process,

$$x(t) = \frac{1}{2\pi}\int_{-\pi}^{\pi} dX_G(\omega)\exp(j\omega t). \tag{1.1}$$

It is analogous to the Fourier integral representation of a finite energy function. We shall show later that $X_G(\omega)$ is the generalized Fourier transform[1] of a stochastic process and $dX_G(\omega)$ is the differential of $X_G(\omega)$.

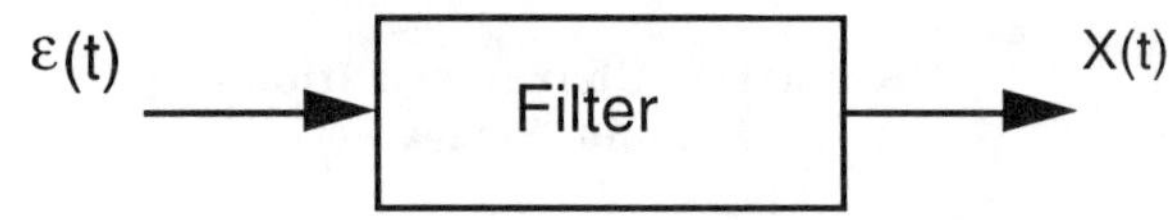

Figure 1.1: Filter model of stochastic time series.

B. Parametric Model

In the second approach we invoke the fundamental decomposition theorem of Wold,[2] which states that a stochastic process can be represented as the sum of a deterministic component and a random component which may be expressed as a weighted average of the past and present random inputs,

$$x(t) = s(t) + \sum_{i=0}^{\infty} b_i \varepsilon(t-i), \tag{1.2}$$

where b_i, $i = 0, 1, 2, \ldots, \infty$ are weighting coefficients and $\varepsilon(t)$ now represent past inputs. When $s(t)=0$ we have

$$x(t) = \sum_{i=0}^{\infty} b_i \varepsilon(t-i), \tag{1.3}$$

Here we have expressed x(t) as a moving average (MA) of current and past random inputs. It is also possible to rewrite (1.3) in a recursive form as

$$x(t) = \sum_{i=1}^{\infty} a_i x(t-i) + \varepsilon(t), \tag{1.4}$$

where a_i, $i = 1, 2, \ldots, \infty$ are a new set of coefficients which can be obtained from the b_i s. The representation shown in (1.4) is known as the autoregressive (AR) model. In Figure 1.2 we have a schematic representation of moving average (MA) and autoregressive (AR) models. In general a time series is best modeled as a combination of MA and AR models, known as the ARMA model. Note that the entire structure of a stochastic time series is contained in a set of parameters $a_1, a_2, a_3 \ldots$ and/or $b_0, b_1 b_2 \ldots$. Such models of time series are called parametric models.

The two approaches, namely, the filter model and parametric model of stochastic time series, are indeed like two faces of a coin. Traditionally, the engineers tended to use the filter model approach because of their interest in the extraction of signal in the presence of noise. On the other hand, the statisticians with interest in prediction and forecasting relied on parametric models. During the past two decades there has been considerable cross-fertilization between these two approaches. This has resulted in a new family of high-resolution spectrum estimation methods.

C. Historical Remarks on Spectrum Analysis

The spectrum of a time series essentially tells us how the energy is distributed as a function of temporal frequency. Often, a time series may be modeled as a sum of sinusoids, to which a certain amount of white noise has been added. From such a time series we naturally

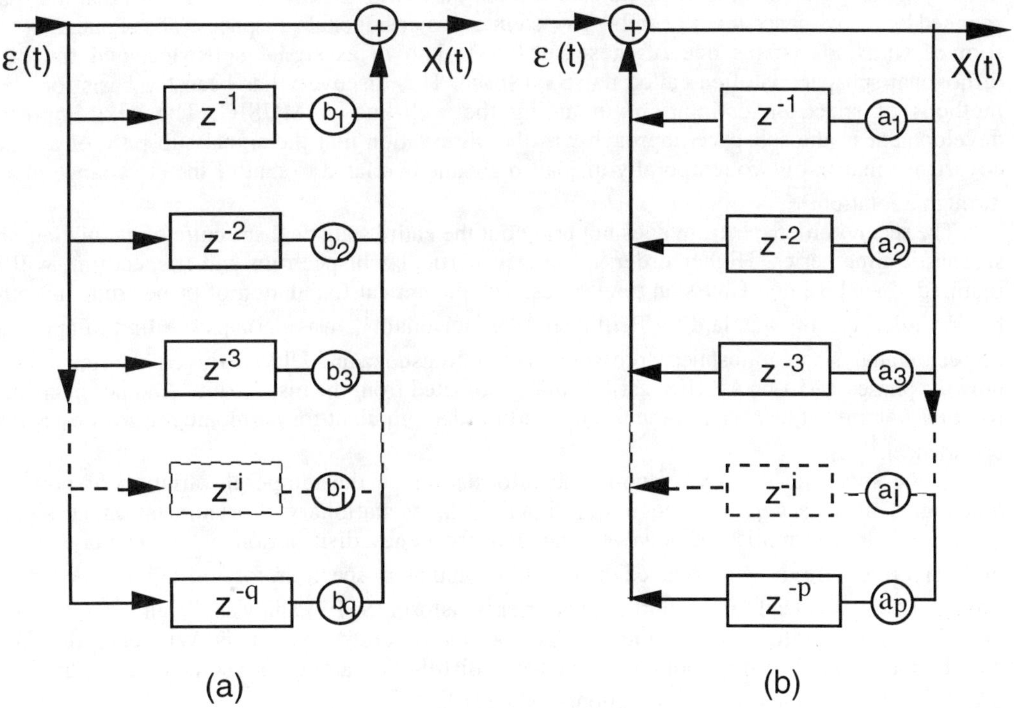

Figure 1.2: A schematic representation of MA and AR models of time series. (a) Moving average (MA) model and (b) Autoregressive (AR) model.

seek to estimate the frequency and amplitude of each sinusoid. Indeed, this has been the goal of spectrum analysis from the days of Schuster[3] who wanted to estimate the periodicity in the geomagnetic field. He simply computed the periodogram, that is, the square of the Fourier coefficients. It was soon found that the periodogram approach was inappropriate when the time series was random in nature. Yule[4] was the first to suggest the use of the AR model for such a situation. He determined the AR coefficients and then the frequencies of sinusoids. Norbert Wiener[1] introduced the concept of generalized harmonic analysis and gave a precise definition and relationship between the covariance function and the spectrum. That these two functions constitute a Fourier transform pair is known as the Wiener-Khintchin theorem. Following this theorem, it became possible to compute the spectrum as a Fourier transform of the autocorrelation function computed from the observed data. Tukey[5] showed how to correctly estimate the spectrum from the empirical autocorrelation function and obtained the statistical distribution of the estimated spectrum. The next major development occurred in 1965 with the discovery of fast Fourier transform (FFT) by Cooley and Tukey.[6] This enabled fast computation of the periodogram and thereby the direct estimation of the spectrum (Welch[7]). Around this time two important ideas were introduced in the field of spectrum estimation. These were the data adaptive window in the maximum likelihood method (MLM) proposed by Capon[8] and extrapolation of autocorrelation function under the constraint of maximum entropy by Burg.[9] Another significant development took place a few years later, in 1972, when Pisarenko[10] found that the frequencies of the sinusoids in the background of white noise may be obtained from the poles of the z-transform of the minimum eigenvector of the covariance matrix. In the recent past, intense research activity centered on a model of sinusoids in the presence of white noise background, which, because of its relevance in array signal processing, is being pursued around the world. Schmidt[11] has shown that the space spanned by a covariance matrix can be divided into two orthogonal subspaces. The signal, that is, a sum of sinusoids, spans one of these subspaces known as signal subspace, and the other orthogonal subspace is often called noise subspace. This discovery has been the basis for many methods of spectrum estimation, including the well-known MUSIC. The next important development in the subspace approach was the observation that the signal subspace of a cross-covariance matrix of two temporally displaced signals is related to that of the covariance matrix through a rotation.[12]

The second order spectrum does not bring out the entire statistical structure of a non-Gaussian stochastic time series. Higher order spectra, in particular bispectrum and trispectrum, will be required to analyze non-Gaussian time series. The theoretical foundation of bispectrum and other higher order spectra was laid by Brillinger.[13] MacDonald[14] was perhaps the first to apply the bispectral analysis to atmospheric pressure records. Matsuoka and Ulrych[15] were the first to show how the phase of MA or AR time series can be estimated from the bispectrum. Another interesting recent development has been in the use of fourth-order cumulant for harmonic retrieval by Swami and Mendel.[16]

In the spectrum density function the information on the temporal variation of power is missing. This is acceptable when the signal is truly stationary. For nonstationary signals (stochastic or deterministic), however, the time-frequency distribution of power plays a very important role. Priestly[17] introduced the idea of evolutionary spectrum for nonstationary stochastic signal, and this has lead to the short time Fourier transform (STFT) analysis. Wigner[18] introduced the idea of quadratic time-frequency distribution function known as Wigner-Ville (WV) distribution. In modern days both STFT and WV distribution along with some other methods are widely used for the analysis of nonstationary signals.[19]

1.2 A Review of Stochastic Process

A. Basic Definitions

Let us visualize a random variable as an outcome of repeated experiments carried out to measure a quantity, say, temperature with an unreliable thermometer which gives a different reading each time an experiment is carried out. Let $\theta_1, \theta_2, \theta_3, \ldots, \theta_r$ be the outcomes of r experiments. We denote the quantity being measured as θ, which we treat as a random variable. The observations may be treated as different realizations of a random variable θ taken from a sample space which may be continuous or discrete. In the case of a continuous sample space the random variable may take any value lying in a certain range, for example, +1 to -1; hence, it is a continuous random variable. In the case of a discrete random space the random variable can take one of a set of discrete values. We shall confine ourselves to the continuous random variable. Formally, the different realizations of a random variable are represented as x_λ, $\lambda = 1, 2, 3, \ldots$.

When a random variable is a function of some independent parameter, say time, we call that function a random function of time or simply a stochastic process. When t is discrete, for example, t=1, 2, 3, ..., we have then a random sequence or a stochastic time series. As an elementary example of a stochastic process consider a sinusoidal function with random phase,

$$x_\phi(t) = \sin(\omega t + \phi),$$

where ϕ is the random phase variable. For each realization of the random phase we obtain one realization of the stochastic process. The random variable ϕ lies in the range $\pm\pi$ and is uniformly distributed. A few realizations of the above stochastic process are sketched in Figure 1.3.

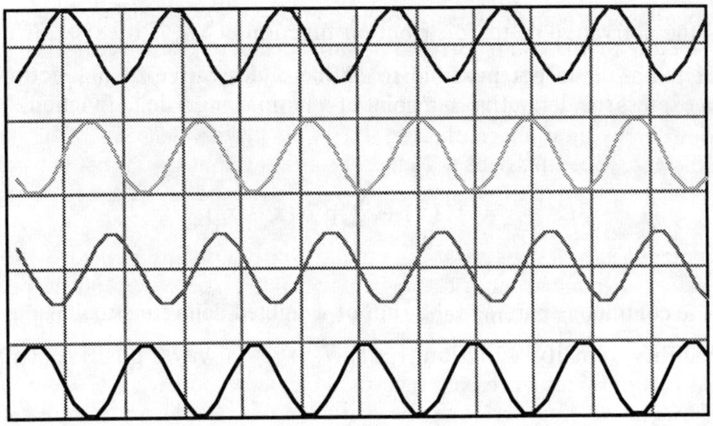

Figure 1.3: Random sinusoids. The frequency and amplitude are the same but the phase is a uniformly distributed random variable.

B. Probability Distribution

A random variable x is characterized by a distribution function which is defined as follows:

$$F(x_0) = \text{Prob}\{ x \le x_0 \}, \tag{1.5}$$

where x_0 is a real number. The distribution function is a positive and nondecreasing function. From physical considerations we must have

$$F(x_0 \to \infty) = 1$$
$$F(x_0 \to -\infty) = 0.$$

From (1.5) we can obtain the probability that x lies between two arbitrary limits a and b as

$$\text{Prob}\{ a < x \le b \} = F(b) - F(a).$$

Let $a = x_0 - \Delta x/2$ and $b = x_0 + \Delta x/2$ with $\Delta x \to 0$

$$\text{Prob}\{ x_0 - \Delta x/2 < x \le x_0 + \Delta x/2 \} = F(x_0 + \frac{\Delta x}{2}) - F(x_0 - \frac{\Delta x}{2})$$
$$= \frac{F(x_0 + \frac{\Delta x}{2}) - F(x_0 - \frac{\Delta x}{2})}{\Delta x} \Delta x$$
$$\Delta x \to 0$$
$$= F'(x_0)\Delta x,$$
$$\Delta x \to 0$$

where $F'(x_0)$ is the derivative of the distribution function at x_0, if it exists. If F(x) contains a finite number of jumps, it is not possible to define a derivative of a function with jumps. However, we can express the derivative at a point of a jump using a delta function,

$$\begin{aligned} F'(x) &= f(x) \\ &= f_0(x) + \sum_i p_i \delta(x - x_i), \end{aligned} \tag{1.6}$$

where $f_0(x)$ is the continuous part, and the sum of weighted delta functions is the discontinuous part. The probability density function (pdf) f(x) is always positive ($f_0(x) \geq 0$ and $p_i \geq 0$, i= 1, 2,...).

In practical applications the Gaussian or normal pdf plays an important role. The Gaussian pdf is given by

$$f(x) = \frac{1}{\sqrt{2\pi Var_x}} \exp\left(-\frac{(x - m_x)^2}{2\, Var_x}\right),$$

where m_x is the mean of the random variable; that is, $m_x = E\{x\}$ and var_x are the variance of the random variable, $Var_x = E\{(x - m_x)^2\}$. Note that $E\{.\}$ stands for expectation or ensemble averaging. A zero mean (m_x =0) and unit variance (Var_x =1.0) Gaussian random variable has the following pdf:

$$f(x) = \frac{1}{\sqrt{2\pi}} \exp\left(-\frac{(x)^2}{2}\right).$$

We have emphasized Gaussian random variables; this, however, does not imply that all random variables that we encounter in practice are Gaussian. In many cases it is possible to approximate an observed random variable as a Gaussian random variable which follows from an important theorem known as the central limit theorem, stated here without proof. Let $x_1, x_2, x_3, \ldots, x_p$ be independent and identically distributed random variables with zero mean and variance σ^2. Let us define a variable $\hat{x} = \frac{1}{p}\sum_{i=1}^{p} x_i$. The central limit theorem states that $\hat{x}$ will be a zero mean Gaussian random variable having a variance of $\frac{\sigma^2}{p}$ for large p. For finite p, $\hat{x}$ is approximately Gaussian; however, a good approximation can be obtained even for small p ($p \approx 50$) if $x_1, x_2, x_3, \ldots, x_p$ are close to being Gaussian.

An arbitrary probability density function can be conveniently expressed in terms of a Gaussian pdf, moments, and Hermite's polynomial.[20] Let x be a normalized zero mean random variable ($m_x = 0$ and Var_x =1.0). We can express an arbitrary pdf as follows:

$$f(x) = \frac{1}{\sqrt{2\pi}} \exp\left(-\frac{x^2}{2}\right) \left[\begin{aligned} &1 + \tfrac{1}{3!}\mu_3 H_3(x) + \tfrac{1}{4!}(\mu_4 - 3) H_4(x) \\ &+ \tfrac{1}{5!}(\mu_5 - 10\mu_3) H_5(x) + \ldots \end{aligned} \right], \tag{1.7}$$

where $\mu_3, \mu_4, \mu_5, \mu_6, \ldots$ etc. are moments of the normalized random variable,

$$\mu_i = E\{x^i\}, \quad i = 3, 4, \ldots$$

and $H_3(x)$, $H_4(x)$, ... are Hermite polynomials defined as follows[21]:

$$\begin{aligned}
H_0(x) &= 1 \\
H_1(x) &= x \\
H_2(x) &= x^2 - 1 \\
H_3(x) &= x^3 - 3x \\
H_4(x) &= x^4 - 6x^2 + 3 \\
H_5(x) &= x^5 - 10x^3 + 15x \\
H_6(x) &= x^6 - 15x^4 + 45x^2 - 15
\end{aligned}$$

in the range $-\infty < x < +\infty$. In this representation of an arbitrary pdf it is possible to ascertain the non-Gaussian part from the moments of the random variable without having to estimate the actual pdf.

C. Pair of Random Variables

Let x_1 and x_2 be two random variables. It is of interest to characterize the dependence of one on the other. For this we now introduce a joint probability density function, $f(x_1,x_2)$, and conditional pdf, $f(x_1|x_2)$, defined as

$$f(x_1,x_2) = \frac{\text{Prob}\{x_1 - \frac{\Delta x_1}{2} < x_1 \le x_1 + \frac{\Delta x_1}{2}; x_2 - \frac{\Delta x_2}{2} < x_2 \le x_2 + \frac{\Delta x_2}{2}\}}{\Delta x_1 \Delta x_2}$$

$$\Delta x_1 \to 0$$
$$\Delta x_2 \to 0.$$

The probability that a pair of random variables will lie inside a small rectangle can be written as a product of the joint probability density function and the area of the rectangle (see Figure 1.4). The conditional pdf is defined as

$$f(x_1|x_2) = \frac{f(x_1,x_2)}{f(x_2)}$$

Two random variables are said to be independent when

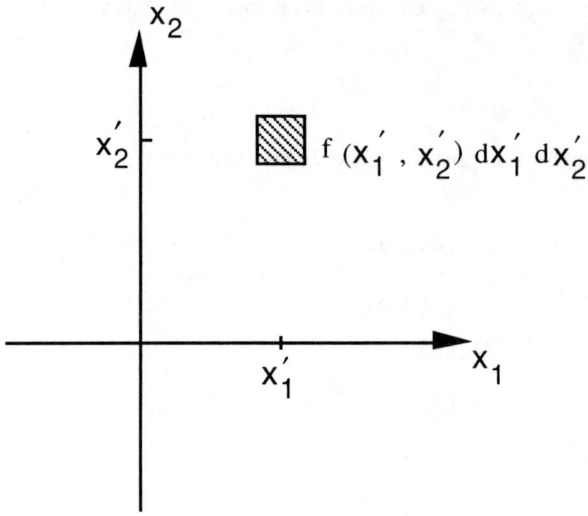

Figure 1.4: Probability that x_1 and x_2 will be inside the hatched box is equal $f(x'_1, x'_2)\,dx'_1\,dx'_2$.

$$f(x_1|x_2) = \frac{f(x_1, x_2)}{f(x_2)} = f(x_1) \quad (1.8)$$

The pdf of two jointly Gaussian random variables is given by

$$f(x_1, x_2) = \alpha \exp[-A(x_1 - m_{x_1})^2 - B(x_1 - m_{x_1})(x_2 - m_{x_2}) - C(x_2 - m_{x_2})^2], \quad (1.9)$$

where

$$\alpha = \frac{1}{2\pi\sqrt{\mathrm{Var}_{x_1}\mathrm{Var}_{x_2} - \mathrm{Var}^2_{x_1x_2}}}$$

$$A = \frac{\mathrm{Var}_{x_2}}{2(\mathrm{Var}_{x_1}\mathrm{Var}_{x_2} - \mathrm{Var}^2_{x_1x_2})}$$

$$B = \frac{\mathrm{Var}_{x_1x_2}}{2(\mathrm{Var}_{x_1}\mathrm{Var}_{x_2} - \mathrm{Var}^2_{x_1x_2})}$$

$$C = \frac{\mathrm{Var}_{x_1}}{2(\mathrm{Var}_{x_1}\mathrm{Var}_{x_2} - \mathrm{Var}^2_{x_1x_2})}.$$

$\mathrm{Var}_{x_1x_2}$, cross-covariance between x_1 and x_2, is defined as

$$\mathrm{Var}_{x_1x_2} = E\{(x_1 - m_{x_1})(x_2 - m_{x_2})\}.$$

Two random variables are said to be uncorrelated when $\mathrm{Var}_{x_1x_2} = 0$. Note that two uncorrelated Gaussian random variables are independent. This may be easily checked by setting B=0 ($\mathrm{Var}_{x_1x_2} = 0$) in (1.9). We obtain $f(x_1, x_2) = f_1(x_1) f_2(x_2)$, a condition for independence (1.8). Uncorrelatedness implies independence only in the case of Gaussian random variables. The converse, namely, independence implies uncorrelatedness, is, however, true for any pair of random variables.

D. Vector Random Variable

A collection of random variables is termed a vector random variable, $\mathbf{x} = \{x_1, x_2\, x_3, \ldots, x_p\}$. To characterize a vector random variable we have to invoke the concept of a joint pdf denoted by $f(x_1, x_2, \ldots, x_p)$. This is defined in the same manner as we have defined a joint pdf for a pair of random variables in the last section. A Gaussian vector random variable (zero mean) has a pdf given by

$$f(x_1, x_2, \ldots x_p) = \frac{1}{\sqrt{2^p \pi^p |\mathbf{Var}_x|}} \exp\left[- \frac{1}{2|\mathbf{Var}_x|} \sum_{\nu=1}^{p} \sum_{\mu=1}^{p} \mathrm{Var}^*_{\nu\mu} x_\nu x_\mu \right], \tag{1.10}$$

where $\mathbf{Var}_x$ is a matrix of variances and cross-variances of the components of the vector random variable,

$$\mathbf{Var}_x = \begin{bmatrix} \mathrm{Var}_{x_1} & \cdots & \mathrm{Var}_{x_px_1} \\ \vdots & & \vdots \\ \mathrm{Var}_{x_1x_p} & \cdots & \mathrm{Var}_{x_p} \end{bmatrix}.$$

$|\mathbf{Var}_x| = \det\{\mathbf{Var}_x\}$, and $\mathrm{Var}^*_{\nu\mu}$ is cofactor of $\mathrm{Var}_{x_\nu x_\mu}$. We can also express (1.10) in a compact matrix form as follows:

$$f(\mathbf{x}) = \frac{1}{\sqrt{(2\pi)^p |\mathbf{Var}_x|}} \exp\left(- \frac{1}{2} \mathbf{x}^T \mathbf{Var}_x^{-1} \mathbf{x} \right). \tag{1.11}$$

We may note that a knowledge of the variance-cross-variance matrix of a Gaussian vector random variable is adequate to completely characterize a vector random variable. This result would be useful in characterizing a stochastic function (next section).

E. Characteristic Function

The characteristic function of a random variable is defined as

$$\Phi_x(u) = E\{\exp(-jux)\} = \int_{-\infty}^{\infty} f(x)\exp(-jux)\,dx \tag{1.12}$$

that is, the Fourier transform of pdf. For example, the characteristic function of a normalized Gaussian random variable is given by $\Phi_x(u) = \exp(-\frac{1}{2}u^2)$. It is easy to show that the moments of a random variable are related to the derivatives of the characteristic function. For example, the mth moment can be related to the mth derivative of $\Phi_x(u)$ at u=0.

$$\frac{d^m\Phi_x(u)}{du^m} = (-j)^m \int_{-\infty}^{\infty} x^m f(x)\exp(-jux)\,dx \; .$$

At u=0 we obtain

$$\frac{d^m\Phi_x(u)}{du^m} = (-j)^m \int_{-\infty}^{\infty} x^m f(x)\,dx = (-j)^m\{m^{th} \text{ moment }\} . \tag{1.13a}$$

The cumulants of a random variable are defined in terms the derivatives of the logarithm of the characteristic function,

$$\left.\frac{d^m \ln \Phi_X(u)}{du^m}\right|_{u=0} = (-j)^m\{\ m^{th} \text{ cumulant } \}. \tag{1.13b}$$

Example 1.1: Using the method of characteristic function we shall evaluate the mean and variance of a Gaussian random variable, whose probability density function and characteristic function are given by

$$\left.\frac{d^m \ln \Phi_X(u)}{du^m}\right|_{u=0} = (-j)^m\{\ m^{th} \text{ cumulant } \}.$$

and

$$\Phi_x(u) = \exp(-jm_x u)\exp(-\frac{var_x}{2}u^2),$$

respectively. The first and second derivatives of the characteristic function at u=0 are easily shown as

$$\left.\frac{d\Phi_x(u)}{du}\right|_{u=0} = m_x$$

and

$$\left.\frac{d^2\Phi_x(u)}{du^2}\right|_{u=0} = m_x^2 + var_x \; .$$

Cumulants of a Gaussian random variable may be computed from (1.13b).

$$\mathrm{cum}_1 = \frac{1}{-j}\frac{d\ln\Phi_X(u)}{du}\bigg|_{u=0} = m_X$$

$$\mathrm{cum}_2 = -\frac{d^2\ln\Phi_X(u)}{d^2u}\bigg|_{u=0} = \mathrm{var}_X .$$

All higher order cumulants are equal to zero. However, all higher order moments are nonzero (End of Example).

Now, for a vector random variable the characteristic function is defined in an analogous fashion.

$$\Phi_X(\mathbf{u}) = \int_{-\infty}^{\infty} f(\mathbf{x})\exp\left(-\frac{1}{2}\mathbf{u}^T\mathbf{x}\right)d\mathbf{x} \quad , \tag{1.14}$$

where $\mathbf{u} = \mathrm{col}\{u_1, u_2, \ldots, u_p\}$. For a Gaussian vector random variable the characteristic function is given by

$$\Phi_X(\mathbf{u}) = \exp\left(-\frac{1}{2}\mathbf{u}^T\mathbf{Var}_x\,\mathbf{u}\right). \tag{1.15}$$

We shall now illustrate how to use the characteristic function of a vector random variable to obtain the first moment of p Gaussian random variables. It is easy to show from (1.14) that

$$E\{X_1, X_2, \ldots, X_p\} = (j)^p \frac{\partial^p}{\partial u_1 \partial u_2, \ldots, \partial u_p}\Phi_X(u_1, u_2, \ldots, u_p)\Big|_{u_1=u_2=\ldots=u_p=0.}$$

To permit differentiation it is convenient to express (1.15) in a series form,[22]

$$\Phi_X(\mathbf{u}) = \sum_{s=0}^{\infty}\frac{\left(-\frac{1}{2}\right)^s}{s!}\sum_{p_1}\sum_{q_1}\ldots\sum_{p_s}\sum_{q_s=1}^{p}\mathrm{Var}_{p_1q_1}\mathrm{Var}_{p_2q_2}\ldots\mathrm{Var}_{p_sq_s}\,u_{p_1}u_{q_1}u_{p_2}u_{q_2}\ldots u_{p_s}u_{q_s}. \tag{1.16}$$

Now we differentiate (1.16) with respect to $u_1, u_2, \ldots, u_p$ and evaluate the derivative at $u_1 = u_2 = \ldots = u_p = 0$. We will find that in the process all terms except the one for s = p/2 will vanish. If p is odd all terms will vanish. We have finally

$$E\{x_1, x_2 \ldots x_p\}$$

$$= (-j)^p\frac{\left(-\frac{1}{2}\right)^s}{s!}\sum_{p_1}\sum_{q_1}^{p}\ldots\sum_{p_s}\sum_{q_s}\mathrm{Var}_{p_1q_1}\mathrm{Var}_{p_2q_2}\ldots\mathrm{Var}_{p_sq_s}\Big|_{S=\frac{P}{2}} ,$$

where the summation is extended over all possible different permutations of p indices, $p_1q_1, p_2q_2, \ldots, p_sq_s$, taken s pairs at a time without duplication. For example, when p=4, that is, s=2, then

$$E\{x_1 x_2 x_3 x_4\} = \frac{1}{8}\sum_{p_1}^{4}\sum_{q_1}\sum_{p_2}^{4}\sum_{q_2} Var_{p_1q_1} Var_{p_2q_2}. \tag{1.17}$$

There will be 256 terms in the summation, of which only 24 terms are without duplication (terms of the type $Var_{12}Var_{23}$, $Var_{12}Var_{14}$, etc. will vanish). Collecting all nonvanishing terms we obtain

$$E\{x_1 x_2 x_3 x_4\} = [Var_{12}Var_{34} + Var_{13}Var_{24} + Var_{14}Var_{23}]. \tag{1.18}$$

1.3 **Stationary Stochastic Process:** Second Order

A random function may be considered as a vector random variable whose components are obtained by sampling the random function at a large number of time instants, ideally infinite. Let a random function x(t) be sampled at time instants $t_1, t_2, t_3, \ldots, t_p$. The samples are $x_1, x_2, x_3, \ldots, x_p$ which go to make up a vector random variable. A joint pdf $f(x_1, x_2, x_3, \ldots; t_1, t_2, t_3, \ldots)$ may be defined to characterize these samples. Ideally, one must use an infinite number of samples, but this leads to extreme analytical complexities. One often considers just two (second order) or three (third order) samples requiring only second- or third-order pdf. We shall first consider the two-sample case and then go on to the three-sample case in the next section.

A. **Covariance Function**

Let x_1 and x_2 be two samples obtained by sampling a stochastic process at time instants, t_1 and t_2. When the joint probability density function depends upon the difference, $t_2 - t_1$, that is, it is of the form $f(x_1,x_2; t_2 - t_1)$, the stochastic process is said to be stationary with respect to the second-order pdf. For such a second-order stochastic process the mean and the variance are independent of time. The covariance function of a zero mean real valued stochastic process is defined as

$$\begin{aligned} C_x(t_2 - t_1) &= E\{x_1 x_2\} \\ &= \int\int_{-\infty}^{\infty} x_1 x_2 f(x_1, x_2; t_2 - t_1) dx_1 dx_2 \end{aligned} \tag{1.19}$$

It is a function of the time lag, $t=t_2 - t_1$. The covariance function plays a very important role in the time series analysis. Often, it is difficult to estimate from a single available time series the probability density function. However, the sample covariance function or the sample autocorrelation function, defined as the lagged product average,

$$\frac{1}{T-\tau}\sum_{i=1}^{T-\tau} x_i x_{i+\tau},$$

is easily computed. We shall now summarize some of the properties of the covariance function:

$$
\begin{aligned}
&\text{i)} \quad C_x(\tau) = C_x(-\tau),\\
&\text{ii)} \quad C_x(0) \ge |C_x(\tau)| \qquad \tau \ge 0,\\
&\text{iii)} \quad \sum\sum C_x(t_2 - t_1)\varphi(t_1)\varphi(t_2) \ge 0,\\
&\qquad \text{(positive semidefinite)}
\end{aligned}
\tag{1.20}
$$

for any function $\varphi(t)$. A covariance function may be written in a matrix form as follows:

$$
\mathbf{C}_x = \begin{bmatrix} C_x(0) & C_x(1) & \dots & C_x(p) \\ C_x(-1) & C_x(0) & \dots & C_x(p-1) \\ \vdots & & \ddots & \vdots \\ C_x(-p) & & \dots & C_x(0) \end{bmatrix}
\tag{1.21}
$$

The covariance matrix defined above is similar to the variance-cross-variance matrix appearing in (1.11), that is, in the joint pdf of a Gaussian random vector with p components, which are obtained by sampling a stationary stochastic process. Thus, a covariance matrix with $p \to \infty$ will completely characterize a Gaussian stationary stochastic process.

On account of the properties listed in (1.20) the covariance matrix will have a special structure: it is symmetric Toeplitz, that is, all elements on any diagonal are equal and it is positive semidefinite, that is, all its eigenvalues are real and nonnegative (including zero eigenvalue). The covariance matrix is of rank equal to the time-bandwidth product. It may be expressed as follows:

$$
\mathbf{C}_x \approx \sum_{p=-L}^{L} S_x(\Delta\omega p)\,\mathbf{a}(p)\mathbf{a}^H(p),
\tag{1.22a}
$$

where

$$
\mathbf{a}(p) = \frac{1}{\sqrt{T}}\operatorname{col}\{1, e^{j\Delta\omega p}, e^{j2\Delta\omega p}, \dots, e^{j(T-1)\Delta\omega p}\},
$$

$$
\Delta\omega = \frac{2\pi}{T},
$$

and $S_x(\omega)$ is the spectrum (see page 16) of the stochastic process. The rank of $\mathbf{C}_X$ will be equal to or less than 2L+1 provided $\mathbf{a}(p)$ $\forall\ p \in -L,\dots,L$ are independent. Let $2f_B$ be the bandwidth of the signal; then $(2L+1)=2f_BT$ (see Exercise 1.13). The approximate representation of the covariance given in (1.22a) becomes exact when the spectrum consists of equispaced lines,

$$
S_x(\omega) = \sum_{i=-L}^{L} S_p\delta(\omega - p\Delta\omega),
$$

where $\Delta\omega = \frac{2\pi}{2L+1}$. Equation (1.22a) becomes

$$\mathbf{C}_x = T \sum_{p=-L}^{L} S_p \mathbf{a}(p)\mathbf{a}^H(p). \tag{1.22b}$$

The covariance matrix in (1.22b) has a special structure, namely, each row is obtained by circularly shifting the previous row by one unit. The resulting covariance matrix is called a circulant matrix. A circulant matrix has some interesting properties, namely, its inverse is also a Toeplitz matrix, its eigenvalues occur in pairs, and the corresponding eigenvectors are real and imaginary parts of $\mathbf{a}(p)$.

Example 1.2: We give below an example of circulant covariance matrix. Let us consider a stochastic process which is a sum of two random sinusoids and unit variance background white noise,

$$x(t) = \sqrt{2}\cos\left(\frac{2\pi}{7}t + \varphi_1\right) + \sqrt{2}\cos\left(\frac{2\pi}{7}3t + \varphi_2\right) + \eta(t).$$

Since in this example $\Delta\omega = \frac{2\pi}{7}$, we shall compute a covariance matrix of size 7×7. The result is as shown below:

$$\mathbf{C}_x = \begin{bmatrix} 3.0 & -0.277 & 0.401 & -1.123 & -1.123 & 0.401 & -0.277 \\ -0.277 & 3.0 & -0.277 & 0.401 & -1.123 & -1.123 & 0.401 \\ 0.401 & -0.277 & 3.0 & -0.277 & 0.401 & -1.123 & -1.123 \\ -1.123 & 0.401 & -0.277 & 3.0 & -0.277 & 0.401 & -1.123 \\ -1.123 & -1.123 & 0.401 & -0.277 & 3.0 & -0.277 & 0.401 \\ 0.401 & -1.123 & -1.123 & 0.401 & -0.277 & 3.0 & -0.277 \\ -0.277 & 0.401 & -1.123 & -1.123 & 0.401 & -0.277 & 3.0 \end{bmatrix}$$

Note the meaning of circular shifting. Take any row and shift it to the right by one unit. The element which is pushed out reappears as the first element of the new row which is indeed the next row of the covariance matrix (End of Example).

A covariance function gives us an idea how rapidly the correlation between two adjacent samples decreases with separation. We define a correlation distance, τ_0, where the covariance function is one tenth of its maximum at zero lag. When the samples are separated by an interval greater than the correlation distance, it is safe to assume that they are uncorrelated. As the correlation distance decreases the stochastic process becomes increasingly uncorrelated. In the limiting case as $\tau_0 \to 0$ we have a completely uncorrelated stochastic process. The covariance function of the uncorrelated process is given by

$$C_x(\tau) = \mathrm{var}_x\,\delta(\tau). \tag{1.23a}$$

For discrete uncorrelated time series

$$C_x(\tau) = \mathrm{var}_x\,\delta_\tau, \tag{1.23b}$$

where δ_τ is the Kronecker delta function,

$$\delta_0 = 1$$
$$\delta_\tau = 0 \qquad |\tau| > 0.$$

The concept of correlation distance, however, cannot be applied to the sum of random sinusoids as its covariance function is nondecreasing.

B. Cross-Covariance Function

When we have a pair of real valued stochastic processes, we define a cross-covariance function,

$$\begin{aligned} C_{xy}(\tau) &= E\{x(t)y(t+\tau)\} \\ &= \int\int x_t y_{t+\tau} f(X_t, Y_{t+\tau}; \tau) dx_t dy_{t+\tau}. \end{aligned} \tag{1.24}$$

It has the following properties:

$$\begin{aligned} C_{xy}(\tau) &= C_{yx}(-\tau) \\ C^2_{xx}(\tau) &\leq C_x(0)C_y(0). \end{aligned} \tag{1.25}$$

Example 1.3: Consider stochastic processes

$$\begin{aligned} x(t) &= ax(t-1) + \eta(t) \\ y(t) &= x(t-\tau_0) + \varepsilon(t) \end{aligned} ,$$

where $a<1$ and $\eta(t)$ and $\varepsilon(t)$ are uncorrelated white noise processes. The covariance function of x(t) is given by

$$C_x(\tau) = \frac{\mathrm{Var}\ \eta}{1-a^2} a^{|\tau|} ,$$

and the cross-covariance functions x(t) and y(t) are given by

$$C_{xy}(\tau) = C_x(\tau - \tau_0)$$

and

$$C_{yx}(\tau) = C_x(\tau + \tau_0).$$

The covariance and cross-covariance functions for a=0.8 and $t_0=4$ are shown in Figure 1.5 (End of Example).

Two stochastic processes are said to be uncorrelated when $C_{xy}(t)=0$ for all t. The importance of the covariance function lies in the fact that when the stochastic process is Gaussian, the joint probability density function of any order involves only the covariance matrix. This was demonstrated in the previous section for a vector random variable. Thus, the covariance function contains all the information that is there in a Gaussian stationary stochastic process. In the case of a non-Gaussian process the covariance function provides only partial information. It does not fully characterize even the second-order probability density function. A stochastic process that is stationary and Gaussian with respect to the second-order joint probability density function may not possess these properties with respect to its higher order probability density functions. However, the converse statement always holds good; that is, a stochastic process which is stationary and Gaussian with respect to a pth order probability density function is stationary and Gaussian with respect to any lower order probability density function.

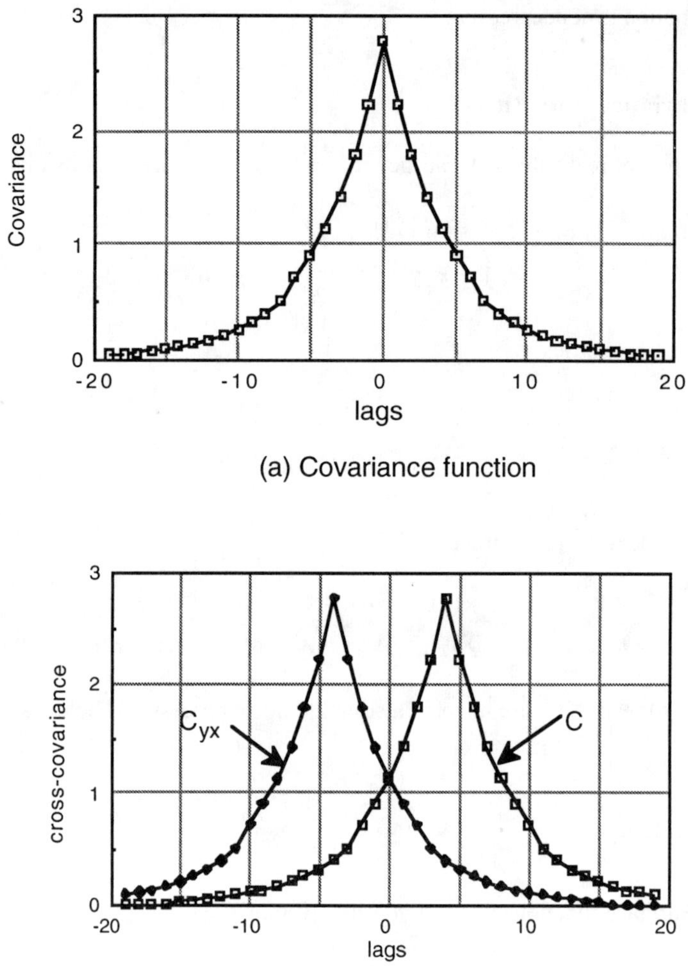

(a) Covariance function

(b) Cross-covariance function

Figure 1.5: Covariance and cross-covariance functions.

C. Spectral Representation of Covariance Function

Wiener established the existence of a bounded nondecreasing function F(w) called the distribution function such that

$$C_x(\tau) = \frac{1}{2\pi}\int_{-\pi}^{\pi} \exp(j\omega\tau)dF(\omega). \tag{1.26}$$

$F(\omega)$ is the power lying between $-\pi$ and ω. $dF(\omega)$ is the differential of $F(\omega)$, that is, the power lying in a small frequency interval $\Delta\omega$ around ω. $F(\omega)$ can be expressed as a sum of two components, the discrete and continuous parts,

$$F(\omega) = F_d(\omega) + F_c(\omega), \tag{1.27}$$

where $F_d(\omega)$, the discrete part, is given by

$$F_d(\omega) = \sum_i p_i U(\omega - \omega_i),$$

where $U(\omega)$ is a unit step function and the continuous part

$$F_c(\omega) = \frac{1}{2\pi}\int_{-\pi}^{\omega} S_X(\omega')d\omega'. \tag{1.28}$$

$S_x(\omega)$ is the spectral density function of a stochastic process with continuous spectrum distribution function. Assuming that $F_d(\omega) = 0$, the spectral representation in (1.26) reduces to

$$C_x(\tau) = \frac{1}{2\pi}\int_{-\pi}^{\pi} S_x(\omega)\exp(j\omega\tau)\,d\omega. \tag{1.29}$$

Note that the unit of $S_x(\omega)$ is power per hertz. It must satisfy, being power, a condition that $S_x(\omega) \geq 0$ (Khintchin's theorem). The converse of (1.29) is

$$S(\omega) = \sum_{\tau=-\infty}^{\infty} C_x(\tau)\exp(-j\omega\tau). \tag{1.30}$$

When the discrete part is not zero, that is, $F_d(\omega) \neq 0$, the covariance function can be expressed as follows:

$$C_x(\tau) = \sum_i p_i \cos(\omega_i\tau) + \frac{1}{2\pi}\int_{-\pi}^{\pi} S_x(\omega)\exp(j\omega\tau)\,d\omega. \tag{1.31}$$

It may be emphasized that the above covariance function does <u>not</u> satisfy the property of square summability, namely,

$$\sum_{\tau=-\infty}^{\infty} |C_x(\tau)|^2 < \infty$$

Corresponding to the discrete and continuous parts of a spectral distribution function a given time series can be expressed as a sum of random sinusoids and a stochastic process with continuous spectrum,

$$x(t) = \sum_i \alpha_i \sin(\omega_i t + \phi_i) + x_c(t), \tag{1.32}$$

where $p_i = \alpha_i^2$ and ϕ_i is random phase. The sinusoidal component is fully predictable from the past samples; hence, it is also sometimes known as the deterministic component.

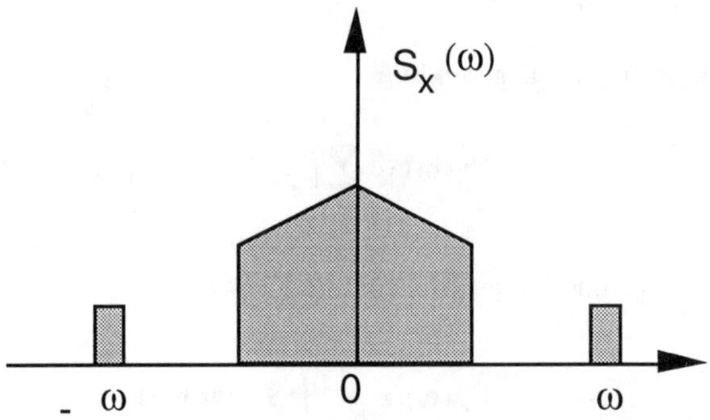

Figure 1.6: An example of a band-limited spectrum.

Example 1.4: To demonstrate the predictability of a sum of sinusoids we consider the following example:

$$x(t)=1.2\cos(0.1\pi t+30^0)+1.8\cos(0.16\pi t+22^0).$$

The linear prediction is obtained from the following equation:

$$\hat{x}(t)=3.6546x(t-1)-5.3334x(t-2)+3.6546x(t-3)-x(t-4),$$

where the coefficients of prediction are obtained from the equations given in Chapter Two. The original time series and its predicted version are given in the Table 1.1 (End of Example).

The spectral density function may be nonzero inside a band of frequencies or bands of frequencies and zero outside (see Figure 1.6) $S_x(\omega) = 0, \quad \omega \neq \Omega$ where Ω is the domain on the ω axis where the spectrum is nonzero. Such a stochastic process is said to be band-limited. No physically realizable stochastic process can have infinite bandwidth. The hypothetical white noise process (the closest example is the radiation from the sun) has infinite bandwidth; that is,

$$\Omega S_x(\omega) = \text{Cons tan t} \quad -\infty < \omega < \infty.$$

In the case of discrete time series, however, the spectrum may be nonzero over the entire bandwidth of $\pm \frac{\pi}{\Delta t}$ where Δt is the sampling interval.

D. Predictable Processes

A stochastic process may be predicted from its past values,

$$\hat{x}(t) = \sum_{i=1}^{\infty} a_i x(t-i),$$

where a_i, i=1, 2, ... are prediction coefficients. The minimum mean square error in the prediction (i.e., optimum prediction coefficients) is given by

Original	Predicted
1.4747	1.4748
0.3432	0.3432
-0.7532	-0.7532
-1.5903	-1.5903
-2.0152	-2.0153
-1.9787	-1.9790
-1.5427	-1.5428

Table 1.1: Error-free prediction of random sinusoids (noise-free) is possible as shown above.

$$\sigma_\varepsilon^2 = \exp\left(\frac{1}{2\pi}\int_{-\pi}^{\pi} \ln S_x(\omega)\,d\omega\right) \tag{1.33}$$

(see Exercise 13 of Chapter Four). From (1.33) we note that the prediction error will be zero whenever $S_x(\omega) = 0$ over some finite domain as

$$\int_{-\pi}^{\pi} S_x(\omega)\,d\omega = -\infty$$

A stochastic process whose spectrum is zero over some finite frequency domain is said to be predictable. Some examples of such process are (a) a sum of random sinusoids and (b) a band-limited process. In the case of a band-limited process the prediction error can be made arbitrarily small.[23]

Predictable processes (continuous in time) are analytic in the sense that a Taylor series expansion is possible. Consequently, given a finite segment of a predictable process, its extrapolation or prediction everywhere outside the segment is, in principle, possible. This, however, is not generally true of band-limited discrete time series. We shall return to this topic later in Chapter Four where we shall examine the possibility of extrapolation of a band-limited time series and thus achieve a higher spectral resolution.

E. **Cross-Spectrum and Coherence**

The cross-spectrum is defined as the Fourier transform of the cross-covariance function,

$$S_{xy}(\omega) = \sum_{\tau} C_{xy}(\tau)\exp(-j\omega\tau). \tag{1.34}$$

Since $C_{xy}(\tau)$ is not symmetric, $S_{xy}(\omega)$ will be a complex function. From the symmetry properties of the cross-covariance function (1.25) it follows that $S_{xy}(\omega) = S^*_{yx}(\omega)$. Let us express it in terms of its real and imaginary parts,

$$S_{xy}(\omega) = P(\omega) + jQ(\omega), \tag{1.35}$$

where $P(\omega)$ is known as cospectrum and $Q(\omega)$ is known as quadrature spectrum. Alternatively, we can also express the cross-spectrum in terms of the amplitude spectrum and the phase spectrum,

$$S_{xy}(\omega) = A(\omega)\exp(j\phi(\omega)),$$

where

$$A(\omega) = \sqrt{P^2(\omega) + Q^2(\omega)}\,,$$

and

$$\phi(\omega) = \tan^{-1}\frac{Q(\omega)}{P(\omega)}.$$

The spectrum and cross-spectrum of a pair of time series can be expressed in a matrix form,

$$\mathbf{S}(\omega) = \begin{vmatrix} S_x(\omega) & S_{yx}(\omega) \\ S_{xy}(\omega) & S_y(\omega) \end{vmatrix}. \tag{1.36}$$

The spectral matrix, **S**(w), is always Hermitian positive semidefinite. Hence, its determinant must be greater than or equal to zero.

$$\det|\mathbf{S}(\omega)| = S_x(\omega)S_y(\omega) - |S_{xy}(\omega)|^2$$
$$= S_x(\omega)S_y(\omega)[1 - \frac{|S_{xy}(\omega)|^2}{S_x(\omega)S_y(\omega)}] \ge 0$$

Hence

$$0 \le \frac{|S_{xy}(\omega)|^2}{S_x(\omega)S_y(\omega)} \le 1. \tag{1.37}$$

We define coherence as

$$\mathrm{Coh}_{xy}(\omega) = \frac{S_{xy}(\omega)}{\sqrt{S_x(\omega)S_y(\omega)}}, \tag{1.38}$$

where $S_x(\omega) \ne 0$ and $S_y(\omega) \ne 0$. Equation (1.37) expressed in terms of $\mathrm{Coh}_{xy}(\omega)$ becomes

$$0 \le |\mathrm{Coh}_{xy}(\omega)|^2 \le 1. \tag{1.39}$$

$\mathrm{Coh}_{xy}(\omega)$ is known as the coherence function. It is a measure of the relationship that exists between the two time series, x(t) and y(t). One way of measuring this relationship between these two time series is to estimate the fraction of energy in the first time series that can be predicted from the knowledge of the second time series. Let $a_{yx}(t)$ be the prediction filter acting on x(t) to estimate the predictable part of y(t). Let $y'(t)$ be the predictable part,

$$y'(t) = \sum_{\tau} a_{yx}(\tau)\, x(t-\tau). \tag{1.40}$$

We shall minimize the energy of the nonpredictable part by optimally selecting the filter function, that is, minimize

$$E\left|y(t) - y'(t)\right|^2 = E\left|y(t) - \sum_{\tau} a_{yx}(\tau)\, x(t-\tau)\right|^2. \tag{1.41}$$

Differentiating (1.41) with respect to $a_{yx}(t)$ and setting the derivative to zero, we obtain

$$C_{yx}(\tau) = \sum_{t} a_{yx}(t) C_x(\tau - t). \tag{1.42}$$

Taking the Fourier transform on both sides of (1.42), we obtain

$$A_{yx}(\omega) = \frac{S_{yx}(\omega)}{S_x(\omega)}, \tag{1.43}$$

where $A_{yx}(\omega)$ is the transfer function of the prediction filter. Since $y'(t)$ and the error, $\varepsilon(t) = y(t) - y'(t)$, are orthogonal, we can write

$$S_y(\omega) = S_{y'}(\omega) + S_{\varepsilon}(\omega),$$

where the spectrum of $y'(t)$ may be obtained from (1.40). For this multiply the left-hand side by $y'(t+\tau)$ and the right-hand side by $\sum_{\tau} a_{yx}(\tau)\, x(t+\tau-\tau)$ and carry out the expectation operation on both sides. We obtain

$$C_{y'}(\tau) = \sum_{\tau''}\sum_{\tau'} a_{yx}(\tau''+\tau')\, a_{yx}(\tau') C_x(\tau - \tau'').$$

On further taking the Fourier transform on both sides of the above equation, we obtain the spectrum of $y'(t)$,

$$S_{y'}(\omega) = \left|A_{yx}(\omega)\right|^2 S_x(\omega).$$

Using (1.43) in the above equation we obtain

$$S_y(\omega) = \frac{\left|S_{yx}(\omega)\right|^2}{S_x(\omega)} + S_{\varepsilon}(\omega). \tag{1.44}$$

Rearranging the terms in (1.44) we obtain

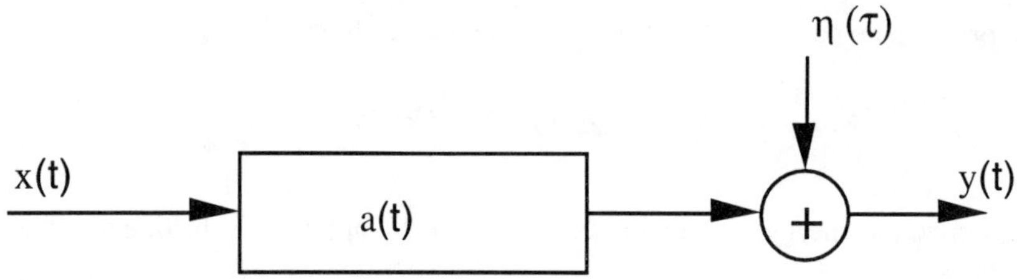

Figure 1.7: A schematic representation of a linear system whose output is corrupted with noise.

$$S_\varepsilon(\omega) = S_y(\omega)[1 - \frac{|S_{yx}(\omega)|^2}{S_x(\omega)S_y(\omega)}]$$

or

$$\frac{S_\varepsilon(\omega)}{S_y(\omega)} = [1 - Coh^2_{yx}(\omega)].$$

Hence, the predictable energy in y(t) can be expressed as a ratio,

$$\frac{S_y(\omega) - S_\varepsilon(\omega)}{S_y(\omega)} = Coh^2_{yx}(\omega). \tag{1.45}$$

The coherence function as shown in (1.45) is clearly a measure of predictability.

The coherence function can also be viewed as a measure of the noise figure of a linear system. Consider a linear system whose output is corrupted with noise as shown in Figure 1.7. The coherence between the output and input may be related to the signal-to-noise ratio (SNR) of the system. We model the system as a linear system with additive noise,

$$y(t) = \sum_\tau a_{yx}(\tau)\, x(t - \tau) + \eta(t), \tag{1.46}$$

where $\eta(t)$ is the system noise which is believed to be uncorrelated with x(t). Multiply both sides of (1.46) with $x(t + \tau)$ and compute the expected value. We obtain

$$C_{yx}(\tau) = \sum_t a_{yx}(t) C_x(\tau + t).$$

Now, taking the Fourier transform on both sides of the above expression, we obtain

$$A_{yx}(\omega) = \frac{S_{yx}(\omega)}{S_x(\omega)}. \tag{1.47}$$

Following the approach used in deriving (1.44), we can show that

$$S_y(\omega) = \left|A_{yx}(\omega)\right|^2 S_x(\omega) + S_\eta(\omega). \tag{1.48}$$

Substituting for $\left|A_{yx}(\omega)\right|^2$, using (1.47), we obtain

$$S_y(\omega) = \frac{\left|S_{yx}(\omega)\right|^2}{S_x(\omega)} + S_\eta(\omega)$$

$$S_y(\omega)[1 - \mathrm{Coh}^2_{yx}(\omega)] = S_\eta(\omega)$$

or

$$\frac{S_y(\omega)}{S_\eta(\omega)} = \frac{1}{[1 - \mathrm{Coh}^2_{yx}(\omega)]}.$$

Define the signal-to-noise ratio, SNR, as a ratio of the output of the linear system divided by the noise power,

$$\mathrm{SNR} = \frac{\left|A_{yx}(\omega)\right|^2 S_x(\omega)}{S_\eta(\omega)} = \frac{\mathrm{Coh}^2_{yx}}{1 - \mathrm{Coh}^2_{yx}}. \tag{1.49}$$

As Coh^2_{yx} varies from 0 to 1, the SNR varies from 0 to ∞.

Example 1.5: For a linear system with no noise the coherence function between the output and input may be expressed in terms of the phase delay introduced by the system,

$$\mathrm{Coh}_{yx}(\omega) = \exp(j\omega\tau_{yx}),$$

where τ_{yx} is the delay introduced in the output with respect to the input. An interesting application of this property of the coherence function is in the measurement of propagation delay (End of Example).

F. **Ergodicity of Stationary Stochastic Process**

We have defined mean, variance and covariance in terms of the expectation operation; for example, the mean of a stochastic process was defined as $E\{x(t)\}$. The expectation operation involves averaging over many realizations at a fixed time instant. In many practical situations we have just one realization; then, it is not possible to carry out the expected operation. For a stationary stochastic process the mean is independent of time. We shall now show that the sample mean defined as

$$\hat{m}_x = \frac{1}{T}\sum_{t=0}^{T-1} x(t)$$
$$T \rightarrow \infty$$

tends to the true mean, m_x, defined using the expectation operation in the least mean square sense, provided the covariance function of x(t) satisfies a certain decay rate.

$$E\{(\hat{m}_x - m_x)^2\} = E\{m_x^2 - 2m_x \frac{1}{T}\sum_{t=0}^{T-1} x(t) + \frac{1}{T^2}\sum_{t=0}^{T-1}\sum_{t'=0}^{T-1} x(t)x(t')\}_{T\to\infty}$$

$$= \frac{1}{T^2}\sum_{t}\sum_{t'} C_x(t-t').$$

$$T \to \infty \tag{1.50}$$

From (1.50) we note that $\hat{m}_x \to m_x$ as $T \to \infty$ only when

$$\sum_{t=1}^{T}\sum_{t'=1}^{T} C_x(t-t') < \infty. \tag{1.51}$$

The stochastic process is said to be ergodic with respect to the mean; that is, the arithmetic mean tends to true mean. For this to take place, the condition given by (1.51) must be satisfied. Ergodicity with respect to variance and covariance can be shown to exist if

$$\sum_{t=1}^{T}\sum_{t'=1}^{T} C_x^2(t-t') < \infty. \tag{1.52}$$

An example of a nonergodic process is a sum of random sinusoids,[22]

$$x(t) = \sum_{i=1}^{p} (\alpha_i \cos(\omega_i t) + \beta_i \sin(\omega_i t)),$$

where α_i and β_i, i=1, 2, ..., p are uncorrelated zero mean random variables with variances, $E\{\alpha_i^2\} = E\{\beta_i^2\} = \sigma^2$. The covariance function of x(t) may be shown to be equal to

$$C_x(\tau) = \sum_{i=1}^{p} \sigma^2 \cos(\omega_i \tau). \tag{1.53}$$

Now, we introduce the autocorrelation function defined as arithmetic average of the lagged product,

$$A_x(\tau) = \frac{1}{T-\tau}\sum_{t=0}^{T-\tau} x(t)x(t+\tau)$$

$$T \to \infty$$

$$= \sum_{i=1}^{p} \frac{(\alpha_i^2 + \beta_i^2)}{2} \cos(\omega_i \tau). \tag{1.54}$$

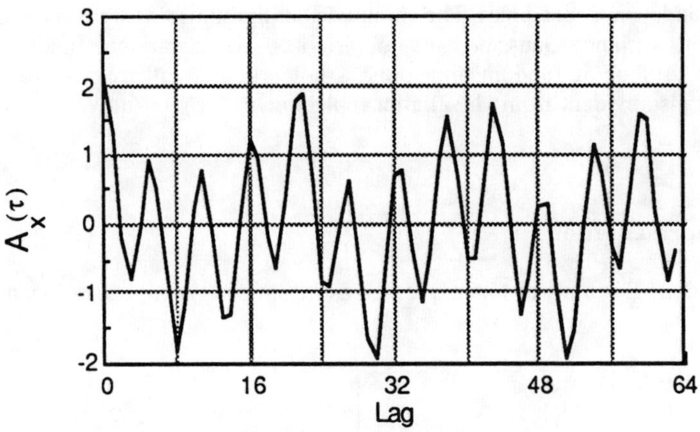

(a) Autocorrelation function

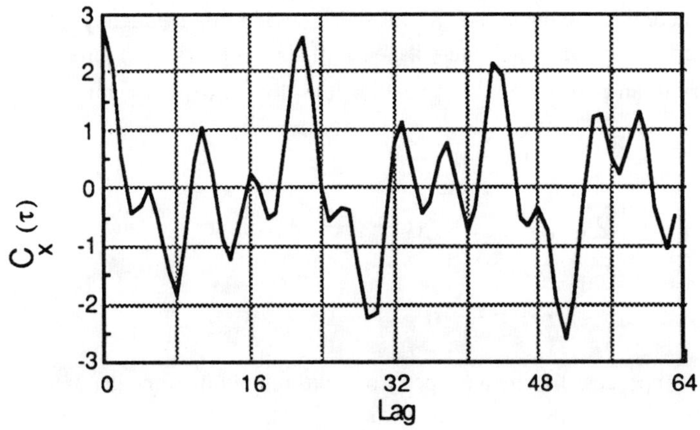

(b) Covariance function

Figure 1.8: A sum of random sinusoids is *not* an ergodic process. Its covariance function obtained through expectation operation is *not* equal to the autocorrelation function obtained through time averaging.

Note that, since $A_x(\tau)$ depends upon the choice of α_i and β_i, i=1, 2,..., p but $C_x(\tau)$ depends only on the variance, the autocorrelation function as defined above is *not* equal to the covariance function.

Example 1.6: The difference between the covariance function and the autocorrelation function of a sum of random sinusoids is demonstrated in Figure 1.8. A time series was generated as a sum of three random sinusoids (random amplitudes $\alpha=\{-1.1864, -0.5711, 1.4338\}$ and $\beta=\{-0.0253, -0.3300, -0.6769\}$ and frequencies=$\{0.1\pi, 0.18\pi, 0.37\pi\}$). The autocorrelation function was computed via time averaging (1.54). The data length was T=2000 samples. The result is shown in Figure 1.8a. Next, the covariance function was evaluated via the expectation operation. Here also we generated 2000 independent samples. The amplitudes of the sinusoids were taken independently from zero mean, unit variance Gaussian random variables. The covariance function is shown in Figure 1.8b. Thus, a sum of random sinusoids is not ergodic with respect to variance and covariance. This is also evident from (1.52) after replacing $C_x(\tau)$ by its value given by (1.53) (End of Example).

1.4 Spectral Representation

A time series of length T can always be expressed as a sum of N complex exponential functions such that

$$\left| x(t) - \sum_{n=1}^{N} X_n \exp(j\omega_n t) \right|^2 \le \varepsilon \qquad -\frac{T}{2} \le t \le \frac{T}{2}, \tag{1.55}$$

where ε is a small positive number. For large T($T \to \infty$) and small error $(\varepsilon \to 0)$ we would require a large number of exponentials $(N \to \infty)$. The frequencies ω_n, n = 1, 2, ..., N would be randomly distributed over the interval $-\pi$ to $+\pi$. Divide this frequency interval into a large number of equal-sized non-overlapping intervals, each of size $\Delta\omega$ (see Figure 1.9). Sum all X_n lying in one interval and assume that $\exp(j\omega_n t)$ does not vary much within a range $\omega + \frac{\Delta\omega}{2}$ to $\omega - \frac{\Delta\omega}{2}$. Let

$$\begin{aligned} \sum_{\Delta\omega} X_n &= \frac{1}{2\pi}\left[X_G\left(\omega + \frac{\Delta\omega}{2}\right) - X_G\left(\omega - \frac{\Delta\omega}{2}\right)\right] \\ &= \frac{1}{2\pi} dX_G(\omega), \end{aligned} \tag{1.56}$$

where X_n is a random process having a property of uncorrelated increments; that is,

$$E\left\{\left(X_G\left(\omega_1 + \frac{\Delta\omega}{2}\right) - X_G\left(\omega_1 - \frac{\Delta\omega}{2}\right)\right)\left(X_G\left(\omega_2 + \frac{\Delta\omega}{2}\right) - X_G\left(\omega_2 - \frac{\Delta\omega}{2}\right)\right)^*\right\} = 0.$$

Using (1.56) in (1.55) we obtain

$$\left| x(t) - \frac{1}{2\pi}\int_{-\pi}^{\pi} dx_G(\omega)\exp(j\omega t) \right|^2 \le \varepsilon, \qquad \begin{matrix} -\infty < t < \infty \\ \varepsilon \to 0 \end{matrix}$$

or

$$x(t) = \frac{1}{2\pi}\int_{-\pi}^{\pi} dx_G(\omega)\exp(j\omega t) \tag{1.57}$$

in the least mean square sense. Equation (1.57) is known as spectral representation.[24] This is also known as Cramer representation.

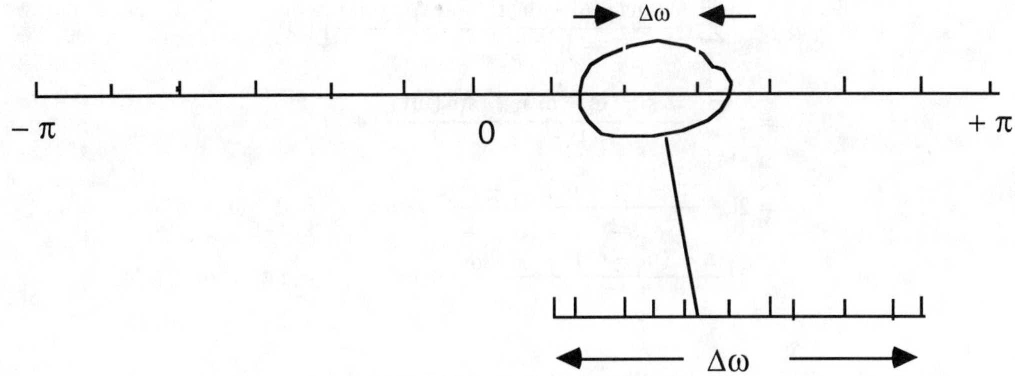

Figure 1.9: Random complex sinusoids in a frequency interval Δw.

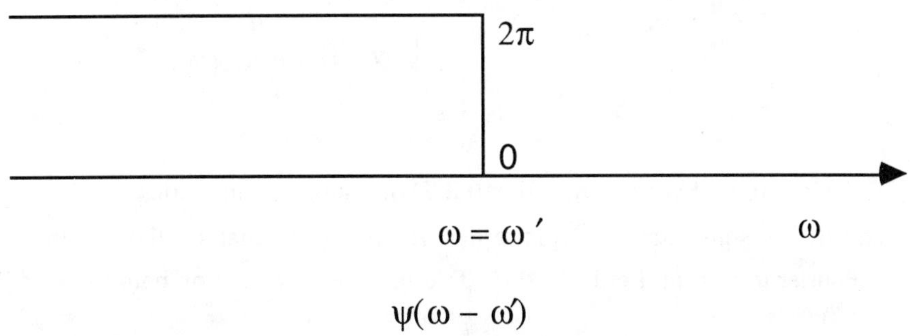

Figure 1.10: Step function.

A. Generalized Fourier Transform

To obtain the inverse of (1.57) multiply both sides by $\frac{\exp(-j\omega' t) - 1}{-jt}$ and sum it over t,

$$\sum_{t=-\infty}^{\infty} x(t) \frac{\exp(-j\omega' t) - 1}{-jt}$$

$$= \frac{1}{2\pi} \int_{-\pi}^{\pi} dX_G(\omega) \sum_{t=-\infty}^{\infty} \left[\frac{\exp(j(\omega - \omega')t)}{-jt} - \frac{\exp(j\omega t)}{-jt} \right]. \tag{1.58}$$

The summation on the right-hand side of (1.58) can be carried out using summation formulas given in Reference 25. Let $\psi(\omega - \omega')$ represent the sum in (1.58),

$$\psi(\omega - \omega') = \sum_{t=-\infty}^{\infty} \left[\frac{\exp(j(\omega - \omega')t)}{-jt} - \frac{\exp(j\omega t)}{-jt}\right]$$

$$= 2\sum_{t=1}^{\infty}\left[-\frac{\sin(\omega - \omega')t}{t} + \frac{\sin(\omega t)}{t}\right] + \omega'$$

$$= 2\left[-\frac{\pi - (\omega - \omega')}{2} + \frac{\pi - \omega}{2}\right] + \omega' \qquad = 0 \quad \omega > \omega'$$

$$= 2\left[\frac{\pi + (\omega - \omega')}{2} + \frac{\pi - \omega}{2}\right] + \omega' \qquad = 2\pi \quad \omega < \omega'.$$

Note that $\psi(\omega - \omega')$ is a step function as illustrated in Figure 1.10.

Making use of the above result in (1.58) we obtain

$$\sum_{t=-\infty}^{\infty} x(t)\frac{\exp(-j\omega' t) - 1}{-jt} = \frac{1}{2\pi}\int_{-\pi}^{\pi} dX_G(\omega)\,\psi(\omega - \omega')$$

$$= \int_{-\pi}^{\omega'} dX_G(\omega) = X_G(\omega') \tag{1.59}$$

$X_G(\omega')$ is the generalized Fourier transform (GFT) of stationary time series.

We shall derive some statistical properties of $dX_G(\omega)$, that is, the differential of the generalized Fourier transform. Let E{x(t)}=0. Taking the expectation on both sides of (1.57), we obtain

$$\int_{-\pi}^{\pi} E\{dX_G(\omega)\}\exp(j\omega t) = 0 \qquad \text{for all } t.$$

Hence, $E\{dX_G(\omega)\} = 0$. Now let us compute

$$E\{x(t)x(s)\} = \frac{1}{4\pi^2}\int\int_{-\pi}^{\pi} E\{dX_G(\omega)\,dX^*_G(\omega')\}\exp(-j(\omega t - \omega' s)). \tag{1.60}$$

Since x(t) is a stationary stochastic process, the left-hand side in (1.60) must be a function of (t – s). For this to be possible the integrand on the right-hand side must be of the type

$$\left.\begin{aligned} E\{\tfrac{1}{2\pi}dX_G(\omega)\tfrac{1}{2\pi}dX^*_G(\omega')\} &= \tfrac{1}{2\pi}dF_x(\omega)\delta(\omega - \omega')d\omega' \\ E\{\left|\tfrac{1}{2\pi}dX_G(\omega)\right|^2\} &= \tfrac{1}{2\pi}dF_x(\omega). \end{aligned}\right\}, \tag{1.61}$$

where $F_x(\omega)$ is the spectral distribution function of x(t). Substitute (1.61) in (1.60) and assume that $F_x(\omega)$ is differentiable; that is, $dF_x(\omega) = S_x(\omega)d\omega$. We obtain the spectral representation of the covariance function (1.29).

The complex analytical representation (CAR) of a stationary stochastic process is expressed as

$$x_c(t) = x(t) + jx_H(t), \tag{1.62a}$$

where $x_H(t)$ is the Hilbert transform of x(t) defined as

$$x_H(t) = \frac{1}{2\pi}\int_{-\pi}^{\pi} -j\mathrm{Sgn}(\omega)\, dX_G(\omega)\exp(j\omega t), \tag{1.62b}$$

where Sgn(ω)=1 when ω >0, and Sgn(ω)= – 1 when ω <0. Using the spectral representation of x(t) and (1.62b) in (1.62a), we get the spectral representation of CAR,

$$x_c(t) = \frac{1}{\pi}\int_0^{\pi} dX_G(\omega)\exp(j\omega t). \tag{1.63}$$

Note that $x_c(t)$ is a complex function. From its phase function we define the instantaneous frequency as its derivative with respect to time,

$$\Omega(t) = \frac{x'_H(t)x(t) - x'(t)x_H(t)}{x^2(t) + x_H^2(t)}, \tag{1.64}$$

where (') stands for differentiation. The derivative of a discrete time series appearing in (1.64) may be computed numerically. The instantaneous frequency given by (1.64) is valid for a single frequency or tone. When there is more than one tone, it would be necessary to separate them first by filtering before estimating the instantaneous frequency. It may be noted that the instantaneous frequency defined above is conceptually different from the usual frequency appearing as an independent variable in the Fourier transform. Yet they are related through weighted mean (see Exercise 1.20).

B. Karhunen-Loeve (KL) Expansion

A general representation theorem of Karhunen[26] states that there exists a complete orthonormal family of uncorrelated zero mean and unit variance random variables,
such that

$$x(t) = \sum_{n=0}^{N} \xi_n \sigma_n u_n(t) \quad 0 \le t \le T-1, \tag{1.65}$$

where $\sigma_n > 0$ and $u_n(t)$, n=0,1,2, ..., are orthonormal functions; that is,

$$\sum_{0}^{T-1} u_n(t)u_m(t) = \delta_{mn},$$

where δ_{mn} =1 when m=n, and $\delta_{mn} = 0$ when $m \ne n$. Using (1.65) we can obtain the covariance function of x(t) in terms of $u_n(t)$.

$$
\begin{aligned}
C_x(t-s) &= E\{x(t)x(s)\} \\
&= \sum_m \sum_n \sigma_m \sigma_n u_m(t) u_n(s) E\{\xi_m \xi_n\} \\
&= \sum_{n=0}^{N} \sigma_n^2 u_n(t) u_n(s).
\end{aligned} \tag{1.66}
$$

Multiply both sides of (1.66) by $u_n(s)$ and sum over s in the range 0 to $T-1$. Using the orthogonality property of $u_n(s)$, we get the following result:

$$
\sum_{s=0}^{T-1} C_x(t-s)\, u_n(s) = \sigma_n^2 u_n(t) \quad 0 \le t \le T-1. \tag{1.67a}
$$

From (1.67a) we note that $u_n(t)$ is an eigenfunction and σ_n^2 is the corresponding eigenvalue of the covariance matrix.

The projection of x(t) on the nth eigenfunction $u_n(t)$ is called the nth principal component (PC) of x(t),

$$
x_n(t) = \sum_{s=0}^{\Delta T-1} x(t+s)u_n(s), \tag{1.67b}
$$

where $\Delta T << T$ is the size of the covariance matrix. The nth PC is a time series with zero mean and variance equal to n^{th} eigenvalue of the covariance matrix. Further, it is uncorrelated (at zero lag only) with all other principal components. The spectrum of a PC has some interesting features. Let us replace x(t+s) by its spectral representation. We obtain after some simplification

$$
x_n(t) = \frac{1}{2\pi} \int_{-\pi}^{\pi} dX(\omega) U_n^*(\omega) e^{j\omega t},
$$

where $U_n(\omega)$ is the Fourier transform of $u_n(t)$. Hence, the spectrum of the nth PC is $S_{x_n}(\omega) = S_x(\omega)\left|U_n(\omega)\right|^2$. The spectrum of an eigenfunction has a sharp peak in the region where sharp peaks in the spectrum of the time series occur. Hence, the power in a PC will be concentrated in a narrow frequency band corresponding to the passband of the eigenfunction. The passbands of the eigenfunctions corresponding to significant eigenvalues are adaptively tuned to the peaks in the spectrum.[27]

Example 1.7: We shall illustrate the above phenomenon through a simple example of a real random sinusoid and white noise. The covariance function is given by

$$
C_x(\tau) = \frac{\alpha^2}{2} \cos(\omega_0 \tau) + \sigma_\eta^2 \delta_\tau,
$$

where α is amplitude of the random sinusoid, ω_0 is angular frequency, and σ_η^2 is variance of the white noise, $\eta(t)$. The covariance matrix (of any size) will have two significant eigenvalues while the remaining eigenvalues are repeated and equal to the noise variance (see Section 4.5 for more details). The first two significant eigenvalues are (see Exercise 4.17)

$$\lambda_1 = \frac{\alpha^2 \Delta T}{4}\left(1 + \frac{\sin(\omega_0 \Delta T)}{\Delta T \sin(\omega_0)}\right) + \sigma_\eta^2$$

$$\lambda_2 = \frac{\alpha^2 \Delta T}{4}\left(1 - \frac{\sin(\omega_0 \Delta T)}{\Delta T \sin(\omega_0)}\right) + \sigma_\eta^2 ,$$

and the corresponding eigenfunctions are

$$\text{Eigenvectors} : \begin{cases} \mathbf{u}_1 = a\{\cos\left(\left(\frac{\Delta T - 1}{2} - k\right)\omega_0\right), & k = 0, 1, \ldots, \Delta T - 1\} \\ \mathbf{u}_2 = b\{\sin\left(\left(\frac{\Delta T - 1}{2} - k\right)\omega_0\right), & k = 0, 1, \ldots, \Delta T - 1\} \end{cases} ,$$

where a and b are normalizing constants. The Fourier transforms of the eigenfunctions will evidently peak at $\omega = \pm\,\omega_0$, and the width of the peak will be proportional to $\frac{1}{\Delta T}$ (End of Example).

1.5 **Stationary Stochastic Process:** Third Order

We have seen in the last section that for a Gaussian stochastic process the joint probability density function of any order is governed by the covariance matrix. Thus, the covariance function or the spectrum completely characterizes a Gaussian stochastic process. However, for a non-Gaussian stochastic process a higher order density function would be required for complete characterization of the process. When the stochastic process is slightly non-Gaussian, the first few terms in the expansion (1.7) would be adequate to describe the non-Gaussian part. The third moment is an important measure of non-Gaussian character of a time series. Sometimes, the fourth moment can also be used as a measure of non-Gaussianness. We shall now introduce bispectrum and trispectrum for the analysis of non-Gaussian time series (More details may be found in a recent text by C. L. Nikias and A. P. Petropulu.[46]). A non-Gaussian time series has another important property, namely, it is time irreversible, by which we mean that a higher order probability density function depends upon whether the time samples are arranged in an increasing or decreasing order,

$$f(x_1, x_2, x_3, \ldots, x_N;\ t_2 - t_1, t_3 - t_1, \ldots, t_N - t_1) \neq$$
$$f(x_N, x_{N-1}, x_{N-2}, \ldots, x_1;\ t_N - t_1, t_{N-1} - t_1, t_{N-2} - t_1, \ldots, t_2 - t_1),$$

where $t_N > t_{N-1} > t_{N-2} > \ldots > t_1$. The only exception to this rule is when a time series is obtained by passing a non-Gaussian white noise process through a filter with zero or linear phase.[28]

A. **Bicovariance Function**

Let $f(x_1, x_2, x_3; t_2 - t_1, t_3 - t_1)$ be a third-order probability density function of a zero mean stationary stochastic process. The third moment, called here as bicovariance, is defined as follows:

$$\begin{aligned} Bc(s,h) &= E\{x(t)x(t+s)x(t+h)\} \\ &= \int\!\!\int\!\!\int x_1 x_2 x_3 f(x_1, x_2, x_3; s,h)\, dx_1 dx_2 dx_3. \end{aligned} \tag{1.68}$$

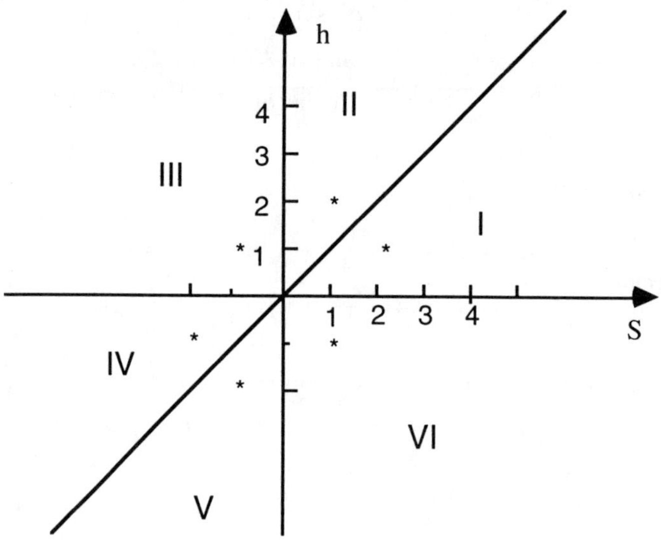

Figure 1.11: Sixfold symmetry of the bicovariance function.

We shall assume throughout the book, unless stated otherwise, that x(t) is a real process. From the definition of the bicovariance function we show the following properties:

1. $$\begin{aligned} Bc(s,h) &= E\{x(t)x(t+s)x(t+h)\} \\ &= E\{x(t)x(t+h)x(t+s)\} \\ &= Bc(h,s), \end{aligned}$$

2. $$\begin{aligned} Bc(s-h,-h) &= E\{x(t)x(t+s-h)x(t-h)\} \\ &= E\{x(t'+h)x(t'+s)x(t')\} \qquad \text{where} \quad t' = t-h \\ &= Bc(s,h), \end{aligned}$$

3. $$\begin{aligned} Bc(-h,s-h) &= E\{x(t)x(t-h)x(t+s-h)\} \\ &= E\{x(t'+h)x(t'+s)x(t')\} \qquad \text{where} \quad t' = t-h \\ &= Bc(s,h), \end{aligned}$$

4. $$\begin{aligned} Bc(h-s,-s) &= E\{x(t)x(t+h-s)x(t-s)\} \\ &= E\{x(t'+s)x(t'+h)x(t')\} \qquad \text{where} \quad t' = t-s \\ &= Bc(s,h), \end{aligned}$$

Thus, we have a sixfold symmetry,

$$\begin{aligned} &Bc(s,h) = Bc(h,s) = Bc(s-h,-h) = \\ &Bc(-h,s-h) = Bc(h-s,-s) = Bc(-s,h-s) \ . \end{aligned} \tag{1.69}$$

The sixfold symmetry is illustrated in Figure 1.11. The (s,h) plane of the bicovariance function is partitioned into six domains such that the bicovariance function given in any one domain can be mapped into all other domains. To understand the mapping operation described by (1.69) consider a point (2,1) in the first domain (shown by * in Figure (1.11)). This point will be mapped using (1.69) onto five other domains as illustrated in the figure by *. Note that $Bc(0,0)=E\{x^3(t)\}$, the third moment, is equal to zero for any symmetrically distributed random variable, including the Gaussian random variable. However, for a Gaussian process the bicovariance function is equal to zero for all s and h. For time reversible process it follows from (1.68) and the property of time reversibility that

$$Bc(s,h) = Bc(-s,-h). \tag{1.70}$$

The bicovariance function defined on the right diagonal, a line passing through the center with slope = +1.0, that is $s=h=\tau$, has a particularly interesting form:

$$\begin{aligned} Bc_r(\tau) &= E\{x(t)x(t+\tau)x(t+\tau)\} \\ &= E\{x(t)x^2(t+\tau)\}. \end{aligned}$$

Applying the symmetry properties, we find that $Bc_r(\tau)=Bc(0,\tau)= Bc(\tau,0)$. Similarly, the bicovariance function defined on the left diagonal, a line passing through the center with slope = – 1.0, that is $s=-h=\tau$, has the following form:

$$Bc_l(\tau) = E\{x(t)x(t+\tau)x(t-\tau)\}.$$

It may be noted that $Bc_r(\tau) \neq Bc_r(-\tau)$, but $Bc_l(\tau) = Bc_l(-\tau)$. The bicovariance function of a third-order white noise process is given by

$$Bc_x(s,h) = \beta_x^3 \delta_{s,h},$$

where β_x^3 is the third moment of the white noise process (zero mean), and $\delta_{s,h} = 1$ for s=h=0; otherwise $\delta_{s,h} = 0$.

B. Bispectrum

The two-dimensional Fourier transform of the bicovariance function gives us the bispectrum,

$$Bs_x(\omega_1,\omega_2) = \sum_s \sum_h Bc(s,h)\exp(-j(\omega_1 s + \omega_2 h)), \tag{1.71}$$

where ω_1 and ω_2 are continuous frequency variables ranging from $-\pi$ to $+\pi$. On account of the symmetry properties of the bicovariance function, the bispectrum will have the following symmetry:

$$\begin{aligned} Bs_x(\omega_1,\omega_2) &= Bs_x(\omega_2,\omega_1) \\ &= Bs_x(-\omega_1-\omega_2,\omega_2) \\ &= Bs_x(\omega_1,-\omega_1-\omega_2). \end{aligned} \tag{1.72}$$

When x(t) is real we have the following two additional symmetries:

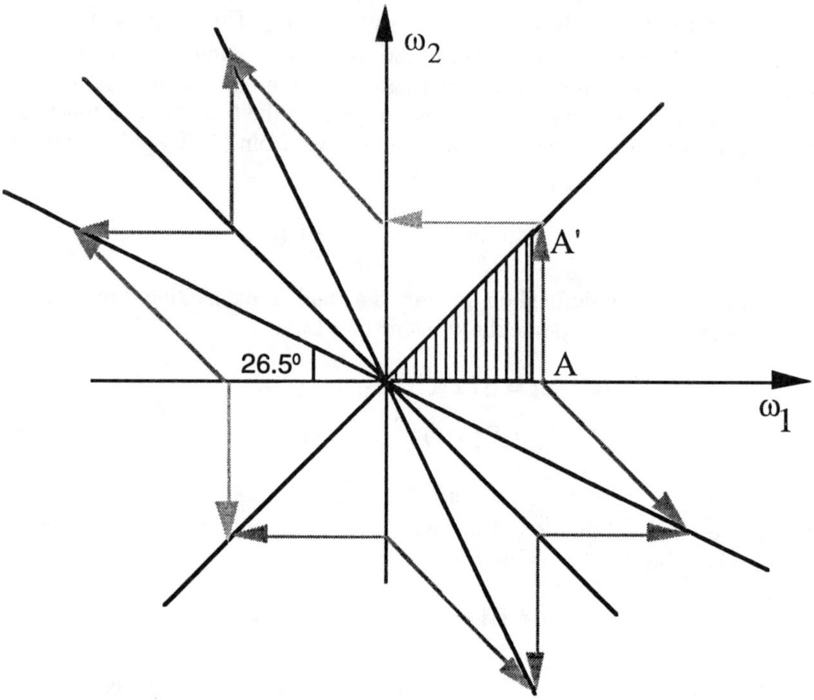

Figure 1.12: Symmetry relations in the bispectrum of real time series.

$$Bs_x(\omega_1, \omega_2) = Bs_x^*(-\omega_1, -\omega_2)$$
$$Bs_x(\omega_1, -\omega_2) = Bs_x^*(-\omega_1, \omega_2). \tag{1.73}$$

These symmetry relations lead to a twelvefold symmetry in the bispectrum plane as illustrated in Figure 1.12. A vector AA' in domain I is mapped onto the remaining eleven domains as shown by dashed arrows.

We can also express the bicovariance function in terms of the generalized Fourier transform of a stationary stochastic process (1.57). For this, replace x(t) in (1.68) by its spectral representation,

$$Bc(s,h) = \frac{1}{8\pi^3}\int\int\int_{-\pi}^{\pi} E\{dX_G(\omega_1)\, dX_G(\omega_2)\, dX_G(\omega_3)\}\, e^{(j(t(\omega_1+\omega_2+\omega_3)+s\omega_2+h\omega_3))} \tag{1.74}$$

The integral on the right-hand side will be independent of t if, and only if,

$$E\{\frac{1}{2\pi}dX_G(\omega_1)\frac{1}{2\pi}dX_G(\omega_2)\frac{1}{2\pi}dX_G(\omega_3)\}$$
$$= \frac{1}{8\pi^3}Bs(\omega_1,\omega_2)d\omega_1 d\omega_2 \delta(\omega_1+\omega_2+\omega_3)\, d\omega_3. \tag{1.75}$$

Then, since

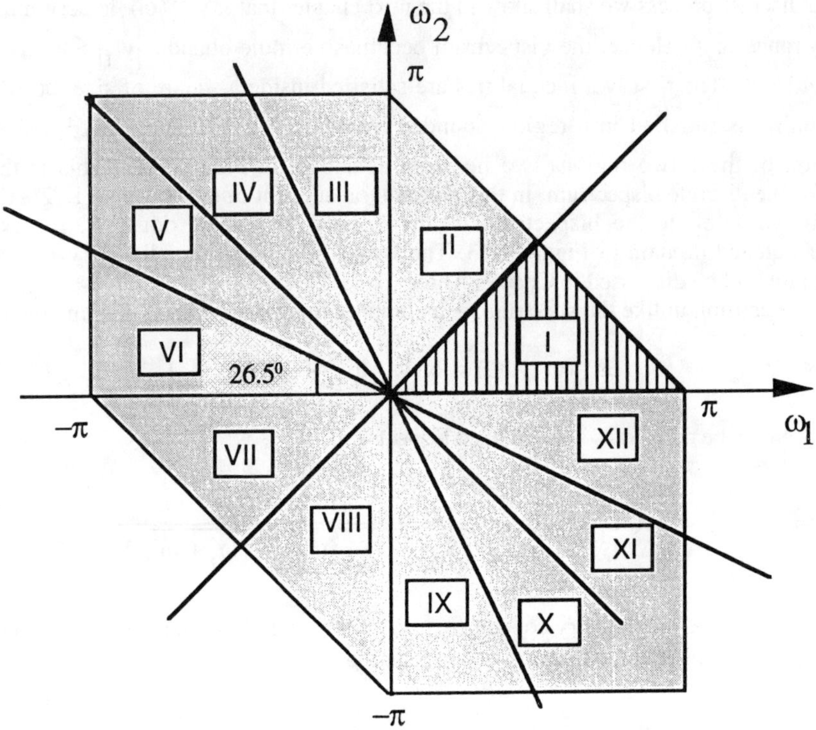

Figure 1.13: Symmetry of digital bispectrum of a real stationary stochastic process. It is enough if the bispectrum is computed in the first region.

$$\frac{1}{2\pi}\int_{-\pi}^{\pi}\delta(\omega_1+\omega_2+\omega_3)\exp(jt(\omega_1+\omega_2+\omega_3)d\omega_3 = 1$$

on a plane,

$$\omega_1+\omega_2+\omega_3 = 0\ ,$$

(1.74) reduces to

$$Bc\,(s,h) = \frac{1}{4\pi^2}\int_{-}\int_{\pi}^{\pi} Bs\,(\omega_1,\omega_2)\exp\,(j\,(s\omega_1+h\omega_2))\,d\omega_1 d\omega_2. \qquad (1.76)$$

This result may be looked upon as a spectral representation of the bicovariance function, analogous to the spectral representation of covariance function (1.29). Integrate both sides of (1.75) with respect to ω_3. We obtain

$$E\{\frac{1}{2\pi}dX_G(\omega_1)\frac{1}{2\pi}dX_G(\omega_2)\frac{1}{2\pi}dX^*_G(\omega_1+\omega_2)\} = \frac{1}{4\pi^2}Bs(\omega_1,\omega_2)d\omega_1 d\omega_2.$$

)

Equation (1.77) provides a convenient means of estimating the bispectrum directly.

For a discrete process we shall show in the next chapter that $dX_G(\omega)$ is periodic outside the frequency range $\pm\pi$. Hence, the bispectrum becomes periodic outside $|\omega_1| \le \pi$, $|\omega_2| \le \pi$ and $|\omega_1 + \omega_2| \le \pi$. The first two inequalities are satisfied inside a square of size $2\pi \times 2\pi$, and the last inequality is satisfied in a region bounded by $\omega_1 + \omega_2 = \pi$ and $\omega_1 + \omega_2 = -\pi$. The intersection of these two regions is a hexagon as shown in Figure 1.13. This is the principal domain for the discrete bispectrum. In this domain, on account of symmetries (1.72 and 1.73), it is enough if we compute the bispectrum in any one of the twelve domains, for example, the triangular hatched domain in Figure 1.13. The issues relating to the aliasing error and required sampling rate will be discussed in Chapter Three (page 220).

The bispectrum, unlike the spectrum, is a complex function with real and imaginary parts,

$$Bs_x(\omega_1, \omega_2) = \alpha(\omega_1, \omega_2) + j\beta(\omega_1, \omega_2).$$

Bicoherence may be defined as a normalized bispectrum,

$$bs(\omega_1, \omega_2) = \frac{Bs_x(\omega_1, \omega_2)}{\sqrt{[S_x(\omega_1) S_x(\omega_2) S_x(\omega_1 + \omega_2)]}}. \tag{1.78}$$

The bicoherence, like the ordinary coherence, may be shown to be equal to the fraction of energy at $(\omega_1 + \omega_2)$ due to interaction of energy at ω_1 and ω_2.

C. Interpretation of Bispectrum

We would now like to a provide a physical meaning to the bispectrum and bicoherence functions. Let us consider three ideal bandpass filters, $h_k(t)$, $h_l(t)$, and $h_m(t)$, centered at frequencies $\pm\omega_k$, $\pm\omega_l$, and $\pm\omega_m$, respectively. The bandwidth is Δ $(\Delta \approx 0)$. A time series, $x(t)$, is passed through each filter (see Figure 1.14). Let the outputs be $x_k(t)$, $x_l(t)$, and $x_m(t)$, where

$$\begin{aligned} x_k(t) &= h_k(t) * x(t), \\ x_l(t) &= h_l(t) * x(t), \\ x_m(t) &= h_m(t) * x(t), \end{aligned}$$

where * denotes convolution. We would like to evaluate the expectation of the triple product,

$$\begin{aligned} E\{x_k x_l x_m\} &= \sum\sum\sum h_k(t')h_l(t'')h_m(t''')E\{x(t-t')x(t-t'')x(t-t''')\} \\ &= \sum\sum\sum h_k(t')h_l(t'')h_m(t''')Bc(t''-t', t'''-t'). \end{aligned} \tag{1.79}$$

Let us replace the filter impulse response functions in (1.79) by their Fourier representation,

$$h_k(t) = \frac{1}{2\pi}\int_{-\pi}^{\pi} H_k(\omega)\exp(j\omega t)\,d\omega \quad ,$$

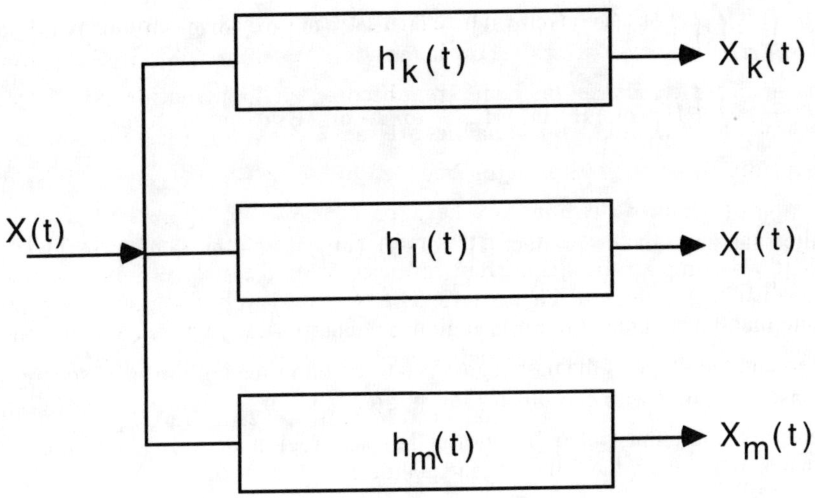

Figure 1.14: A time series is passed through three ideal bandpass filters.

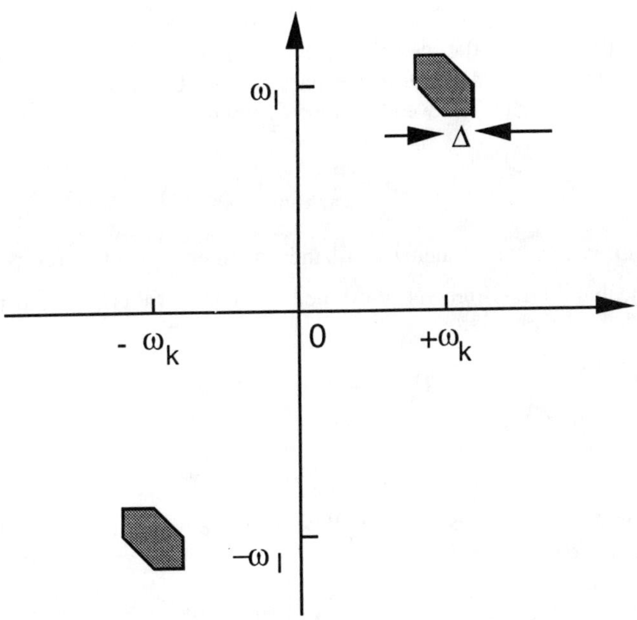

Figure 1.15: Passband of the product $H_k(\omega_1)H_l(\omega_2)H_m(-\omega_1-\omega_2)$.

and so on. We obtain

$$E\{X_k X_l X_m\}$$

$$= \left(\frac{1}{2\pi}\right)^3 \int\int\int H_k(\omega_1) H_l(\omega_2) H_m(\omega_3) Bs_X(\omega_1,\omega_2)\delta(\omega_1,\omega_2,\omega_3)\, d\omega_1 d\omega_2 d\omega_3$$

$$= \left(\frac{1}{2\pi}\right)^2 \int\int H_k(\omega_1) H_l(\omega_2) H_m(-\omega_1,-\omega_2) Bs_X(\omega_1,\omega_2)\, d\omega_1 d\omega_2. \tag{1.80}$$

The filter passband of the product $H_k(\omega_1)H_l(\omega_2)H_m(-\omega_1 - \omega_2)$ in (1.80) is shown in Figure 1.15.

Note that the area under the pass region is equal to $\frac{3}{4}\Delta^2$, where Δ is the passbandwidth of the filter, and the shape of the pass region is a hexagon instead of the usual square shape. To show this, consider a point with coordinates $\omega_1 + \delta\omega_1$ and $\omega_2 + \delta\omega_2$ lying inside the pass region. Further, this point must also lie within the pass region of individual filters; therefore, the coordinates must satisfy the following inequalities:

$$-\frac{\Delta}{2} \le \delta\omega_1 \le \frac{\Delta}{2}$$
$$-\frac{\Delta}{2} \le \delta\omega_2 \le \frac{\Delta}{2}$$
$$-\frac{\Delta}{2} \le \delta\omega_1 + \delta\omega_2 \le \frac{\Delta}{2}.$$

This requires that the shape of the pass region be as shown in Figure 1.15. We shall assume that the bispectrum is almost constant within the pass region. Using the symmetry properties of the bispectrum of a real process (1.73), we obtain the following result:

$$E\{x_k x_l x_m\} = \frac{3}{2}\Delta^2 \operatorname{Re}\{Bs_x(\omega_k,\omega_l)\}, \tag{1.81}$$

where ω_k and ω_l are center frequencies of bandpass filters h_k and h_l, respectively.

If one of the bandpass filters is combined with the Hilbert transform having a transfer function,

$$\begin{aligned} H(\omega) &= +j & \omega > 0 \\ &= 0 & \omega = 0 \\ &= -j & \omega < 0, \end{aligned}$$

the response of the product of three filters will be negative in the third quadrant. Consequently, in place of (1.81) we have the following result:

$$E\{x_k^H x_l x_m\} = -\frac{3}{2}\Delta^2 \operatorname{Im}\{Bs_x(\omega_k,\omega_l)\}, \tag{1.82}$$

where x_k^H stands for the Hilbert transform of $x_k(t)$.

D. Coefficient of Interaction

Consider the following model for interaction of power at frequencies ω_k and ω_l to produce a third frequency, $\omega_m = -\omega_k - \omega_l$. Let $x_k(t)$, $x_l(t)$, and $x_m(t)$, the three time series as defined earlier, satisfy a bilinear relationship,

$$x_m(t) = a_{k,l} x_k(t) x_l(t) + \varepsilon(t)$$
$$x_m^{\perp}(t) = b_{k,l} x_k^H(t) x_l(t) + \eta(t), \tag{1.83}$$

where $x_m^{\perp}(t)$ is the quadrature component of $x_m(t)$, and $\varepsilon(t)$ and $\eta(t)$ are error series. We would like to minimize the error power with respect to the interaction coefficients $a_{k,l}$ and $b_{k,l}$ in (1.83). We obtain the following result:

$$a_{k,l} = \frac{E\{x_k x_l x_m\}}{E\{x_k^2 x_l^2\}}$$

and

$$b_{k,l} = \frac{E\{x_k^H x_l x_m^{\perp}\}}{E\{x_k^{H^2} x_l^2\}}. \tag{1.84}$$

The fraction of power in $x_m(t)$ that can be predicted from $x_k(t)$ and $x_l(t)$, assuming the interaction model as given in (1.83), is

$$r = \frac{E\{x_k x_l x_m\}^2}{E\{x_k^2 x_l^2\} E\{x_m^2\}} + \frac{E\{x_k^H x_l x_m^{\perp}\}^2}{E\{x_k^{H^2} x_l^2\} E\{x_m^2\}}. \tag{1.85}$$

We shall assume for the sake of simplicity that

$$E\{x_k^2 x_l^2\} = E\{x_k^2\} E\{x_l^2\}$$

and

$$E\{x_k^{H^2} x_l^2\} = E\{x_k^{H^2}\} E\{x_l^2\},$$

a result that is strictly valid only if x_k and x_l are independent. Further, we note that $E\{x_k^2\} = 2\Delta S_x(\omega_k)$ and similarly for other time series. Using (1.81) and (1.82) in (1.85), we obtain

$$r = \frac{\frac{9}{4}\Delta^4 \left|Bs_x(\omega_k, \omega_l)\right|^2}{8\Delta^3 S_x(\omega_k) S_x(\omega_l) S_x(\omega_k + \omega_l)}$$
$$= \frac{9}{32}\Delta \left|bs(\omega_k, \omega_l)\right|^2. \tag{1.86}$$

The ratio appears to depend upon the filter passband. Thus, $r \to 0$ as $\Delta \to 0$. This may be explained[14] as the effect of the length of the bandpass filter. As $\Delta \to 0$ the filter length increases, and, therefore, its output tends to be Gaussian, as a consequence of the central limit theorem. The average of the triple product in (1.81) or in (1.82) would naturally tend to zero as Δ goes to zero. The non-Gaussian structure appears to have been destroyed in the process of filtering.

Example 1.8: A test signal is generated as follows:

$$x(t) = \cos(0.22\pi t + \theta_k) + \cos(0.375\pi t + \theta_l) + \frac{1}{2}\cos(0.595\pi t + \theta_m)$$
$$+ \cos(0.22\pi t + \theta_k)\cos(0.375\pi t + \theta_l) + \eta(t).$$

The phases θ_k, θ_l, and θ_m are taken from random numbers in the range of $-\pi$ to $+\pi$. $\eta(t)$ is the background Gaussian white noise (SNR=20 dB). The power spectrum of the above test signal showed four peaks at frequencies 0.155π, 0.22π, 0.375π, and 0.595π. The fourth term will give rise to two sinusoids at the sum frequency, 0.595π, and difference frequency, 0.155π. The bicoherence at $\omega_1=0.22\pi$ and $\omega_2=0.375\pi$ was found to be 0.48, and that at $\omega_1=-0.22\pi$ and $\omega_2=0.375\pi$ was found to be 0.99. The higher bicoherence is due to the fact that the power at the difference frequency, that is, at $\pm 0.155\pi$, is entirely due to the interaction. This theoretical model formed the basis for understanding the nonlinearly coupled waves in plasma reported in Reference 29 from where the above example is taken (End of Example).

E. Trispectrum

A fourth-order probability density function may be used to obtain additional information on the non-Gaussian character of a time series. A fourth-order moment of a zero mean random process, defined as

$$E\{x(t)x(t+s)x(t+h)x(t+g)\}$$
$$= \int\int\int\int x_1 x_2 x_3 x_4 f(x_1, x_2, x_3, x_4; t, t+s, t+h, t+g)\, dx_1 dx_2 dx_3 dx_4 \tag{1.87}$$

is called the tricovariance function, Tc(s,h,g), analogous to the bicovariance function introduced earlier. Note that when the process is Gaussian the tricovariance function can be written in terms of the covariance function (use (1.18)),

$$Tc_g(s,h,g) = C_x(s)C_x(h) + C_x(s)C_x(g) + C_x(h)C_x(g).$$

Often a fourth-order cumulant is defined by subtracting Tc_g from Tc; that is,

$$T_0(s,h,g) = Tc(s,h,g) - Tc_g(s,h,g). \tag{1.88}$$

The fourth-order cumulant has many useful properties when it is evaluated on the diagonal s=h=g in (1.87) and (1.88),

$$T_0(s) = E\{x(t)x^3(t+s)\} - 3C_x^2(s).$$

It is easy to show that $T_0(s) = 0$ for s=0 and $T_0(s) \to 0$ $|s| \to \infty$. The latter follows under the assumption that x(t) and x(t+s), $|s| \to \infty$ are independent. For some useful applications of $T_0(s)$ refer to Chapter Four.

The trispectrum is the Fourier transform of the tricovariance function,

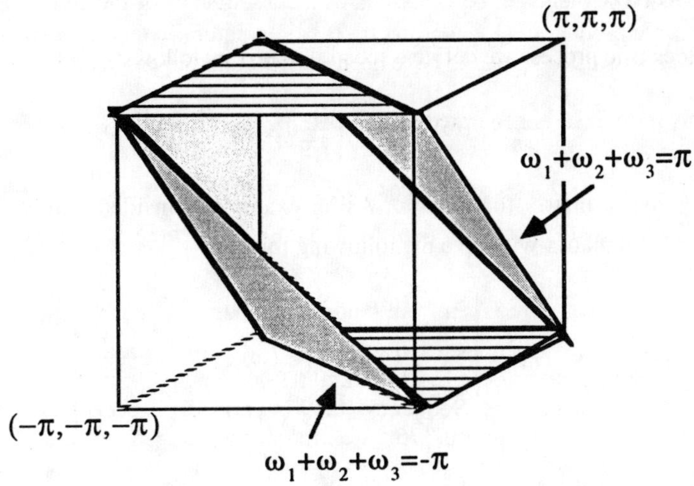

Figure 1.16: Unit volume of trispectrum, a octohedron with eight triangular faces.

$$Ts(\omega_1, \omega_2, \omega_3) = \frac{1}{8\pi^3} \sum_s \sum_g \sum_h Tc\,(s,h,g)\exp\,(-j(\omega_1 s + \omega_2 h + \omega_3 g)).$$

Like the bispectrum we find that the trispectrum is also periodic outside the basic unit volume defined by $|\omega_1| \le \pi$, $|\omega_2| \le \pi$, $|\omega_3| \le \pi$, and $|\omega_1 + \omega_2 + \omega_3| \le \pi$. The first three inequalities define a cube, and the last inequality defines a space bounded by two planes, $\omega_1 + \omega_2 + \omega_3 = \pi$ and $\omega_1 + \omega_2 + \omega_3 = -\pi$. The intersection of the cube and the space bounded by above planes is the unit volume of trispectrum. It is an octahedron (a solid with eight triangular faces). The construction of the unit volume is illustrated in Figure 1.16

1.6 Vector Stochastic Process

A collection of time series, often representing a single phenomenon, is a vector stochastic process. Let **x(t)** represent a vector time series with components $x_0(t), x_1(t), \dots, x_{N-1}(t)$, that is,

$$\mathbf{x}(t) = \text{col}\,\{x_0(t), x_1(t), \dots, x_{N-1}(t)\}, \tag{1.89}$$

where col{...} stands for column vector. As in the case of a single or scalar time series we shall assume that **x(t)** is a stationary stochastic process with a spectral representation,

$$\mathbf{x}(t) = \frac{1}{2\pi}\int_{-\pi}^{\pi} d\mathbf{X}(\omega)\exp\,(j\omega t), \tag{1.90}$$

where $d\mathbf{X}(\omega) = \text{col}\,\{dX_0(\omega), dX_1(\omega), \dots, dX_{N-1}(\omega)\}$, and $dX_i(\omega)$ is a differential of the generalized Fourier transform of the ith time series.

A. Spectral Matrix

For a vector stochastic process we define a spectral matrix as follows:

$$\frac{1}{2\pi}\mathbf{S}_{\mathbf{x}}(\omega)\,d\omega = E\{\frac{1}{2\pi}d\mathbf{X}(\omega)\frac{1}{2\pi}d\mathbf{X}^{H}(\omega)\}. \tag{1.91}$$

Note that $\mathbf{S}_{\mathbf{X}}(\omega$ is a matrix for fixed ω. As an example, consider a three-component vector process. The spectral matrix will have the following form:

$$\mathbf{S}_{\mathbf{x}}(\omega) = \begin{bmatrix} S_{x_0 x_0}(\omega) & S_{x_1 x_0}(\omega) & S_{x_2 x_0}(\omega) \\ S_{x_0 x_1}(\omega) & S_{x_1 x_1}(\omega) & S_{x_2 x_1}(\omega) \\ S_{x_0 x_2}(\omega) & S_{x_1 x_2}(\omega) & S_{x_2 x_2}(\omega) \end{bmatrix}. \tag{1.92}$$

Note that each entry on the diagonal represents the spectrum of individual time series, and the off-diagonal entries represent the cross spectra. Since

$$S_{x_1 x_2}(\omega) = S^{*}_{x_2 x_1}(\omega)$$

(see page 19), the spectral matrix becomes Hermitian. Further, the spectral matrix is always positive definite (or semidefinite). To see this property of the spectral matrix consider the following quantity:

$$E\{\left|\mathbf{A}^{H}\frac{1}{2\pi}d\mathbf{X}\right|^{2}\} \geq 0, \tag{1.93}$$

where **A** is an arbitrary filter vector. After carrying out the indicated operations, the inequality in (1.93) may be simplified to

$$\begin{aligned} E\{\mathbf{A}^{H}\frac{1}{2\pi}d\mathbf{X}\frac{1}{2\pi}d\mathbf{X}^{H}\mathbf{A}\} &= \mathbf{A}^{H}E\{\frac{1}{2\pi}d\mathbf{X}\frac{1}{2\pi}d\mathbf{X}^{H}\}\mathbf{A} \\ &= \frac{1}{2\pi}\mathbf{A}^{H}\mathbf{S}_{\mathbf{x}}(\omega)\mathbf{A} \geq 0. \end{aligned} \tag{1.94}$$

A matrix is said to be positive semidefinite whenever (1.94) is valid for any nonzero vector. Unlike a covariance matrix the spectral matrix is of generally non-Toeplitz form (see Example 1.9 for a specific instance of the spectral matrix becoming Toeplitz.). The fact that the spectral matrix is Hermitian positive semidefinite implies that all its eigenvalues are real and positive, or possibly zero. In case of a vector process we have three different types of coherences.[30,31]

B Ordinary Coherence

Let $x_k(t)$ and $x_l(t)$ be two components of a vector process. The ordinary coherence is defined as below:

$$\mathrm{Coh}_{k,l}(\omega) = \frac{S_{x_k x_l}(\omega)}{\sqrt{S_{x_k}(\omega) S_{x_l}(\omega)}}. \tag{1.95}$$

Note that we have N(N – 1)/2 ordinary coherence functions from an N-component vector process.

C. Multiple Coherence

It is a fraction of power in, say, the ith component that can be predicted from the knowledge of all or a subset of remaining components. Consider a component $x_i(t)$ and a subvector defined as

$$x_{\mathbf{m}}(t) = \text{col}\{ x_{m_1}(t), x_{m_2}(t), \ldots, x_{m_q}(t)\},$$

where $m_1, m_2, \ldots, m_q$ are integers lying between 0 and N – 1 but not equal to i, the index of the selected component. Let $x_{i\cdot\mathbf{m}}(t)$ represent the predicted time series using the subvector, $x_{\mathbf{m}}(t)$, where $\mathbf{m} = \{m_1, m_2, \ldots, m_q\}$. The multiple coherence is defined as a ratio of the predicted power to the total power,

$$\text{Coh}^{M}_{i\cdot\mathbf{m}}(\omega) = \frac{S_{x_{i\cdot m}}(\omega)}{S_{x_i}(\omega)}. \tag{1.96}$$

D. Partial Coherence

It is an ordinary coherence between two time series from which the predictable power has been subtracted. Let $u_k(t)$ and $u_l(t)$ be the residual time series defined as

$$u_k(t) = (x_k(t) - x_{k\cdot\mathbf{m}}(t))$$
and
$$u_l(t) = (x_l(t) - x_{l\cdot\mathbf{m}}(t)). \tag{1.97}$$

Note that the subvector **m** does not include the kth and l^{th} components. We define the partial coherence as

$$\text{Coh}^{P}_{k,l}(\omega) = \frac{S_{u_k u_l}(\omega)}{\sqrt{S_{u_k}(\omega) S_{u_l}(\omega)}}. \tag{1.98}$$

It is a complex number whose magnitude lies between 0 and 1.

Since the residual series is uncorrelated with the predicted component, that is, $u_k(t) \perp x_{k.\mathbf{m}}(t)$, we can write

$$S_{x_k}(\omega) = S_{u_k}(\omega) + S_{x_{k.m}}(\omega) \tag{1.99}$$

and similarly for the lth component,

$$S_{x_l}(\omega) = S_{u_l}(\omega) + S_{x_{l.m}}(\omega). \tag{1.100}$$

Further, let us assume that the residual in the kth time series is uncorrelated with the predicted component in the lth time series, that is, $u_k \perp x_{l.\mathbf{m}}$. We can then show that

$$S_{u_k u_l}(\omega) = S_{x_k x_l}(\omega) - S_{x_{k.m} x_{l.m}}(\omega). \tag{1.101}$$

Using Equations (1.99 to 1.101) in (1.98) we obtain

$$\mathrm{Coh}^{p}_{k,l}(\omega) = \frac{\left(1 - \dfrac{S_{x_{k.m} x_{l.m}}}{S_{x_k x_l}}\right)}{\sqrt{(1 - \mathrm{coh}^{M}_{k.m})(1 - \mathrm{coh}^{M}_{l.m})}} \mathrm{coh}_{k,l}(\omega). \tag{1.102}$$

Let $H_{k.m}(\omega)$ be a vector $(q \times 1)$ transfer function of the prediction filter which acts on q components, $x_{m_1}(t), x_{m_2}(t), \ldots, x_{m_q}(t)$ of the vector time series and produces as close a replica of $x_k(t)$ as possible. The output of the filter is given by

$$dX_{k.m}(\omega) = \mathbf{H}^{H}_{k.m}(\omega)\, dX_{\mathbf{m}}(\omega), \tag{1.103}$$

where $dX_{\mathbf{m}}(\omega)$ is the generalized Fourier transform of $x_{\mathbf{m}}(t)$. In order that

$$E\{(x_k(t) - x_{k.\mathbf{m}}(t))^2\} = \min,$$

the prediction filter (assumed to be two-sided) must be of the form given by

$$\mathbf{H}_{k.m}(\omega) = \mathbf{S}^{-1}_{x_m}(\omega)\mathbf{S}_{x_k x_m}(\omega), \tag{1.104}$$

where $\mathbf{S}_{x_k x_m}(\omega)$ is a column vector of cross-spectra between $x_k(t)$ and subvector $x_{\mathbf{m}}(t)$. The spectrum of the residual error is given by

$$S_{u_k}(\omega) = S_{x_k}(\omega) - \mathbf{S}^{H}_{x_k x_m}(\omega)\mathbf{S}^{-1}_{x_m}(\omega)\mathbf{S}_{x_k x_m}(\omega), \tag{1.105}$$

and the spectrum of the predicted component is given by

$$S_{x_{k.m}}(\omega) = \mathbf{S}^{H}_{x_k x_m}(\omega)\mathbf{S}^{-1}_{x_m}(\omega)\mathbf{S}_{x_k x_m}(\omega). \tag{1.106}$$

Using Equation (1.106) in (1.96) the multiple coherence may be written as follows:

$$\mathrm{Coh}^{M}_{k.m}(\omega) = \frac{\mathbf{S}^{H}_{x_k x_m}(\omega)\mathbf{S}^{-1}_{x_m}(\omega)\mathbf{S}_{x_k x_m}(\omega)}{S_{x_k}(\omega)}. \tag{1.107}$$

The fraction of power in $x_k(t)$ which cannot be predicted from $x_{\mathbf{m}}(t)$ is given by

$$\frac{S_{u_k}(\omega)}{S_{x_k}(\omega)} = [1 - \mathrm{Coh}^{M}_{k.m}(\omega)]. \tag{1.108}$$

Similarly, the partial coherence may be expressed as

$$\mathrm{Coh}^{P}_{k,l}(\omega) = \frac{S_{u_k u_l}(\omega)}{\sqrt{S_{u_k}(\omega) S_{u_l}(\omega)}}$$

$$= \frac{[1 - \mathrm{Coh}^{M}_{k.m,l.m}(\omega)]}{\sqrt{1 - \mathrm{Coh}^{M}_{k.m}(\omega)}\sqrt{1 - \mathrm{Coh}^{M}_{l.m}(\omega)}} \mathrm{Coh}_{k,l}(\omega), \tag{1.109}$$

where

$$\mathrm{Coh}^{M}_{k.m,l.m}(\omega) = \frac{\mathbf{S}^{H}_{x_k x_m}(\omega)\mathbf{S}^{-1}_{x_m}(\omega)\mathbf{S}_{x_l x_m}(\omega)}{S_{x_k x_l}(\omega)}$$

is a fraction of the common power in $x_k(t)$ and $x_l(t)$ which may be predicted from $x_{\mathbf{m}}(t)$.

Example 1.9: We shall compute the multiple coherence and partial coherence in a system where there is one input and many outputs. A simple example of such a system is an array of sensors for finding the direction of arrival (DOA) of a plane wavefront. Consider a linear array of p sensors. Let k=1 and $x_{\mathbf{m}}=$ col$\{x_2(t), x_3(t), \ldots, x_p(t)\}$. Then, $\mathbf{S_m}$ is a matrix formed from the spectral matrix of the array signal by dropping the first column and the first row. The vector $\mathbf{S}_{x_k x_m}(\omega)$ is the first column without the first element. This is shown in Figure 1.17.

For concreteness consider a four-element array, p=4 (see Figure 1.18). Let

$$\begin{aligned}
x_1(t) &= x(t) + \eta_1(t),\\
x_2(t) &= x(t-p) + \eta_2(t),\\
x_3(t) &= x(t-q) + \eta_3(t),\\
x_4(t) &= x(t-r) + \eta_4(t),
\end{aligned}$$

where p, q, and r are delays relative to the first sensor, and $\eta_1(t)$, $\eta_2(t)$, $\eta_3(t)$, and $\eta_4(t)$ are uncorrelated background noise terms with equal power, σ^2_η. We would like to compute the multiple coherence of $x_1(t)$ with respect to $x_2(t)$ and $x_3(t)$ and similarly for $x_4(t)$ with respect to $x_2(t)$ and $x_3(t)$. In (1.107) the required vectors and matrix are easily computed as follows:

$$\mathbf{S}_{x_1 x_m}(\omega) = \{S_x(\omega)e^{-j\omega p}, S_x(\omega)e^{-j\omega q}\}^T,$$

$$\mathbf{S}_{x_m}(\omega) = \begin{bmatrix} S_x(\omega) + \sigma^2_\eta & S_x(\omega)e^{j\omega(q-p)} \\ S_x(\omega)e^{-j\omega(q-p)} & S_x(\omega) + \sigma^2_\eta \end{bmatrix},$$

where $S_x(\omega)$ is the spectrum of x(t). After some simple algebra we obtain

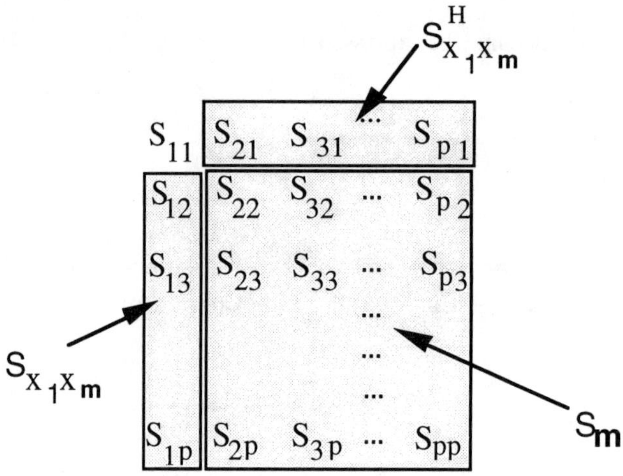

Figure 1.17: Spectral matrix of the output of a p sensor array is partitioned into a square matrix $\mathbf{S_m}$ and column matrix $S_{x_1 x_m}(\omega)$ for computation of multiple coherence.

$$\mathrm{Coh}^{M}_{1.\{2,3\}}(\omega) = \frac{2S_x^2(\omega)}{(S_x(\omega)+\sigma_\eta^2)(2S_x(\omega)+\sigma_\eta^2)} \underset{\sigma_\eta^2 \to 0}{=} 1.$$

Similarly for $x_4(t)$,

$$\mathrm{Coh}^{M}_{4.\{2,3\}}(\omega) = \frac{2S_x^2(\omega)}{(S_x(\omega)+\sigma_\eta^2)(2S_x(\omega)+\sigma_\eta^2)} \underset{\sigma_\eta^2 \to 0}{=} 1.$$

Next let us compute the partial coherence between sensors 1 and 4 after removing the combined effect of sensors 2 and 3. For this we need to compute $\mathrm{Coh}^{M}_{1.\{2,3\},4.\{2,3\}}(\omega)$, which is defined in (1.109), and is evaluated in the same manner as (1.107) was evaluated. We need an additional vector,

$$\mathbf{S}_{x_4 x_{2,3}}(\omega) = \{S_x(\omega)e^{-j\omega(p-r)}, S_x(\omega)e^{-j\omega(q-r)}\}.$$

We give the final result,

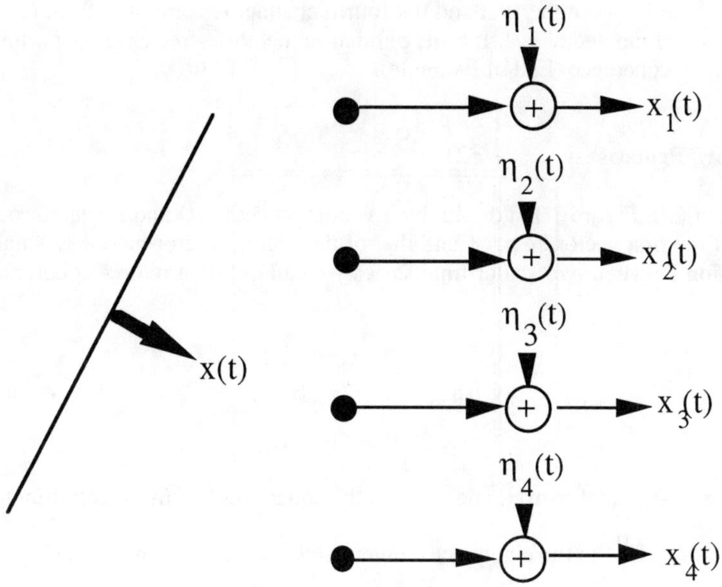

Figure 1.18: Four-element linear array of sensors. A plane wavefront with temporal waveform x(t) is incident on the array. To each sensor output a random white noise is added. The array output consists of four time series.

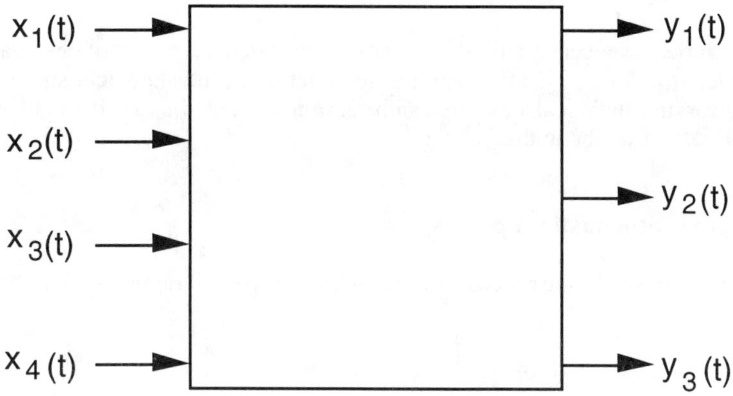

Figure 1.19: Vector input and vector output.

$$\mathrm{Coh}^{M}_{1.\{2,3\},4.\{2,3\}}(\omega) = \frac{2S_x^2(\omega)e^{j\omega r}}{(S_x(\omega)e^{j\omega r} + \sigma_\eta^2)(2S_x(\omega) + \sigma_\eta^2)}$$

$$= 1.$$

$$\sigma_\eta^2 \to 0$$

The partial coherence between the first and the fourth channel is computed from (1.109) using the multiple coherences computed above. It turns out that in the noise-free case the partial coherence is equal to the ordinary coherence (End of Example).

E. Two-Vector Processes

Consider a system (see Figure 1.19) driven by a vector process **x(t)** and its vector output **y**(t). Let the dimension of the first vector be $p \times 1$ and that of the second vector be $q \times 1$. Analogous to the coherence function between two scalar time series we can define a matrix coherence function as follows:

$$\mathbf{Coh}_{\mathbf{xy}} = \mathbf{S}_{\mathbf{x}}^{-\frac{1}{2}} \mathbf{S}_{\mathbf{xy}} \mathbf{S}_{\mathbf{y}}^{-\frac{1}{2}}, \tag{1.110}$$

where $\mathbf{S}_{\mathbf{xy}}$ is a cross-spectral matrix, and $\mathbf{S}_{\mathbf{x}}^{-\frac{1}{2}}$ is the square root of the spectral matrix defined as follows. Let $\mathbf{S}_{\mathrm{x}} = \mathbf{U}\Lambda\mathbf{U}^{\mathrm{H}}$ be the eigenvector/eigenvalue decomposition. We shall assume that the spectral matrix is positive definite. We define $\mathbf{S}_{\mathbf{x}}^{-\frac{1}{2}}$ as $\mathbf{S}_{\mathbf{x}}^{-\frac{1}{2}} = \mathbf{U}\Lambda^{-\frac{1}{2}}\mathbf{U}^{\mathrm{H}}$. Note that, because of positive definiteness of the spectral matrix, $\Lambda^{-\frac{1}{2}}$ is assumed to exist.

1.7 Nonstationary Process

No real-life time series is indeed strictly stationary over an extended period of observation. A more appropriate model would be to incorporate a time variation in the statistical structure of a time series. A slowly varying statistical structure can be accounted with the existing mathematical tools which we shall briefly describe in this section.

A. Time-Varying Stochastic Process

A discrete time-varying stochastic process has the following spectral representation:

$$x(t) = \frac{1}{2\pi} \int_{-\pi}^{\pi} H(t,\omega)\, d\varepsilon(\omega) \exp(j\omega t), \tag{1.111}$$

where $H(t,\omega)$ is a time-varying transfer function of a filter which is driven by a white noise process $\varepsilon(t)$ (see Figure 1.20). A component of x(t) at a specific frequency ω_0

$$x(t,\omega_0) = H(t,\omega_0) e^{j\omega_0 t} d\bar{\varepsilon}(\omega_0)$$

may be considered as a complex exponential of frequency ω_0 with time-varying and random amplitude and phase such that

$$E\{|x(t,\omega_0)|^2\} = S_x(t,\omega_0) d\omega_0.$$

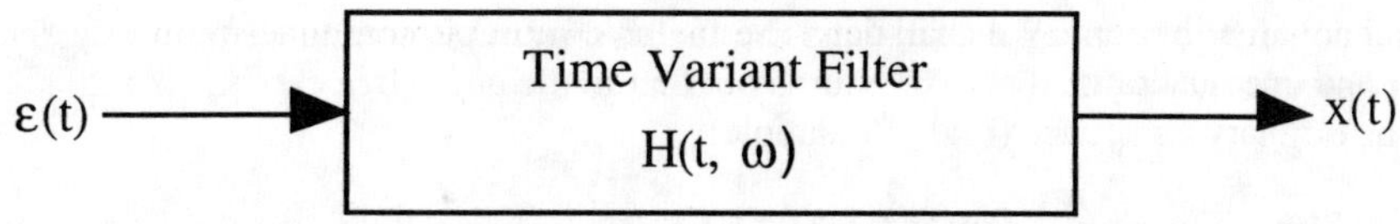

Figure 1.20: A time-variant linear system driven by stationary white noise processes.

We define a time-varying covariance function as follows:

$$C_x(t,\tau) = E\{x(t+\tfrac{\tau}{2})x(t-\tfrac{\tau}{2})\}. \tag{1.112a}$$

From the above equation we note that $C_x(t,\tau)$ is real and symmetric,

$$C_x(t,\tau) = C_x(t,-\tau).$$

Now, using the spectral representation (1.111) in (1.112a) we obtain

$$C_x(t,\tau) = \frac{\sigma_\varepsilon^2}{2\pi}\int_{-\pi}^{\pi} H(t+\tfrac{\tau}{2},\omega)H^*(t-\tfrac{\tau}{2},\omega)\exp(j\omega\tau)d\omega,$$

which may be considered as a spectral representation of the time-varying covariance function. However, there is no unique definition of a time-varying spectrum. Bendat and Piersol[32] proposed the following definition of a time-varying spectrum:

$$S_x(t,\omega) = \sum_{\tau=-\infty}^{\infty} E\{x(t+\tfrac{\tau}{2})x(t-\tfrac{\tau}{2})\}\exp(-j\omega\tau). \tag{1.112b}$$

Another type of time-varying spectrum is the evolutionary spectrum introduced by Priestly,[17]

$$S_x(t,\omega) = |H(t,\omega)|^2 \geq 0. \tag{1.113}$$

A time-varying cross-covariance function between two nonstationary time series may be defined as

$$C_{xy}(t,\tau) = E\{x(t+\tfrac{\tau}{2})y(t-\tfrac{\tau}{2})\} \tag{1.114a}$$

and the time-varying cross-spectrum as

$$S_{xy}(t,\omega) = \sum_{\tau=-\infty}^{\infty} C_{xy}(t,\tau)\exp(-j\omega\tau). \tag{1.114b}$$

A complete statistical characterization of a nonstationary signal is difficult, given just one sample of the stochastic process. It may be more appropriate to treat the available time series as a deterministic signal which is then mapped into the time-frequency plane. The map shows us how the various frequency components are distributed in the time-frequency plane. The physical quantity measured as a function of time and frequency is related to the energy, but this relationship is usually involved. Yet, a considerable insight into the structure of the signal can be gained from the time-frequency distributions. We shall describe the three most popular time-frequency distributions, two linear and one quadratic.

B. Short Time Fourier Transform (STFT)

When the time-varying filter, $H(t, \omega)$, is a slowly varying function of time, it may be approximated by the Fourier transform of a short segment of data, assuming that the filter transfer function does not vary much over the duration of the segment. The short-time Fourier transform (STFT)[33] is defined as

$$X(t,\omega) = \sum_{\tau=-\infty}^{\infty} x(\tau) w(t-\tau) \exp(-j\omega\tau), \tag{1.115}$$

where w(t) is a lowpass filter. Equation (1.115) may be looked upon as an output of a lowpass filter (centered at $\omega = 0$ and the bandwidth $= \Delta\omega$) with input $x(t)\exp(-j\omega t)$ (see Figure 1.21).

The STFT $X(t,\omega)$ is a complex function even though the actual signal is real. Further, $X(t,\omega)$ is discrete with its time and frequency intervals determined by the effective bandwidth and the effective length of the lowpass filter. There is, however, no unique definition of an effective bandwidth or an effective length. We shall define these quantities as follows: Let w(t) be an infinite length lowpass filter and W(w) be its Fourier transform. The *effective* length of a signal and its effective bandwidth may be defined as

$$T_{eff} = 2\sqrt{\frac{\sum_{t=-\infty}^{\infty} t^2 w(t)^2}{\sum_{t=-\infty}^{\infty} w(t)^2}}$$

and

$$\Delta\omega_{eff} = 2\sqrt{\frac{\int_{-\pi}^{\pi} \omega^2 |W(\omega)|^2 d\omega}{\int_{-\pi}^{\pi} |W(\omega)|^2 d\omega}},$$

respectively. It is known that $T_{eff}\Delta\omega_{eff} \geq 2\sqrt{2}$ (uncertainty principle[34]), where the sign of equality holds for a Gaussian window. STFT with Gaussian window is called a Gabor transform.[35] The STFT will be sampled in time with interval equal to T_{eff} and in frequency with interval equal to $\Delta\omega_{eff}$.

It is interesting to note that the STFT can be inverted. For this, we need to pass the STFT through an inverse filter (yet to be defined) and then compute its inverse Fourier transform,

$$x(t) = \frac{1}{2\pi}\int_{-\pi}^{\pi}\left\{\sum_{\tau=-\infty}^{\infty} X(\tau,\omega) w^{INV}(t-\tau)\right\} e^{j\omega t} d\omega.$$

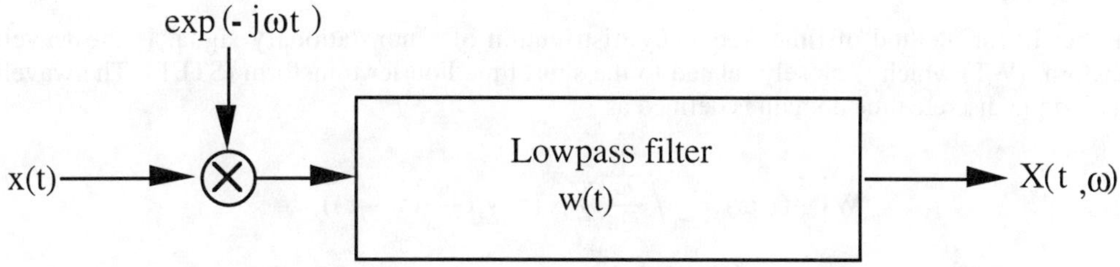

Figure 1.21 Short time Fourier transform is shown as an output of a lowpass filter whose input is a frequency shifted signal.

For this purpose the inverse filter must satisfy the condition,

$$\sum_{t=-\infty}^{\infty} w(t) w^{INV}(t) = 1$$

One simple choice is $w(t) = w^{INV}(t)$ with proper normalization. Before going back to the signal domain it is possible to modify the STFT, as one would modify the Fourier transform of a signal for the purpose of filtering.

Next we show a relationship between squared STFT and evolutionary spectrum. In (1.115) substitute for x(t) using (1.111). We get

$$X(t,\omega) = \frac{1}{2\pi}\int_{-\pi}^{\pi} d\bar{\varepsilon}(\omega')\exp(j(\omega' - \omega)t) \sum_{\tau=-\infty}^{\infty} w(\tau) H(t-\tau,\omega')\exp(-j(\omega'-\omega)\tau). \tag{1.116}$$

We assume that $H(t - \tau,\omega) \approx H(t,\omega)$ over the effective width of the lowpass filter then (1.116) reduces to

$$X(t,\omega) \approx H(t,\omega) e^{-j\omega t} \frac{1}{2\pi} \int_{\omega - \frac{\Delta\omega}{2}}^{\omega + \frac{\Delta\omega}{2}} d\bar{\varepsilon}(\omega') e^{j\omega' t}. \tag{1.117}$$

It is now possible to relate the STFT to the evolutionary spectrum defined in (1.112b). For this evaluate

$$E\left\{|X(t,\omega)|^2\right\} \approx |H(t,\omega)|^2 \frac{\sigma_\varepsilon^2}{2\pi}\Delta\omega$$
$$= S_x(t,\omega)\frac{\sigma_\varepsilon^2}{2\pi}\Delta\omega. \tag{1.118}$$

Equation (1.118) forms the basis for estimation of the time-varying spectrum. The simplest approach, therefore, is to use a bank of narrow bandpass filters whose outputs are squared and averaged over a moving window. The result is often known as a spectrogram. The method works well when the time variations are slow.

C. Wavelet Transform

Another linear method of time-frequency distribution of a nonstationary signal is the wavelet transform (WT) which is closely related to the short time Fourier transform (STFT). The wavelet transform in discrete time domain is defined as

$$WT_x(t,\omega) = \sqrt{\frac{\omega}{\omega_0}} \sum_{\tau} x(\tau)\,\gamma^*\!\left(\frac{\omega}{\omega_0}(\tau - t)\right),$$

where $\gamma(t)$ is a bandpass filter centered at frequency $\omega_0 \neq 0$. The bandwidth of the filter, however, does not remain fixed but linearly increases with the frequency. Let $\Gamma(\Omega)$ be the Fourier transform of G(t). Being a bandpass filter centered at ω_0, it is expected that $\Gamma(\Omega)$ will have a narrow peak at ω_0. The Fourier transform of $\sqrt{\frac{\omega}{\omega_0}}\,\gamma(\frac{\omega}{\omega_0}t)$ is $\sqrt{\frac{\omega_0}{\omega}}\,\Gamma(\frac{\omega_0}{\omega}\Omega)$, which has a maximum at $\Omega = \omega$ as desired, but its bandwidth increases with increasing frequency. Consider an ideal bandpass filter response,

$$\begin{aligned} \Gamma(\Omega) &= 1 \qquad \omega_0 - \frac{\Delta}{2} \leq \Omega \leq \omega_0 + \frac{\Delta}{2}, \\ &= 0 \qquad \text{otherwise} \;, \end{aligned}$$

The Fourier transform of the scaled bandpass filter, $\gamma(\frac{\omega}{\omega_0}t)$, is given by

$$\begin{aligned} \Gamma(\frac{\omega_0}{\omega}\Omega) &= 1 \qquad \omega_0 - \frac{\Delta}{2} \leq \frac{\omega_0}{\omega}\Omega \leq \omega_0 + \frac{\Delta}{2}, \\ &= 0 \qquad \text{otherwise} \;, \end{aligned}$$

whose center frequency is ω and bandwidth is $\Delta' = \frac{\omega}{\omega_0}\Delta$, which is linearly increasing with center frequency as shown in Figure 1.22. We express the Q factor of a filter as a ratio of the center frequency to bandwidth, $Q = \frac{\omega}{\Delta'}$. This filter's quality factor turns out to be a constant equal to $\frac{\omega_0}{\Delta}$.

The wavelet transform, like STFT, can be inverted provided the wavelet satisfies the so-called admissibility condition,

$$\Gamma(0) = 0 \,,$$

or in the time domain its mean value must be zero,

$$\sum_{t=-\infty}^{\infty} \gamma(t) = 0,$$

which will be satisfied by any bandpass filter whose center frequency is not at zero (see Reference 36, Chapter 5). Note that this is a necessary condition. Thus, g(t) is an impulse response of a bandpass filter, and since its mean is zero, it must oscillate, hence the name wavelet. A popular example of a wavelet function is the second derivative of a Gaussian function,

$$\gamma(t) = (1 - t^2)e^{\frac{-t^2}{2}} \leftrightarrow \Gamma(\Omega) = \sqrt{2\pi}\,\Omega e^{\frac{-\Omega^2}{2}} .$$

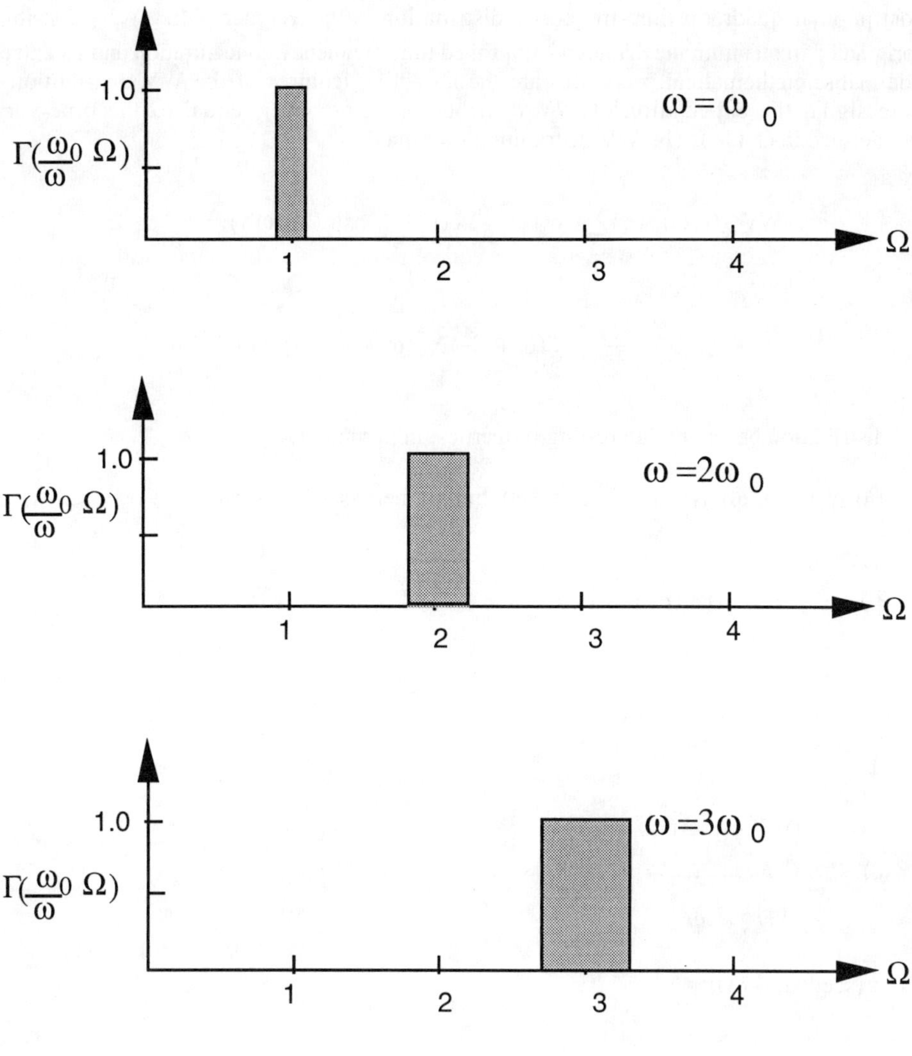

Figure 1.22: The bandwidth of the bandpass filter linearly increases with the center frequency.

This function is found to possess excellent property of localization in time and frequency, and it satisfies the admissibility condition.

The important difference between STFT and WT lies in the bandwidth of the filter. In STFT the filter is a lowpass filter of a fixed bandwidth, but in WT the filter is a bandpass filter with increasing passband, that is, with constant Q. At high frequencies WT achieves better time resolution but poor frequency resolution. As in STFT it is possible to invert WT and retrieve the original signal, possibly after some modification, if desired. The square of the wavelet transform is often known as a scalogram, which, like the spectrogram, gives a distribution of power as a function of time and scale (note: scale is inversely related to the frequency) but with a difference; that is, the bandpass filter used in the transformation possesses increasing bandwidth or constant Q.

D. Wigner-Ville Distribution

The most popular quadratic time-frequency distribution is the Wigner-Ville (WV) distribution which originated in quantum mechanics.[18] Improved time-frequency concentration and an extensive list of desirable mathematical properties are the attractive features of the WV distribution. For stochastic signal the expectation of WV distribution is, however, equal to the time-varying spectrum defined in (1.114). The WV distribution is defined as

$$WV_x(t,\omega) = \sum_{\tau=-\infty}^{\infty} x(t+\tfrac{\tau}{2})x(t-\tfrac{\tau}{2})\exp(-j\omega\tau)$$

$$= \frac{1}{2\pi}\int_{-\pi}^{\pi} X(\omega+\tfrac{\omega'}{2})X^*(\omega-\tfrac{\omega'}{2})\exp(j\omega t)\,d\omega'. \tag{1.119}$$

The WV distribution has many interesting properties, in particular,

(a) $WV_x(t,\omega)$ is real for all (t,ω) but not necessarily positive,

(b) $\frac{1}{2\pi}\int_{-\pi}^{\pi} WV_x(t,\omega)\,d\omega = |x(t)|^2$,

(c) $\sum_{t=-\infty}^{\infty} WV_x(t,\omega) = |X(\omega)|^2$,

(d) $\dfrac{\sum_{t=-\infty}^{\infty} tWV_x(t,\omega)}{\sum_{t=-\infty}^{\infty} WD_x(t,\omega)} = \text{group delay}$

(see page 78)),

(e) $WV_{x_0}(t,\omega) = WV_x(t-t_0,\omega)$,

where $x_0(t) = x(t-t_0)$,

and

(f) $WV_{x_1}(t,\omega) = WV_x(t,\omega-\omega_1)$,

where $x_1(t) = x(t)\exp(j\omega_1 t)$.

Example 1.10: Let us consider a few simple examples of WV distributions. Let $x(t) = \exp(j\omega_0 t)$ and compute the WV using its definition in (1.119).

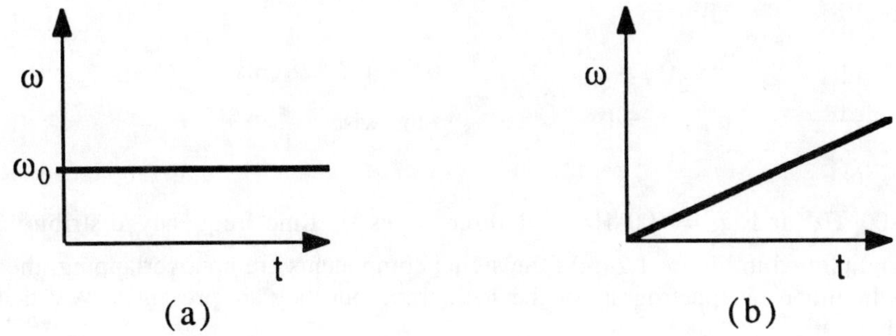

Figure 1.23: WV distribution. (a) Complex sinusoid. (b) Complex chirp.

$$WV_x(t,\omega) = \sum_{\tau=-\infty}^{\infty} \exp(j\omega_0(t+\tfrac{\tau}{2}))\exp(-j\omega_0(t-\tfrac{\tau}{2}))\exp(-j\omega\tau)$$

$$= \sum_{\tau=-\infty}^{\infty} \exp(j(\omega_0-\omega)\tau) = \delta(\omega_0-\omega)$$

In (t,ω) plane the WV distribution of a simple sinusoid is a horizontal line parallel to the time axis (see Figure 1.23a). Let us take another example of a chirp signal, $x(t) = \exp(-j\beta t^2)$, where β is a constant.

$$WV_x(t,\omega) = \sum_{\tau=-\infty}^{\infty} \exp(j\beta(t+\tfrac{\tau}{2})^2)\exp(-j\beta(t-\tfrac{\tau}{2})^2)\exp(-j\omega\tau)$$

$$= \sum_{\tau=-\infty}^{\infty} \exp(j2\beta t\tau - j\omega\tau) = \delta(2\beta t - \omega)$$

The WV distribution of a chirp signal is a radial line in the (t, ω) plane as shown in Figure 1.23b (End of Example).

The WV distribution suffers from an important drawback, namely, the presence of interference terms, even when the components of the signals are nonoverlapping in the time-frequency plane. This is best illustrated in the following numerical example. (Many other examples and applications of the WV and related distributions are given in Reference 19). The spectrogram as well as the scalogram suffer from a similar drawback. But there is one important difference; namely, the cross terms are absent when the signal components are nonoverlapping.

Example 1.11: Consider a two-component discrete complex signal

$$x(t) = x_1(t) + x_2(t),$$

where

$$x_1(t) = e^{j2\pi f_1 t} \qquad 0 \le t \le 32 \text{ ms}$$

$$= 0 \qquad \text{otherwise} \quad ;$$

$$x_2(t) = e^{j2\pi f_2 t} \qquad 64 \le t \le 128 \text{ ms}$$
$$= 0 \qquad \text{otherwise}$$

Let $f_1 = 200$ Hz and $f_2 = 400$ Hz. All three types of time-frequency distributions were computed and plotted in Figure 1.24. As the signal components are nonoverlapping, there are no cross terms in either the spectrogram or the scalogram, but they are present in WV distribution (End of Example).

Another type of quadratic time-frequency distribution of a signal is the ambiguity function defined as

$$A(\Omega,\tau) = \sum_{t=-\infty}^{\infty} x(t+\tfrac{\tau}{2})x^*(t-\tfrac{\tau}{2})e^{-j\Omega t}. \tag{1.120}$$

It is easy to show that the Fourier transform of $A(\Omega,\tau)$ is equal to WV_X; that is,

$$WV_x(t,\omega) = \frac{1}{2\pi}\int_{-\pi}^{\pi} d\Omega \sum_{\tau=-\infty}^{\infty} A(\Omega,\tau)e^{j(\Omega t-\omega\tau)}. \tag{1.121}$$

Before computing the Fourier transform of $A(\Omega,\tau)$, one often uses a weighting or a filter function to remove the cross terms in the WV distribution. By selecting different types of filter functions we obtain a class of WV distributions (Cohen class). For example, the following WV distributions belong to Cohen class:

$$WV_x(t,\omega) = \frac{1}{2\pi}\int_{-\pi}^{\pi} d\Omega \sum_{\tau=-\infty}^{\infty} A(\Omega,\tau)\Phi(\Omega,\tau)e^{j(\Omega t-\omega\tau)},$$

where

$\Phi(\Omega,\tau) = 1$:Wigner – Ville distribution ,

$\Phi(\Omega,\tau) = e^{-\frac{(\tau\omega)^2}{\sigma}}$:Choi – Williams distribution ,

$\Phi(\Omega,\tau) = \dfrac{1}{1+(\frac{\tau}{\tau_0})^{2M}+(\frac{\Omega}{\Omega_0})^{2N}}$:Butterworth distribution .

E. Cyclostationary Process

We shall now consider a class of nonstationary processes where the temporal variation of the second-order statistical structure is periodic. They are useful in modeling time series generated by modulating a periodic process by a stochastic process. Many modern-day communication signals

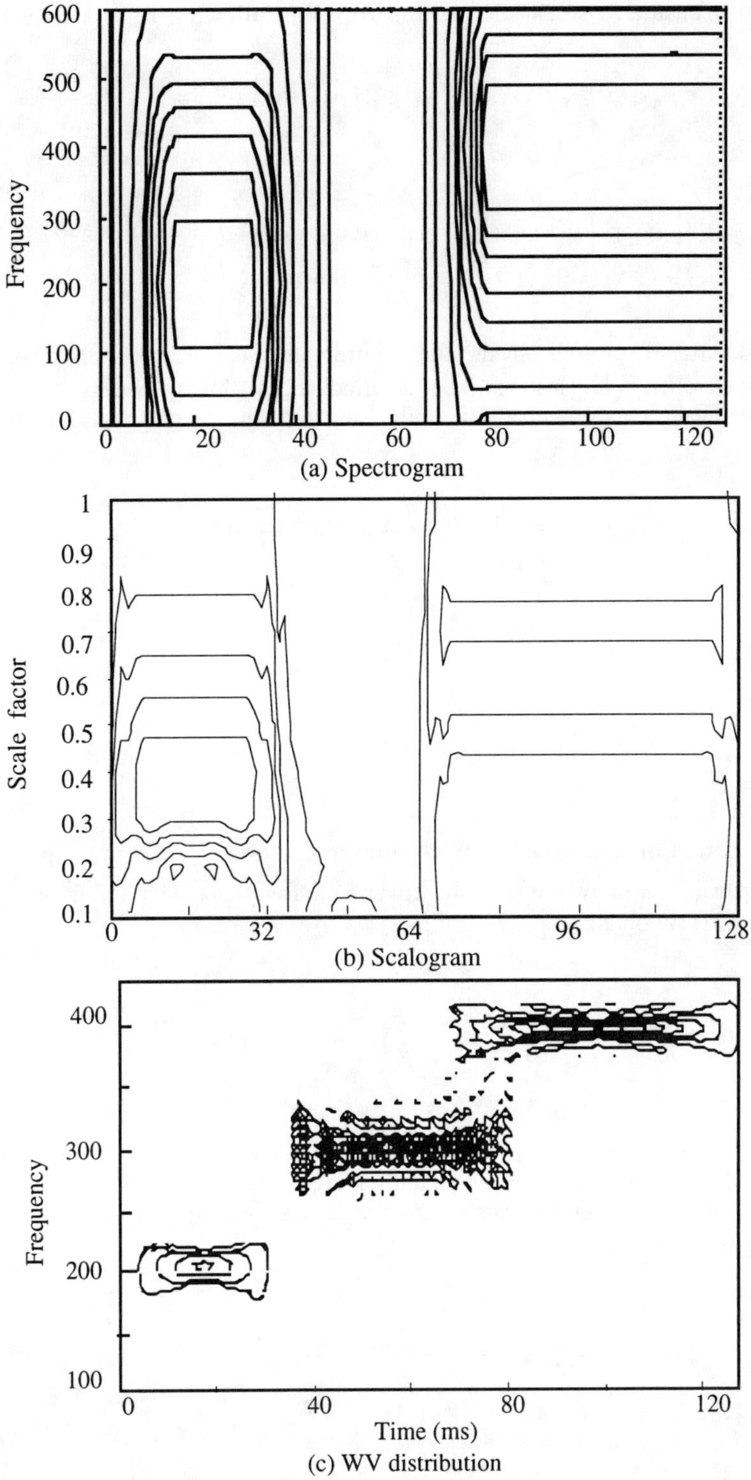

Figure 1.24: Three different types of time-frequency distributions: (a) Spectrogram.(b) Scalogram. (c) WV distribution. Note the absence of cross terms in (a) and (b) but their prominence in (c).

fall in this group. Consider a stochastic process with the following first- and second-order characteristics:

$$E\{x(t)\} = m_x(t) = m_x(t + pT_0), \quad p = 0, 1, 2, \ldots$$

and

$$E\{[x(t+\tfrac{\tau}{2}) - m_x(t+\tfrac{\tau}{2})][x(t-\tfrac{\tau}{2}) - m_x(t-\tfrac{\tau}{2})]\}$$
$$= C_x(t,\tau) = C_x(t + pT_0, \tau) \qquad p = 0, 1, 2, \ldots \tag{1.122}$$

The covariance function is periodic in the time coordinate. Such a process has been termed *cyclostationary* by Gardner.[37] It is also known as a periodically stationary process.

Since the covariance function is periodic in the time variable, we can expand it in a Fourier series,

$$C_x(t,\tau) = \sum_{p=-\infty}^{\infty} C_x(\frac{2\pi p}{T_0}, \tau) \exp(j\frac{2\pi p}{T_0}t) \quad , \tag{1.123a}$$

with inverse,

$$C_x(\frac{2\pi p}{T_0}, \tau) = \frac{1}{T}\sum_{-\frac{T}{2}}^{\frac{T}{2}} C_x(t,\tau)\exp(-j\frac{2\pi p}{T_0}t), \tag{1.123b}$$

where T_0 is the period of repetition of the covariance function in t. Each component in the Fourier series expansion is known as a cyclic covariance function. Substituting in (1.123) for $C_x(t,\tau)$ from (1.122) we obtain

$$C_x(\frac{2\pi p}{T_0}, \tau) = E\{\frac{1}{T}\sum_{-\frac{T}{2}}^{\frac{T}{2}} x(t+\frac{\tau}{2})x(t-\frac{\tau}{2})\exp(-j\frac{2\pi p}{T_0}t)\}. \tag{1.124}$$

The summation in (1.124) may be evaluated using just a single observed time series. There is no need for the statistical operation, E{.}, which requires a large number of realizations. Analogous to the autocorrelation defined in (1.54) using a single time series, we shall introduce a cyclic autocorrelation function,[38]

$$\hat{C}_x(\frac{2\pi p}{T_0}, \tau) = \frac{1}{T}\sum_{-\frac{T}{2}}^{\frac{T}{2}} x(t+\frac{\tau}{2})x(t-\frac{\tau}{2})\exp(-j\frac{2\pi p}{T_0}t). \quad T \to \infty \tag{1.125}$$

A cyclic covariance function reduces to a simple covariance function for p=0 as can be seen from (1.124), where the exponential term is equal to one, and the rest of the expression is the standard covariance function. Again from (1.124) it follows that for a stationary process the cyclic covariance function is equal to zero, except for p=0, when it is equal to the covariance function of the stationary process. The magnitude of the cyclic covariance function for $p \neq 0$ may be used as a measure of the degree of cyclostationarity. We give a different explanation to (1.124). For this purpose let us rewrite (1.124) as

$$C_x\left(\frac{2\pi p}{T_0},\tau\right) = E\left\{\frac{1}{T}\sum_{t=0}^{T-1} x\left(t+\frac{\tau}{2}\right)\exp\left(-j\frac{\pi p}{T_0}\left(t+\frac{\tau}{2}\right)\right)x\left(t-\frac{\tau}{2}\right)\exp\left(-j\frac{\pi p}{T_0}\left(t-\frac{\tau}{2}\right)\right)\right\}$$

$$= E\left\{\frac{1}{T}\sum_{t=0}^{T-1} u_p\left(t+\frac{\tau}{2}\right)v_p^*\left(t-\frac{\tau}{2}\right)\right\}, \tag{1.126}$$

where

$$u_p(t) = x(t)\exp\left(-j\frac{\pi p}{T_0}t\right),$$

$$v_p(t) = x(t)\exp\left(+j\frac{\pi p}{T_0}t\right).$$

We can consider $u_p(t)$ and $v_p(t)$ as frequency shifted versions of x(t), by an amount equal to $-\frac{\pi p}{T_0}$ and $+\frac{\pi p}{T_0}$, respectively. The cyclic covariance function is a cross-covariance between $u_p(t)$ and $v_p(t)$.

Next let us introduce the cyclic spectrum as a Fourier transform of the cyclic covariance function.

$$S_x\left(\frac{2\pi p}{T_0},\omega\right) = \sum_{-\infty}^{\infty} C_x\left(\frac{2\pi p}{T_0},\tau\right)\exp(-j\omega\tau). \tag{1.127}$$

Both the cyclic covariance function and the cyclic spectrum share many of the properties of the covariance and spectrum of a stationary process. The cyclic spectrum may also be looked upon as a cross-spectrum of $u_p(t)$ and $v_p(t)$ defined in (1.126). Using (1.126) in (1.127) we obtain

$$S_x\left(\frac{2\pi p}{T_0},\omega\right) = \sum_{\tau=-T}^{T} E\left\{\frac{1}{T}\sum_{t=0}^{T-1} u_p\left(t+\frac{\tau}{2}\right)e^{-j\left(t+\frac{\tau}{2}\right)\omega} v_p^*\left(t-\frac{\tau}{2}\right)e^{j\left(t-\frac{\tau}{2}\right)\omega}\right\}$$

$$= E\left\{\frac{1}{T}\sum_{t=\frac{\tau}{2}}^{T-1+\frac{\tau}{2}} u_p(t)e^{-j\omega t}\sum_{t'=-\frac{\tau}{2}}^{T-1-\frac{\tau}{2}} v_p^*(t')e^{+j\omega t'}\right\}$$

$$= E\left\{\frac{1}{T}\ {}_p(\omega)\bar{V}_p^*(\omega)\right\} = S_{u_p v_p}(\omega). \tag{1.128}$$

$S_{u_p v_p}(\omega)$ is the correlation between the frequency components $\omega + \frac{\pi p}{T_0}$ and $\omega - \frac{\pi p}{T_0}$. Note that for stationary stochastic process, $S_{u_p v_p}(\omega) = 0$.

F. Communication Signals

The significance of cyclostationarity rests on the facts that many communication signals can be modeled as cyclostationary processes and that, since the cyclic covariance function of a stationary process is identically zero for $p \neq 0$, it is at least in principle possible to remove the noise or the interference which is more likely to be a stationary process unlike a communication signal. This has been the basis for current interest in the cyclostationary process.[37,38] We shall briefly examine here how the communication signals are modeled as a cyclostationary process. The signal to be communicated is used to modulate a carrier. The signal may be looked upon as a pure stationary stochastic process, but the carrier is a periodic wave form, for example, a sinuoidal carrier whose amplitude or phase or frequency are varied according to the signal. The dominant effect of the modulation is to introduce a correlation between two Fourier coefficients at frequencies separated by twice the carrier frequency. This effect will be illustrated through an example of amplitude modulation of a sinusoidal carrier. Note that for a purely stationary process there is *no* correlation between the two differentials of the generalized Fourier transform over nonoverlapping intervals (1.61).

Example 1.12: Consider an amplitude modulated (AM) sinusoidal carrier signal (some more examples of communication signals may be found in References 37,38),

$$x(t) = x_0(t)\cos(\omega_0 t + \theta),$$

where $x_0(t)$ is a stationary process and q is the phase of the carrier. We can represent the signal as a sum of up and down frequency-shifted components (like $u_p(t)$ and $v_p(t)$ in (1.126)),

$$x(t) = \frac{1}{2}x_0(t)e^{j(\omega_0 t+\theta)} + \frac{1}{2}x_0(t)e^{-j(\omega_0 t+\theta)},$$

and use it in (1.124) for computation of the cyclic covariance function. We obtain, after some algebraic simplification, the following result:

$$\begin{aligned} C_x\left(\frac{2\pi p}{T_0},\tau\right) &= \frac{1}{4}C_x(\tau)\,e^{\pm j2\theta} && \text{for } p = \pm 2, \\ &= \frac{1}{2}C_x(\tau)\cos(\omega_0\tau) && \text{for } p = 0, \\ &= 0 && \text{otherwise} \end{aligned}$$

The cyclic spectrum, obtained by Fourier transforming the cycliccovariance function is given by

$$\begin{aligned} S_x\left(\frac{2\pi p}{T_0},\omega\right) &= \frac{1}{4}S_x(\omega)\,e^{\pm j2\theta} && \text{for } p = \pm 2, \\ &= \frac{1}{4}\{S_x(\omega+\omega_0) + S_x(\omega-\omega_0)\} && \text{for } p = 0, \\ &= 0 && \text{otherwise}. \end{aligned}$$

In Figure 1.25 the cyclic spectrum is shown as a function two frequencies, ω and $\frac{2\pi p}{T_0}$, where $p = 0, \pm 2$ (End of Example).

The most interesting feature in the plot is the presence of a signal spectrum at $\frac{2\pi p}{T_0} = \pm 2\omega_0$ which represents the common signal power between two frequency-shifted components of the

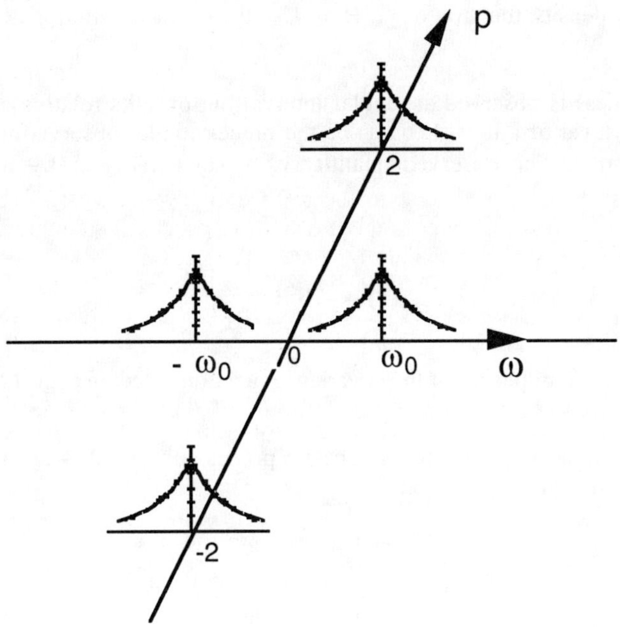

Figure 1.25: Spectrum of amplitude modulated signal. The message is a first-order AR process. ω_0 is the carrier frequency. For simplicity the phase is assumed to be zero.

signal. For a pure stationary process this phenomenon is not observed. It is therefore possible to devise a filtering scheme exploiting the above unique property of the cyclostationary process for the purpose of extracting a communication signal in the presence of noise or interference which is more likely to be a pure stationary process.[38] Note the important role of the phase of the carrier. If we treat the phase as unknown or random, a uniformly distributed random variable, we find that the averaged cyclic spectrum will no longer contain the term representing the signal power at $\frac{2\pi p}{T_0} = \pm 2\omega_0$ (since $E\{\exp(j2\theta)\} = 0$). The process behaves like a pure stationary process Therefore, a knowledge of the phase or the timing information is mandatory. It is a basic property of a cyclostationary process that when shifted by a random time interval, which is uniformly distributed over the period of repetition, T_0, the cyclostationary process becomes a stationary process.[26]

1.8 **Exercises**

1. Let x_1 and x_2 be two independent random variables. Define another random variable, $y=x_1+x_2$. Let the probability density functions of x_1 and x_2 be given as

$$f_1(x_1) = \frac{1}{2}\delta(x_1 - 2) + \frac{1}{2}\delta(x_1 - 2.5)$$

$$f_2(x_2) = \frac{1}{2} \quad -1 \le x_2 \le 1$$

$$= 0 \quad \text{otherwise} \;.$$

Find the probability density function of y. [Hint: Use the characteristic function approach to find the pdf of y.]

2. A stochastic process is observed at regular interval but only its relative change is noted, for example, the water level of a lake. Let x(t) be the process under observation and h be the time interval of observation. The observed quantity is $x_1(t) = x(t) - x(t-h)$. Show that for small h,

$$C_{x_1}(\tau) \approx -h^2 \frac{\partial^2 C_x(\tau)}{\partial \tau^2}.$$

Does $C_{x_1}(\tau)$ satisfy all properties of the covariance function listed on page 12?

3. Now consider a situation where the stochastic process observed is an integral over a time interval h. The observed quantity is given by

$$x_1(t) = \int_{t-h}^{t} x(t)dt \qquad h > 0,$$

for example, the rainfall over a duration of time h. Show that

$$C_{x_1}(\tau) = \frac{1}{2\pi} \int_{-\infty}^{\infty} S_x(\omega) \sin c^2 \frac{\omega h}{2} \exp(j\omega\tau) d\omega.$$

Discuss the effect of integration in terms of filtering.

4. Let a random process x(t) be given by $x(t) = a\cos(\omega t + \phi)$, where ω is a random frequency with a symmetric pdf given by $f(\omega)$ and ϕ is a random phase uniformly distributed over an interval $0 < \phi < 2\pi$ and independent of ω. Show (a) the random process is stationary (second order) and (b) its spectrum is given by $\frac{a^2}{2} f(\omega)$.[39]

5. Let $y(t) = bx(t) + x^3(t)$, where x(t) is a zero mean Gaussian stochastic process. Compute the covariance function of y(t) in terms of the covariance function of x(t). Show that the coherence function between x(t) and y(t) is *not* equal to one. In fact, it can be shown to be equal to zero if $b = -3\sigma^2$, where σ^2 is the variance of x(t).[26]

6. A triple correlation of a deterministic real valued function is defined as

$$Bc(s,h) = \int_{-\infty}^{\infty} x(t)x(t+s)x(t+h)\,dt$$

and bispectrum as a Fourier transform of the triple correlation,

$$Bs_x(\omega_1,\omega_2) = \sum_{-\infty}^{\infty}\sum_{-\infty}^{\infty} Bc(s,h)\exp(-j(s\omega_1 + h\omega_2))$$
$$= X(\omega_1)X(\omega_2)X(-\omega_1, -\omega_2).$$

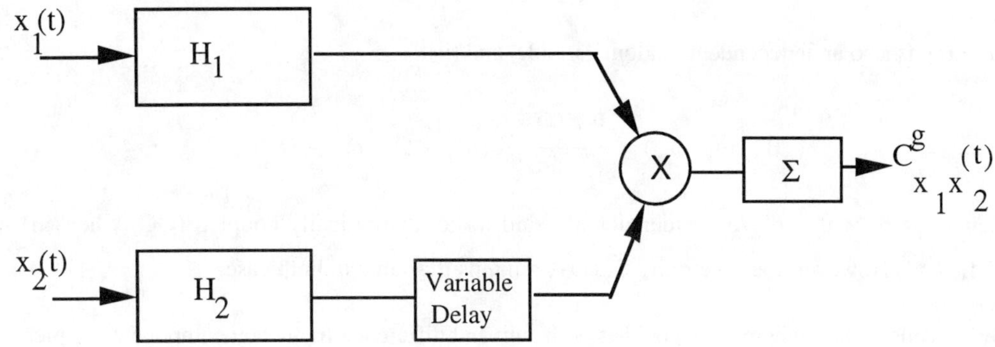

Figure 1.26: A generalized cross-covariance function.

Sketch the triple correlation function of x(t)= rect[t/ Δ], that is x(t)=1, for $|t| \leq \frac{\Delta t}{2}$, otherwise equal to zero. Also, sketch the bispectrum when $X(\omega) = \mathrm{Rect}\,[\frac{\omega}{\Delta\omega}]$. (For details see Reference 40).

7. Let $\varepsilon(t)$ be a discrete time series. We form another time series y(t) as follows:

$$y(t) = (-1)^t \varepsilon(t).$$

Find the properties of the new time series; in particular, is it stationary? Derive these relations assuming $\varepsilon(t)$ is a stationary white noise sequence.

8. A scheme for generalized cross-covariance (GCC) is shown in Figure 1.26. Let H_1 and H_2 be whitening filters. Show that the generalized cross-covariance is the Fourier transform of the coherence function. GCC has found application in the estimation of time delay. How would you measure the time delay? [Reference 41 gives a good review of the work on coherence and time delay estimation.]

9. A linear system is driven by a non-Gaussian white noise process. Given the output process, one can compute the phase of the transfer function using the bispectrum. Show that the bispectrum of x(t) may be given by

$$Bs_x(\omega_1, \omega_2) = H(\omega_1)H(\omega_2)H^*(\omega_1 + \omega_2)\beta_\varepsilon^3,$$

where β_ε^3 is the third moment of $\varepsilon(t)$. Rosenblatt[42] gave a method for estimating the phase of H(w) from the bicoherence function (see Section 2.6 for more details).

10. Let

$$x(t) = \alpha_1 \cos(\Omega_1 t + \theta_1) + \alpha_2 \cos(\Omega_2 t + \theta_2) + \alpha_3 \cos((\Omega_1 + \Omega_2)t + \theta_3),$$

where θ_1 and θ_2 are independent uniformly distributed random variables and $\Omega_2 > \Omega_1$. Show that (i)

$$B_x(\omega_1, \omega_2) = 0$$

when θ_3 is also an independent random variable, and (ii)

$$B_x(\omega_1, \omega_2) = \frac{\alpha_1 \alpha_2 \alpha_3}{8} \delta(\omega_1 - \Omega_1, \omega_2 - \Omega_2)$$

when $\theta_3 = \theta_1 + \theta_2$. (Two frequencies are said to be quadratically coupled (QC) when (ii) is satisfied.[43]) However, the spectrum, $S_x(\omega)$, remains the same in both cases.

Now consider a case where the signal has both sum and difference frequency components (typically caused by nonlinear interaction),

$$x(t) = \alpha_1 \cos(\Omega_1 t + \theta_1) + \alpha_2 \cos(\Omega_2 t + \theta_2) + \alpha_3 \cos((\Omega_1 + \Omega_2)t + \theta_3) + \alpha_4 \cos((\Omega_2 - \Omega_1)t + \theta_4),$$

where $\theta_4 = \theta_2 - \theta_1$. Show that there is an additional peak at $(\Omega_1, \Omega_2 - \Omega_1)$ which can cause some confusion as regard to fundamental and QC frequencies, but it can be eliminated by modifying the definition of the bispectrum as suggested in Reference 44.

11. Prove the property (d) of Wigner-Ville (WV) distribution on page 54 of the current chapter. To prove this result it is convenient to start with the definition of WV distribution in the frequency plane (see (1.119)). The proof is outlined in Reference 45, but it is educative to derive all intermediate steps.

12. A cross-spectrum is a measure of common power between two time series, say, $x_k(t)$ and $x_l(t)$. Consider another vector time series, $x_{\mathbf{m}}(t)$, influencing both $x_k(t)$ and $x_l(t)$. We define a ratio of the cross-spectrum between the predicted parts in $x_k(t)$ and $x_l(t)$ and the cross-spectrum between $x_k(t)$ and $x_l(t)$. Show that this ratio is equal to

$$\mathrm{Coh}^{M}_{k.\mathbf{m},l.\mathbf{m}}(\omega) = \frac{S^H_{x_m x_k} S^{-1}_{x_m} S_{x_m x_l}}{S_{x_k x_l}}$$

13. Let $\mathbf{x} = \mathrm{Col}\{x(0), x(1), \ldots, x(T-1)\}$, where x(t), t=0,1, ..., $T-1$ is a stationary stochastic time series. Show that the covariance matrix $\mathbf{C}_x$ of the time series may be expressed as follows:

$$\mathbf{C}_x = \frac{1}{2\pi} \int_{-\pi}^{\pi} S_x(\omega) \mathbf{h}(\omega) \mathbf{h}^H(\omega) d\omega,$$

where $\mathbf{h}(\omega) = \mathrm{col}\{1 \;\; e^{-j\omega} \;\; e^{-j2\omega} \;\; \ldots \;\; e^{-j(T-1)\omega}\}$ (frequency vector). [Hint: Express each component of **x** in terms of its spectral representation.]

14. Next, we will generalize the result proved in Exercise 1.13 to vector time series. Let us define a vector by stacking all components as follows:

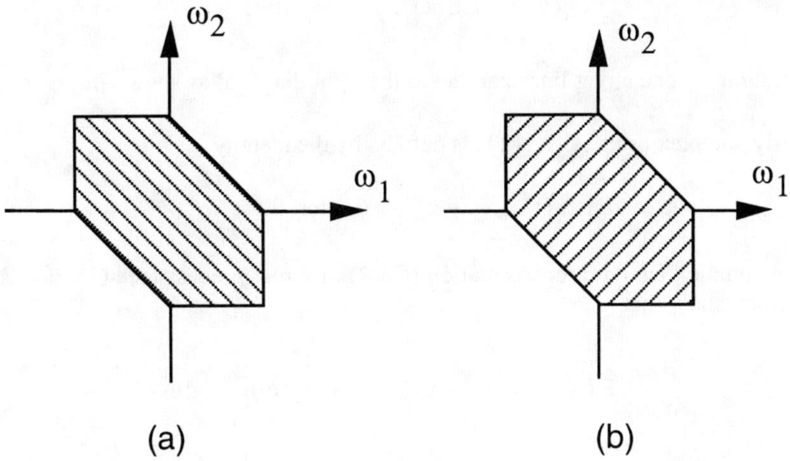

Figure 1.27: Integration of bispectrum along (a) left diagonal and (b) right diagonal.

$$\mathbf{x} = \text{Col}\{\ x_0(0)\ x_0(1)\ \ldots\ x_0(T-1)\ x_1(0)\ x_1(1)\ \ldots\ x_1(T-1)\ \ldots\ x_{N-1}(0)\ x_{N-1}(1)\ \ldots\ x_{N-1}(T-1)\}.$$

Show that

$$\mathbf{C}_x = E\{\mathbf{x}\mathbf{x}^H\} = \frac{1}{2\pi}\int_{-\pi}^{\pi} \mathbf{S}_{\mathbf{x}}(\omega) \otimes [\mathbf{h}(\omega)\mathbf{h}^H(\omega)]d\omega,$$

where $\mathbf{S}_{\mathbf{x}}(\omega)$ is the spectral matrix of the vector time series, and $\otimes$ stands for Kronecker product of two matrices.

15. Show that the bicovariance function on the right and left diagonals (s=h=t and s=– h=t, see page 33) may be expressed in terms of the bispectrum,

$$D_{right}(\tau) = \int_{-\pi}^{\pi} B_r(\omega)\exp(j\omega\tau)d\omega,$$

where

$$B_r(\omega) = \int B(\omega_1, \omega - \omega_1)d\omega_1,$$

and the integration is done along lines parallel to the left diagonal as shown in Figure 1.27a. Similarly,

$$D_{left}(\tau) = \int_{-\pi}^{\pi} B_l(\omega)\exp(j\omega\tau)d\omega,$$

where

$$B_1(\omega) = \int B(\omega_1, \omega + \omega_1)\, d\omega_1 ,$$

and the integration is done along lines parallel to the right diagonal as shown in Figure 1.27b.

16. A circularly complex process, $x_c(t)$, is defined by the identity

$$E\{x_c(t)x_c(t+\tau)\} = 0 \quad \text{for all } \tau .$$

Show that the complex analytic representation (CAR) of a real process (see (1.62a)) is circularly complex. Further, show that

$$E\{x_c(t)x_c^*(t+\tau)\} = \frac{2}{\pi}\int_0^{\pi} S_x(\omega)e^{j\omega\tau}\, d\omega,$$

where $S_x(\omega)$ is the spectrum of the real process.

17. A Fourier transform of a nonstationary time series can be defined without bringing in the concept of the generalized Fourier transform. For this it is sufficient if the time-varying filter transfer function (see (1.111)) satisfies a condition, $H(t,\omega) \underset{t \to \infty}{\to} 0$. Show that the Fourier transform is given by

$$X(\Omega) = \frac{1}{2\pi}\int_{-\pi}^{\pi} d\varepsilon(\omega)\bar{H}(\omega - \Omega, \omega),$$

where $\bar{H}(\Omega, \omega)$ is the Fourier transform of $H(t,\omega)$. Further, show that correlation between the Fourier coefficients at two frequencies is

$$Q_x\left(\omega + \frac{\Omega}{2}, \omega - \frac{\Omega}{2}\right) = \frac{1}{2\pi}\sum_{-\infty}^{\infty} S_x(t,\omega)\exp(j\Omega t),$$

where $Q_x(\omega + \frac{\Omega}{2}, \omega - \frac{\Omega}{2}) = E\left\{X(\omega + \frac{\Omega}{2})X^*(\omega - \frac{\Omega}{2})\right\}$. Note that for a stationary process, $Q_x(\omega + \frac{\Omega}{2}, \omega - \frac{\Omega}{2}) = 0$ for $\Omega \neq 0$.

18. In STFT the window function may be chosen so as to improve the localization of the signal component in the time-frequency plane. Consider a window which is matched to the signal. Let $w(t) = x(-t)$. Show that STFT now reduces to the ambiguity function (which is a 2D Fourier transform of WV distribution); that is,

$$X(t,\omega) = \exp\left(-j\omega\frac{t}{2}\right)A(\omega,t).$$

19. A decimator is a device that takes an input time series and produces an output series by retaining samples at integer multiples of M. An M-fold decimator is given by

$$x_D(t) = x(tM).$$

Let x(t) be a stationary stochastic process. Sketch the spectrum of the decimated time series in terms of the spectrum of x(t).

20. The instantaneous frequency (1.64) and the frequency appearing in the Fourier transform are conceptually different,[47] where it is called Fourier frequency. Define an average Fourier frequency,

$$\omega_{ave} = \frac{\int_0^\infty \omega \left| F_c(\omega) \right|^2 d\omega}{\int_0^\infty \left| F_c(\omega) \right|^2 d\omega},$$

where $F_c(\omega)$ is the Fourier transform of complex analytic representation, $f_c(t)$, of a continuous time signal, f(t). Show that ω_{ave} is related to the instantaneous frequency W(t) through the following relation:

$$\omega_{ave} = \frac{\int_{-\infty}^\infty \Omega(t) \left| f_c(t) \right|^2 dt}{\int_{-\infty}^\infty \left| f_c(t) \right|^2 dt}.$$

21. The time-varying spectrum and cross-spectrum between two nonstationary time series were introduced in (1.114). The next step is to introduce the time-varying coherence function as

$$\mathrm{Coh}_{xy}(t, \omega) = \frac{S_{xy}(t, \omega)}{\sqrt{S_x(t, \omega) S_y(t, \omega)}}.$$

While $\mathrm{Coh}_{xy}(t, \omega)$ shares many properties with the coherence function of stationary time series, it differs in one important way; namely, the coherence magnitude may <u>not</u> be equal to one even when the two time series are linearly related.[48] In particular, let

$$y(t) = \sum_{\tau = 0}^{\infty} h(\tau) x(t - \tau);$$

show that the time-varying coherence function is given by

$$\mathrm{Coh}_{xy}(t, \omega) = \frac{S_x(t,\omega) * 2h^*(2t) \exp(2j\omega t)}{\sqrt{S_x(t, \omega) S_x(t, \omega) * WV_h(t,\omega)}},$$

where $WV_h(t,f)$ is the Wigner-Ville distribution of $h(\tau)$, and convolution (*)is with respect to t.

References

1. N. Wiener, *Extrapolation, Interpolation, and Smoothing of Stationary Time Series*, M.I.T. Press, 1949.

2. H. Wold, *A Study in the Analysis of Stationary Time Series*, Stockholm University, Stockholm, Sweden, Almqvist and Wiksell, 1954.

3. A. Schuster, On the investigation of hidden periodicities with application to a supposed 26 day period of meteorological phenomenon, *Terra Magnet*, 3, 13-41, 1898.

4. G. V. Yule, On a method of investigating periodicities in disturbed series, with special reference to Wolfer's sunspot numbers, *Philos. Trans. R. Soc., London*, 226, 269-278, 1927.

5. J. W. Tukey, The sampling theory of power spectrum estimates, *Proc. Symp. Applied Autocorrelation Analysis in Physics Problems*, US Office of Naval Research (Navex Ds-p-725), 1949.

6. J. Cooley and J. W. Tukey, An algorithm for the machine calculation of complex Fourier series, *Math. Comput.*, 19, 297-301, 1965.

7. P. D. Welch, A direct digital method of power spectrum estimation, *IBM J. Res. Dev.*, 5, 141-156, 1961.

8. J. Capon, High resolution frequency-wave number spectrum analysis, *Proc IEEE*, 57, 1408-1418, 1969.

9. J. P. Burg, Maximum entropy spectrum analysis. Annual Meeting of Society of Exploration Geophysicists, Oklahoma City, 1967.

10. V. F. Pisarenko, The retrieval of harmonics from a covariance function, *R. Astronomical Soc., Geophys. J.*, 33, 247-266, 1973.

11. R. O. Schmidt, Multiple emitter location and signal parameter estimation, in Proc. RADC Spectrum Estimation Workshop, Griffiths AFB, NY, 243-258, 1979.

12. R. Roy, A. Paulraj, and T. Kailath, ESPRIT A subspace rotation approach to estimation of parameters of cisoids in noise, *IEEE Trans.*, ASSP-34, 1340-1342, 1986.

13. D. R. Brillinger, An introduction to polyspectra, *Ann. Math. Stat.*, 36, 1351-1374, 1965.

14. G. J. F. MacDonald, The bispectrum of atmospheric pressure records, in *Spectral Analysis of Time Series,* B. Harris, (Ed) John Wiley & Sons, New York, NY, 247-264, 1967.

15. T. Matsuoka and T. J. Ulrych, Phase estimation using bispectrum, *Proc. IEEE*, 72(11), 1403-1411, 1984.

16. A. Swami and J. M. Mendel, Cumulant-based approach to the harmonic retrieval problem. *IEEE. Proc., ICASSP*-88, 2264-2267, 1988.

17. M. B. Priestly, Evolutionary spectra and nonstationary processes, *J. R. Stat. Soc. B.*, 27, 204-239, 1965.

18. E. Wigner, On the quantum correction for thermodynamic equilibruim, *Phys. Rev.*, 40, 749-759, 1932.

19. F. Hlawatsch and G. F. Boudreaux-Bartels, Linear and quadratic time-frequency signal representations, *IEEE Signal Processing Magazine*, 21-67, April 1992.

20. V. V. Ol'shevskii, *Characteristics of Sea Reverberations*, Consultant Bureau, New York, NY, 32, 1967.

21. E. Jahnke and F. Emde, *Tables of Functions*, Dover Publication, New York, 32, 1945.

22. V. S. Pugachev, *Theory of Random Functions and Its Application to Control Problems*, Pergamon Press, Oxford, 1965.

23. P. P. Vaidyanathan, On predicting a band limited signal based on the past sample values, *Proc. IEEE,* 75, 1125-1127, 1987.

24. A. H. Yaglom, *Introduction to Theory of Stationary Random Functions*, Prentice-Hall, Englewood Cliffs, 1962, chap two.

25. J. B. W. Jolley, *Summation of Series*, Dover Publications, Inc., New York, 96, 1961.

26. A. Papoulis, *Probability, Random Variables, and Stochastic Processes*, Third edition, McGraw-Hill International Book Co, New York, NY, 413, 1991.

27. R. Vautard and M. Ghil, Singular spectrum analysis in nonlinear dynamics, with application to paleoclimatic time series, *Physica D*, 35, 395-424, 1989.

28. M. Hallin, C. Lefevre, and M. Puri, On time reversibility and uniqueness of MA representation for non-Gaussian stationary time series, *Biometrika*, 170-171, 1988.

29. Y. C. Kim and E. J. Powers, Digital bispectral analysis and its applications to nonlinear wave interactions, *IEEE Trans. Plasma Sci.*, PS-7, 120-131, 1979.

30. L. H. Koopmans, *The Spectral Analysis of Time Series*, Academic Press, New York, NY, 153, 1974.

31. J. S. Bendat and A. G. Piersol, *Engineering Applications of Correlation and Spectral Analysis*, Wiley-Interscience, New York, 1980.

32. J. S. Bendat and A. G. Piersol, *Random Data: Analysis and Measurement Procedures*, Wiley-Interscience, New York, 1971.

33. M. R. Portnoff, Time-frequency representations of digital signals and systems based on short time Fourier analysis, *IEEE Trans.*, ASSP., 28, 55-69, 1980.

34. A. Papoulis, *Signal Analysis,* McGraw-Hill, New York, 1977.

35. D. Gabor, Theory of communication, *J. IEE*, 93, 429-457, 1947.

36. A. N. Akansu and R. A. Haddad, *Multiresolution Signal Decomposition*, Academic Press, Inc., San Diego, CA, 1992.

37. W. A. Gardner, Exploitation of spectral redundancy in cyclostationary signals, *IEEE Signal Processing Magazine*, 14-36, April 1991.

38. W. A. Gardner, The spectral correlation theory of cyclostationary time series, *Signal Process*, 11, 13-86, 1986.

39. W. B. Davenport, Jr. and W. L. Root, *Random Signals and Noise*, McGraw-Hill, New York, 110, 1958.

40. A. W. Lohmann and B. Wirnitzer, Triple correlation, *Proc. IEEE*, 72, 889-901, 1984.

41. G. C. Carter, Coherence and time delay estimation, *Proc. IEEE*, 75, 236-255, 1987.

42. M. Rosenblatt, Linear process and bispectra, *J. Appl. Probabil.*, 17, 265-271, 1980.

43. M. R. Raghuveer, Time domain approaches to quadratic phase coupling estimation, *IEEE Trans.*, AC-35, 48-56, 1990.

44. T. Ning and S. M. Gao, Detection of amplitude using bispectra, *IEEE Proc., ICASSP-92*, V469-472, 1992.

45. T. A. C. M. Claasen and W. F. G. Mecklenbrauker, The wigner distribution — A tool for time frequency signal analysis, Part II, *Philips J. Res.*, 35, 276-300, 1980.

46. C. L. Nikias and A. P. Petropulu, *Higher-order Spectra Analysis, A Nonlinear Signal Processing Framework,* PTR Prentice-Hall Inc., New York, 1993.

47. M. Sun and R. J. Sclabassi, Discrete-time instantaneous frequency and its computation, *IEEE Trans., SP*, 41, 1867-1880, 1993.

Chapter Two

Mathematical Models of Time Series

The stochastic time series may be further modeled as an output of a filter driven by a white noise process or as a solution of a difference equation driven by white noise process. The filter output may be related to the input through their Fourier transforms (or since stochastic processes are involved, generalized Fourier transform will be used) and the filter transfer function. The difference equations which we shall consider here are linear and time invariant, driven by Gaussian or non-Gaussian white noise. But there are many other possibilities which we will not consider. However, we shall consider the chaotic model of time series involving nonlinear difference equations which are highly sensitive to initial conditions.

2.1 Time Series Models

In Chapter One we mentioned two approaches to time series analysis, one involving the concept of filter and the other involving the difference equation with random inputs. The impulse response function of a filter is ideally an infinite function, whereas the difference equation requires a finite number of parameters; therefore, in order to characterize a time series a complete filter function must be found in the first approach, but only a few parameters need to be estimated in the second approach. From a finite length of time series it is obviously easier to estimate a few parameters reliably than a filter function of infinite duration. Thus, from the point of estimation, the difference equation approach appears to have some advantage over the filter approach. However, the two approaches are not independent; in fact, they are interchangeable. A filter model can be expressed as a difference equation model or vice-versa. This complete equivalence gives us a choice of one of the two models depending upon the data and the purpose of analysis, such as extraction of a signal buried in noise or prediction of time series, computational cost, etc. The time series modeling must be based on the physical nature of the generating mechanism. An inappropriate model applied to perfectly good data can be just as bad as an appropriate model applied to poor data. Gutowski[1] has demonstrated, by means of synthetic data, the consequence of an inappropriate model on spectrum estimation. In Figure 2.1, taken from the above reference, a 4th-order AR process is fitted with MA and ARMA models, and the resulting spectra are superimposed on the exact spectrum.

The spectrum of a time series can be approximated to any degree of accuracy by means of a rational model,

$$S_x(\omega) = \frac{\sigma^2}{2\pi}\left|\frac{\sum_{i=0}^{q} b_i \exp(ji\omega)}{\sum_{i=0}^{p} a_i \exp(ji\omega)}\right|^2 \tag{2.1}$$

with $a_0=1$ and $a_1, a_2, ..., a_p$; $b_0, b_1, b_2, ..., b_q$; and σ^2 as unknown parameters[2], where p and q are integers. For sufficiently large p and q the approximation can be made as close as desired. However, in practice, the model order p and q can at the most be of moderate size. Determination of the spectrum model order is a basic problem in time series modeling and must be viewed in conjunction with the physical nature of the time series. Even the question, which of the time series models (AR, MA, or ARMA) is appropriate for a given time series, is by no means easy.

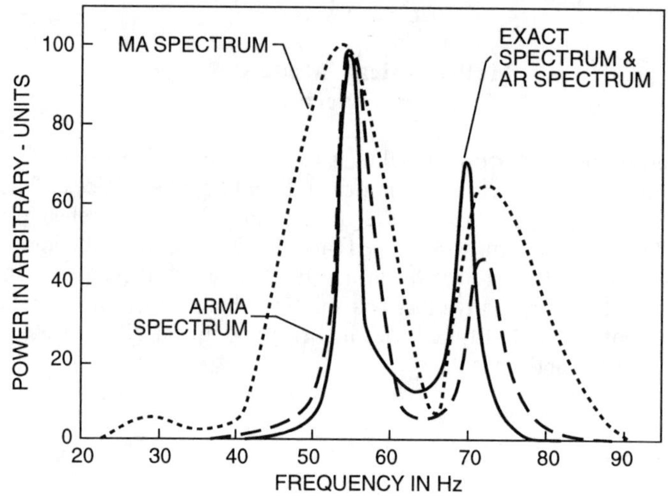

Figure 2.1: Result of fitting an AR(4) process (Example 2.9) with MA and ARMA models. (Taken from Reference 1. © 1978 IEEE)

2.2 **Filter Model**

A stationary stochastic process may be modeled as an output of a filter driven by white noise, $\varepsilon(t)$.

$$x(t) = \sum_{\tau=0}^{\infty} \varepsilon(t-\tau)h(\tau), \tag{2.2}$$

where $h(\tau)$ is the impulse response function of the filter, that is, the output to a delta function input.

A. Causal Filter

We would like to bring out some of the basic properties of the filter and relate them to the spectrum of x(t). To start with we shall impose the practical constraint that all dynamic systems must be causal and must be of finite energy; hence, h(t)=0 for t<0 and

$$\sum_{t=0}^{\infty} |h(t)|^2 < \infty.$$

Let us now look at its z-transform,

$$H(z) = \sum_{t=0}^{\infty} h(t)z^{-t}.$$

H(z) is analytic in the z-plane in the region bounded by the unit circle and the infinite circle (Figure 2.2). A function ceases to be analytic at a point of singularity. Let us assume that there is a singularity of order one at z = a, $|a| > 1$ and then show that the inverse z-transform of such a function is unstable. The inverse z-transform,

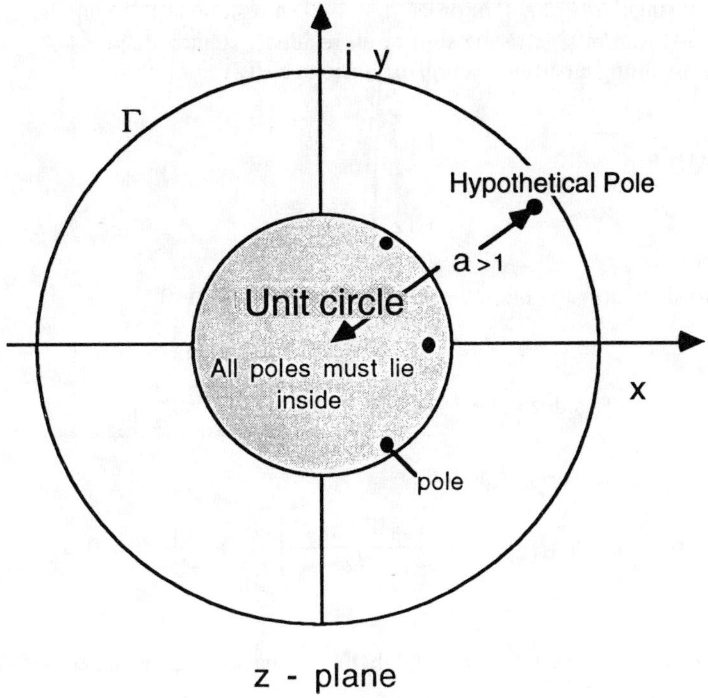

Figure 2.2: All poles of a stable filter must lie inside the unit circle. In the event a pole lies outside the unit circle, it is shown in the text how the impulse response function becomes unstable.

$$h(t) = \frac{1}{2\pi j}\oint_{\Gamma} H(z)z^{t-1}dz$$

$$= \text{Residue of } \{H(z)z^{t-1}\}, \tag{2.3}$$

where Γ is a closed path as shown in Figure 2.2. Let $H(z) = \frac{H_0(z)}{(z-a)}$, where $H_0(z)$ is analytic in the region enclosed by Γ. Note that the residue of a function f(t) (its z-transform is F(z)) having a pole of order m at z=a is given by

$$\text{Residue} = \frac{1}{(m-1)!}\lim_{z\to a}\frac{d^{m-1}}{dz^{m-1}}[(z-a)^m F(z)].$$

From (2.3) we obtain

$$h(t) = H_0(a)\,a^{t-1}. \tag{2.4}$$

It is easy to see that h(t) as given in (2.4) will not satisfy the finite energy constraint unless $|a| < 1$. Hence, all poles of H(z) must lie inside the unit circle. For $t \le -1$, $H(z)z^{t-1}$ will have an additional singularity at z=0 of order $|t-1|$. The residue at this singularity may be shown to be equal but opposite in sign to the sum of all residues. Hence h(t)=0, $t \le -1$. This may be easily shown by assuming a partial fraction expansion of H(z).

Example 2.1: Let

$$H(z) = \frac{\alpha}{z-a}.$$

We would like to show how to obtain the residue of $H(z)z^{t-1}$ for t<0.

For $t = -1$

$$\text{Residue of}\{H(z)z^{-2}\} = \left.\frac{d}{dz}\frac{\alpha}{z-a}\right|_{z=0} + \left.\frac{\alpha}{z^2}\right|_{z=a}.$$

For $t = -2$

$$\text{Residue of}\{H(z)z^{-3}\} = \left.\frac{1}{2}\frac{d^2}{dz^2}\frac{\alpha}{(z-a)}\right|_{z=0} + \left.\frac{\alpha}{z^3}\right|_{z=a} = 0.$$

similarly for other values of t<0. For $t \ge 0$, $h(t)$ will have to be evaluated individually (End of Example).

B. Inverse Filter

An inverse filter of a given stochastic process is one whose output is white noise when driven by the given stochastic process.

$$\varepsilon(t) = \sum_{\tau=0}^{\infty} x(t-\tau)h_{INV}(\tau), \tag{2.5}$$

where $h_{INV}(\tau)$ is the impulse response function of the inverse filter. Figure 2.3 illustrates the role of an inverse filter. Using (2.5) in (2.2) we get

$$x(t) = \sum_{\tau=0}^{\infty} h(\tau) \sum_{\tau'=0}^{\infty} x(t-\tau-\tau')h_{INV}(\tau').$$

Let $p = \tau + \tau'$ or $\tau = p - \tau'$ and rewrite the above as

$$x(t) = \sum_{p=0}^{\infty} \sum_{\tau'=0}^{p} h(p-\tau')h_{INV}(\tau')x(t-p)$$

$$= \sum_{p=0}^{\infty} o(p)x(t-p).$$

For this result to be valid at all t we must have $o(\tau) = \delta(\tau)$; that is,

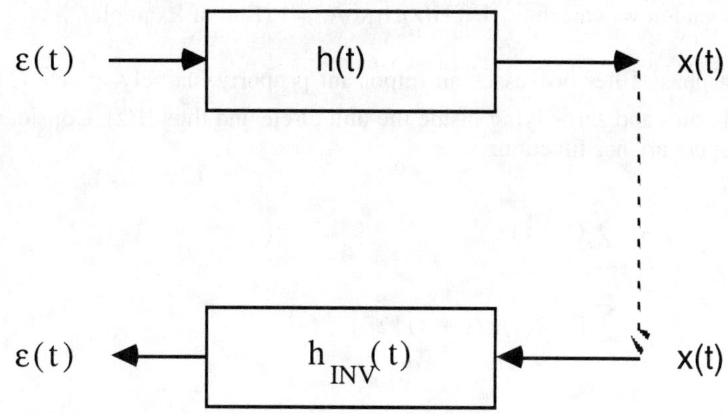

Figure 2.3: The role of inverse filter.

$$o(p) = \sum_{\tau'=0}^{p} h(p-\tau')h_{INV}(\tau') = \delta(p).$$

Taking the z-transform on both sides of the above equation, we obtain

$$H(z).H_{INV}(z) = 1$$

$$H_{INV}(z) = H(z)^{-1}. \tag{2.6}$$

On account of (2.6) all zeros of H(z) will become poles of $H_{INV}(z)$, which necessarily have to lie within the unit circle in z-plane. Note that $h_{INV}(t)$ is also a causal filter of finite energy. As a result all zeros as well as poles of H(z) must lie inside the unit circle. A causal filter, which is also invertible in the sense of (2.6), is called a minimum phase filter.

Example 2.2: This example shows how an inverse filter can be obtained. Consider a filter with one pole and its complex conjugate:

$$H(z) = \frac{z^2}{(z-b)(z-b^*)}$$

$$= [1 + (b+b^*)z^{-1} + (b^2 + b^{*2} + bb^*)z^{-2} + (b^3 + b^{*3} + b^2b^* + b^{*2}b)z^{-3}$$
$$+ (b^4 + b^{*4} + b^3b^* + b^{*3}b + b^2b^{*2})z^{-4}$$
$$+ \ldots].$$

$$H_{INV}(z) = \frac{(z-b)(z-b^*)}{z^2}$$
$$= (1 - bz^{-1})(1 - b^*z^{-1})$$
$$= 1 - (b+b^*)z^{-1} + bb^*z^{-2}.$$

By actual multiplication we can show that $H(z)H_{INV}(z)=1$ (End of Example).

A minimum phase filter possesses an important property; namely, given $C_x(\tau)$, one can uniquely find all poles and zeros lying inside the unit circle and thus H(z). Consider the two-sided z-transform of the covariance function:

$$\begin{aligned}
C_x(z) &= \sum_{\tau=-\infty}^{\infty} C_x(\tau) z^{-\tau} \\
&= \sum_{\tau=-\infty}^{\infty} E\{x(t)x(t+\tau)\} z^{-\tau} \\
&= \sum_{\tau} E\{\sum_{\tau_1=0}^{\infty} \varepsilon(t-\tau_1)h(\tau_1) \sum_{\tau_2=0}^{\infty} \varepsilon(t+\tau-\tau_2)h(\tau_2)\} z^{-\tau} \\
&= \sigma_\varepsilon^2 \sum_{\tau} z^{-\tau} \sum_{\tau_1}\sum_{\tau_2} h(\tau_1)h(\tau_2)\delta(\tau-\tau_2+\tau_1) \\
&= \sigma_\varepsilon^2 \sum_{\tau=-\infty}^{\infty} z^{-\tau} \sum_{\tau_1=0}^{\infty} h(\tau_1)h(\tau+\tau_1) \\
&= \sigma_\varepsilon^2 H(z)H(z^{-1}).
\end{aligned} \tag{2.7}$$

$C_x(z)$ may be factored, and all its poles and zeros lying inside the unit circle are combined into the desired transfer function, H(z). Alternatively, given the spectrum we can obtain the phase of the transfer function by Hilbert transform of the log spectrum.[3] A minimum phase filter will have much of its energy concentrated at the beginning, and its length is the minimum possible.[4] This property may be interpreted as the shortest possible range of influence; that is, the current output is influenced by as few immediate past inputs as possible. Distant past inputs have little influence.

Example 2.3: Given an exponentially decreasing covariance function, find H(z). Let

$$C_x(\tau) = a^{|\tau|}, \quad |a| < 1$$

and compute its two-sided z-transform.

$$\begin{aligned}
C_x(z) &= \sum_{\tau=-\infty}^{\infty} a^{|\tau|} z^{-\tau} \\
&= \sum_{\tau=0}^{\infty} a^{\tau} z^{-\tau} + \sum_{\tau=0}^{\infty} a^{\tau} z^{\tau} - 1 \\
&= \frac{1}{1-az^{-1}} + \frac{1}{1-az^{+1}} - 1 \\
&= \frac{1-a^2}{(1-az^{-1})(1-az^{+1})}.
\end{aligned}$$

It is easy to show that the first pole is at z=a, inside the unit circle and the second pole at $z = \frac{1}{a^*}$, outside the unit circle. The transfer function of the minimum phase filter will be $H(z) = \frac{1}{1 - az^{-1}}$, and the variance of the noise input will be (1-a^2) (End of Example).

C. Spectral Representation of Filter Output

Let x(t) be the output of a filter with impulse response function given by h(t). Let the input be white stationary noise, e(t). Hence,

$$x(t) = \sum_{t'=0}^{\infty} h(t')\varepsilon(t - t'). \tag{2.8}$$

Using the spectral representation of $\varepsilon(t)$ in (2.8) we obtain

$$\begin{aligned} x(t) &= \frac{1}{2\pi}\sum_{t'=0}^{\infty} h(t')\int_{-\pi}^{\pi} d\bar{\varepsilon}_G(\omega)\exp(j\omega(t - t')) \\ &= \frac{1}{2\pi}\int_{-\pi}^{\pi} d\bar{\varepsilon}_G(\omega)\exp(j\omega t)\sum_{t'=0}^{\infty} h(t')\exp(-j\omega t'). \end{aligned} \tag{2.9}$$

Let

$$H(\omega) = \sum_{0}^{\infty} h(t')\exp(-j\omega t')$$

be the frequency response function of the filter. In terms of the filter response, (2.9) may be written as

$$x(t) = \frac{1}{2\pi}\int_{-\pi}^{\pi} d\bar{\varepsilon}_G(\omega)H(\omega)\exp(j\omega t). \tag{2.10}$$

Equation (2.10) is the spectral representation of the filter output. Writing

$$x(t) = \frac{1}{2\pi}\int_{-\pi}^{\pi} dX_G(\omega)\exp(j\omega t)$$

and comparing it with (2.10) we obtain

$$dX_G = H(\omega)\,d\bar{\varepsilon}_G(\omega). \tag{2.11}$$

The spectrum of the filter output may be obtained from (2.10) using the properties of the generalized Fourier transform of a stationary stochastic process (Section 1.4, (1.61)),

$$S_x(\omega) = \sigma_\varepsilon^2|H(\omega)|^2,$$

where σ_ε^2 is the variance of the random input to the filter. The phase of the filter does not contribute to the spectrum. Conversely, it is not possible to completely characterize the filter from the spectrum of the output except when the filter is of minimum phase type. For a given spectrum the corresponding minimum phase may be obtained by computing the Hilbert transform of the log spectrum,[3]

$$\phi_{min}(\omega) = \frac{1}{2\pi j}\int_{-\pi}^{\pi} \frac{\ln S_x(\omega')}{\omega - \omega'} d\omega'. \tag{2.12}$$

D. **Group Delay**

A filter will introduce a delay at each frequency, and it is defined as the negative derivative of the phase of the transfer function with respect to the frequency. Let $\tau_p(\omega)$ stand for the group delay defined from the phase function,

$$\tau_p(\omega) = -\frac{d\phi_{min}(\omega)}{d\omega}. \tag{2.13a}$$

The group delay is a characteristic of a filter, and it may be related to the spectrum or to the covariance function of the output of the filter which is driven by white noise input. We shall assume that the filter is a minimum phase filter. In (2.12) use the Fourier series expansion of $\ln S_x(w)$, which is a symmetric periodic function (period being 2π). We obtain

$$\phi_{min}(\omega) = \frac{1}{2\pi j}\int_{-\pi}^{\pi} \frac{\frac{K(0)}{2} + \sum_{\tau=1}^{\infty} K(\tau)\cos(\omega\tau)}{\omega - \omega'} d\omega'$$

$$= -\sum_{\tau=1}^{\infty} K(\tau)\sin(\omega\tau), \tag{2.13b}$$

where $K(\tau)$, $\tau = 0, 1, 2, \ldots$ are known as cepstrum coefficients.[5] Using (2.13b) in (2.13a) we obtain[6,7]

$$\tau_p(\omega) = \sum_{\tau=1}^{\infty} \tau K(\tau)\cos(\omega\tau). \tag{2.14a}$$

It is also possible to obtain the group delay in terms of covariance and spectrum. The derivation is left out as an exercise (see Exercise 2.17), but the result is reproduced here:

$$\tau_p(\omega) = \frac{\frac{1}{\sigma_\varepsilon^2}\sum_{\tau=0}^{\infty}\sum_{\tau'=0}^{\infty} \tau C_x(\tau) C_x(\tau')\cos(\omega(\tau - \tau'))}{S_x(\omega)}. \tag{2.14b}$$

The group delay of a filter can also be defined in terms of the Wigner-Ville distribution of the filter impulse response function (see page 53).

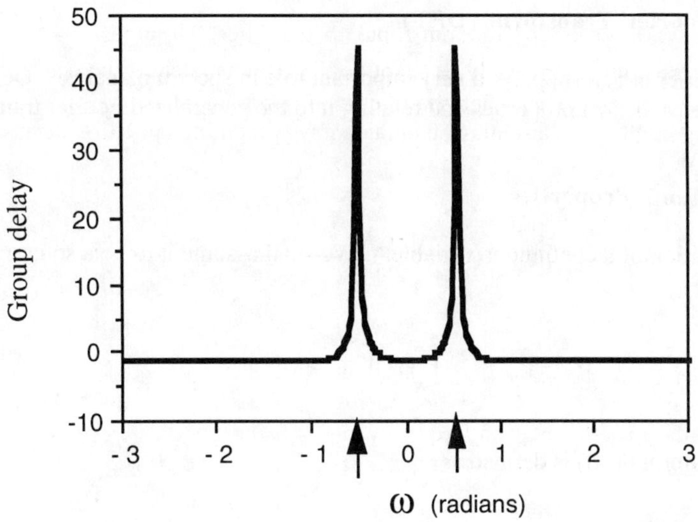

Figure 2.4: Group delay (in seconds) of the transfer function of AR(2). The poles of the filter are at $0.98e^{\pm j30^0}$. The peaks in the group delay correctly show the pole positions.

Example 2.4: In this example we shall consider a simple second-order filter which was earlier considered in Example 2.2. The transfer function is

$$H(z) = \frac{z^2}{(z-b)(z-b^*)}.$$

The phase term is obtained from the imaginary part of lnH(z). Therefore, the group delay is given by

$$\tau_p(\omega) = -\,\mathrm{Im}\left\{\frac{d\ln H(z)\big|_{z=e^{j\omega}}}{d\omega}\right\}$$

$$= -\,\mathrm{Im}\left\{ j\left(1 - \frac{e^{j\omega}}{e^{j\omega}-b} + 1 - \frac{e^{j\omega}}{e^{j\omega}-b^*}\right)\right\}.$$

Simplification of the above expression leads to the following result:

$$\tau_p(\omega) = -\left[\frac{r^2 - r\cos(\omega-\theta)}{1+r^2-2r\cos(\omega-\theta)} + \frac{r^2 - r\cos(\omega+\theta)}{1+r^2-2r\cos(\omega+\theta)}\right].$$

where

$$b = re^{j\theta}.$$

A plot of the group delay shows two peaks at the corresponding pole positions (see Figure 2.4) (End of Example).

2.3 Discrete Fourier Transform (DFT)

The discrete Fourier transform plays a very important role in spectrum analysis. Our emphasis will be on studying its statistical properties and relating it to the generalized Fourier transform.

A. Definition and Properties

Let f(t) be a function of a continuous variable, t. We shall assume it to be a square integrable; that is,

$$\int_{-\infty}^{\infty} |f(t)|^2 dt < \infty \quad .$$

The Fourier transform of f(t) is defined as

$$F(\omega) = \int_{-\infty}^{\infty} f(t) \exp(-j\omega t) dt \, . \tag{2.15a}$$

Analogously, we define the Fourier transform of a discrete sequence or time series of finite length. Remember that a finite sequence of finite magnitude is always of finite energy, and hence its discrete Fourier transform exists.

$$X(k) = \sum_{t=0}^{T-1} x(t) \exp\left(-j\frac{2\pi k}{T} t\right), \qquad k = 0,1,2, \ldots, T-1. \tag{2.15b}$$

Note that X(k) is also discrete; hence it is known as discrete Fourier transform (DFT). Since $F(\omega)$ is continuous, the Fourier transform defined by (2.15a) is sometimes called a continuous Fourier transform (CFT). The inverse of (2.15b) exists and is given by

$$x(t) = \frac{1}{T} \sum_{k=0}^{T-1} X(k) \exp\left(+j\frac{2\pi}{T} kt\right). \tag{2.16}$$

The DFT coefficients, X(k), possess several interesting properties which are listed below:

$$\left.\begin{aligned} &X(nT + k) = X(k) && \text{periodicity,} \\ &X(-k) = X(T-k) && \text{mapping,} \\ &X(-k) = X^*(k) && \text{for real time series.} \end{aligned}\right\} \tag{2.17}$$

At k=0

$$X(0) = \sum_{t=0}^{T-1} x(t) \quad = T \times \text{mean of the sequence} \quad ,$$

and at $k = \dfrac{T}{2}$

$$X(\frac{T}{2}) = \sum_{t=0}^{T-1} (-1)^t x(t).$$

It is important to note that the DFT of a finite sequence implicitly assumes that the sequence is periodic; that is, for all p (integer),

$$x(pT + t) = x(t). \tag{2.18}$$

The DFT operation is linear, for it may be easily shown that for $y(t)=a_1x_1(t)+a_2x_2(t)$, $DFT\{y(t)\}=a_1DFT\{x_1(t)\}+a_2DFT\{x_2(t)\}$, where a_1 and a_2 are any constants. The DFT of a (circularly) shifted sequence is equal to the DFT of the sequence before shifting multiplied by a phase factor. This is known as the shift property of DFT.

$$\begin{aligned} DFT\{x(t-\tau)\} &= \sum_{t=0}^{T-1} x(t-\tau)\exp(-j\frac{2\pi k}{T}t) \\ &= \sum_{t'=-\tau}^{T-\tau-1} x(t')\exp(-j\frac{2\pi k}{T}(t'+\tau)) \\ &= \exp(-j\frac{2\pi k\tau}{T}) \sum_{t'=-\tau}^{T-\tau-1} x(t')\exp(-j\frac{2\pi k}{T}t') \\ &= X(k)\exp(-j\frac{2\pi k}{T}\tau). \end{aligned} \tag{2.19}$$

Because of the implied periodicity of x(t) and the periodicity of $\exp(-j\frac{2\pi k}{T}t')$, the summation in (2.19) may be replaced by X(k). Next, we consider the DFT of the output of a filter,

$$y(t) = \sum_{\tau=0}^{T-1} h(\tau)x(t-\tau).$$

Carry out the DFT operation on both sides of the above equation. Using (2.19) we obtain

$$\begin{aligned} Y(k) &= \sum_{\tau=0}^{T-1} h(\tau)X(k)\exp(-j\frac{2\pi k\tau}{T}) \\ Y(k) &= H(k)X(k), \end{aligned} \tag{2.20}$$

where H(k) is equal to $DFT\{h(\tau)\}$. Equation (2.20) is the discrete frequency domain equivalent of the convolution theorem.

B. **Statistical Properties**

Let x(t), t=0, 1, 2, ..., $T-1$ be a finite segment of a stochastic process. We would like to evaluate the statistical properties of the Fourier coefficients.

Mean:

$$E\{X(k)\} = \sum E\{x(t)\}\exp(-j\frac{2\pi k}{T}t)$$

$$= \sum m_x \exp(j\frac{2\pi}{T}kt)$$

$$= m_x T \quad \text{for } k = 0,$$

$$= 0 \quad \text{for } k \neq 0.$$

Except the first coefficient, all other DFT coefficients are zero mean. To remove this asymmetry, it is a common practice to subtract the mean from the time series before computing its Fourier transform.

Covariance:

$$E\{X(k)X^*(l)\} = \sum_t \sum_s E\{x(t)x(s)\}\exp(-j\frac{2\pi kt}{T} + j\frac{2\pi ls}{T})$$

$$= \sum_t \sum_s C_x(t-s)\exp(-j\frac{2\pi k}{T}(t-s)\exp(-j\frac{2\pi}{T}(k-l)s).$$

$$= \sum_{\tau=-(T-1)}^{(T-1)} C_x(\tau)\exp(-j\frac{2\pi k\tau}{T})\frac{\sin[\frac{\pi}{T}(k-l)(T-|\tau|)]}{\sin[\frac{\pi}{T}(k-l)]} \exp(-j\frac{\pi(k-l)}{T}(T-|\tau|-1)). \tag{2.21}$$

See Figure 2.5 for details on the evaluation of the sum in (2.21). For $T >> \tau_0$ (data length is much longer than the correlation distance),

$$\frac{\sin\frac{\pi}{T}(k-l)(T-|\tau|)}{\sin\frac{\pi}{T}(k-l)} \approx T \quad \text{when } k = l$$

$$\approx 0 \quad \text{when } k \neq l.$$

Hence,

$$E\{X(k)X^*(l)\} = T\sum_{\tau=-(T-1)}^{(T-1)} C_x(\tau)\exp(-j\frac{2\pi k\tau}{T})\,\delta_{kl}\,. \tag{2.22}$$

From (2.22) two facts emerge; first, X(k) and X(l) are uncorrelated for large data lengths (much larger than the correlation distance), and second,

$$\frac{1}{T}E\{|X(k)|^2\} = \sum_{\tau=-(T-1)}^{\tau=(T-1)} C_x(\tau)\exp(-j\frac{2\pi k}{T}\tau). \tag{2.23}$$

The right-hand side of (2.23) is by definition the spectrum of a time series. Thus, (2.23) provides a basis for estimation of the spectrum directly from the DFT coefficients.

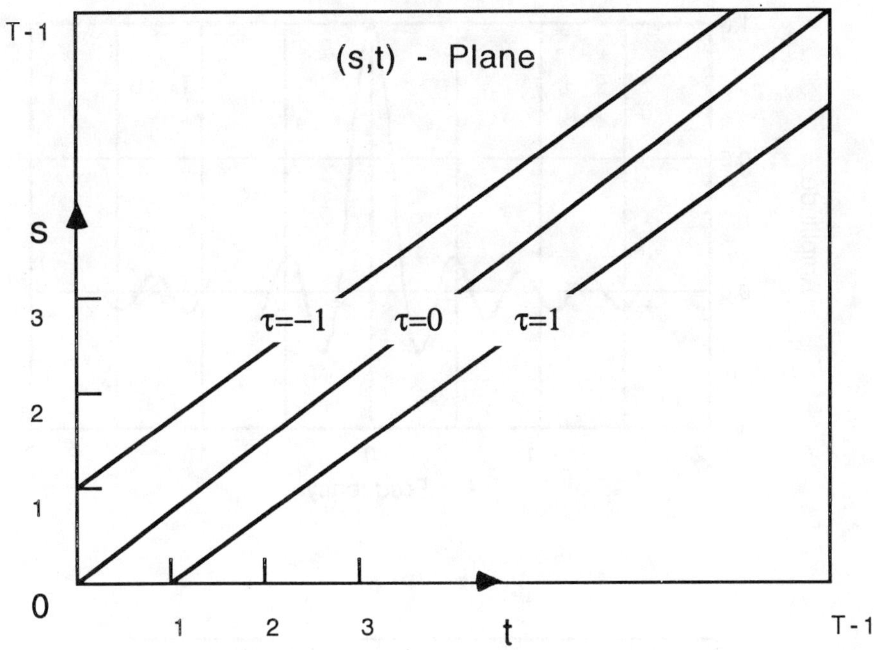

Figure 2.5: The double sum in (2.21) is evaluated first along a diagonal for fixed (t-s) and next over all diagonals.

Finally, we shall show that the joint pdfs of $\{x(t), t=0,1, \ldots, T-1\}$ and $\{X(k), k=0,1, \ldots, T-1\}$ are equal to each other. To show this note that $f_X = \frac{f_x}{|\mathbf{J}|}$[8], where $|\mathbf{J}|$ is determinant of the Jacobian **J** of the transformation from time to frequency space,

$$\mathbf{J} = \begin{bmatrix} \exp\left(-j\frac{2\pi kt}{T}\right) \\ k,t = 0,1,\ldots,T-1 \end{bmatrix}.$$

It is easy to show that $|\mathbf{J}|=1$ (see Reference 9); hence $f_X = f_x$. As a corollary to this result, any non-Gaussian character present in the time series would be preserved in the DFT coefficients. In particular, it is possible to obtain the bispectrum from the DFT coefficients.

C. DFT Coefficients and *Discrete* GFT

We shall now establish a relationship between the DFT coefficients and generalized Fourier transform (GFT) of a discrete process. In the DFT equation,

$$X(k) = \sum_{t=0}^{T-1} x(t)\exp\left(-j\frac{2\pi k}{T}t\right),$$

we replace x(t) by its spectral representation (1.57). We obtain

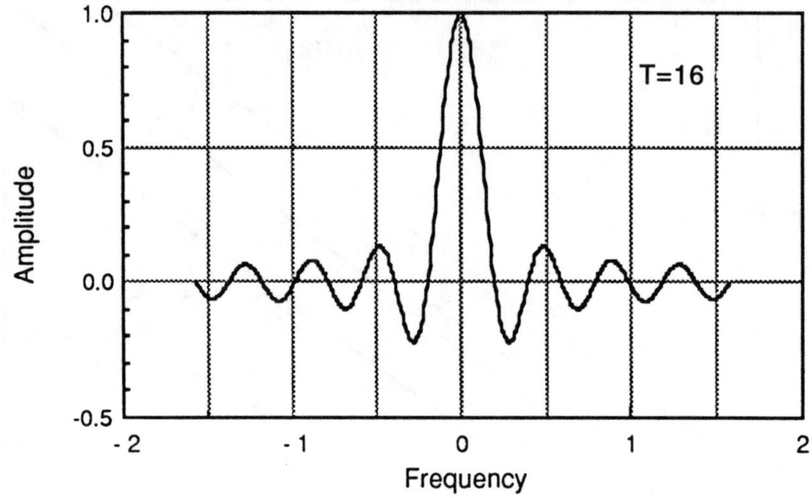

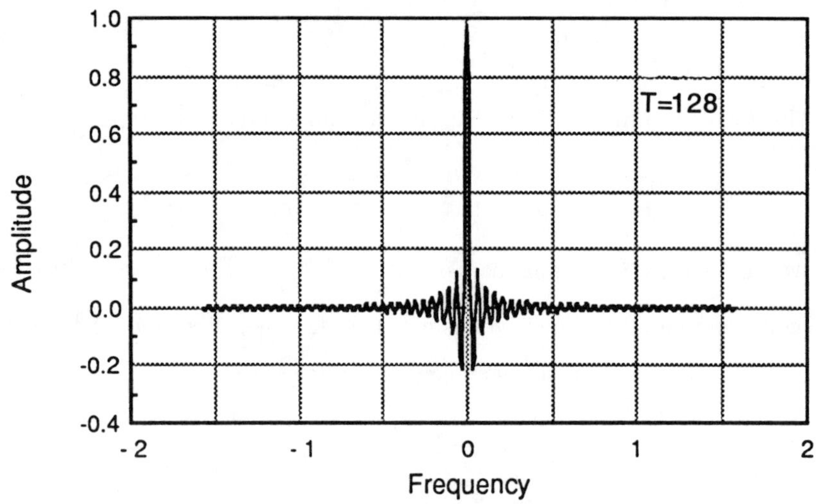

Figure 2.6: A DFT coefficient of a finite stochastic time series is the weighted average of the generalized Fourier transform of an ideal infinite time series. The weighting function depends upon the data length as shown above.

$$X(k) = \frac{1}{2\pi}\int_{-\pi}^{\pi} dX_G(\omega) \sum_{t=0}^{T-1} \exp\left(j\left(\omega - \frac{2\pi k}{T}\right)t\right)$$

$$= \frac{1}{2\pi}\int_{-\pi}^{\pi} dX_G(\omega) \frac{\sin\left(\omega - \frac{2\pi k}{T}\right)\frac{T}{2}}{\sin\left(\omega - \frac{2\pi k}{T}\right)/2} \exp\left(j\left(\omega - \frac{2\pi k}{T}\right)\frac{T-1}{2}\right). \tag{2.24}$$

where

$$\frac{\sin\left(\omega - \frac{2\pi k}{T}\right)\frac{T}{2}}{\sin\left(\omega - \frac{2\pi k}{T}\right)/2}$$

is Dirichlet function. It has a peak at $\omega = \frac{2\pi k}{T}$ in the frequency range $-\pi \le \omega \ge \pi$. This peak becomes sharper as $T \to \infty$, tending to a delta function (see Figure 2.6). In this limiting case

$$\frac{1}{T}X(k) \to \frac{1}{2\pi} dX_G(\omega)\Big|_{\omega=\frac{2\pi k}{T}}. \qquad T \to \infty \tag{2.25}$$

The significance of this result is that for large data length, we can in all practical computations involving real-life data replace $dX_G(\omega)$ by the corresponding DFT coefficient. We also note that (2.24) is an integral equation connecting the DFT coefficient to the generalized Fourier transform. However, it cannot be solved through a least squares local solution as has been suggested.[10]

D. DFT Coefficients and *Continuous* GFT

Let $x^c(t)$ (superscript c denotes continuous) be a stationary stochastic process of continuous parameter, t, having a spectral representation analogous to that of a discrete process given by (1.57). Let us introduce a discrete version of $x^c(t)$. It may be expressed as

$$x(t) = x^c(t)\sum_{n=0}^{T-1}\delta(t - n\Delta t), \tag{2.26}$$

where Δt is the sampling interval. x(t) is a function consisting of T spikes with amplitude equal to $x^c(t)$ at $t = n\Delta t, n = 0, 1, \ldots, T-1$ at other values of t, x(t)=0. Notice that x(t) is a comb function whose teeth are of variable magnitude, controlled by $x^c(t)$ (see Figure 2.7). As $T \to \infty$, x(t) will represent an infinite duration discrete time series. Multiply both sides of (2.26) by $\exp(-j\omega t)$ and integrate with respect to t.

$$\int_0^{\Delta t\, T} x(t)\exp(-j\omega t)\,dt = \int_0^{\Delta t\, T} x^c(t)\sum_{n=0}^{T-1}\delta(t - n\Delta t)\exp(-j\omega t)\,dt. \tag{2.27}$$

Replace $x^c(t)$ by its spectral representation. The right-hand side of (2.27) becomes

$$\frac{1}{2\pi}\int_{-\infty}^{\infty} dX_G^c(\omega')\int_0^{\Delta t\, T}\sum_0^{T-1}\delta(t - n\Delta t)\exp(-j(\omega - \omega')t)\,dt$$

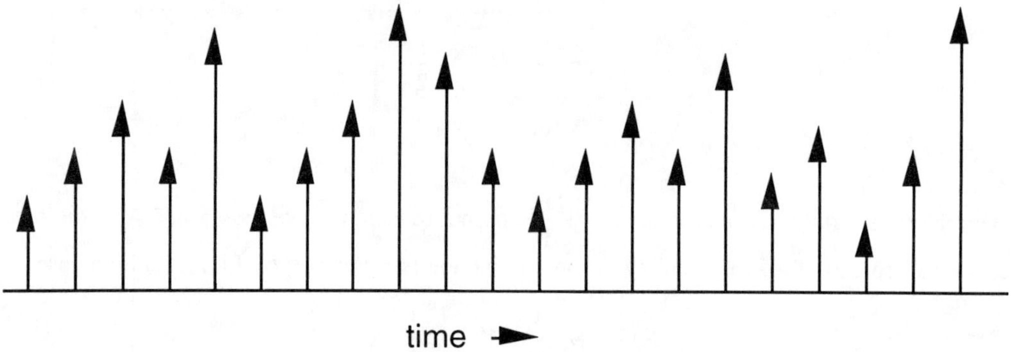

Figure 2.7: Sampled version of stochastic process of a continuous parameter, t.

$$= \frac{1}{2\pi}\int_{-\infty}^{\infty} dX_G^c(\omega')\frac{\sin(\omega-\omega')\frac{T\Delta t}{2}}{\sin(\omega-\omega')\frac{\Delta t}{2}}\exp(-j(\omega-\omega')\frac{T-1}{2}\Delta t).$$

But, by straightforward evaluation of (2.27) we obtain

$$\int_0^{\Delta t\, T} x(t)\exp(-j\omega t)\,dt = \sum_0^{T-1}\int_0^{\Delta t\, T} x^c(t)\delta(t-n\Delta t)\exp(-j\omega t)\,dt$$

$$= \sum_0^{T-1} x(n\Delta t)\exp(-j\omega n\Delta t)$$

$$= X(k),$$

where we have set $\omega = \frac{2\pi k}{\Delta t T}$. Hence,

$$X(k) = \frac{1}{2\pi}\int_{-\infty}^{\infty} dX_G^c(\omega)\frac{\sin(\omega-\frac{2\pi k}{\Delta t T})\frac{T\Delta t}{2}}{\sin(\omega-\frac{2\pi k}{\Delta t T})\frac{\Delta t}{2}}\exp(j(\omega-\frac{2\pi k}{\Delta t T})\frac{T-1}{2}\Delta t). \tag{2.28}$$

Equation (2.28) relates a DFT coefficient to the continuous GFT. Observe the close similarity between (2.24) and (2.28).

E. Continuous and Discrete GFT

Next we shall relate the GFT of a continuous and discrete stochastic process. For this, divide both

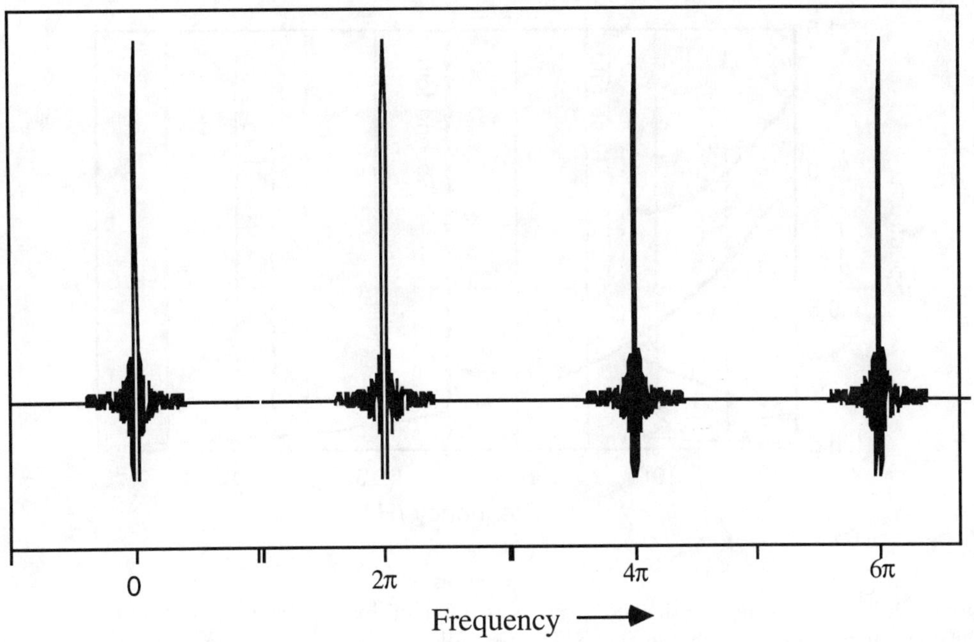

Figure 2.8: Periodic Dirichlet function for T=128: a sharp peak rising above compressed sidelobes. As data length increases ($T \to \infty$) all sidelobes collapse into the mainlobe which is now a spike. The Dirichlet function reduces to a train of spikes.

sides of (2.28) by T and compute the right-hand side as $T \to \infty$. Note that

$$\frac{1}{T}\frac{\sin\left(\omega T\frac{\Delta t}{2}\right)}{\sin\left(\omega\frac{\Delta t}{2}\right)},$$

which is known as a Dirichlet function, is periodic with a period $\frac{2\pi}{\Delta t}$. As $T \to \infty$ the peaks become sharper and sharper, and finally, in the limiting case the peaks become spikes (Figure 2.8),

$$\frac{1}{T}\frac{\sin \omega T\frac{\Delta t}{2}}{\sin \frac{\omega \Delta t}{2}} \underset{T\to\infty}{\to} \sum_{n=-\infty}^{\infty} \Delta\left(\omega - \frac{2\pi}{\Delta t}n\right) , \tag{2.29}$$

where $\Delta(\omega) = 1$ when $\omega=0$; otherwise it is zero. Using (2.25) and (2.29) in (2.28) we obtain

$$\frac{1}{T}X(k) \underset{T\to\infty}{\to} \frac{1}{2\pi}dX_G(\omega) = \frac{1}{2\pi}\sum_{-\infty}^{\infty} dX_G^c\left(\omega - \frac{2\pi n}{\Delta t}\right). \tag{2.30}$$

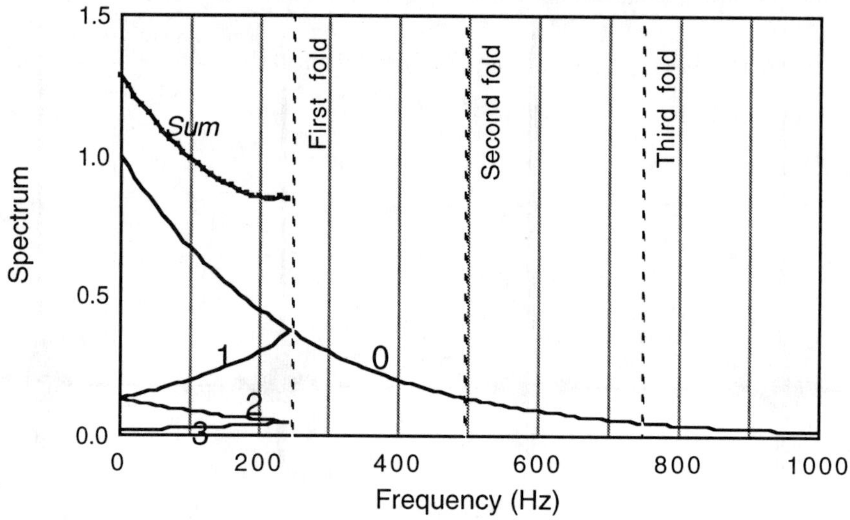

Figure 2.9: The spectrum of a discrete process is obtained by successively folding the spectrum of the continuous process as shown above. The first fold is at half the sampling frequency.

Equation (2.30) states that the differential of a *discrete* GFT is equal to the periodically extended continuous GFT. The periodicity of extension is $\frac{2\pi}{\Delta t}$.

F. Spectrum of Discrete and Continuous Process

Next we relate the spectrum of a discrete process to that of a continuous process. For this consider the case of an infinite sequence. On squaring both sides of (2.30) and carrying out the expectation, we obtain, using the properties of the differential GFT (Chapter One, (1.57)),

$$dF_X(\omega) = \sum_{n=-\infty}^{\infty} dF_{X^c}(\omega - \frac{2\pi}{\Delta t}n). \tag{2.31}$$

Assume that the spectral distribution function is continuous; then (2.31) reduces to

$$S_X(\omega) = \sum_{n=-\infty}^{\infty} S_{X^c}(\omega - \frac{2\pi}{\Delta t}n). \tag{2.32}$$

The mathematical operation on the right-hand side of (2.32) is often known as folding of the spectrum as illustrated in Figure 2.9. The spectral power lying outside the range of $\pm\frac{\pi}{\Delta t}$ is folded back into the same range. Further, $S_X(\omega)$ is periodic with period $\frac{2\pi}{\Delta t}$. When $S_{X^c}(\omega)$ is band limited to $\pm\frac{\pi}{\Delta t}$, there is no reflected power into the fundamental period $-\frac{\pi}{\Delta t} \le \omega \le \frac{\pi}{\Delta t}$. Consequently, the discrete spectrum $S_X(\omega)$ is said to be alias free.

We shall now show that the variance of X(k), the discrete Fourier transform of a finite sequence, is related to the spectrum. From (2.28) and the fact that $\frac{\sin(\omega T\frac{\Delta t}{2})}{T\sin(\omega\frac{\Delta t}{2})}$ is periodic we obtain

$$E\{|X(k)|^2\} = \frac{1}{2\pi}\int_{-\frac{\pi}{\Delta t}}^{\frac{\pi}{\Delta t}} \sum_p S_{x^c}(\omega - \frac{2\pi p}{\Delta t})\left|\frac{\sin(\omega - \frac{2\pi k}{\Delta t T})T\frac{\Delta t}{2}}{\sin(\omega - \frac{2\pi k}{\Delta t T})\frac{\Delta t}{2}}\right|^2 d\omega. \tag{2.33}$$

Using (2.32) in (2.33) we get

$$E\{|X(k)|^2\} = \frac{1}{2\pi}\int_{-\frac{\pi}{\Delta t}}^{\frac{\pi}{\Delta t}} S_x(\omega)\left|\frac{\sin(\omega - \frac{2\pi k}{\Delta t T})T\frac{\Delta t}{2}}{\sin(\omega - \frac{2\pi k}{\Delta t T})\frac{\Delta t}{2}}\right|^2 d\omega \quad . \tag{2.34}$$

Divide by T^2 on both sides of (2.34):

$$S_x(k,T) = \frac{E\{|X(k)|^2\}}{T^2} = \frac{1}{2\pi}\int_{-\frac{\pi}{\Delta t}}^{\frac{\pi}{\Delta t}} S_x(\omega)\left|\frac{\sin(\omega - \frac{2\pi k}{\Delta t T})T\frac{\Delta t}{2}}{T\sin(\omega - \frac{2\pi k}{\Delta t T})\frac{\Delta t}{2}}\right|^2 d\omega. \tag{2.35}$$

The left-hand side of (2.35) is the discrete spectrum of a time series of length T. Thus, (2.35) provides a link between what can be observed, $S_x(k,T)$, from finite data and what cannot be observed, namely, $S_x(\omega)$. This would form a basis for discussion of windows in Chapter Three.

G. z-Transform and DFT

We define z-transform of a sequence x(t), t=0,1, ... , T − 1 as follows:

$$X(z) = \sum_{t=0}^{T-1} x(t) z^{-t}, \tag{2.36}$$

where z is a complex number. DFT is a special case of z-transform when it is evaluated for $z = e^{j\frac{2\pi}{T}k}$, that is, on a unit circle in complex z-plane. In general, z-transform maps the s-plane to the z-plane using a transformation $z = e^s$. The left half s-plane is mapped into a unit circle in the z-plane as shown in Figure 2.10. Similarly, a line in the left half s-plane parallel to the imaginary axis is mapped into a circle inside the unit circle. Thus, the entire left half s-plane is mapped inside the unit circle.

We can also evaluate the z-transform on other types of contours, in particular, on a circle concentric with the unit circle. Such a circle may be defined by an equation, $|z| = \exp(\sigma_0)$. The z-transform defined in Equation (2.36) is now evaluated at

$$z = \exp(\sigma_0 + j\frac{2\pi k}{T}), \; k = 0,1,\ldots T-1.$$

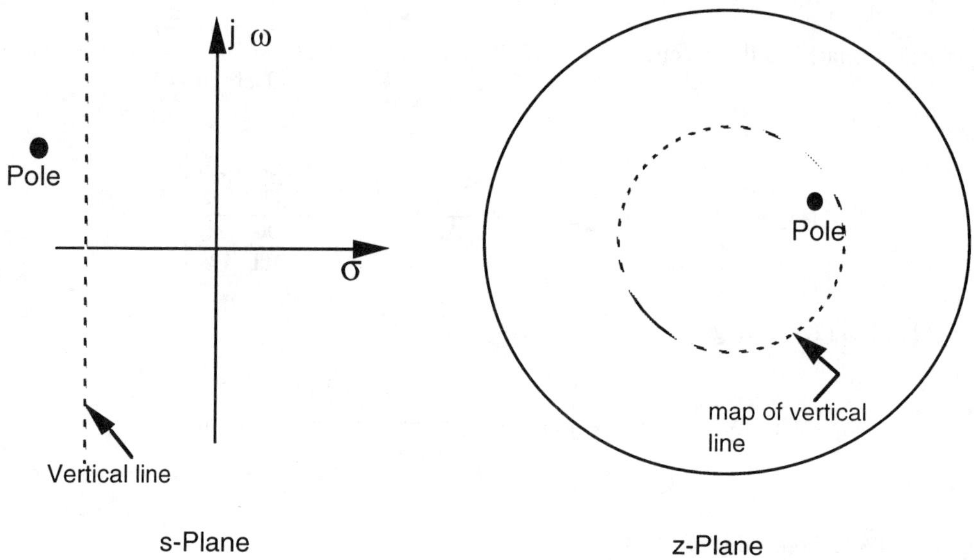

Figure 2.10: z-Transform maps the s-plane to the z-plane. As an example of mapping take a vertical line in the s-plane. It is mapped into a circle in the z-plane.

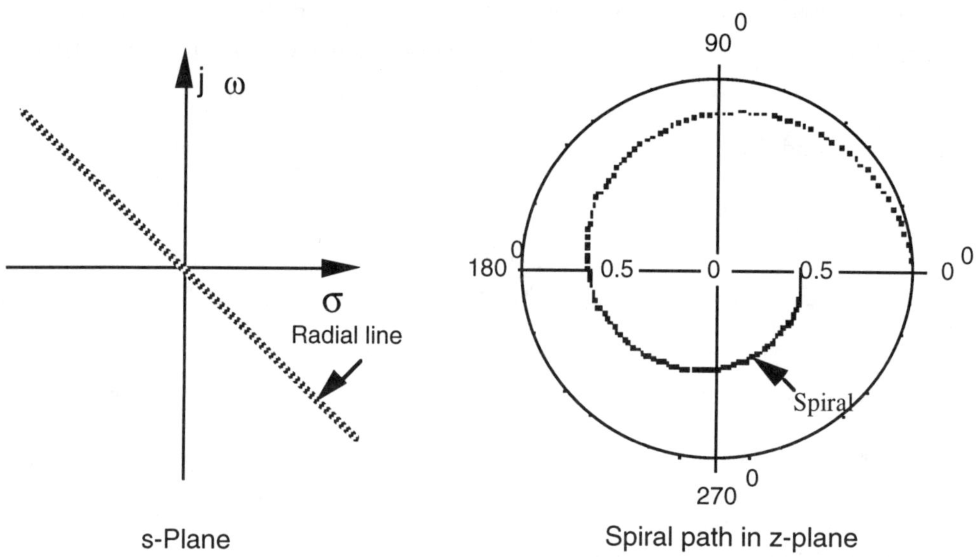

Figure 2.11: A radial line in the s-plane is mapped into a spiral in the z-plane.

Note that $\sigma_0 = 0$ is for unit circle and $\sigma_0 > 0$ for a concentric circle outside the unit circle. The z-transform on a concentric circle, $|z| = \exp(\sigma_0)$, is given by

$$X_{\sigma_0}(z) = \sum x(t) z^{-t}$$

$$= \sum_{t=0}^{T-1} (x(t) e^{-\sigma_0 t}) \exp(-j\frac{2\pi k}{T} t). \tag{2.37}$$

We notice from (2.37) that the z-transform on a $|z| = \exp(\sigma_0)$ concentric circle is simply the DFT of the original sequence multiplied by $e^{-\sigma_0 t}$.

There is yet another type of contour, the spiral contour, over which the z-transform can be evaluated. The z-transform is now evaluated for

$$z = \exp(\sigma_0 k + j\frac{2\pi k}{T}).$$

As k varies from 0 to $T-1$, z varies along a spiral as shown in Figure 2.11 (where $\sigma_0 = -\frac{2\pi}{T}$). Here also we can express z-transform as a DFT. We rewrite (2.36) as

$$X(z_k) = \sum_{t=0}^{T-1} x(t) \exp(-\sigma_0 tk - j\frac{2\pi k}{T} t). \tag{2.38}$$

By using the identity,

$$tk = \frac{(t^2 + k^2 - (k-t)^2)}{2},$$

in (2.38), we can express it as a convolution.

$$X(z_k) = \exp(-\frac{k^2}{2}(\sigma_0 + j\frac{2\pi}{T})) \sum_{t=0}^{T-1} x(t) \exp[-\frac{1}{2}(\sigma_0 + j\frac{2\pi}{T})(t^2 - (k-t)^2]$$

$$= \exp(-\frac{k^2}{2}(\sigma_0 + j\frac{2\pi}{T})) \sum_{t=0}^{T-1} h(t-k) x'(t), \tag{2.39}$$

where

$$x'(t) = x(t) \exp(-\frac{1}{2}(\sigma_0 + j\frac{2\pi}{T}) t^2)$$

$$h(t) = \exp(-\frac{1}{2}(\sigma_0 + j\frac{2\pi}{T}) t^2).$$

Equation (2.39) is a convolution sum which may be computed via Fourier domain, that is, expressing the convolution sum as a DFT product and then computing the inverse DFT of the product.

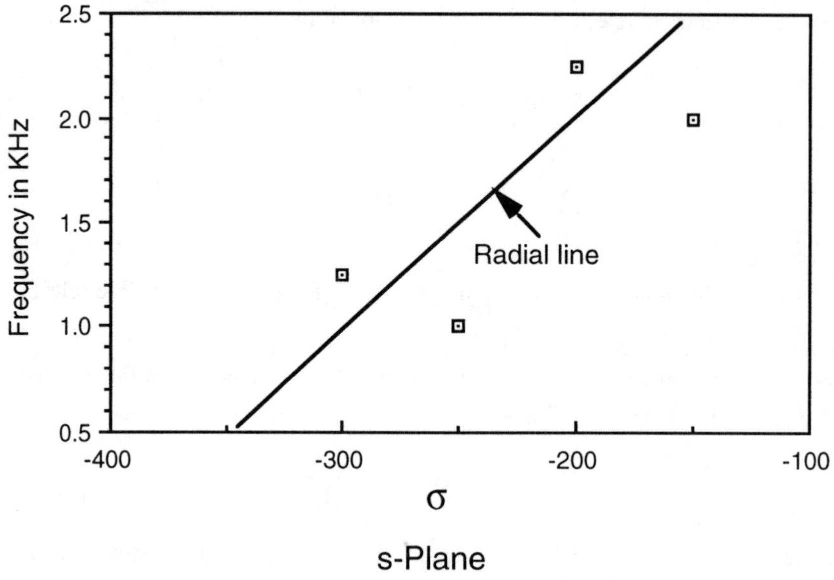

Figure 2.12: Poles of damped sinusoids in the s-plane. z-Transform is evaluated over a spiral which corresponds to a radial line as shown in the figure.

The need for evaluating z-transform over a contour other than the unit circle arises when it is required to approach a pole lying away from the unit circle. Tseng and Sarkar[11] have defined a modified DFT chirp z-transform for enhancement of poles in spectral analysis.

Example 2.5: Consider the following damped sinusoids:

$$x(t) = \sum_{i=1}^{4} a_i e^{\sigma_i t} \sin(2\pi f_i t + \phi_i),$$

where

$$a_1 = 0.1,\ a_2 = 1.0,\ a_3 = 1.0,\ a_4 = 0.1$$

$$\sigma_1 = -250,\ \sigma_2 = -300,\ \sigma_3 = -150,\ \sigma_4 = -200$$

$$f_1 = 1.0,\ f_2 = 1.25,\ f_3 = 2.0,\ f_4 = 2.25\ \text{kHz}$$

$$\phi_i,\ i = 1,\ldots,4 \ : \text{random phase.}$$

There are four pairs of poles in the s-plane as shown in Figure 2.12.The z-transform is evaluated along a spiral which corresponds to a radial line. The spectrum computed on this path is shown in Figure 2.13b, and it is compared with the spectrum on the unit circle shown in Figure 2.13a. All four poles are well reflected in the computed spectrum; however, there is significant error in the indicated pole position; in particular, the error for the first pole, being quite deep inside and of low power, is large, approximately, 6 bins (1 bin=20.8333 Hz) (End of Example).

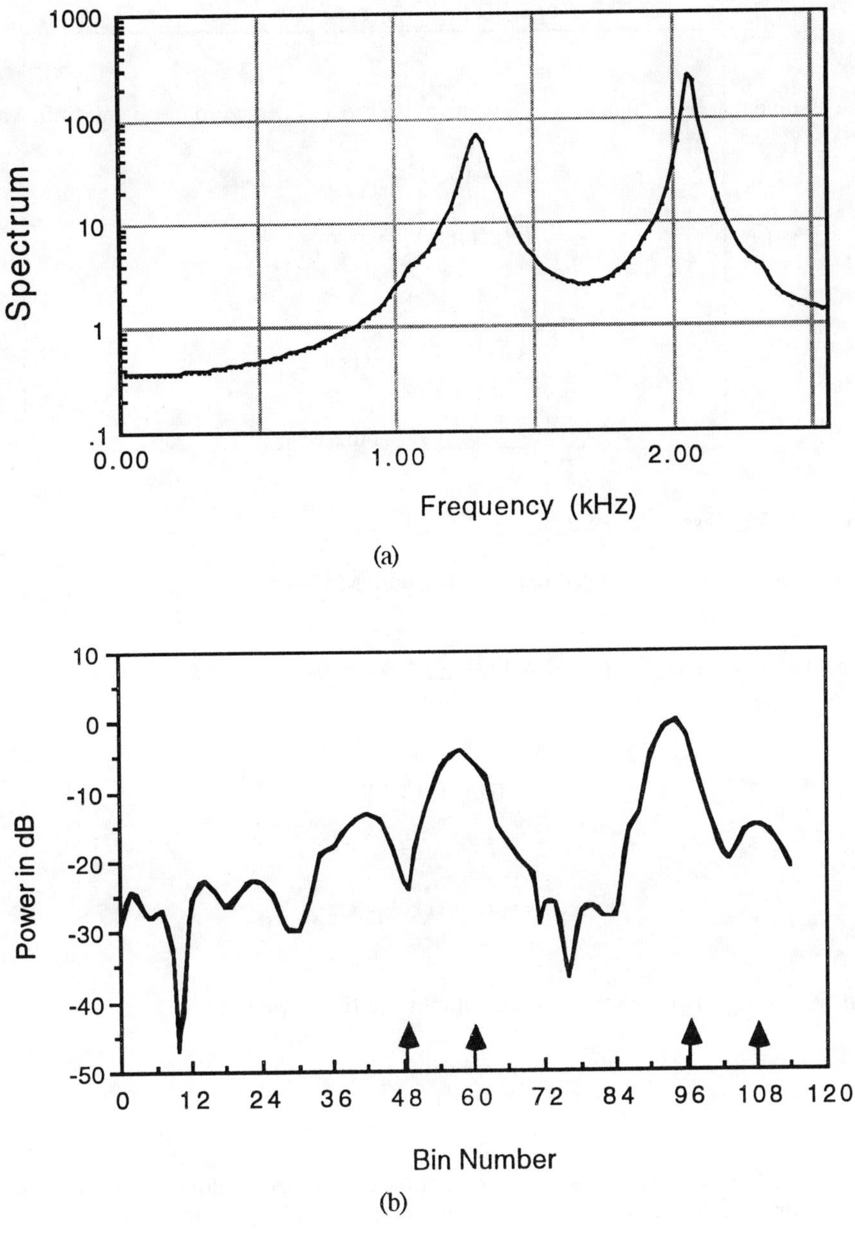

Figure 2.13: (a) The spectrum computed on the unit circle and (b) spiral path corresponding to the radial line shown in Figure 2.12.

2.4 Parametric Models: MA/AR

Following the fundamental theorem of decomposition of time series,[12] it is possible to model any time series as a sum of a deterministic component, s(t), and a random component having a one-sided moving average (MA) representation,

$$x(t) = s(t) + \sum_{i=0}^{\infty} b_i \varepsilon(t - i), \tag{2.40}$$

where we assume that $b_0=1$ and $\varepsilon(t)$ is a white noise process with zero mean and finite variance,

$$E\{\varepsilon(t)\} = 0$$

$$E\{\varepsilon(t)\varepsilon(s)\} = \sigma_\varepsilon^2 \delta_{t,s}, \tag{2.41}$$

where

$$\delta_{t,s} = 1 \qquad \text{when } t = s,$$

$$\delta_{t,s} = 0 \qquad \text{otherwise.}$$

A. **Pure MA Process**

When s(t)=0 the time series modeled in (2.40) is a pure MA process,

$$x(t) = \sum_{i=0}^{\infty} b_i \varepsilon(t - i). \tag{2.42}$$

Note that

$$E\{x(t)\} = 0$$

and

$$\mathrm{Var}_x = E\{x(t)^2\} = \sigma_\varepsilon^2 \sum_{i=0}^{\infty} b_i^2 .$$

For stability, that is, for finite input the output must be finite, we require

$$\sum_{i=0}^{\infty} b_i^2 < \infty. \tag{2.43}$$

x(t) may be looked upon as an output of a filter excited by random input. Let us define the transfer function of an MA filter as the z-transform of the model parameters,

$$B(z) = \sum_{i=0}^{\infty} b_i z^{-i}. \tag{2.44}$$

Equation (2.43) implies that if B(z) has poles, they must be inside the unit circle ($|z| < 1$). We had shown earlier on page 73 that for a stable filter all poles must lie inside the unit circle. Since the poles of B(z) must lie inside the unit circle, we express it as a product of two factors,

$$B(z) = A(z)B_0(z), \tag{2.45}$$

where A(z) has only poles inside the unit circle and $B_0(z)$ has only zeros. An MA process is said to be an all-zero model when all the poles of A(z) are at the center. Note that a complex function which is not equal to a constant everywhere must possess poles lying somewhere in the complex plane. Even for all-zero MA, model poles do exist, but at the center and hence of no consequence. We shall hereafter consider only the all-zero model. The spectral representation of x(t) can be written as follows:

$$x(t) = \frac{1}{2\pi}\int_{-\pi}^{\pi}\sum_n b_n e^{-jn\omega}\, d\varepsilon(\omega)\exp(j\omega t)$$
$$= \frac{1}{2\pi}\int_{-\pi}^{\pi} dX_G(\omega)\exp(j\omega t), \tag{2.46}$$

where

$$dX_G(\omega) = \sum_n b_n e^{-jn\omega}\, d\varepsilon(\omega).$$

Referring to the properties of the generalized Fourier transform (GFT), we can obtain the spectrum of MA process from (2.46):

$$S_X(\omega) = \left|\sum_n b_n \exp(-j\omega n)\right|^2 \sigma_\varepsilon^2$$
$$= \left|B(z = e^{j\omega})\right|^2 \sigma_\varepsilon^2. \tag{2.47}$$

Since all poles of B(z) lie at the center of the unit circle, B(z) will be free of singularities over the rest of the unit circle; hence, $S_X(w)$ will not possess sharp peaks. Thus, the spectrum of a moving average (MA) process will be a smooth and continuous function. If there are zeros in B(z) close to the unit circle, the spectrum will show nulls at the corresponding positions of the zeros. In fact, B(z) can be completely characterized by the position of its zeros. It is not necessary, except for invertibility, that the zeros must lie inside the unit circle.

Next we shall take a look at the covariance function of an MA process. Using the definition of a covariance function we can write

$$C_X(\tau) = E\{x(t)x(t+\tau)\}$$

$$= \sum_i\sum_k b_i b_k E\{\varepsilon(t-i)\varepsilon(t+\tau-k)\}$$

$$= \sigma_\varepsilon^2 \sum_i\sum_k b_i b_k \delta(\tau - k + i)$$

$$= \sigma_\varepsilon^2 \sum_{k=\tau}^{\infty} b_k b_{k-\tau}, \quad \text{or} \quad = \sigma_\varepsilon^2 \sum_{i=0}^{\infty} b_i b_{i+\tau}. \tag{2.48}$$

It is interesting to note that if the MA process is of finite order, that is, $b_i = 0$ for $i > q$, (2.48) reduces to ,

C_x	2.7770	-1.8074	0.9025	0.0	0.0
Lag	0	1	2	3	4

Table 2.1: The covariance function of MA(2) process.

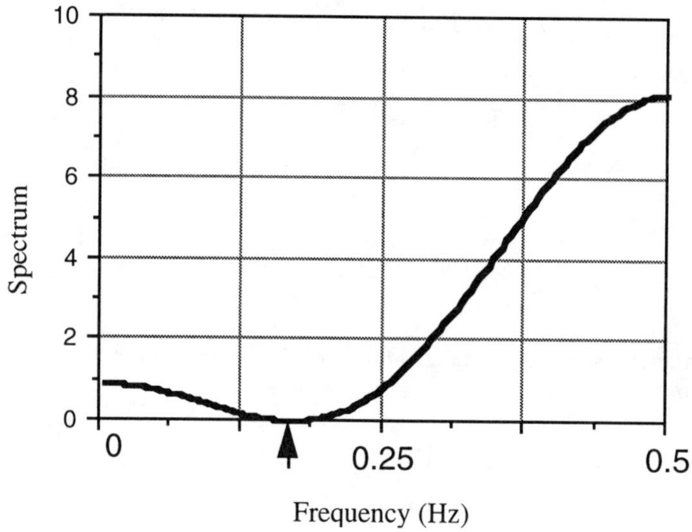

Figure 2.14: Spectrum of MA(2) process. There is zero at w=0.1666p. Notice that the spectrum around zero is gradual.

$$C_x(\tau) = \sigma_\varepsilon^2 \sum_{i=0}^{q-\tau} b_i b_{i+\tau} \qquad \tau \le q$$
$$= 0 \qquad \tau > q \; . \tag{2.49}$$

The covariance function is nonzero only for lags less than or equal to the order of the MA process. This fact is useful in rapid estimation of the order of the MA process.

The covariance function of a qth-order MA process may be written in a matrix form. For this, define a matrix,

$$\mathbf{B} = \underbrace{\begin{bmatrix} b_q & b_{q-1} & \cdots & b_1 & b_0 & & & & \\ & b_q & b_{q-1} & \cdots & b_1 & b_0 & & \mathbf{0} & \\ & & b_q & b_{q-1} & \cdots & b_1 & b_0 & & \\ & \mathbf{0} & & \ddots & \ddots & & \ddots & \ddots & \\ & & & & b_q & b_{q-1} & \cdots & b_1 & b_0 \end{bmatrix}}_{(q+1)\times(2q+1)} .$$

We can now express the covariance matrix in terms of **B**:

$$\mathbf{C}_x = \sigma_\varepsilon^2 \mathbf{B}\mathbf{B}^T . \tag{2.50}$$

Example 2.6: Consider an MA process (zeros are at $0.95e^{\pm j\frac{\pi}{3}}$) given by

$$x(t) = \varepsilon(t) - 0.95\varepsilon(t-1) + 0.9025\,\varepsilon(t-2).$$

The covariance function is tabulated (Table 2.1), and its spectrum is shown in Figure 2.14. The position of the zero is shown by an arrow.

The **B** matrix defined as above turns out to be

$$\mathbf{B} = \begin{bmatrix} 0.9025 & -0.95 & 1.0 & & \\ & 0.9025 & -0.95 & 1.0 & \\ & & 0.9025 & -0.95 & 1.0 \end{bmatrix}.$$

By actual multiplication we can show that

$$\mathbf{B}\mathbf{B}^T = \begin{bmatrix} 2.7770 & -1.8074 & 0.9025 \\ -1.8074 & 2.7770 & -1.8074 \\ 0.9025 & -1.8074 & 2.7770 \end{bmatrix},$$

which is the covariance matrix (covariance function arranged in a matrix form) of the MA process (End of Example).

Finally, note that from the spectrum of an MA process we can obtain the magnitude of its transfer function only. Since the zeros of an MA process can lie outside the unit circle, we cannot determine the phase of the transfer function given its magnitude, and hence the model parameters cannot be estimated from the spectrum. However, from the bispectrum we shall see later (page 132) how to determine the parameters of a model which is driven by a non-Gaussian input.

B. **Pure Autoregressive (AR) Process**

An autoregressive process is defined as follows:

$$x(t) + a_1 x(t-1) + a_2 x(t-2) + \ldots + a_p x(t-p) = \varepsilon(t), \tag{2.51}$$

where $\varepsilon(t)$ is zero mean white noise (not necessarily Gaussian) driving a feedback system as shown in Figure 1.2. Note that

$$E\{X(t)\varepsilon(s)\} = 0 \quad s > t$$

imposes a constraint that $\varepsilon(t)$ be a white noise process.[13] We can also think of a two-sided AR process for spatial series,

$$x(n) = \sum_{\substack{m=-N \\ m \neq n}}^{N} h(m)x(n-m) + \varepsilon(n) \quad -\infty < n < \infty,$$

where $E\{x(n)\varepsilon(m)\} = c\delta_{mn}$; that is, $\varepsilon(m)$ is uncorrelated with all x(m) actually used in the estimate of x(n).

Equation (2.51) may be looked upon as a difference equation driven by $\varepsilon(t)$. Naturally, it has a transient output which is observed soon after it is excited. The transient output may be obtained by solving the difference equation with $\delta(t)$ as input,

$$x(t) + a_1 x(t-1) + a_2 x(t-2) + \ldots + a_p x(t-p) = \delta(t), \tag{2.52}$$

whose solution is a sum of a complex exponential,

$$x(t) = \sum_{i=1}^{p} \alpha_i \exp(\gamma_i t) \quad t \geq 0,$$

where γ_i, i = 1,2, ... p are given by the roots of the polynomial,

$$A(z) = 1 + a_1 z^{-1} + a_2 z^{-2} + \ldots + a_p z^{-p} = 0 \quad , \tag{2.53}$$

whose i^{th} root, $z_i = \exp(\gamma_i)$. For the transient to decay fast, the roots of (2.53) must lie well inside the unit circle. If any one of the roots were to lie close to the unit circle, the decay rate would be very slow. As a result the mean of AR process will become time dependent, and thus the resulting solution will be nonstationary.

The covariance function of an AR process can be expressed in terms of its parameters,

$$\begin{aligned} C_x(\tau) &= E\{x(t)x(t+\tau)\} \\ &= E\{x(t)[-a_1 x(t+\tau-1) - a_2 x(t+\tau-2) \ldots - a_p x(t+\tau-p) + \varepsilon(t+\tau)] \\ &= -a_1 C_x(\tau-1) - a_2 C_x(\tau-2) \ldots - a_p C_x(\tau-p) + E\{x(t)\varepsilon(t+\tau)\}. \end{aligned}$$

Since x(t) is uncorrelated with all future inputs but correlated with the current and the past inputs,

$$E\{x(t)\varepsilon(t+\tau)\} = \sigma_\varepsilon^2 \delta(\tau) \qquad \tau \geq 0.$$

Hence

$$C_x(\tau) + a_1 C_x(\tau-1) + a_2 C_x(\tau-2) + \ldots + a_p C_x(\tau-p) = \sigma_\varepsilon^2 \delta(\tau), \quad \tau \geq 0. \tag{2.54}$$

Equation (2.54) may be solved using the z-transform approach. Taking the z-transform on both sides of (2.54) we obtain

$$\tilde{S}_x(z) = \frac{\sigma_\varepsilon^2}{[1 + a_1 z^{-1} + a_2 z^{-2} + \ldots + a_p z^{-p}]}, \tag{2.55}$$

where

$$\tilde{S}_x(z) = C_x(0) + \sum_{i=1}^{\infty} C_x(i) z^{-i}$$

is the one-sided z-transform of the covariance function. Note that the spectrum is related to the one-sided z-transform of the covariance function,

$$S_x(\omega) = 2\text{Re}\{\bar{S}_x(z)\Big|_{z=e^{j\omega}}\} - C_x(0)$$

The denominator polynomial in (2.55) may be factored. Let $z_1, z_2, \ldots, z_p$ be the roots of the polynomial and the corresponding residues be $\alpha_1, \alpha_2, \ldots, \alpha_p$, where

$$\alpha_i = \frac{(z_i)^{p-1}}{\prod_{i \neq j}^{p} (z_i - z_j)}.$$

Equation (2.55) reduces to

$$S_x(z) = \sigma_\varepsilon^2 \left[\frac{\alpha_1 z}{(z - z_1)} + \frac{\alpha_2 z}{(z - z_2)} + \ldots + \frac{\alpha_p z}{(z - z_p)}\right]. \tag{2.56}$$

The inverse z-transform of (2.56) is given by

$$\hat{C}_x(\tau) = \sigma_\varepsilon^2 \sum_{i=1}^{p} \alpha_i \exp(\gamma_i \tau) \qquad \tau \geq 0, \tag{2.57a}$$

where $z_i = \exp(\gamma_i)$. For $\hat{C}_x(\tau)$ to be a covariance function it has to be made symmetric by reflecting it onto the $\tau < 0$ axis. Consider the following expression:

$$C_x(\tau) = \sigma_\varepsilon^2 \sum_{i=1}^{p} \Omega_i \exp(\gamma_i |\tau|), \tag{2.57b}$$

where Ω_i, i=1, 2, ..., p are constants to be determined by insisting that (2.57b) satisfies (2.54).

Example 2.7: For p=1 the covariance function is of the form,

$$C_x(\tau) = \sigma_\varepsilon^2 \Omega_1 \exp(\gamma_1 |\tau|) \quad ,$$

where $\exp(\gamma_1) = -a_1$, the only root of (253). Next, to obtain Ω_i we substitute for the covariance function in (2.54) for $\tau = 0$. The result is $\Omega_i = \frac{1}{1-a^2}$. For p=2, the covariance function is of the form

$$C_x(\tau) = \sigma_\varepsilon^2 \{\Omega_1 e^{\gamma_1 \tau} + \Omega_2 e^{\gamma_2 \tau}\} \quad ,$$

where e^{γ_1} and e^{γ_2} are the roots of the coefficient polynomial (2.53). These may be expressed in terms of the radial distance and the angular position of the roots, that is, $r_0 e^{\pm j\omega_0}$. In terms of the AR coefficients,

$$a_1 = -2r_0 \cos \omega_0$$

and

$$a_2 = r_0^2 .$$

Next, to obtain Ω_1 and Ω_2 we substitute for the covariance function in (2.54) for $\tau = 0$ and $\tau = 1$ and solve the resulting pair of linear equations. We obtain

$$\Omega_1 = \Omega_2 = \frac{0.5}{1-2r_0^2 \cos^2(\omega_0)+r_0^4 \cos(2\omega_0)} .$$

Finally, the covariance function of an AR(2) process is given by

$$C_X(\tau) = \frac{\sigma_\varepsilon^2}{1-2r_0^2 \cos^2(\omega_0)+r_0^4 \cos(2\omega_0)} r_0^{|\tau|} \cos(\omega_0 \tau) .$$

(End of Example).

Note that for the output to be finite we must have

$$C_x(0) = \sigma_\varepsilon^2 \sum_{i=1}^{p} \Omega_i < \infty .$$

Further, for $C_x(\tau) \to 0$ as $|\tau| \to \infty$, $\mathrm{Re}\{\gamma_i\} < 0$ for all i, which implies that the roots of the polynomial, (2.53), must lie inside the unit circle. It may be observed that the transient solution and the covariance function satisfy the same difference equation (2.52 or 2.54) but with different initial conditions.

An AR model can be completely specified by means of its poles inside the unit circle. For this reason the AR model is sometimes known as the all-pole model. An AR model can be reversed to obtain a stable MA model and conversely an MA model can also be reversed provided all its zeros lie inside the unit circle. There is, however, an important difference. A low-order AR model if reversed will give rise to a high-order MA model or vice versa.

Example 2.8: Consider a first-order AR model, $A(z) = (1 - az^{-1})$. We shall reverse the AR(1) model to obtain an MA model,

$$B(z) = \frac{1}{A(z)} = (1 - az^{-1})^{-1}$$
$$= [1 + az^{-1} + (az^{-1})^2 + (az^{-1})^3 + \dots],$$

which is a transfer function of an infinite-order MA model having coefficients, $\{1, a, a^2, a^3, a^4, \dots\}$. Since $|a| < 1$ the sequence is convergent (End of Example).

For a general AR model the equivalent MA model may be obtained from the transient solution of (2.52). Let the solution of (2.52) be given by $b(0), b(1), b(2), \ldots$ with

$$b(0) = 1 \text{ and } b(-1) = b(-2) = \ldots = 0.$$

The solution of the AR equation can be written in terms of its transient solution,

$$x(t) = \sum_{i=0}^{t} b(i)\varepsilon(t-i). \tag{2.58}$$

Let us compute the covariance function of x(t) given by (2.58):

$$\begin{aligned} C_x(\tau) = E\{x(t)x(t+\tau)\} &= \sum_{i=0}^{t}\sum_{l=0}^{t+\tau} b(i)b(l)\delta(\tau+i-l) \\ &= \sigma_\varepsilon^2 \sum_{i}^{t} b(i)\,b(\tau+i). \end{aligned} \tag{2.59}$$

From (2.59) we see that the covariance function will be independent of t if and only if the transient solution decays at a sufficiently fast rate (see Exercise 2.4 for an illustration).

C. Yule-Walker Equations

Equation (2.54) may be written in a matrix form as follows:

$$\underbrace{\begin{bmatrix} C_x(0) & C_x(1) & C_x(2) & \cdots & C_x(p) \\ C_x(1) & C_x(0) & C_x(1) & \cdots & C_x(p-1) \\ C_x(2) & C_x(1) & C_x(0) & \cdots & C_x(p-2) \\ \vdots & & & \vdots & \\ C_x(p) & C_x(p-1) & C_x(p-2) & \cdots & C_x(0) \end{bmatrix}}_{(p+1)\times(p+1)} \begin{bmatrix} 1 \\ a_1 \\ a_2 \\ \vdots \\ a_p \end{bmatrix} = \begin{bmatrix} \sigma_\varepsilon^2 \\ 0 \\ \vdots \\ 0 \end{bmatrix}. \tag{2.60}$$

This is the well-known Yule-Walker equation which enables us to compute the AR coefficients given the covariance matrix. This is a forward solution. We can also obtain the reverse Yule-Walker equation where given the AR coefficients we can compute the covariance vector.

$$\begin{bmatrix} 1 & a_1 & a_2 & \cdots & & a_p \\ a_1 & (1+a_2) & a_3 & \cdots & a_p & 0 \\ a_2 & (a_1+a_3) & (1+a_4) & \cdots & a_p\ 0 & 0 \\ \vdots & & & & \therefore & \\ a_i & (a_{i-1}+a_{i+1}) & (a_{i-2}+a_{i+2}) & \cdots\ (a_{i-p}+a_{i+p}) & 0\ \cdots & 0 \\ \vdots & & & & & \\ a_p & a_{p-1} & a_{p-2} & \cdots & a_1 & 1 \end{bmatrix} \begin{bmatrix} C_x(0) \\ C_x(1) \\ \\ \vdots \\ \\ C_x(p) \end{bmatrix} = \begin{bmatrix} \sigma_\varepsilon^2 \\ 0 \\ \\ \vdots \\ \\ 0 \end{bmatrix}. \tag{2.61}$$

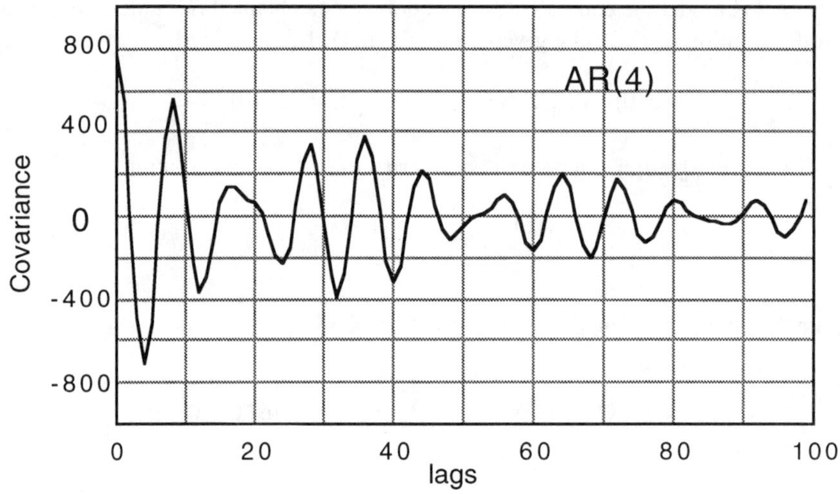

Figure 2.15: Covariance function of an AR(4) process.

Covariance function					
C_x	761.78	545.8	27.14	-487.70	-705.30
Lag	0	1	2	3	4

Table 2.2: Covariance function for lags 0 to 4 of an AR(4) process.

In the above matrix the ith row is given by

$$\text{row}\,[a_i \quad a_{i+1}+a_{i-1} \quad a_{i+2}+a_{i-2} \quad \cdots \quad a_{i+p}+a_{i-p}],$$

where we note that $a_m = 0$ whenever m>p and m<0, and $a_0 = 1$. The reverse Yule-Walker equations are useful for computing the theoretical covariance function given the AR coefficients and variance of the driving stochastic process.

Example 2.9: A covariance function of an AR process of order four with coefficients -2.7607, 3.8106, -2.6535, and 0.9238 and driven by a unit variance stochastic process is computed using the reverse Yule-Walker equations. First, the covariance function for lags 0 to 4 are computed using (2.61). The numerical values are shown in Table 2.2. These values are used in (2.54) with $\tau > 0$ for computation of the covariance function at other lags. The results are shown in Figure 2.15 (End of Example)

Dropping the first equation of (2.54) corresponding to $\tau=0$ we get the following system of equations:

$$\underbrace{\begin{bmatrix} C_x(1) & C_x(0) & C_x(1) & \ldots & C_x(p-1) \\ C_x(2) & C_x(1) & C_x(0) & \ldots & C_x(p-2) \\ C_x(3) & C_x(2) & C_x(1) & \ldots & C_x(p-3) \\ \vdots & & & & \vdots \\ C_x(p+1) & C_x(p) & C_x(p-1) & \ldots & C_x(1) \\ \vdots & & & & \vdots \\ C_x(N) & C_x(N-1) & C_x(N-2) & \ldots & C_x(N-p) \end{bmatrix}}_{N\times(p+1)} \begin{bmatrix} 1 \\ a_1 \\ a_2 \\ \vdots \\ a_p \end{bmatrix} = \begin{bmatrix} 0 \\ \vdots \\ 0 \end{bmatrix}, \tag{2.62}$$

Or in a matrix form

$$\mathbf{C}_x^e \mathbf{A} = \mathbf{0}, \tag{2.63}$$

where $\mathbf{C}_x^e$ is an $N\times(p+1)$ extended covariance matrix, $\mathbf{A}$ is a vector of the AR coefficients, and $\mathbf{0}$ is a zero vector (all entries are zeros). In order that $\mathbf{A}$ has nontrivial solution, $\mathbf{C}_x^e$ must *not* be full rank. On squaring both sides of (2.63) we obtain

$$\mathbf{A}^T \mathbf{C}_x^{e\,T} \mathbf{C}_x^e \mathbf{A} = \mathbf{0}. \tag{2.64}$$

Note that $\mathbf{C}_x^{e\,T} \mathbf{C}_x^e$ is $(p+1)\times(p+1)$ matrix of rank p (the rank of $\mathbf{C}_x^e$ is equal to p; see Exercise 2.7). The AR parameter vector $\mathbf{A}$ is equal to the minimum eigenvector of $\mathbf{C}_x^{e\,T} \mathbf{C}_x^e$. The first element has to be normalized to unity. By interchanging the rows of the extended covariance matrix $\mathbf{C}_x^e$ (standard form) in (2.62), we can form many (N!) equivalent extended matrices.

D. Order of an AR Process

The rank of an extended covariance matrix is a good measure of the order of the AR process. For this purpose it is recommended to form an oversized extended covariance matrix and then estimate the rank of $\mathbf{C}_x^{e\,T} \mathbf{C}_x^e$. That the order of an AR model can be effectively estimated from a sample covariance matrix is illustrated in the example given below.

Example 2.10: A fourth-order process with parameters $a_1=-2.7607$, $a_2=3.8106$, $a_3=-2.6555$, and $a_4=0.9238$ was simulated on computer. White Gaussian noise of zero mean and unit variance was used to drive the AR filter. The first two thousand samples were rejected as they were likely to contain the transient component. A data vector of eight samples was formed, and sixteen such data vectors were used to compute the covariance matrix,

$$\hat{\mathbf{C}}_x = \sum_{i=1}^{16} \mathbf{x}_i \mathbf{x}_i^H,$$

2539.5	217.3	96.8	10.1	0.91	0.03	0.01	0.0
1962.4	1073.2	65.2	43.8	1.33	0.08	0.02	0.0
1917.8	1067.1	82.4	58.2	0.77	0.04	0.01	0.0
1816.6	466.5	48.8	21.1	1.48	0.07	0.01	0.0
393.2	309.2	25.8	14.5	0.78	0.07	0.01	0.0
1715.9	503.6	95.3	20.7	1.14	0.07	0.01	0.0
2698.8	1823.2	104.4	63.6	1.16	0.11	0.01	0.0
1295.9	561.9	45.3	15.2	1.55	0.08	0.01	0.0

Significant Eigenvalues

Table 2.3: The eigenvalues of the estimated covariance matrix arranged along a row. Results of eight independent experiments are shown. The significant eigenvalues (clear background) and insignificant eigenvalues (speckled background) are also marked.

where $\mathbf{x}_i$ is the ith data vector. The covariance matrix was subjected to eigenvalue decomposition. A set of eight samples of computations are shown in Table 2.3 (End of Example).

Order estimation requires a means to distinguish significant eigenvalues from insignificant ones. An iterative method based on threshold bounds on the estimated eigenvalues has been reported in Reference 14 and was demonstrated to perform as well as or better than the well-known order estimation techniques such as AIC and MDL, which are discussed in Chapter Five (page 367).

The inverse of a covariance matrix is often useful in many applications. Given the AR coefficients we can directly compute the inverse of the covariance function using Gohberg-Semenual relation[15]

$$\mathbf{C}_x^{-1} = \frac{1}{\sigma_\varepsilon^2}(\mathbf{L}_1\mathbf{L}_1^T - \mathbf{L}_2\mathbf{L}_2^T),$$

where

$$\mathbf{L}_1 = \begin{bmatrix} 1 & & & & \\ a_1 & 1 & & \mathrm{O} & \\ a_2 & a_1 & 1 & & \\ \vdots & & & \ddots & \\ a_p & a_{p-1} & \cdots & a_1 & 1 \end{bmatrix} \qquad \mathbf{L}_2 = \begin{bmatrix} 0 & & & & \\ a_p & 0 & & \mathrm{O} & \\ a_{p-1} & a_p & 0 & & \\ \vdots & & \ddots & \ddots & \\ a_1 & a_2 & \cdots & a_p & 0 \end{bmatrix}.$$

E. Spectrum of an AR Process

To evaluate the spectrum of an AR process we shall use the spectral representation of a stationary stochastic process (see Chapter One, (1.57)). We obtain after substituting for x(t) in (2.51),

$$\int_{-\pi}^{\pi} dX_G(\omega)\exp(j\omega t)\left[1+\sum_{i=1}^{p} a_i \exp(-j\omega i)\right] = \int_{-\pi}^{\pi} d\bar{\varepsilon}_G(\omega)\exp(j\omega t).$$

Hence,

$$dX_G(\omega) = \frac{d\bar{\varepsilon}_G(\omega)}{A(\omega)},$$

where

$$A(\omega) = \left[1+\sum_{i=1}^{p} a_i \exp(-j\omega i)\right].$$

Using the relation between the spectrum and GFT (1.61) we obtain

$$\begin{aligned} 2\pi S_x(\omega)d\omega &= E\{dX_G dX_G^H\} \\ &= \frac{2\pi S_\varepsilon(\omega)d\omega}{|A(\omega)|^2}, \end{aligned}$$

where, since $\varepsilon(t)$ is a white noise process, $S_\varepsilon(\omega)$ is a constant equal to σ_ε^2. Therefore,

$$S_x(\omega) = \frac{\sigma_\varepsilon^2}{|A(\omega)|^2}. \tag{2.65}$$

We note that

$$A(\omega) = A(z)\big|_{z=e^{j\omega}}.$$

Therefore, the roots of $A(\omega)$, which may be treated as a polynomial in $\exp(j\omega)$ or in z if we replace $\exp(j\omega)$ by z, are the same as the roots of A(z). We shall now express the spectrum of the AR process in terms of its poles (assuming p distinct poles).

$$S_x(\omega) = \frac{\sigma_\varepsilon^2}{|(z-z_1)(z-z_2)(z-z_3)\ldots(z-z_p)|^2}\Bigg|_{z=e^{j\omega}}.$$

A peak in the spectrum is a consequence of the pole, which has a great bearing upon the physical process responsible for the generation of time series. But often the spectral peak does not give the exact angular position of the pole, except when the pole is close to the unit circle.

Example 2.11: Consider an AR(2) process having two poles at $r_0 e^{\pm j\omega_0}$ and with the corresponding AR coefficients,

$$a_1 = -2r_0\cos(\omega_0)$$

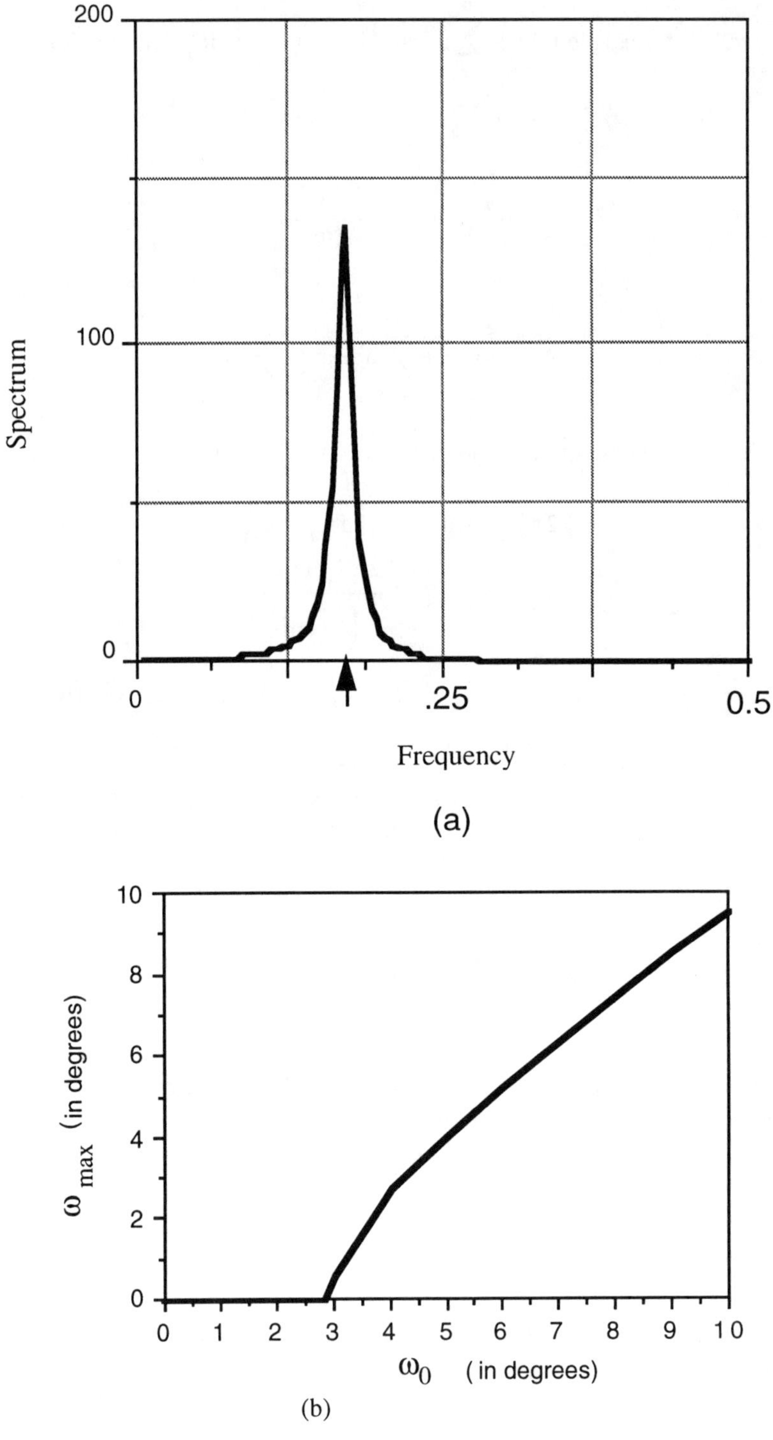

Figure 2.16: (a) AR spectrum for r_0 =0.95 and $\omega_0 = \frac{\pi}{3}$. Pole position is shown by an arrow. (b) Spectrum peak position as a function of pole position. The radial distance of the pole is 0.95, and the corresponding critical angle is ω_c=2.94^0.

and

$$a_2 = r_0^2.$$

The AR spectrum is given by

$$S_x(\omega) = \frac{\sigma_\varepsilon^2}{[(1 - r_0^2)^2 \sin^2(\omega_0) + 4 r_0^2 \{\cos(\omega) - \frac{1+r_0^2}{2r_0}\cos(\omega_0)\}^2]}.$$

and displayed in Figure 2.16a. The position of the spectral peak, ω_{max}, and the pole position, ω_0, are related as

$$\omega_{max} = \cos^{-1}\{\frac{1 + r_0^2}{2r_0}\cos(\omega_0)\} \quad \text{for } \omega_0 > \omega_c,$$

$$\omega_{max} = 0 \qquad \text{for } \omega_0 \le \omega_c,$$

where

$$\cos(\omega_c) = \frac{2r_0}{1 + r_0^2}$$

(see Figure 2.16b).[16] The peak height also depends upon the pole position,

$$S_x(\omega_{max}) = \frac{\sigma_\varepsilon^2}{(1 - r_0^2)^2 \sin^2(\omega_0)} \qquad \omega_0 > \omega_c$$

$$= \frac{\sigma_\varepsilon^2}{[(1 - r_0^2)^2 \sin^2(\omega_0) + 4r_0^2\{1 - \frac{1+r_0^2}{2r_0}\cos(\omega_0)\}^2]} \qquad \omega_0 < \omega_c$$

Finally, the 3 dB width (full width) of the peak is given by

$$= \frac{\sigma_\varepsilon^2}{[(1 - r_0^2)^2 \sin^2(\omega_0) + 4r_0^2\{1 - \frac{1+r_0^2}{2r_0}\cos(\omega_0)\}^2]} \qquad \omega_0 < \omega_c,$$

where we have assumed that $\Delta\omega << 1$. It is observed that the spectral peak broadens as the pole moves away from the circumference but practically remains independent of ω_0. This effect is more dramatic in the case of an AR(4) process (End of Example).

F. **Resolution of Poles from Spectrum**

Consider an AR(4) process with pole pair, $r_1 e^{\pm j\omega_1}$ and $r_2 e^{\pm j\omega_2}$, and with the corresponding AR coefficients

$$a_1 = -2(r_1 \cos \omega_1 + r_2 \cos \omega_2),$$

$$a_2 = r_1^2 + r_2^2 + 4r_1 r_2 \cos \omega_1 \cos \omega_2,$$

$$a_3 = -2r_1 r_2 \{r_2 \cos \omega_1 + r_1 \cos \omega_2\},$$

$$a_4 = r_1^2 r_2^2.$$

The AR spectrum is obtained by substituting the above values of the coefficients expressed in terms of position of poles in (2.65). To locate the extreme of the spectrum it is necessary to solve a cubic equation which results from the differentiation of the expression for the spectrum. We shall avoid the tedious algebra but give an outline of the method and the final results (for details see Reference 17). The points of extrema are the solutions of the following equation:

$$\sin(\omega)\{8p_4 \cos^3(\omega) - 4p_3 \cos^2(\omega) + (2p_2 - 4p_4)\cos(\omega) + p_3 - p_1, \qquad (2.66)$$

where

$$p_1 = -a_1 - \sum_{k-1}^{3} a_k a_{k+1},$$

$$p_2 = 2(a_2 + \sum_{k-1}^{2} a_k a_{k+2}),$$

$$p_3 = 3(-a_3 - a_1 a_4),$$

and

$$p_4 = 4a_4.$$

The trivial roots of (2.66) are at $\omega = 0$ and $\pm\pi$, and other roots are obtained by solving the cubic equation (inside the curly brackets) using Cardan's formula.[18] The roots are

$$\cos(\omega) = 2\sqrt{-\frac{s}{3}} \cos\left(\frac{\psi + 2\pi k}{3}\right) - \frac{p}{3} \qquad k = 0, 1, 2,$$

where

$$s = \frac{(3q - p^2)}{3},$$

$$t = \frac{2p^3 - 9pq + 27r}{27},$$

and

$$\psi = \cos^{-1}\left(-\frac{t}{2}\sqrt{-\frac{s^3}{27}}\right).$$

Further, p, q, and r are defined as

$$p = -4p_3/8p_4,$$

$$q = (2p_2 - 4p_4)/8p_4,$$

and

$$r = (p_3 - p_1)/8p_4.$$

The extrema at k=0,1 correspond to frequencies ω_1 and ω_2, respectively, and the one at k=2 corresponds to the valley in between. It is found that the spectral peaks do not coincide with the pole positions. There is a small shift, of the order one degree or less, depending upon the pole angle and the radial distance.

Example 2.12: As an example we consider two poles at a radial distance of 0.96 and separated by 9 degrees. The position of the spectral peaks was computed using Cardan's formula for solving the cubic equation (2.66). The results are plotted in Figure 2.17 and tabulated in Table 2.4 for angles close to the real axis where the change is rapid. The shift decreases as one moves away from the real axis, but it does not become zero except when the poles are on the unit circle. However, the spectrum at these points of extreme displays an interesting dependence on the radial distance of the poles. This is shown in Figure 2.18. The amplitude of the spectral peaks decreases very rapidly as the poles recede from the circumference while the floor of the valley remains almost at a constant level. As a result, by the time the poles move 0.04 inside the unit circle, the amplitude of the spectral peak has fallen to the level of the valley, at which point there are not two spectral peaks but only one peak, as if there is only one pole. This loss of resolution takes place very close to the circumference (End of Example).

2.5 **Parametric Models**: ARMA

An AR model of a time series is an all-pole model while the MA model is an all-zero model (except for the poles at the center). There are, however, situations requiring pole zero configuration. Such a model of time series is called the ARMA model, where the current value of the process is expressed as the weighted sum of the past values and the weighted sum of the past random inputs. An ARMA model of order (p,q) is written as follows:

$$x(t) + \sum_{k=1}^{p} a_k x(t-k) = \sum_{k=0}^{q} b_k \varepsilon(t-k), \tag{2.67}$$

where a_k, k=1,2, ..., p and b_k, k=0,1,2, ..., q are parameters of the ARMA model, and p and q are integers.

A. **Transfer Function**

Let us take the z-transform on both sides of (2.67),

$$X(z) + \sum_{k=1}^{p} a_k z^{-k} X(z) = \sum_{k=0}^{q} b_k z^{-k} \bar{\varepsilon}(z),$$

ω_1 (degs)	ω_{max_1} (degs)	ω_{max_2} (degs)
0.0	2.69	10.23
1.0	2.85	10.63
2.0	3.42	11.38
3.0	4.17	12.28
4.0	5.00	13.24
5.0	5.91	14.20

Table 2.4: The position of the spectral peaks is compared with the pole position. The poles are at a radial distance of 0.96. The first pole is at ω_1, and the second pole is at $\omega_2 = \omega_1 + 9 \cdot$ The first spectrum peak is at ω_{max_1} and the second peak at ω_{max_2}.

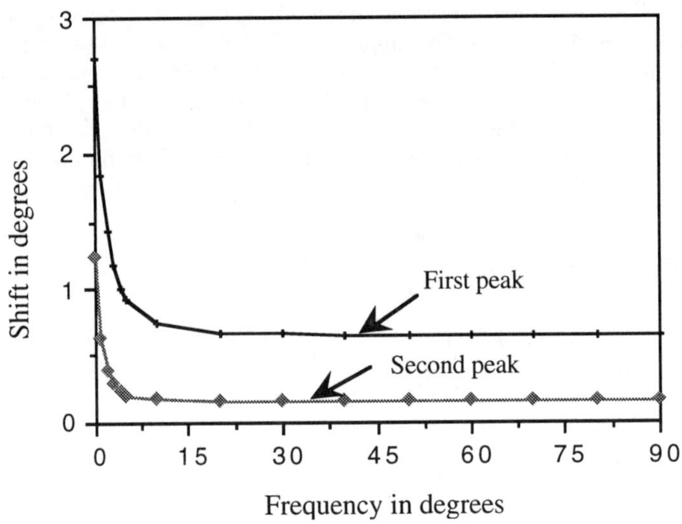

Figure 2.17: Shift in the spectral peak position as a function pole angle (ω_1). The two poles are at a radial distance of 0.96 and separated by 9 degrees.

$$X(z) = \frac{\sum_{k=0}^{q} b_k z^{-k}}{1 + \sum_{k=1}^{p} a_k z^{-k}} \bar{\varepsilon}(z) = H(z)\bar{\varepsilon}(z). \tag{2.68}$$

H(z) is the transfer function of the model. Let h_τ, $\tau = 0, 2, \ldots$ be the impulse response or the transient response of the ARMA model; that is,

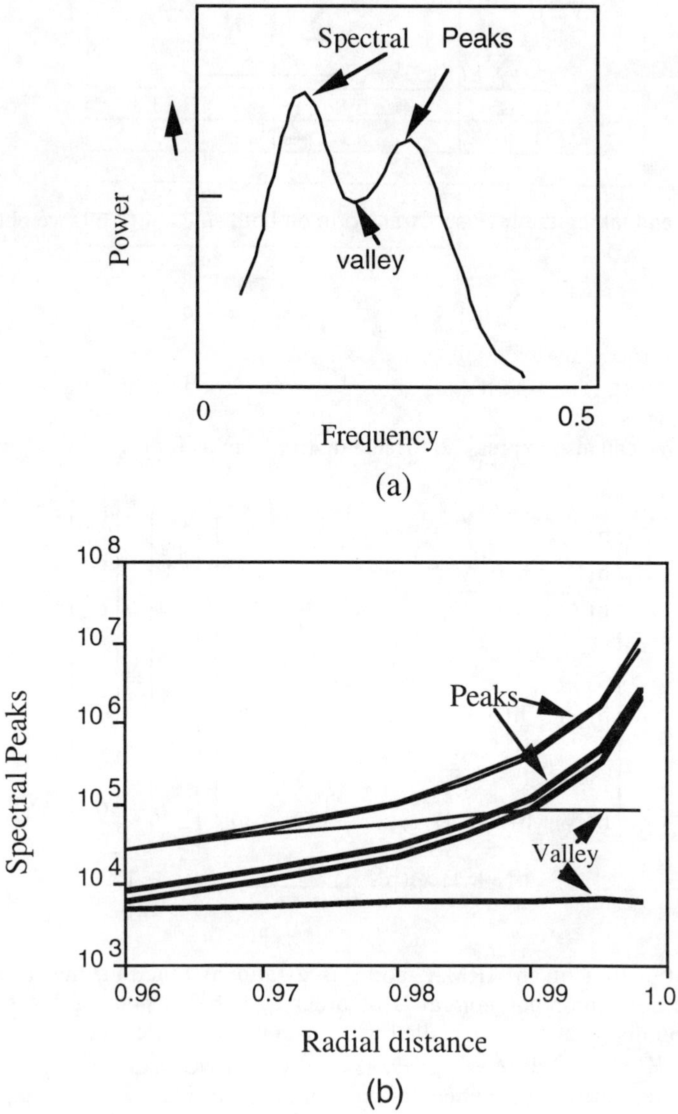

Figure 2.18: (a) A typical spectrum caused by a pair of poles. (b) Spectral peak height corresponding to two poles, 5 degrees apart (thin lines) and 9 degrees apart (thick lines) and the floor of the valley in between as a function of the radial distance of the poles (Note: $\omega_1 = 40$ degrees).

$$H(z) = \sum_{\tau=0}^{\infty} h(\tau)\, z^{-\tau}.$$

We can relate the impulse response to the parameters of the ARMA model. Note that

$$\sum_{\tau=0}^{\infty} h(\tau) z^{-\tau} = \frac{\sum_{k=0}^{q} b_k z^{-k}}{1 + \sum_{k=1}^{p} a_k z^{-k}}. \tag{2.69}$$

Cross multiplying and taking the inverse z-transform on both sides of (2.69) we obtain

$$\begin{aligned} \sum_{i=0}^{p} h(k-i) a_i &= b_k \qquad 0 \le k \le q \\ &= 0 \qquad\quad k > q, \end{aligned} \tag{2.70}$$

where $a_0 = 1.0$. We can also express (2.70) in a matrix form as follows:

$$\underset{(N+1)\times(p+1)}{\begin{bmatrix} h(0) & & & \\ h(1) & h(0) & & \mathbf{0} \\ h(2) & h(1) & & \\ \vdots & & \ddots & \\ h(p) & h(p-1) & \dots & h(0) \\ \vdots & & & \vdots \\ h(N) & h(N-1) & \dots & h(N-p) \end{bmatrix}} \underset{(p+1)\times 1}{\begin{bmatrix} 1 \\ a_1 \\ a_2 \\ \vdots \\ a_p \end{bmatrix}} = \begin{bmatrix} b_0 \\ b_1 \\ b_2 \\ \vdots \\ b_q \\ 0 \\ 0 \end{bmatrix}. \tag{2.71}$$

The transfer function of an ARMA model is a rational function, that is, a ratio of two polynomials. The numerator polynomial is of order q. It has a pole of order q at z=0. The denominator polynomial is of order p, and all its zeros must lie inside the unit circle. It has a pole of order p at z=0. $H(z) \to 0$ as $z \to 0$ only if p>q, which we shall assume. Note that H(z) is invertible provided the zeros of the numerator polynomial lie inside the unit circle; in other words, H(z) must be a minimum phase filter.

B. Covariance Function of an ARMA Process

To relate the covariance function to the parameters of an ARMA model we multiply both sides of (2.67) by $x(t-\tau)$ and evaluate the expected value,

$$C_x(\tau) + \sum_{k=1}^{p} a_k C_x(\tau - k) = \sum_{k=0}^{q} b_k E\{\varepsilon(t-k) x(t-\tau)\}. \tag{2.72}$$

The right-hand side of (2.72) may be evaluated by treating $\varepsilon(t)$ as the input and x(t) as the output of a linear causal system. The cross-covariance function is equal to the impulse response function,

Covariance Function					
C_x	2185.0	1539.6	4.2	– 1479.4	– 2033.5
Lag	0	1	2	3	4

Table 2.5: Covariance function for lags 0 to 4 of an ARMA(4,2) process.

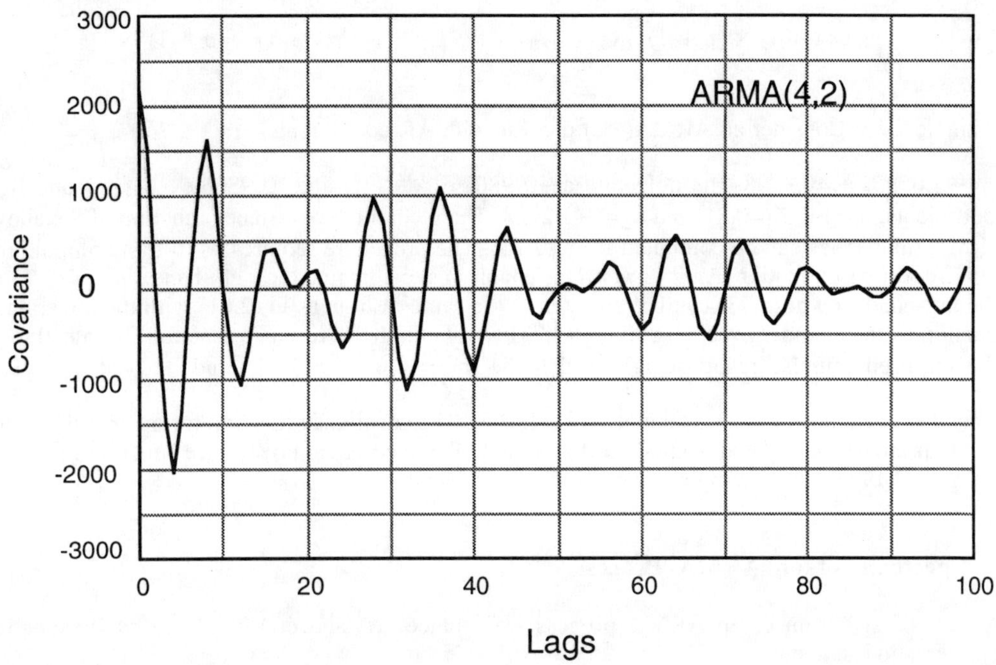

Figure 2.19: Covariance function of an ARMA(4,2) process.

$$E\{\varepsilon(t-k)x(t-\tau)\} = \sigma_\varepsilon^2 h(k-\tau) \qquad k \geq \tau \text{ and } h(0) = 1$$
$$= 0 \qquad k < \tau.$$

Note that when $k \geq \tau$ the output follows the input, and when $k < \tau$, the output precedes the input. Finally, the covariance function of an ARMA process satisfies the following Yule-Walker equation:

$$\sum_{k=0}^{p} a_k C_x(\tau - k) = \sigma_\varepsilon^2 \sum_{k=0}^{q} b_k h(k-\tau) \qquad \tau = 0,1.2,\ldots, q$$
$$= 0 \qquad \tau > q. \tag{2.73a}$$

We can also express (2.73a) in a matrix form,

$$\begin{bmatrix} C_X(0) & C_X(1) & C_X(2) & \cdots & C_X(p) \\ C_X(1) & C_X(0) & C_X(1) & \cdots & C_X(p-1) \\ \vdots & & & & \vdots \\ C_X(p) & C_X(p-1) & & \cdots & C_X(0) \\ C_X(p+1) & C_X(p) & & \cdots & C_X(1) \\ \vdots & & & & \vdots \\ C_X(\tau) & C_X(\tau-1) & & \cdots & C_X(\tau-p) \end{bmatrix}_{(\tau+1)\times(p+1)} \begin{bmatrix} 1 \\ a_1 \\ a_2 \\ a_3 \\ \vdots \\ a_p \end{bmatrix} = \sigma_\varepsilon^2 \begin{bmatrix} h(0) & h(1) & \cdots & h(q) \\ 0 & h(0) & \cdots & h(q-1) \\ \vdots & & \ddots & \\ 0 & 0 & & h(0) \\ \vdots & \vdots & & \vdots \\ 0 & 0 & & 0 \\ 0 & 0 & & 0 \end{bmatrix}_{(\tau+1)\times(q+1)} \begin{bmatrix} b_0 \\ b_1 \\ b_2 \\ \vdots \\ b_q \end{bmatrix}. \tag{2.73b}$$

Example 2.13: Consider an ARMA(4,2) process with AR coefficients, $a_1=-2.7607$, $a_2=3.8106$, $a_3=-2.6535$, $a_4=0.9238$ (the poles are at $0.9802\, e^{\pm j39.58^0}$, and $0.9805\, e^{\pm j50.41^0}$) and MA coefficients, $b_0=1.0$, $b_1=0.81$, and $b_2=-1.2728$. The theoretical covariance function of the above ARMA time series can be computed from (2.73). The procedure requires us to first compute the impulse response function using (2.71). This enables us to compute the right-hand side of (2.73b). The left-hand side of (2.73) can be rewritten in the same fashion as in (2.61) with the covariance function as the unknown vector. Solving (2.73) we obtain the covariance function for lags 0 to p. The required impulse response is found to be $h_0 = 1.0, h_1 = 3.53$, and $h_2 = 4.77$. The computed covariance vector is tabulated (Table 2.5). The covariance function for other lags greater than four are next computed from (2.73) for $\tau > 4$. The theoretical covariance function is shown in Figure 2.19 (End of Example).

C. Spectrum of an ARMA Process

To get the spectrum of an ARMA process we replace x(t) and e(t) in (2.67) by its spectral representation and obtain

$$dX_G(\omega) = \frac{\sum_{k=0}^{q} b_k \exp(-jk\omega)}{\sum_{k=0}^{p} a_k \exp(-jk\omega)} d\bar{\varepsilon}_G(\omega). \tag{2.74}$$

On squaring both sides of (2.74) and evaluating the expected value we obtain

$$S_x(\omega) = \sigma_\varepsilon^2 \left| \frac{\sum_{k=0}^{q} b_k \exp(-jk\omega)}{\sum_{k=0}^{p} a_k \exp(-jk\omega)} \right|^2 = \sigma_\varepsilon^2 |H(z)|^2 \Big|_{z=e^{jk\omega}} \tag{2.75}$$

It is interesting to examine the type of ambiguity one faces in recovering the transfer function H(z) from the spectrum. For the time being we assume that we are able to express $S_x(\omega)$ as a ratio of two polynomials in $\exp(j\omega)$ and also that we are able to determine the zeros of the

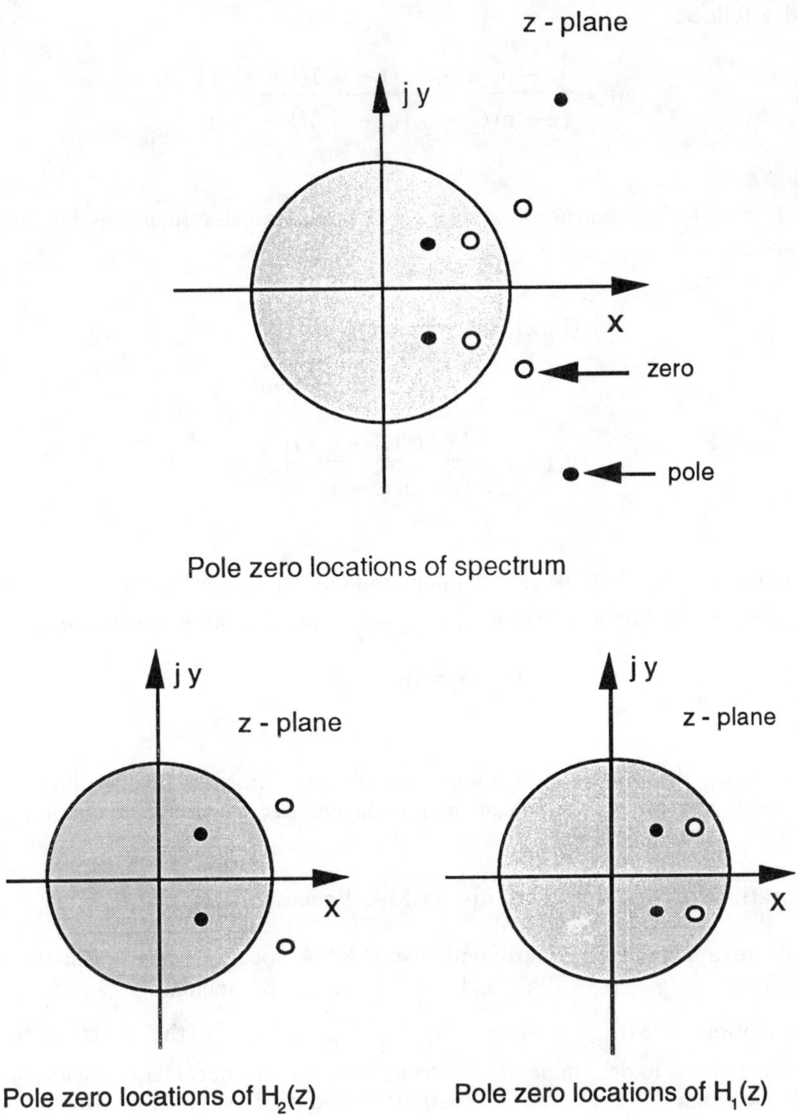

Figure 2.20: Given the pole and zero locations from the spectrum it is not possible to uniquely specify the transfer function without invoking the minimum phase property.

numerator and denominator polynomials. The zeros will lie both inside and outside the unit circle. For every pair of zeros (a zero and its complex conjugate) inside the unit circle there will be another pair outside the unit circle at reciprocal distance. When the system is known to be minimum phase, it is possible to uniquely determine the transfer function; simply select all zeros of the numerator and denominator polynomials lying inside the unit circle to build up the transfer function. However, when the system is nonminimum phase we are free to select the zero pairs lying either inside or outside the unit circle without affecting the amplitude of the transfer function or the spectrum. For q zeros there are $2^{\frac{q}{2}}$ possible alternatives, hence, $2^{\frac{q}{2}}$ different transfer functions.

Example 2.14: Consider an ARMA(2,2) process. Assume that the spectrum of the process has been factored as follows:

$$S_x(\omega) = \left.\frac{(z-a)(z-a^*)(1-az)(1-a^*z)}{(z-b)(z-b^*)(1-zb)(1-zb^*)}\right|_{z=e^{j\omega}},$$

where $|a|<1, |b|<1$. We can now define two different transfer functions having the same amplitude response.

$$H_1(z) = \left.\frac{(z-a)(z-a^*)}{(z-b)(z-b^*)}\right|_{z=e^{j\omega}}$$

$$H_2(z) = \left.\frac{(1-az)(1-a^*z)}{(z-b)(z-b^*)}\right|_{z=e^{j\omega}}$$

$H_1(z)$ is a minimum phase, but $H_2(z)$ is a nonminimum phase transfer function. The pole/zero position for each filter is shown in Figure 2.20. By actual multiplication we can verify that

$$|H_1(z)| = |H_2(z)|\Big|_{z=e^{j\omega}}.$$

The ambiguity in the determination of a nonminimum phase transfer function may be resolved through a study of the bispectrum, to be taken up in the next section (End of Example).

D. **Determination of Spectrum of an ARMA Process**

A straightforward approach for determination of ARMA spectrum is to estimate the model parameters, namely, $a_1, a_2, a_3, \ldots, a_p$ and $b_0, b_1, b_3, \ldots, b_q$ including the order p and q. Of these, determination of MA parameters, $b_0, b_1, b_3, \ldots, b_q$, is the most difficult task. Fortunately, if the aim is to determine the spectrum only, it is *not* necessary to determine the MA parameters. It is enough if we have the AR coefficients and the covariance function defined over a range, $\tau = 0$ to $\tau = p + q$. We shall now describe a method of determining the spectrum from the above-mentioned inputs.[19,20]

Let us express (2.67) in a matrix form,

$$\mathbf{A}^T \mathbf{x}_t = \mathbf{B}^T \varepsilon_t, \tag{2.76}$$

where

$$\mathbf{A} = [1, a_1, a_2, \ldots, a_p]^T,$$

$$\mathbf{x}_t = [x(t), x(t-1), x(t-2), \ldots, x(t-p)]^T,$$

$$\mathbf{B} = [b_0, b_1, b_2, \ldots, b_q]^T,$$

and

$$\varepsilon_t = [\varepsilon(t), \varepsilon(t-1), \varepsilon(t-2), \ldots, \varepsilon(t-q)]^T.$$

Postmultiply the left-hand side of (2.76) by $\mathbf{x}_{t'}^T\mathbf{A}$ and the right-hand side by $\varepsilon_{t'}^T\mathbf{B}$ and evaluate the expected value of the products,

$$\mathbf{A}^T E\{\mathbf{x}_t \mathbf{x}_{t'}^T\}\mathbf{A} = \mathbf{B}^T E\{\varepsilon_t \varepsilon_{t'}^T\}\mathbf{B}. \tag{2.77}$$

Note that

$$E\{\mathbf{x}_t \mathbf{x}_{t'}^T\} = \underbrace{\begin{bmatrix} C_X(k - l + t' - t) \end{bmatrix}}_{(p+1)\times(p+1)} \qquad 0 \le k,l \le p.$$

and

$$E\{\varepsilon_t \varepsilon_{t'}^T\} = \underbrace{\begin{bmatrix} C_\varepsilon(k - l + t' - t) \end{bmatrix}}_{(q+1)\times(q+1)}.$$

Since $\varepsilon(t(\,))$ is a white noise process,

$$C_\varepsilon(k - l + t' - t) = \sigma_\varepsilon^2 \delta(k - l + t' - t).$$

Hence, $E\{\varepsilon_t \varepsilon_{t'}^T\}$ will be a matrix with σ_ε^2 on a subdiagonal $\tau = (t' - t)$ units away from the principal diagonal. For example, for q=5 and τ=3, $E\{\varepsilon_t \varepsilon_{t'}^T\}$ will have the following form:

$$E\{\varepsilon_t \varepsilon_{t'}^2\} = \sigma_\varepsilon^2 \begin{bmatrix} 0 & & \cdots & & & 0 \\ 0 & & & & & \\ 0 & & \ddots & & & \vdots \\ 1 & & & & & \\ 0 & 1 & & & & \\ 0 & 0 & 1 & 0 & 0 & 0 \end{bmatrix}. \tag{2.78}$$

Using (2.78) in (2.77) the right-hand side of (2.77) becomes

$$\sigma_\varepsilon^2 \sum_{k=\tau}^{q} b_k b_{k-\tau},$$

which is indeed equal to the covariance function of an MA process (2.45). The left-hand side of (2.77) can be reduced to

$$\sum_{k=0}^{p}\sum_{l=0}^{p} a_k a_l C_X(k-l+\tau), \quad a_0 = 1.$$

Finally, (2.77) reduces to

$$\sum_{k=0}^{p}\sum_{l=0}^{p} a_k a_l C_x(k-l+\tau) = \sigma_\varepsilon^2 \sum_{k=\tau}^{q} b_k b_{k-\tau} \qquad -q \le \tau \le q,$$
$$= 0 \qquad \text{otherwise} .$$

(2.79

In the above equation we replace $C_x(k-l+\tau)$ by its spectral representation (see (1.26)),

$$C_X(k-l+\tau) = \frac{1}{2\pi}\int_{-\pi}^{\pi} S_X(\omega)\exp(j\omega(k-l+\tau))\,d\omega ,$$

and obtain

$$\frac{1}{2\pi}\int_{-\pi}^{\pi} S_x(\omega)\left|\sum_{k=0}^{p} a_k \exp(j\omega k)\right|^2 \exp(j\omega\tau)\,d\omega = C_{MA\ part}(\tau). \tag{2.80}$$

Taking the inverse Fourier transform on both sides of (2.80) we obtain

$$S_x(\omega) = \frac{\sum_{\tau=-q}^{q} C_{MA\ part}(\tau)\exp(-j\omega\tau)}{\left|\sum_{k=0}^{p} a_k \exp(j\omega k)\right|^2}$$

$$= \frac{\sum_{\tau}\sum_{k=0}^{p}\sum_{l=0}^{p} a_k a_l C_x(k-l+\tau)\exp(-j\omega\tau)}{\left|\sum_{k=0}^{p} a_k \exp(j\omega k)\right|^2}. \tag{2.81}$$

Thus, the spectrum of ARMA process could be written purely in terms of the AR coefficients. The numerator of (2.81) may be treated as a spectrum of the output of AR filter excited by ARMA time series. To show this let

$$x_r(t) = x(t) + a_1 x(t-1) + \ldots + a_p x(t-p) ,$$

or in the matrix notations let

$$x_r(t) = \mathbf{A}^T \mathbf{x}_t$$

be the output of AR filter. On squaring both sides and evaluating the expected operation we obtain the numerator of (2.81). This approach may also be used for practical computation of the ARMA spectrum without having to compute the MA coefficients. The output of the filter, being the MA component of ARMA, will have a smooth spectrum without any peaks. It is then appropriate to use one of the low-resolution methods described in the next chapter.

A moving average model with all its zeros inside the unit circle can be expressed in terms of a very high order AR model (see Example 2.7). Let $A_{equi}(z)$ be the equivalent AR model to the MA(q) model with a z-transform B(z),

$$A_{equi}(z)B(z) = 1.$$

or in the time domain,

$$\sum_{n=0}^{q} \hat{a}_{m-n} b_n = \delta_m ,$$

which may be expressed as a system of linear equations, where $\hat{a}_m$, $m = 0, 1, \ldots, p >> q$ are the AR coefficients of an equivalent AR model. These coefficients are determined from the residual time series, $x_r(t)$.

The AR coefficients of an ARMA model can be estimated independent of MA coefficients using an extended covariance matrix. For this let us look at (2.73),

$$\sum_{k=0}^{p} a_k C_x(\tau - k) = 0 \qquad \tau > q.$$

In the matrix form the above system of equations may be expressed as

$$\begin{bmatrix} C_x(q+1) & C_x(q) & C_x(q-1) & \ldots & C_x(q+1-p) \\ C_x(q+2) & C_x(q+1) & C_x(q), & \ldots & C_x(q+2-p) \\ \vdots & & & \ldots & \vdots \\ C_x(\tau) & C_x(\tau-1) & C_x(\tau-2) & \ldots & C_x(\tau-p) \end{bmatrix} \begin{bmatrix} 1 \\ a_1 \\ a_2 \\ \vdots \\ a_p \end{bmatrix} = \begin{bmatrix} 0 \\ \vdots \\ 0 \end{bmatrix}, \tag{2.82a}$$

or in a compact form,

$$\mathbf{C}^e_x \mathbf{A} = 0. \tag{2.82b}$$

Let $\tau - q > p + 1$; then the rank of C^e_x will be equal to p (see Exercise 2.7). On squaring both sides of (2.82) we obtain

$$\mathbf{A}^H \mathbf{C}^{e\,H}_x \mathbf{C}^e_x \mathbf{A} = 0 .$$

Note that $\mathbf{C}_x^{e\,H}\mathbf{C}_x^e$ is a square matrix of size (p+1)(p+1), but it has rank p. Therefore, $\mathbf{A}$ will correspond to the null eigenvector of $\mathbf{C}_x^{e\,H}\mathbf{C}_x^e$. In practice, on account of finite data errors in the covariance function, $\mathbf{C}_x^{e\,H}\mathbf{C}_x^e$ will not be a singular matrix; that is, its minimum eigenvalue will not be equal to zero but will be a finite value.

We may also express (2.82) as

$$\underbrace{\begin{bmatrix} C_x(q) & C_x(q-1) & \dots & C_x(q+1-p) \\ C_x(q+1) & C_x(q) & \dots & C_x(q+2-p) \\ \vdots & & & \vdots \\ C_x(p+q-1) & & \dots & C_x(q) \end{bmatrix}}_{p\times p} \underbrace{\begin{bmatrix} a_1 \\ a_2 \\ \vdots \\ a_p \end{bmatrix}}_{p\times 1} = - \underbrace{\begin{bmatrix} C_x(q+1) \\ C_x(q+2) \\ \vdots \\ C_x(p+q) \end{bmatrix}}_{p\times 1}, \tag{2.83}$$

or in a compact matrix form

$$\mathbf{C}_x(q)\tilde{\mathbf{A}} = \mathbf{R}_x(q+1).$$

Note that $\mathbf{C}_x(q)$ is a nonsymmetric Toeplitz covariance matrix whose elements differ from those of the standard symmetric covariance by q lags. $\mathbf{C}_x(q)$ is a full rank matrix; hence its inverse exists. $\tilde{\mathbf{A}}$ and $\mathbf{R}_x(q+1)$ are vectors as shown in (2.83).

E. Order of an ARMA Process

The order of the AR part of an ARMA process may be estimated by examining the eigenvalues of the extended covariance matrix in (2.82), assuming a sufficiently large value for q. The steps involved are similar to those used in the order estimation of an AR process earlier. However, the question of order estimation of the MA part is more subtle. Let us examine the rank property of a covariance matrix derived from the extended covariance matrix of (2.82) by including an extra row corresponding to $\tau = q$ in (2.73),

$$\mathbf{C}_x^{ee} = \underbrace{\begin{bmatrix} C_x(q) & C_x(q-1) & \dots & C_x(q-p) \\ C_x(q+1) & C_x(q) & \dots & C_x(q+1-p) \\ C_x(q+2) & C_x(q+1) & \dots & C_x(q+2-p) \\ \vdots & & & \vdots \\ C_x(\tau) & C_x(\tau-1) & \dots & C_x(\tau-p) \end{bmatrix}}_{(\tau-q+1)\times(p+1)},$$

q'	λ_1	λ_2	λ_3	λ_4	λ_5	λ_6
1	6.242	5.274	1.180	**0.834**	0.516	0.232
2	6.050	4.938	1.137	**0.639**	0.272	0.220
3	5.587	4.870	1.021	**0.285**	0.240	0.175
4	5.108	4.848	0.637	**0.258**	0.239	0.176

Table 2.6: Estimation of the order of the MA component from the singular values (SVD) of $\mathbf{C}_x^{ee}(q')$ for different values of q'. The order of the AR component is known from the SVD of $\mathbf{C}_x^{e}$. Notice that the fourth singular value undergoes a significant change as q' is changed from 2 to 3.

All elements of the first column except the very first element can be computed by linearly combining the elements of the p remaining columns (use (2.73a) for $\tau > q$). The first element of the first column, however, cannot be obtained by linearly combining the remaining elements of the first row. Indeed, this element is given by

$$C_x(q) = -a_1 C_x(q-1) - a_2 C_x(q-2) \ldots - a_p C_x(q-p) + \sigma_\varepsilon^2 \sum_{k=0}^{q} b_k h(k-q),$$

where the last term on the right-hand side is the new additional quantity required. Thus, all p+1 columns of $\mathbf{C}_x^{ee}$ are independent, and therefore the rank of $\mathbf{C}_x^{ee}$ will be equal to p+1.[21] We shall now exploit this property of $\mathbf{C}_x^{ee}$ for estimation of the order of the MA component. For this define a matrix,

$$\mathbf{C}_x^{ee}(q') = \underbrace{\begin{bmatrix} C_x(q') & C_x(q'-1) & \ldots & C_x(q'-p) \\ C_x(q'+1) & C_x(q') & \ldots & C_x(q'+1-p) \\ C_x(q'+2) & C_x(q'+1) & \ldots & C_x(q'+2-p) \\ \vdots & & & \vdots \\ C_x(\tau) & C_x(\tau-1) & \ldots & C_x(\tau-p) \end{bmatrix}}_{(\tau - q' + 1) \times (p+1)},$$

where we assume the AR order is known. The rank of $\mathbf{C}_x^{ee}(q')$ depends upon q'

$$\text{Rank}\,\{\mathbf{C}_x^{ee}(q')\} = p+1 \quad \text{for } q' \le q, \qquad = p \quad \text{for } q' > q.$$

The change of rank of $\mathbf{C}_x^{ee}(q')$ can be monitored for estimating the order of the MA component. A numerical example below demonstrates this possibility.

Example 2.15: Consider the time series taken from Reference 21,

$$x(t) - 2.0x(t-1) + 1.7x(t-2) - 0.5x(t-3)$$
$$= \varepsilon(t) - 1.5\varepsilon(t-1) + 0.8\varepsilon(t-2)$$

where $\varepsilon(t)$ is zero mean and unit variance white noise input. Using the autocorrelation estimates provided by the author,[21] $C_X^{ee}(q')$ matrix was formed for different values of q', but p was set to three, the order of the AR component. The singular value decomposition (SVD) for different values of q' gave the results shown in Table 2.6 (End of Example).

F. Sinusoids Plus Noise

We shall next consider an important time series model, namely, sinusoids in the presence of background noise, which we represent as a qth-order MA process:

$$x(t) = \sum_{k=1}^{p} \alpha_k \sin(\omega_k t + \phi_k) + \sum_{i=0}^{q} b_i \varepsilon(t-i), \tag{2.84}$$

where α_k and ϕ_k are random amplitude and phase of the kth sinusoid. For the time being let us consider the noise-free case; that is, $b_0 = b_1 = \ldots = b_q = 0$,

$$\begin{aligned} s(t) &= \sum_{k=1}^{p} \alpha_k \sin(\omega_k t + \phi_k) \\ &= \frac{1}{2j}\Big[\sum_{k=1}^{p} \alpha_k \exp(j(\omega_k t + \phi_k)) - \alpha_k \exp(-j(\omega_k t + \phi_k))\Big]. \end{aligned} \tag{2.85}$$

Define

$$\sum_{i=1}^{p'} a_i s(t-i) = \frac{1}{2j}\Big[\sum_{k=1}^{p} \begin{aligned} &\alpha_k \exp(j(\omega_k t + \phi_k)) \sum_{i=1}^{p'} a_i \exp(-j\omega_k i) \\ &- \alpha_k \exp(-j(\omega_k t + \phi_k)) \sum_{i=1}^{p'} a_i \exp(j\omega_k i)\Big], \end{aligned} \tag{2.86}$$

where a_i, $i = 1, 2, \ldots, p'$ are coefficients satisfying the following equations:

$$\begin{aligned} &\sum_{i=1}^{p'} a_i \exp(-j\omega_k i) = -1 \\ &\sum_{i=1}^{p'} a_i \exp(+j\omega_k i) = -1 \quad k = 1, 2, \ldots, p. \end{aligned} \tag{2.87}$$

The right-hand side of (2.86) simply is equal to s(t). Thus,

$$\sum_{i=1}^{p'} a_i s(t-i) = -s(t). \tag{2.88}$$

Note that (2.87) is indeed a system of 2p equations when we consider its real and imaginary parts separately. Further, we shall require that the coefficients a_i must be real; then the second equation of (2.87) automatically follows from the first equation. We can uniquely determine 2p coefficients (hence $p'=2p$) given p frequencies, w_k, k=1,2, ..., p. The system of equations is as follows:

$$\underset{(2p)\times(2p)}{\begin{bmatrix} \cos(\omega_1) & \cdots & \cos(2p\omega_1) \\ \cos(\omega_2) & \cdots & \cos(2p\omega_2) \\ \vdots & & \vdots \\ \cos(\omega_p) & \cdots & \cos(2p\omega_p) \\ \cdots & \cdots & \cdots \\ \sin(\omega_1) & \cdots & \sin(2p\omega_1) \\ \sin(\omega_2) & \cdots & \sin(2p\omega_2) \\ \vdots & & \vdots \\ \sin(\omega_p) & \cdots & \sin(2p\omega_p) \end{bmatrix}} \underset{(2p)\times 1}{\begin{bmatrix} a_1 \\ a_2 \\ \\ \\ \vdots \\ \\ \\ \\ a_{2p} \end{bmatrix}} = \underset{(2p)\times 1}{\begin{bmatrix} -1 \\ \vdots \\ \\ -1 \\ \cdots \\ 0 \\ \vdots \\ \\ 0 \end{bmatrix}},$$

or in compact form

$$\mathbf{F}\mathbf{A} = \mathbf{D}. \tag{2.89}$$

The matrix **F** has full rank as shown by Chan,[22] and hence its inverse exists. The coefficient vector **A** is given by $\mathbf{A} = \mathbf{F}^{-1}\mathbf{D}$. Conversely, when the coefficient vector is known, for example, estimated from the data, we would like to estimate the frequencies. For this, consider the following polynomial:

$$A(z) = 1 + a_1 z^{-1} + a_2 z^{-2} + a_3 z^{-3} + \ldots + a_{2p} z^{-2p}. \tag{2.90}$$

Following (2.87) it is easy to show that the polynomial will have roots at $z = \exp(j\omega_k)$, $k = 1,2,\ldots,2p$, that is, on the unit circle. We may also rewrite (2.90) as a product of simple factors,

$$\prod_{i=1}^{p} (z - z_i)(z - z_i^*) = 0, \tag{2.91}$$

where $z_i = e^{j\omega_i}$, i=1,2, ..., p are the roots of (2.90). Equation (2.91) is also useful in obtaining the coefficient, given the frequencies, in a straightforward manner. This is illustrated in the next example.

Example 2.16:

<u>p=1</u>

$$(z - z_1)(z - z_1^*) = z^2 - (z_1 + z_1^*)z + 1 = 0,$$

$$a_0 = 1.0 \quad a_1 = -(z_1 + z_1^*) = -2\cos(\omega_1) \quad a_2 = 1.0.$$

p=2

$$(z - z_1)(z - z_1^*)(z - z_2)(z - z_2^*) = 0,$$

$$a_0 = 1,$$
$$a_1 = -(z_1 + z_1^*) - (z_1 + z_1^*) = -2[\cos(\omega_1) + \cos(\omega_2)],$$
$$a_2 = 2 + (z_1 + z_1^*)(z_1 + z_1^*) = 2[1 + 2\cos(\omega_1)\cos(\omega_2)],$$
$$a_3 = -(z_1 + z_1^*) - (z_1 + z_1^*) = -2[\cos(\omega_1) + \cos(\omega_2)],$$
$$a_4 = 1.$$

In this manner it is easy to compute the coefficients as long as the number of frequencies involved is small. The coefficients possess symmetry properties, namely, $a_0 = a_{2p} = 1$ and $a_i = a_{2p-i}$.[22] (End of Example).

Let us now go back to (2.84) and rewrite as follows:

$$s(t) = x(t) - \sum_{i=0}^{q} b_i \varepsilon(t - i),$$

and on substituting for s(t) in (2.88) we obtain

$$x(t) - \sum_{i=0}^{q} b_i \varepsilon(t - i) = - \sum_{k=0}^{2p} a_k [x(t - k) - \sum_{i=0}^{q} b_i \varepsilon(t - k - i)] \quad . \tag{2.92}$$

After rearranging the terms in (2.92) we obtain the ARMA representation for x(t),

$$x(t) + \sum_{k=1}^{2p} a_k x(t - k) = \sum_{l=0}^{2p+q} \beta_l \varepsilon(t - l), \tag{2.93}$$

where

$$\beta_0 = b_0$$
$$\beta_1 = b_1 + b_0 a_1$$
$$\beta_2 = b_2 + b_1 a_1 + b_0 a_2$$
$$\beta_3 = b_3 + b_2 a_1 + b_1 a_2 + b_0 a_3$$
$$\vdots$$

$$\beta_1 = b_1 + \sum_{i=1}^{1} b_{1-i} a_i$$

$$\vdots$$

$$\beta_{2p+q} = a_{2p} b_q .$$

Let $\beta(z)$ be the z-transform of the MA coefficients in (2.93). It may be noted that

$$\beta_1 = b_1 + \sum_{i=1}^{1} b_{1-i} a_i$$

is a convolution product of two finite sequences, b_i, i=0, 1, 2, ... , q and a_k, k=0, 1, 2, ... , 2p (a_0=1.0). Therefore,

$$\beta(z) = B(z)A(z), \tag{2.94}$$

where A(z) is defined in (2.90) and

$$B(z) = b_0 + b_1 z^{-1} + b_2 z^{-2} + \ldots + b_q z^{-q}.$$

Thus, a sum of p sinusoids and a qth order MA process, as a background noise, can be modeled as an ARMA process of order (2p,2p+q). An interesting fact emerges from this representation; that is, the poles corresponding to p sinusoids also appear as zeros of the MA part.

Example 2.17: Consider a situation where we have two sinusoids with normalized frequencies, 0.1 and 0.15, and a second-order MA process as a background noise:

$$x(t) = \alpha_1 \sin(0.1\, t + \phi_1) + \alpha_2 \sin(0.15\, t + \phi_2)$$
$$+ \varepsilon(t) + 0.81\varepsilon(t-1) + 1.2728\, \varepsilon(t-2) .$$

The AR coefficients of the ARMA model of x(t) can be computed from Example 2.16. We obtain

$$a_1 = -3.6840,$$
$$a_2 = 5.3893$$
$$a_3 = -3.6840$$
$$a_4 = 1.0.$$

Next, to obtain the MA coefficients we shall first write down the b(z) polynomial,

$$\beta(z) = (1 + 0.81z^{-1} + 1.2728\, z^{-2})$$
$$(1 - 3.684\, z^{-1} + 5.3839\, z^{-2} - 3.6840\, z^{-3} + z^{-4})$$

$$= 1 - 2.874\, z^{-1} + 3.6780\, z^{-2} - 4.0076\, z^{-3}$$
$$+ 4.8755\, z^{-4} - 3.8789\, z^{-5} + 1.2728\, z^{-6}$$

The MA coefficients are

$$\beta_0 = 1.0, \beta_1 = -2.8740, \beta_2 = 3.6780$$
$$\beta_3 = -4.0070, \beta_4 = 4.8755,$$
$$\beta_5 = -3.8789, \beta_6 = 1.2728$$

(End of Example).

When the background noise is a simple white noise (then, $b_1 = b_2 = \ldots = 0$), (2.93) takes the following form:

$$x(t) + \sum_{k=1}^{2p} a_k x(t-k) = \varepsilon(t) - \sum_{l=1}^{2p} a_l \varepsilon(t-l). \tag{2.95}$$

The ARMA process given in (2.93) or (2.95) is known as a special ARMA process as every pole is accompanied by a zero.

G. AR Process Plus White Noise

Consider another useful model where an AR process is accompanied by uncorrelated observation noise. We would like to show that such a process can also be modeled as an ARMA(p,p) process.[23,24] Let

$$x(t) = \zeta(t) + \eta(t), \tag{2.96}$$

where $\zeta(t)$ is an AR(p) process given by

$$\sum_{i=0}^{p} a_i \zeta(t-i) = \varepsilon(t), \quad a_0 = 1.0,$$

and $\eta(t)$ is uncorrelated with $\varepsilon(t)$ and hence with $\zeta(t)$. Define a new process y(t) as follows:

$$y(t) = \sum_{i=0}^{p} a_i x(t-i). \tag{2.97}$$

The covariance function of y(t) can be written as

$$C_y(\tau) = \sum_{i=0}^{p} \sum_{k=0}^{p} a_i a_k C_x(\tau - k + i). \tag{2.98}$$

But $C_x(\tau) = C_\zeta(\tau) + \sigma_\eta^2 \delta(\tau)$, which we substitute in (2.98) and obtain

$$C_y(\tau) = \sum_{i=0}^{p} \sum_{k=0}^{p} a_i a_k C_\zeta(\tau - k + i) + \sigma_\eta^2 \sum_{i=0}^{p-\tau} a_i a_{i+\tau}. \tag{2.99}$$

Since $\zeta(t)$ is an AR process, it is easy to show that

$$\sum_{i}^{p}\sum_{k=0}^{p} a_i a_k C_\zeta(\tau - k + i) = \sigma_\varepsilon^2 \delta(\tau). \tag{2.100}$$

Using (2.100) in (2.99) we obtain

$$\begin{aligned} C_y(\tau) &= \sigma_\varepsilon^2 \delta(\tau) + \sigma_\eta^2 \sum_{i=0}^{p-\tau} a_i a_{i+\tau} \qquad && \tau \le p, \\ &= 0 && \tau > p. \end{aligned} \tag{2.101}$$

Clearly, since the covariance function is zero beyond a finite number of lags, y(t) must be a moving average process of order p. Hence, formally we can write

$$y(t) = \sum_{i=0}^{p} b_i \xi(t - i). \tag{2.102}$$

where $\xi(t)$ is a white noise process and $b_0, b_1, \ldots, b_p$ are new MA coefficients. Equations (2.97) and (2.102) together yield the desired result,

$$\sum_{i=0}^{p} a_i x(t - i) = \sum_{i=0}^{p} b_i \xi(t - i) \tag{2.103}$$

which is an ARMA(p,p) model.

In the discussion on the ARMA spectrum it was noted that explicit determination of the MA coefficients was not necessary (see page 118). The AR coefficients determined from the extended covariance matrix and the covariance function defined over a range, 0 to p+q, are sufficient to evaluate the covariance function of the MA part and then of the ARMA spectrum (see (2.79)). Therefore, we shall not pursue any further the question of determining the MA coefficients in (2.103). An example of calculation of MA coefficients of the equivalent ARMA(2,2) process of an AR(2) signal in white noise is given in Reference 25.

The spectrum of AR process plus uncorrelated white noise is simply given by

$$\begin{aligned} S_x(\omega) &= S_\zeta(\omega) + \sigma_\eta^2, \\ &= \frac{\sigma_\varepsilon^2}{\left|A(z = e^{j\omega})\right|^2} + \sigma_\eta^2, \end{aligned}$$

where σ_ε^2 is the variance of noise driving the AR process, and σ_ε^2 is the variance of the background noise. They may be determined using (2.101). For $\tau = 0$ and $\tau = p$ we can write from (2.101)

$$\begin{aligned} C_y(0) &= \sigma_\varepsilon^2 + \sigma_\eta^2 \sum_{i=0}^{p} a_i^2, \\ C_y(p) &= \sigma_\eta^2 a_p, \end{aligned}$$

which may be solved for σ_ε^2 and σ_η^2. We obtain

$$\sigma_\varepsilon^2 = C_y(0) - \frac{C_y(p)}{a_p}\sum_{i=0}^{p} a_i^2,$$

$$\sigma_\eta^2 = \frac{C_y(p)}{a_p}.$$

H. State Space Model

A time series model may be treated as a state space model wherein we consider all past inputs as different states of the system generating the time series. We define a state vector at time t as follows:

$$\mathbf{x}(t) = \{x(t), x(t-1), x(t-2), \dots, x(t-p+1)\}^T,$$

where $x(t), x(t-1), x(t-2), \dots, x(t-p+1)$ are different states of the system. Similarly, we can define another state vector $\mathbf{x}(t-1)$ at $t-1$. The components of the state vectors are related through the ARMA model, (2.67). Define the following matrices:

$$\phi = \underset{p\times p}{\begin{bmatrix} -a_1, -a_2, -a_3 & \dots & -a_p \\ 1\ 0\ 0\ 0 & \dots & 0 \\ 0\ 1\ 0\ 0 & \dots & 0 \\ \vdots & \ddots & \vdots \\ 0\ 0\ 0 & \dots\ 0\ 1 & 0 \end{bmatrix}}$$

and

$$\psi = \underset{p\times(q+1)}{\begin{bmatrix} b_0, & b_1 & b_2 & \dots & b_q \\ 0 & 0 & 0 & \dots & 0 \\ 0 & 0 & 0 & \dots & 0 \\ \vdots & & & & \vdots \\ 0 & 0 & 0 & \dots & 0 \end{bmatrix}}.$$

We are now able to express the ARMA model in terms of the state vectors and matrices Φ and Ψ,

$$\mathbf{x}(t)=\Phi\mathbf{x}(t-1)+\Psi\mathbf{e}(t), \tag{2.104}$$

$$x(t) = \mathbf{h}^T\mathbf{x}(t),$$

where

$$\varepsilon(t)= \text{col}\{\varepsilon(t), \varepsilon(t-1), \dots, \varepsilon(t-q)\}$$

is a random vector input and

$$\mathbf{h} = \text{col}\{1, 0, 0, \ldots, 0\}.$$

A general solution of the state equation (2.104) is given below:

$$\mathbf{x}(t) = \phi^t \mathbf{x}(0) + \sum_{i=1}^{t} \phi^{t-i} \psi\, \varepsilon(t-i). \tag{2.105}$$

For large t, $\phi^t \to 0$ (eigenvalues of ϕ are assumed to lie inside the unit circle); hence we can ignore the transient component. Using (2.105) without the transient component we compute the covariance matrix,

$$\mathbf{C}_x = \sum_{k=1}^{t}\sum_{i=1}^{t} \phi^{t-i} \psi E\{\varepsilon(t-i)\varepsilon(t-k)^H\} \psi^H {\phi^{t-k}}^H$$

$$= \sigma_\varepsilon^2 \sum_{i=1}^{t} \phi^{t-i} \psi\psi^H {\phi^{t-i}}^H \qquad t >> 1. \tag{2.106a}$$

One of the appealing aspects of the state space representation is that it is easy to account for time-varying ARMA model parameters. Both matrices in (2.105) become functions of time, $\phi(t)$ and $\psi(t)$.We will not further pursue this aspect of time-varying system modeling. Interested readers are advised to refer to Reference 26.

The state space models of sinusoids or damped sinusoids are of great interest as they enable us to efficiently estimate the frequencies and the damping coefficients of the sinusoids straight from the data. A state space model for a sum of damped sinusoids is given by (2.104) with $b_0 = b_1 = b_2 = \ldots = b_q = 0$,

$$\mathbf{x}(t) = \phi\, \mathbf{x}(t-1)$$
$$x(t) = \mathbf{h}^T \mathbf{x}(t). \tag{2.106b}$$

The eigenvalues of the state feedback matrix ϕ are related to the frequency and damping factor. To show this, consider the characteristic equation of ϕ,

$$\det\,(\phi - z\mathbf{I}) = z^p\Big(1 + \sum_{i=1}^{p} a_i z^{-i}\Big) = 0.$$

The eigenvalues of ϕ are therefore equal to the roots of the polynomial

$$\Big(1 + \sum_{i=1}^{p} a_i z^{-i}\Big) = 0,$$

whose roots, as noted in (2.53), are complex, the ith root being $z_i = \exp(\gamma_i)$, where $\gamma_i = \sigma_i + j\omega_i$ and $\sigma_i \le 0$. It is important to note that the eigenvalues of the feedback matrix remain invariant under a similarity transformation, for example, $\Phi \rightarrow Q\Phi Q^{-1}$, where Q is any nonsingular matrix. Thus, the eigenvalues of the feedback matrix remain unchanged in any minimal realization where the size of the state vector is equal to the number of exponentials with distinct frequencies. Later, in Chapter Five, we shall show how the feedback matrix may be estimated straight from the data matrix using the robust SVD approach.

2.6 Parametric Bispectral Model

In this section we shall examine the bispectrum and the trispectrum of AR/MA time series driven by a non-Gaussian stochastic process. The aim is to show what additional information can be derived from non-Gaussian AR/MA time series models. It turns out that the phase of the transfer function can be estimated from the bispectrum without having to bring in the minimum phase concept. This is not possible for Gaussian time series using the spectrum.

A. Filter Model of Non-Gaussian Time Series

In Section 2.2 we have shown that a time series can be modeled as an output of a filter driven by a white noise process, which we now assume to be non-Gaussian. Following (2.11) we shall write its spectral representation as follows:

$$x(t) = \frac{1}{2\pi}\int_{-\pi}^{\pi} d\varepsilon_G(\omega)H(\omega)\exp(j\omega t).$$

The bicovariance function can be expressed in terms of the generalized Fourier transform (Chapter One, (1.74)) of a stochastic process,

$$Bc(s,h) = \frac{1}{8\pi^3}\int\int\int_{-\pi}^{\pi} H(\omega_1)H(\omega_2)H(\omega_3)E\{d\varepsilon(\omega_1)d\varepsilon(\omega_2)d\varepsilon(\omega_3)\} e^{(j(t(\omega_1+\omega_2+\omega_3)+s\omega_1+h\omega_2))}$$

$$= \frac{\beta_\varepsilon^3}{4\pi^2}\int\int_{-\pi}^{\pi} H(\omega_1)H(\omega_2)H^*(\omega_1+\omega_2)e^{j(s\omega_1+h\omega_2)}d\omega_1 d\omega_2. \tag{2.107a}$$

Hence the bispectrum of a stochastic process in terms of the filter transfer function is given by

$$Bs(\omega_1,\omega_2) = \beta_\varepsilon^2 H(\omega_1)H(\omega_2)H^*(\omega_1+\omega_2). \tag{2.107b}$$

Let $\Psi(\omega_1,\omega_2)$ and $\Phi(\omega)$ be the phase of the bispectrum and filter transfer functions, respectively. From (2.107b) it is easy to show that

$$\Psi(\omega_1,\omega_2) = \Phi(\omega_1)+\Phi(\omega_2)-\Phi(\omega_1+\omega_2). \tag{2.108}$$

It may be noted from (2.108) that linear phase variation in the filter transfer function will not affect the bispectrum phase. As a corollary, any delay in the filter output will not affect its bispectrum.

The projection of a bicovariance function onto one of the axes is proportional to the covariance function. To verify this statement we project the bicovariance function onto the s axis, that is, sum the bicovariance function with respect to h. We obtain from (2.107a)

$$\sum_{h=-\infty}^{\infty} Bc\,(s,h) = \frac{\beta_\varepsilon^3}{4\pi^2}\int_{-\pi}^{+\pi}\!\!\int H(\omega_1)H(\omega_2)H^*(\omega_1+\omega_2)e^{js\omega_1}\sum_{-\infty}^{\infty} e^{jh\omega_2}\,d\omega_1\,d\omega_2$$

$$= \frac{\beta_\varepsilon^3 H(0)}{2\pi}\int_{\infty}^{\infty} H(\omega_1)H^*(\omega_1)e^{js\omega_1}\,d\omega_1$$

$$= \frac{\beta_\varepsilon^3 H(0)}{2\pi\sigma_\varepsilon^2}C_x(s).$$

A similar result can be arrived at if we project the bicovariance function onto the h axis.

B. Non-Gaussian MA Process

Let

$$x(t) = \sum_{i=0}^{q} b_i\varepsilon(t-i) \qquad ,$$

where $\varepsilon(t)$ is a non-Gaussian white noise process with zero mean and unit variance. We shall compute the bicovariance of x(t):

$$Bc\,(s,h) = E\{x(t)x(t+s)x(t+h)\}$$

$$= \sum_i\sum_j\sum_k b_i b_j b_k E\{\varepsilon(t-i)\varepsilon(t+s-j)\varepsilon(t+h-k)\}$$

$$= \sum_i\sum_j\sum_k b_i b_j b_k \beta_\varepsilon^3\delta_{s-j+i,\,h-k+i}$$

$$= \beta_\varepsilon^3\sum_{i=0}^{q} b_i b_{i+s} b_{i+h} \qquad\qquad s,h \in D,$$

$$= 0 \qquad\qquad \notin D, \qquad (2.109)$$

where D stands for a domain shown in Figure 2.21. This property of an MA process is encountered in other analogous statistics, viz. covariance, bicovariance, and tricovariance. We have assumed that the bicovariance function of $\varepsilon(t)$ is $Bc_\varepsilon(s,h) = \beta_\varepsilon^3\delta_{s,h}$.

Example 2.18: In this example we shall illustrate how to compute the bicovariance function of a second-order MA process,

$$x(t) = \varepsilon(t) + 0.81\varepsilon(t-1) - 1.2728\,\varepsilon(t-2).$$

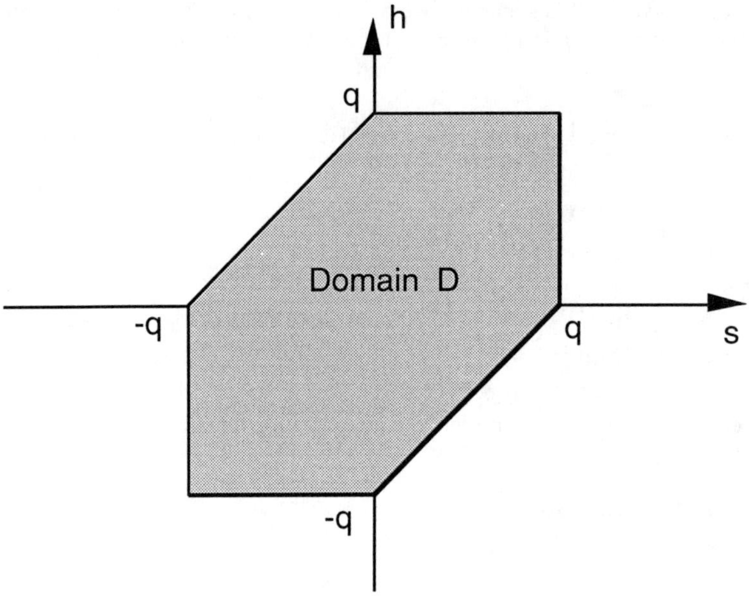

Figure 2.21:The bicovariance function of a qth order MA process is finite only within a speckled domain shown above, and it is zero outside.

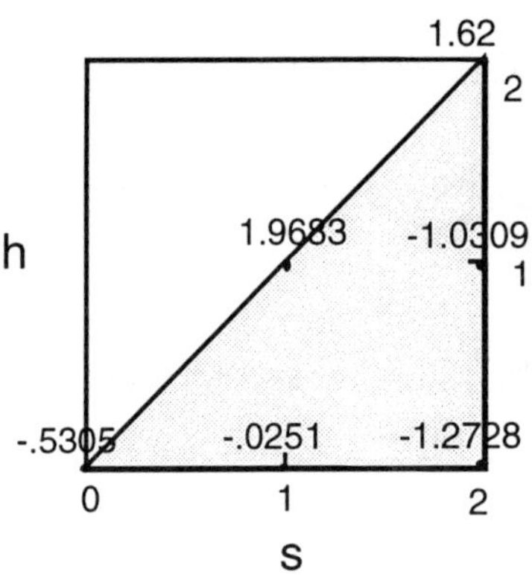

Table 2.7: Bicovariance function of an MA(2) non-Gaussian time series.

We shall compute the bicovariance function in region I of Figure 1.11, that is, inside a triangle defined by $0 \le s \le 2$ and $0 \le h \le s$. Note that $b_0 = 1$, $b_1 = 0.81$ and $b_2 = -1.2728$. Evaluate (2.109) keeping in mind that the product is zero whenever the index of the coefficient is other than 0, 1 or 2. The numerical results are shown in Table 2.7 (End of Example).

Take the DFT on both sides of (2.109). After some simple manipulations we obtain

$$Bs(\omega_1, \omega_2) = \beta_\varepsilon^3 H(\omega_1) H(\omega_2) H^*(\omega_1 + \omega_2), \tag{2.110}$$

where

$$H(\omega) = \sum_{i=0}^{q} b_i \exp(j\omega i).$$

Since the MA coefficients are real, $H(\omega)$ will be real at $\omega = 0$ and π, and hence its phase $\Phi(\omega) = \pm k\pi$, k=0, 1, ... at $\omega = 0$ and π. We shall now show how to determine the phase of H(w) given the phase of the bispectrum.[27] Equation (2.108) is treated as a system of linear equations for a given set of discrete frequencies, $\omega_1 = m\Delta\omega$, $m = n, n+1, \ldots N - n$, and $\omega_2 = n\Delta\omega$, $n = 0, 1, 2, \ldots, \frac{N}{2}$ as shown in Figure 2.22, where $\Delta\omega = \frac{\pi}{N}$. The digital bispectrum needs to be evaluated over a grid of points in the triangular region (Figure 2.22). Equation (2.108)

evaluated at the grid points becomes

$$\begin{aligned} \Psi(m,n) &= \Phi(m) + \Phi(n) - \Phi(m+n) \\ m &= n, n+1, \ldots N - n \\ n &= 0, 1, 2, \ldots \frac{N}{2}, \end{aligned} \tag{2.111}$$

where we have dropped $\Delta\omega$. From (2.108) we observe that $\Psi(m,0) = \Phi(0)$ for all m; hence we can assume that $\Phi(0)$ is known.

There are $\frac{N^2}{4}$ equations for $N - 1$ unknowns. A typical set of equations is shown below:

$$\begin{aligned} &\Psi(1,1) = 2\Phi(1) - \Phi(2) \\ &\Psi(2,1) = \Phi(1) + \Phi(2) - \Phi(3), \\ &\Psi(3,1) = \Phi(1) + \Phi(3) - \Phi(4), \\ &\ldots \\ &\Psi(2,2) = 2\Phi(2) - \Phi(4), \\ &\Psi(3,2) = \Phi(2) + \Phi(3) - \Phi(5), \\ &\ldots. \\ &\Psi(\tfrac{N}{2}, \tfrac{N}{2}) = 2\Phi(\tfrac{N}{2}) - \Phi(N) \qquad (\text{ N is even }) \end{aligned} \tag{2.112}$$

We can express the above system of equations in a compact form. For this, define the following vectors and matrices:

$$\underset{N \times 1}{\phi = \{\Phi(1)\ \Phi(2)\ \Phi(3)\ \ldots\ \Phi(N)\}^T},$$

$$\underset{\frac{N^2}{4} \times 1}{\psi = \{\Psi(1,1)\ \ldots\ \Psi(N-1,1)\ \Psi(2,2)\ \ldots\ \Psi(N-2,2)\ \ldots\ \Psi(\tfrac{N}{2}, \tfrac{N}{2})\}^T}.$$

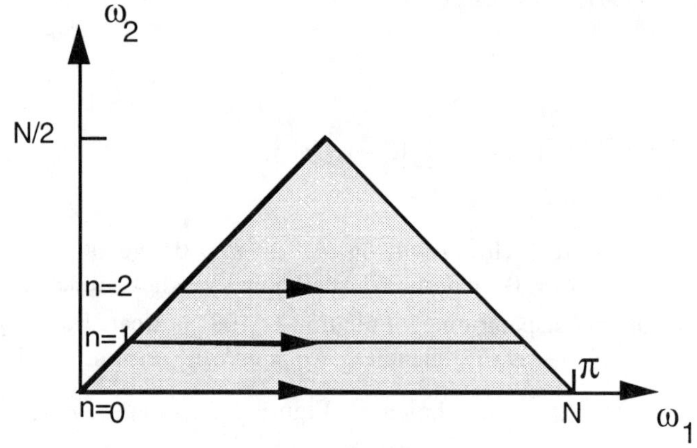

Figure 2.22: The digital bispectrum needs to be evaluated over a grid of points in the triangular region (shaded triangle).

$$
\mathbf{A} = \begin{bmatrix}
2 & -1 & 0 & 0 & 0 & 0 & 0 & 0 \\
1 & 1 & -1 & 0 & 0 & 0 & 0 & 0 \\
1 & 0 & 1 & -1 & 0 & 0 & 0 & 0 \\
1 & 0 & 0 & 1 & -1 & 0 & 0 & 0 \\
1 & 0 & 0 & 0 & 1 & -1 & 0 & 0 \\
1 & 0 & 0 & 0 & 0 & 1 & -1 & 0 \\
1 & 0 & 0 & 0 & 0 & 0 & 1 & -1 \\
0 & 2 & 0 & -1 & 0 & 0 & 0 & 0 \\
0 & 1 & 1 & 0 & -1 & 0 & 0 & 0 \\
0 & 1 & 0 & 1 & 0 & -1 & 0 & 0 \\
0 & 1 & 0 & 0 & 1 & 0 & -1 & 0 \\
0 & 1 & 0 & 0 & 0 & 1 & 0 & -1 \\
0 & 0 & 2 & 0 & 0 & -1 & 0 & 0 \\
0 & 0 & 1 & 1 & 0 & 0 & -1 & 0 \\
0 & 0 & 1 & 0 & 1 & 0 & 0 & -1 \\
0 & 0 & 0 & 2 & 0 & 0 & 0 & -1
\end{bmatrix}
\qquad N = 8, \quad \text{Size} = \frac{N^2}{4} \times N.
$$

In a matrix form, (2.112) may be written as

$$\psi = \mathbf{A}\phi. \tag{2.113}$$

Unfortunately, matrix **A** is not full rank. In fact it has a rank of $N-1$. To show this, let $\mathbf{a}_1, \mathbf{a}_2, \mathbf{a}_3, \ldots, \mathbf{a}_N$ be columns of **A**. It is easily verified that

$$\sum_{i=1}^{N} i\mathbf{a}_i = 0$$

and that there is only one such relation. Hence, the rank of **A** is N − 1. To overcome this problem, we shall assume that $\Phi(N)$ is known and eliminate the last column of **A**. Define

$$\psi_0 = \psi - \mathbf{a}_N \phi(N),$$

and $\mathbf{A}_{N-1}$ is **A** matrix without the last column, and hence it is a full rank matrix. We can now solve (2.113):

$$\phi = (\mathbf{A}_{N-1}^T \mathbf{A}_{N-1})^{-1} \mathbf{A}_{N-1}^T \psi_0. \tag{2.114}$$

C. Non-Gaussian AR Process

We now consider an AR process driven by non-Gaussian white noise,

$$x(t) + \sum_{i=1}^{p} a_i x(t-i) = \varepsilon(t), \tag{2.115}$$

where $\varepsilon(t)$ is non-Gaussian white noise with zero mean, σ_ε^2, the variance and β_ε^3, the third moment. x(t) would naturally be a zero mean stationary stochastic process. To compute the bicovariance of x(t) multiply both sides of (2.115) by $x(t-s)x(t-h)$ and evaluate the expected value of the product.

$$\begin{aligned} Bc(-s,-h) + \sum_{i=1}^{p} a_i Bc(-s+i,-h+i) &= E\{\varepsilon(t)x(t-s)x(t-h)\} \\ &= \beta_\varepsilon^2\, \delta_{s,h} \qquad s,h \geq 0. \end{aligned} \tag{2.116}$$

On account of the symmetry properties of the bicovariance function (see Section 1.5) it is enough if we obtain the bicovariance function in region (I) in Figure (1.11), that is, for $s \geq 0$ and $s \geq h \geq 0$.

The diagonal slice, that is, the bicovariance, is defined on the right diagonal (see Figure 2.23) where $s = h \geq 0$ is of great interest. On this diagonal we can express (2.116) in a matrix form, which is similar in appearance to the Yule-Walker equations involving the covariance function,

$$\mathbf{Bc}\,\mathbf{A} = \beta_\varepsilon^3 \delta, \tag{2.117}$$

where

$$\mathbf{Bc} = \begin{bmatrix} Bc(0,0) & Bc(1,1) & \cdots & Bc(p,p) \\ Bc(-1,-1) & Bc(0,0) & \cdots & Bc(p-1,p-1) \\ \vdots & & & \vdots \\ Bc(-p,-p) & Bc(-p+1,-p+1) & \cdots & Bc(0,0) \end{bmatrix},$$

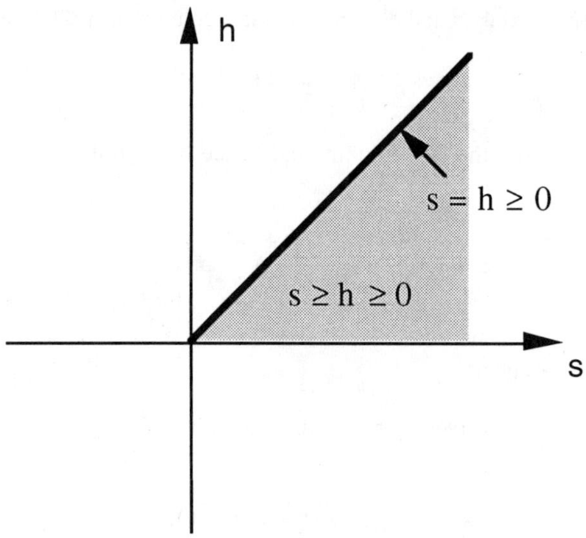

Figure 2.23: The bicovariance function computed on the diagonal as shown above is of considerable interest.

$$\mathbf{A} = \{1\ a_1\ a_2\ \dots\ a_p\}^T \quad \text{and} \quad \delta = \{1\ 0\ 0\ \dots\ 0\}^T.$$

Note that **Bc** is a nonsymmetric Toeplitz bicovariance matrix. To solve for **A** we must ensure that **Bc** is full rank. It is, however, not known whether the **Bc** will always be full rank (see Reference 28 for more discussion on this issue).

The bispectrum of an AR process driven by non-Gaussian white noise may be derived using the approach used for the filter model; that is, an AR process is an output of a filter with a transfer function,

$$H(\omega) = \sum_{i=1}^{p} a_i e^{j\omega i},$$

and driven by non-Gaussian white noise.

$$Bs(\omega_1, \omega_2) = \frac{\beta_\varepsilon^3}{H(\omega_1)H(\omega_2)H^*(\omega_1 + \omega_2)}. \tag{2.118}$$

The phase of the bispectrum is related to the phase of the AR transfer function as in (2.111) but with opposite sign,

$$\Psi(m,n) = -\Phi(m) - \Phi(n) + \Phi(m+n)$$

$$m = n, n+1, \dots, N-n$$

$$n = 0, 1, 2, \dots, \frac{N}{2}.$$

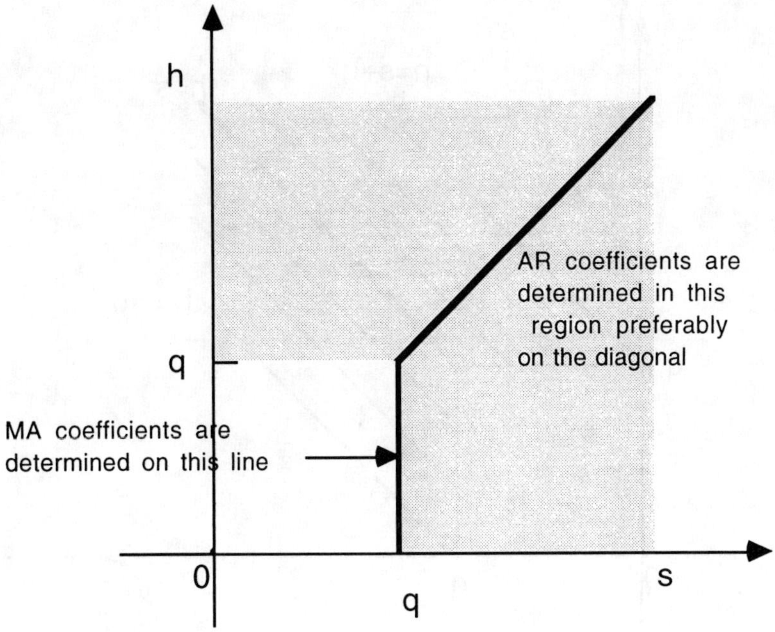

Figure 2.24: The AR and MA coefficients are determined from different parts of the bicovariance plane.

D. **Non-Gaussian ARMA Model**

We shall finally consider an ARMA model driven by a non-Gaussian process, $\varepsilon(t)$:

$$\sum_{i=0}^{p} a_i x(t-i) = \sum_{k=0}^{q} b_k \varepsilon(t-k), \qquad a_0 = 1.$$

Multiply both sides by $x(t-s)x(t-h)$ and then evaluate the expected value of the product.

$$\sum_{i=0}^{p} a_i E\{x(t-i)x(t-s)x(t-h)\} = \sum_{k=0}^{q} b_k \varepsilon(t-k)x(t-s)x(t-h)\}$$

$$\sum_{i=0}^{p} a_i Bc(-s+i,-h+i)$$

$$= \sum_{k=0}^{q} b_k \sum_m \sum_n h_m h_n E\{\varepsilon(t-k)\varepsilon(t-m-s)\varepsilon(t-n-h)\}$$

$$= \sum_{k=0}^{q} b_k \sum_m \sum_n h_m h_n \beta_\varepsilon^3 \delta_{k-m-s,k-n-h},$$

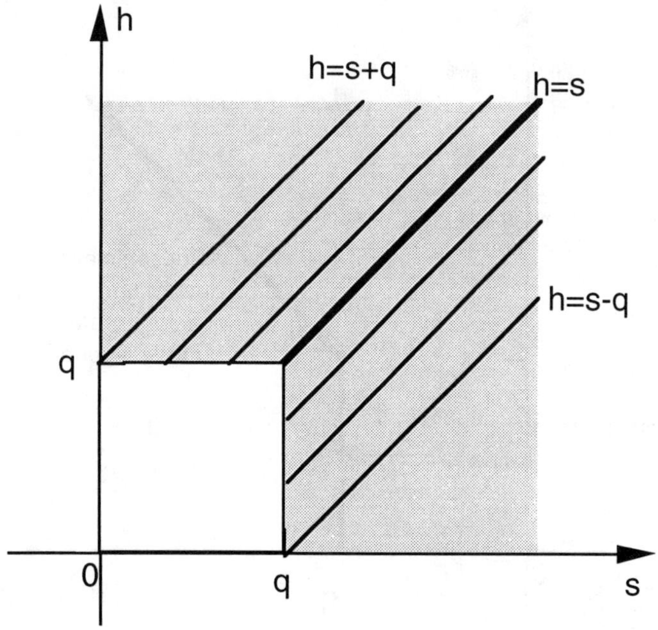

Figure 2.25: Consistent estimates of the AR parameters are possible by considering different diagonal slices as shown in the figure.

where h_m, m=0, 1, 2, ..., is the impulse response function of the ARMA model (see page 112). Note that h_m =0 for m< 0.

$$\sum_{i=0}^{p} a_i Bc(-s+i,-h+i) = \beta_\varepsilon^3 \sum_{k=0}^{q} b_k h_{k-s} h_{k-h} \qquad s \text{ and } h \le q,$$
$$= 0 \qquad s \text{ or } h > q. \tag{2.119}$$

The diagonal slice, given by s=h>q, is shown in Figure 2.24. We can write (2.119) in a compact matrix form,

$$\mathbf{Bc}^e\mathbf{A} = \mathbf{0}. \tag{2.120}$$

where

$$\mathbf{Bc}^e = \begin{bmatrix} Bc(-q-1,-q-1) & Bc(-q,-q) & \dots & Bc(-q-1+p,-q-1+p) \\ Bc(-q-2,-q-2) & & \dots & Bc(-q-2+p,-q-2+p) \\ \vdots & & & \vdots \\ Bc(-q-L,-q-L) & & \dots & Bc(-q-L+p,-q-L+p) \end{bmatrix}_{L\times(p+1)},$$

$\mathbf{A} = \{1 \quad a_1 \quad a_2 \ \dots \ a_p\}.$

Note that $\mathbf{Bc}^e$ is the extended bicovariance matrix similar to the extended covariance matrix defined on page 103. Equation (2.120) may be solved for **A** in the same manner that we solved (2.82); however, the question of the rank of $\mathbf{Bc}^e$ is undecided. In Reference 28 it is argued that by concatenating p+1 extended matrices obtained on p+1 diagonal slices (see Figure 2.25), a consistent estimate of AR parameters is possible.

Having determined AR coefficients we shall now show how to determine the MA coefficients. In (2.119) we let s=h=q and s=q but h<q and then obtain the following pair of equations:

$$\sum_{i=0}^{p} a_i Bc(-q+i,-q+i) = \beta_\varepsilon^3 b_q h_0^2 \tag{2.121}$$

and

$$\sum_{i=0}^{p} a_i Bc(-q+i,-q+i) = \beta_\varepsilon^3 b_q h_0^2 \tag{2.122}$$

Dividing (2.122) by (2.121) we obtain h_{q-h}, h=0,1,2, ... , q

$$h_{q-h} = \frac{\sum_{i=0}^{p} a_i Bc(-q+i,-h+i)}{\sum_{i=0}^{p} a_i Bc(-q+i,-q+i)} \tilde{h}_0. \tag{2.123}$$

Thus, we can determine q coefficients of the impulse response. Given the AR parameters and from q coefficients of the impulse response, we can determine q parameters of the MA part of the ARMA model by making use of (2.71). When p=0, that is, when it is a pure MA process, we determine the parameters of the MA model by letting p go to 0 in (2.123):

$$b_{q-h} = h_{q-h} = \frac{Bc(-q,-h)}{Bc(-q,-q)} h_0. \tag{2.124}$$

Equations (2.123) and (2.124) along with the symmetry properties of the bicovariance function (1.69) can be reduced to those given in Reference 28.

Example 2.19: In this example we shall demonstrate the use of the bicovariance function in the estimation of ARMA parameters. We shall consider an ARMA time series (nonminimum phase) with p=2 and q=1,

$$x(t) - 0.9x(t-1) + 0.81x(t-2) = \varepsilon(t) - 2.0\varepsilon(t-1),$$

where $\varepsilon(t)$ is a zero mean exponentially distributed white noise process with skewness, $\frac{\beta_\varepsilon^2}{\sigma_\varepsilon^2}$ =1.63. For the above ARMA model there are two poles at $0.5\angle \pm 60^0$ and one zero at 2.0. Five thousand samples of the ARMA(2,1) process were generated. The skewness of the ARMA

Pole mag.	Skewness	AR Coeffs	MA Coeffs
0.20	1.06	1.0, -0.1915,0.0477 (1.0, -0.2000, 0.0400)	1.0, -1.94 (1.0, -2.0)
0.50	0.828	1.0, -0.5189, 0.2424 (1.0, -0.5000, 0.2500)	1.0, -2.18 (1.0, -2.0)
0.80	0.580	1.0, -0.8109, 0.6456 (1.0, -0.8000, 0.6400)	1.0, -2.16 (1.0, -2.0)
0.90	0.376	1.0, -0.8731, 0.7984 (1.0, -0.9000, 0.8100)	1.0, -1.80 (1.0, -2.0)
0.99	0.029	1.0, -0.8302, 0.7165 (1.0, -0.9900, 0.9801)	1.0, -1.66 (1.0, -2.0)

Table 2.8: The deterioration in the estimate of ARMA parameters sets in at pole magnitude 0.9 and above. The numerical results are averages of six independent experiments of the type described in Example 2.18. The numbers inside brackets represent actual ARMA parameters.

time series was found to be 0.3368. The bicovariance matrix obtained from the bicovariance function on the main diagonal (s=h) is given by

$$\hat{B}c^{e} = \begin{bmatrix} 0.122 & -0.784 & -1.000 \\ 0.753 & 0.122 & -0.784 \\ 0.514 & 0.753 & 0.122 \\ -0.115 & 0.514 & 0.753 \\ -0.523 & -0.115 & 0.514 \\ -0.535 & -0.523 & -0.115 \\ 0.004 & -0.535 & -0.523 \end{bmatrix}.$$

Similar bicovariance matrices were generated on s=h+1 and s=h-1 diagonals and were concatenated into a single matrix of size (21×3). A least squares solution of (2.120) yielded the following AR coefficients: (1.0, -0.8754, 0.8175). Finally, the MA coefficients, obtained from Equations (2.123) and (2.71), are $(1, -1.6323)$ (End of Example).

In the above example it may be noted that the skewness of the ARMA time series is far less than that of the input. Indeed, as the pole approaches the unit circle, on account of the increased duration of the impulse response function of the ARMA model and the central limit theorem, the output will become increasingly Gaussian even though the input is non-Gaussian. This will affect our ability to estimate the ARMA parameters. The phenomenon is demonstrated in Table 2.8.

The parameters of a linear time series model based on the second-order statistical structure, for example, the covariance matrix, may differ from those based on the third-order statistical structure, for example, the bicovariance matrix.[29] This may be illustrated by considering a process which is the sum of two uncorrelated processes, of which one is Gaussian.

Let

$$x(t) = x_G(t) + x_{NG}(t),$$

where $x_G(t)$ is the Gaussian component, and $x_{NG}(t)$ is the non-Gaussian component. $x_G(t)$ and $x_{NG}(t)$ are uncorrelated. It is evident that

$$\mathbf{C}_X = \mathbf{C}_{X_G} + \mathbf{C}_{X_{NG}},$$

where $\mathbf{C}_X$ stands for the covariance matrix of x(t) and, similarly, the other variables. However, the bicovariance matrix,

$$\mathbf{Bc}_X = \mathbf{Bc}_{X_{NG}},$$

has a contribution from the non-Gaussian component only. Thus, whenever a time series can be written as a sum of Gaussian and non-Gaussian components, the time series models derived from the second-order and third-order structures will differ.

2.7 **Deterministic Chaos**

Nonlinear dynamic systems differ in many ways from linear systems; in particular, a nonlinear system can produce self-sustained oscillatory output, often known as limit cycles. A nonlinear system when driven by an external periodic input may produce a complex set of harmonics and subharmonics which results in a quasi-periodic behavior. The spectrum consists of a series of lines. Some nonlinear dynamic systems produce highly complex outputs exhibiting random behavior. A typical spectrum is made up of discrete lines as well as a continuous broadband spectrum. We call such a time series a chaotic time series, which is generated by a completely deterministic, though nonlinear, system; hence the term deterministic chaos. A simple example of deterministic chaos is the pseudorandom numbers generated on a computer. In this section we shall learn how to characterize a chaotic time series. (An excellent tutorial coverage on this topic is available in Reference 30.)

A. **Nonlinear Dynamic System**

An nth-order dynamic system is characterized by a state equation,

$$\dot{\mathbf{x}}(t) = f(\mathbf{x}, t), \tag{2.125}$$

where $\dot{\mathbf{x}}(t) = \frac{d\mathbf{x}(t)}{dt}$ and f(**x**,t) is a vector field. **x**(t) is a state vector, defined as

$$\mathbf{x}(t) = \text{col}\,\{x_0(t), x_1(t), x_2(t), \ldots, x_{n-1}(t)\},$$

where $x_0(t), x_1(t), x_2(t), \ldots, x_{n-1}(t)$ are different states of the system at time t. The state vector at a given time instant is represented by a point in n-dimensional space known as phase space, and as the time progresses it executes a trajectory. The solution of (2.125) with the initial condition $\mathbf{x}(t = t_0) = \mathbf{x}_0$ is called a trajectory. We represent the solution by $\phi_t(\mathbf{x}_0, t_0)$ as a function of initial condition, $\mathbf{x}_0$ at $t=t_0$. The solution does not depend upon the initial time when the vector field is independent of time.

Steady-state behavior, that is, as $t \to \infty$, of a nonlinear system is of four types, the simplest being equilibrium points. Here the solution of (2.125) is a constant for $t \to \infty$. The solution in the phase space is represented by a point called limit point onto which all trajectories converge. There can be more than one limit point, each with its own basin of attraction. If the point representing the initial conditions of a dynamic system were to fall inside one of the basins of attraction, the system would, as $t \to \infty$, reach the limit point of that basin. In Figure 2.26 we show phase space trajectories of a simple damped pendulum (Figures 2.26, 2.27 and 2.28 are adapted from Reference 30. The two limit points along with their basin of attraction are shown in the same figure. Nonlinear systems often produce self-sustained oscillations. The steady-state behavior of such a system in the phase space is a closed curve or surface known as limit cycle on which all trajectories are located. The classic example of a limit cycle is found in van der Pol's equation,

$$\dot{x}(t) = y(t)$$
$$\dot{y}(t) = (1 - x^2(t))y(t) - x(t)\,. \qquad (2.126)$$

In the phase space, which in this case is a two-dimensional space spanned by a vector, $\mathbf{x}(t) = \text{col}\{x(t), \dot{x}(t)\}$, the limit cycle is a closed curve as shown in Figure 2.27a. The output time series is a periodic function (see Figure 2.27b) with its spectrum, shown in Figure 2.27c, consisting of the fundamental and its harmonics. Now, if we add a sinusoidal forcing term to (2.126), the trajectories are confined to an annular-like region or a torus in the phase space, and they are uniformly distributed in the region (see Figure 2.27d). The output is quasi-periodic (see Figure 2.27e) and has a spectrum as shown in Figure 2.27f. It displays tightly spaced sidebands due to the forcing function.

B. Chaos

The fourth type of steady-state behavior of a nonlinear dynamic system is chaos. It is characterized by the absence of any pattern. It has the appearance of a stochastic process, yet it is an output of a completely deterministic, albeit nonlinear, system. An example of deterministic chaos is shown in Figure 2.28. It was obtained by solving the Duffing equation[30] with a forcing term. The spectrum, shown in Figure 2.28c, is almost featureless. From the spectrum we get no information about the nonlinear system or about the forcing function.

An alternate approach is required. This is precisely what we aim at in this section. We list some of the important features of a chaotic system:

(i) The spectrum of a chaotic time series has a continuous broadband nature over which discrete lines are superposed.

(ii) No matter how precisely the initial condition is known, the long-term behavior of a chaotic system can never be predicted.

(iii) The trajectories of a chaotic system do not lie on a well-defined geometrical object like a circle or torus. They exhibit complex behavior.

Lyapunov exponents are used to determine the stability of any type of steady-state behavior including quasi-periodic and chaotic systems. The Lyapunov exponents are defined as follows:

$$\lambda_i = \lim_{t\to\infty} \frac{1}{t} \ln\left|m_i(t)\right|,$$

where $m_i(t)$ is the eigenvalue of $\phi_t(\mathbf{x}_0)$. The Lyapunov exponents tell us about the evolution of initial condition $\mathbf{x}_0$. In the neighborhood of an equilibrium, point (2.125) is linearized, and its steady-state solution is written in terms of the eigenvectors and eigenvalues,

$$\phi_t(\mathbf{x}_0) = \sum_{i=1}^{n} c_i \eta_i e^{\lambda_i t}$$

Figure 2.26: Phase-space trajectories of a simple damped pendulum. Two limit points along with their basin of attraction are shown. (Taken from Reference 30. © 1987 IEEE)

where η_i and λ_i, i=1, ... , n are eigenvectors and eigenvalues, respectively. The constants c_i, i=1,2, ..., n are determined from the initial conditions. For stable equilibrium points $\lambda_i < 0$ for all i. For the stable limit cycle $\lambda_1 = 0$ and $\lambda_i < 0$ for i =2, 3, ..., n and for a stable torus $\lambda_1 = 0, \lambda_2 = 0$ and $\lambda_i < 0$ for i=3, 4, ..., n. For a chaotic system at least one eigenvalue becomes positive, $\lambda_1 > 0$; however, for overall stability we must have

$$\sum_{i=1}^{n} \lambda_i < 0.$$

The distribution of eigenvalues λ_i, i=1, ..., n, often called the Lyapunov-exponent spectrum, provides a quantative measure of the sensitivity to initial conditions. It is the most useful dynamical diagnostic for chaotic systems.

As an example of chaos from the practical world we consider fluid motion between two rotating concentric cylinders (Couette-Taylor system[31]). The spectrum of a radial component of velocity for different dynamical regimes, in particular, the transition from periodic to quasi-periodic to chaotic states, is shown in Figure 2.29. At low Reynolds number ($\frac{R}{R_c} = 9.6$, where R_c is the critical Reynolds number for the Taylor instability) the spectrum shows a single line (the first fundamental frequency), which corresponds to the azimuthal wave passing through the point

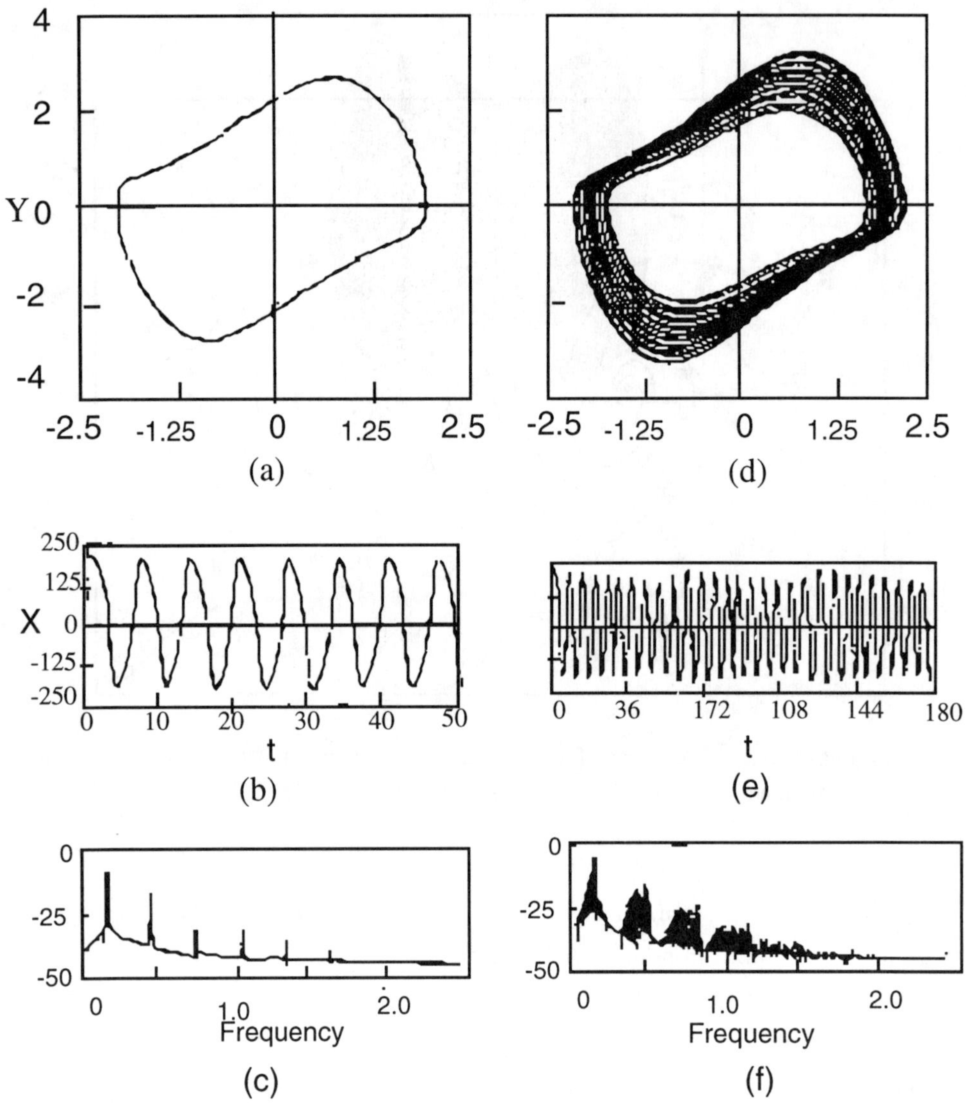

Figure 2.27: Limit-cycle phenomenon in van der Pol's equation (2.126). (a) limit cycle, (b) output time series, (c) spectrum, (d) torus in phase space when the van der Pol's equation is driven by sinusoidal input, (e) quasi-periodic output, and (f) spectrum displays tightly spaced sidebands. (Taken from Reference 30. © 1987 IEEE)

of observation (see Figure 2.29a where we also show a sample of time series). When $\frac{R}{R_c}$ is increased to 15.1, the spectrum consists of many lines; two of them are fundamental frequencies ω_1 and ω_3, and the rest are integer combinations thereof (Figure 2.29b). In addition, we observe a broadband spectrum. The appearance of a broadband spectrum is an indication of the onset of a chaotic state. However, only at a high Reynolds number, that is, at $\frac{R}{R_c}=43.9$, is the flow strongly chaotic in the sense that the entire spectral energy is broadband (see Figure 2.29c).

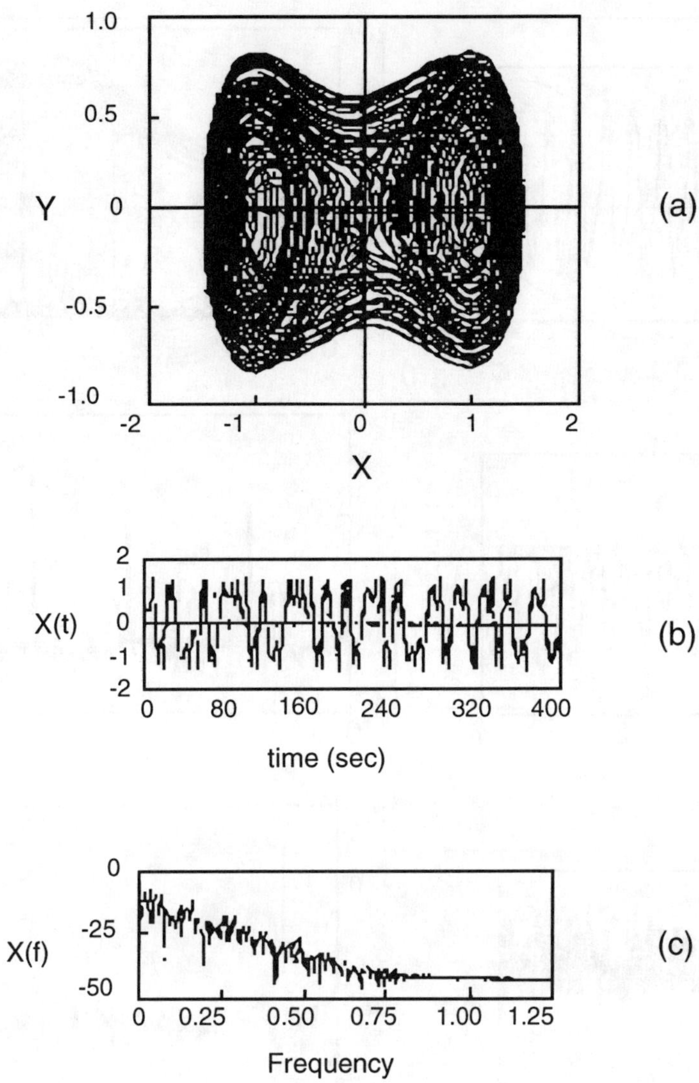

• **Figure** 2.28: Chaotic process. (a) Phase space trajectories of the Duffing equation with a forcing term, (b) output time series, and (c) almost featureless spectrum. (Taken from Reference 30. © 1987 IEEE)

The spectrum is completely devoid of line structure found at lower Reynolds numbers (see Figure 2.29b).

C. **Dimension of Chaos**

We classify an attractor using the concept of dimension, which may be generally taken as degrees of freedom enjoyed by a trajectory in the phase space. An equilibrium point into which all trajectories seem to coalesce is considered to have zero dimension. A limit cycle, or any closed curve in the phase space, is of dimension one, and a torus is of dimension two with two frequencies

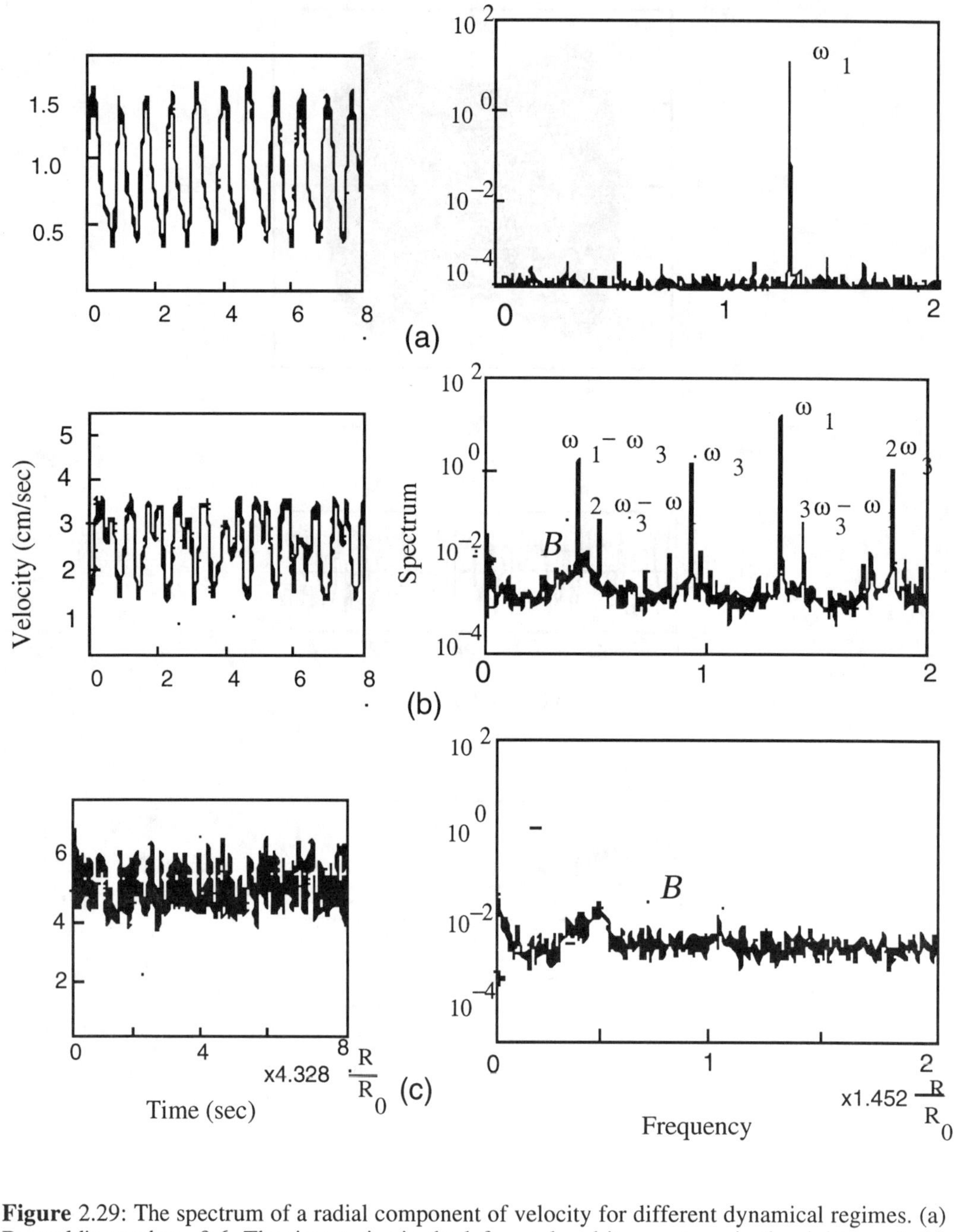

Figure 2.29: The spectrum of a radial component of velocity for different dynamical regimes. (a) Reynold's number= 9.6. The time series in the left panel and its spectrum in the right panel. (b) Reynold's number=15.1. Apart from the line spectrum we have broadband spectrum, which is an indication of onset of chaotic state. (c) Reynold's number= 43.9. Complete chaotic state. (Taken from Reference 31. © Cambridge University Press)

interacting nonlinearly. Similarly, D linearly independent frequencies, that is, those whose linear combination with integer coefficients is nonzero, interacting nonlinearly will produce an attractor of dimension D. In all of the above cases the dimension is an integer. In the case of chaos the steady-state orbits do not lie on a simple geometrical object. The neighborhood of any point of an attractor of a chaotic system, also known as strange attractor, has a fine structure consisting of layers within layers as in some fine French pastry. This complex structure, when viewed through varying magnification, consists of intricate folds of different sizes. The neighborhood of such an attractor does not resemble any Euclidean space. Hence, the strange attractors are not manifolds and do not have integer dimension. Such a strange geometrical set of points has been christened as fractal by Mandelbrot. The dimension of the strange attractor is noninteger, and is said to be of a "fractal" dimension.

D. **Correlation Dimension**

The phase space of a dynamic system is spanned by all components of the state vector. However, the phase space may also be reconstructed from just one component of the state vector. The reconstructed attractors share the same qualitative features as those of the original; in particular, the dimension of the reconstructed attractor is same as that of the original.[32] Using a single time series the phase space is reconstructed as follows. Define data vectors $\mathbf{x}_0, \mathbf{x}_1, \mathbf{x}_2, \ldots, \mathbf{x}_{N-1}$ as

$$\mathbf{x}_0 = \text{col}\,\{x(0)\ \ x(1) \ldots\ x(p-1)\},$$

$$\mathbf{x}_1 = \text{col}\,\{x(1)\ \ x(2) \ldots\ x(p)\},$$

$$\mathbf{x}_2 = \text{col}\,\{x(2)\ \ x(3) \ldots\ x(p+1)\},$$

$$\vdots$$

$$\mathbf{x}_{N-1} = \text{col}\,\{x(N-1)\ \ x(N) \ldots\ x(N+p-2)\},$$

where $N=T-p+1$ and p is an integer (>2D) representing the dimension of the constructed space. We represent the N data vectors in p-dimensional space as N points. Next, we measure the closeness of points by counting the number of pairs of points which are closer than a specified threshold, r. A correlation function, a measure of closeness, is defined as follows:

$$C_p(r) = \lim_{N\to\infty} \frac{1}{N^2} \sum_{i=1}^{N}\sum_{j=1}^{N} U(r - |\mathbf{x}_i - \mathbf{x}_j|), \tag{2.127}$$

where $U(x)=1$ for $x \geq 0$ and $=0$ for $x<0$ is a step function. Observe that $C_p(r)=U(r)$ implies that all points are clustered at r=0. The behavior of $C_p(r)$ at $r \to 0$ is most important. It is believed that $C_p(r) \approx r^D$ for large D and small r. Hence, the correlation dimension of an attractor is defined as

$$D_c = \lim_{\substack{r\to 0 \\ p\to\infty}} \frac{\log C_p(r)}{\log r}. \tag{2.128}$$

There are other types of measures of dimension, but the correlation dimension is the easiest to compute and the most commonly used measure. Yet there are difficulties in its estimation, particularly when the data length is short.

E. **Fractal Dimension**

The degree of roughness of a geometrical object over multiple scales is characterized by fractal dimension. Consider a coastline whose roughness is estimated by measuring its length using a ruler of length, ε . Let the measured length of the coastline be $L(\varepsilon)$. It is observed that

$$L(\varepsilon) \approx k\varepsilon^{-D_f} \quad \text{as} \quad \varepsilon \to 0, \tag{2.129}$$

where D_f is the fractal dimension of the curve, and k is some constant. A curve with noninteger D_f is said to be a fractal curve which is everywhere continuous but nondifferentiable. For analytic curve D_f =0 or equal to an integer depending upon the dimension of the manifold. An important property of a fractal curve is self-similarity; that is, if x(t) is a chaotic process with a fractal dimension, then, $a^{-\nu} x(at + b)$ is also a chaotic process with the same fractal dimension, where ν $(0 < \nu < 1)$ is a scaling exponent or a self-similarity parameter, and a>0 and b are any constants. The scaling exponent n is related to the fractal dimension, $D_f = 2-\nu$.[33]

The fractal dimension in principle may be computed as a limiting value,

$$D_f = \lim_{\varepsilon \to 0} \frac{\log L(\varepsilon)}{-\log \varepsilon}. \tag{2.130}$$

In practice, however, the fractal dimension of a time series may be computed using the method of box counting. The method consists of covering a plot of the time series with square boxes as shown in Figure 2.30 and counting the number of boxes required to cover the entire plot. A plot of this number versus the size of the box on a log-log graph sheet is a straight line whose slope is equal to - D_f (for more details see Reference 34).

F. **Rank of Data Matrix**

The data vectors defined earlier are now arranged in a matrix form:

$$\mathbf{x} = \begin{bmatrix} x(0) & x(1) & \cdots & x(p-1) \\ x(1) & x(2) & \cdots & x(p) \\ x(2) & x(3) & \cdots & x(p+1) \\ \vdots & & & \vdots \\ x(N-1) & x(N) & \cdots & x(N+p-2) \end{bmatrix}.$$

x is also known as a trajectory matrix in the literature on chaos.[35] The rank of the trajectory matrix, obtained from its singular value decomposition, gives an upper bound to the dimensionality of the subspace spanned by the trajectory.[35] It is also known as statistical dimension.[36] An important difference between the statistical dimension and the dynamic dimension is that the latter is invariant of the process of embedding; in particular, it does not depend on the dimension of the embedding space or on the sampling interval or on the length of the time series. The statistical dimension, unlike the dynamic dimension, is easily estimated, even in the presence of measurement noise, and more importantly it does not require excessively large data. A plot of the singular values of the trajectory matrix (or eigenvalues of the correlation matrix) against their order is called a singular spectrum from which the effective rank or the statistical dimension is estimated.

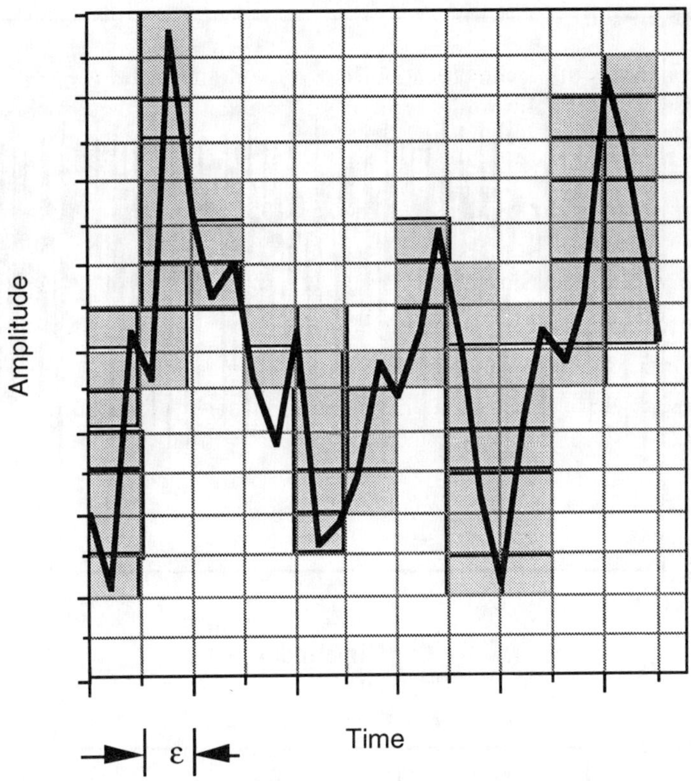

Figure 2.30: A plot of time series is covered with square boxes of size $\varepsilon \times \varepsilon$ as shown above. The number of boxes required to cover the time series is determined.

Example 2.20: Let us consider a simple example of a chaotic process known as the Henon map,

$$x(t + 1) = 1.0 - 1.4x^2(t) + 0.3x(t - 1)\ ,$$

whose correlation dimension is known to be 1.28.[30] We shall compute its statistical dimension by examining the rank of the data matrix. A sample of 256 values of the Henon map was generated by solving the above nonlinear difference equation. A plot of the waveform is shown in Figure 2.31a and that of the singular spectrum in Figure 2.31b. As seen from Figure 2.31 it is difficult to estimate the rank of the trajectory matrix. The eigenvalues continue to decrease slowly. Perhaps, an effective rank of three appears to be a reasonable figure which agrees with the claim that the statistical dimension (effective rank of a trajectory matrix) provides an upper bound to the correlation dimension[35](End of Example).

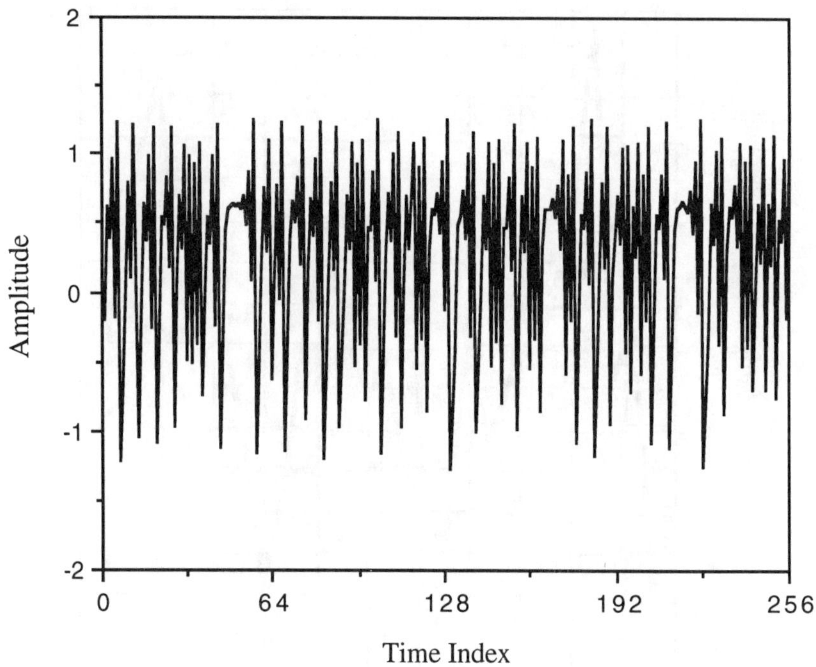

(a)

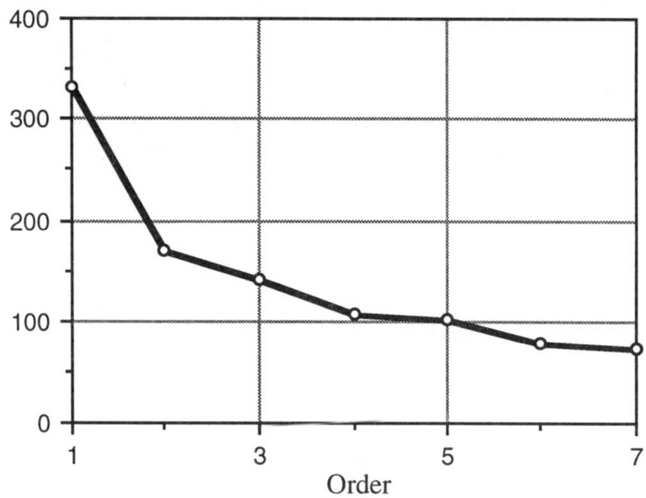

(b)

Figure 2.31: (a) Henon map obtained by solving the nonlinear equation with zero initial conditions. The plot shows the waveform starting from 1500 to 1755 time index. (b) A plot of the eigenvalues of $\mathbf{x}\mathbf{x}^H$ where $\mathbf{x}$ is a data matrix (7×249).

G. Effect of Filtering on Dimension

The output of a linear system driven by a chaotic signal will also be a chaotic signal, possibly with a different structure, i.e., with a different Lyapunov exponent spectrum and fractal dimension. For example, the (fractal) dimension of a first-order AR process driven by chaotic process x(t),

$$y(t) + ay(t-1) = x(t),$$

has been shown to be greater than that of the driving chaotic process, provided $0 \geq \ln(a) > \lambda_{min}$ where λ_{min} is the minimum Lyapunov exponent.[37] For small a, that is, when Y(t) is wideband, the dimension of the filter output remains unaltered. For a higher order AR process when impulse response function of the AR process satisfies the condition,

$$\sum_{t=0}^{\infty} |h(t)| e^{-\lambda_{min} t} < \infty,$$

that is, $|h(t)|$ decays faster than $e^{-\lambda_{min} t}$, the filtering operation has no effect on the fractal dimension.[38] Note that the above condition is satisfied for all FIR (finite impulse response) filters. The effect of filtering on the attractor of a chaotic process in the phase space is to generally make it blurred and thus increase the volume occupied by the attractor.

H. **Chaos as a Signal Model**

In many physical systems the signal source can be highly nonlinear, the most prominent examples of this class being turbulence in fluids, laser instabilities, brain activity, speech generation by the human vocal cord, etc. (see Reference 39 for many examples of deterministic chaos). It is important to remember that even a relatively low-dimensional nonlinear system can produce signals which look like random signals, yet there is no intrinsic randomness in the system. The first question one would like to ask is how to distinguish a deterministic chaos from a stochastic time series. This question may be answered at least partially by computing the dimension of a given time series. In a phase space of dimension greater than the dimension of the chaos, that is p>2D, the trajectories of chaos would occupy a space of dimension D. On the other hand, the stochastic process would, over sufficiently long time, fill the entire phase space of dimension p. Therefore, it is expected that the dimension of the attractor of a pure stochastic process, would be equal to p, that is, always equal to the dimension of the embedding space. However, that this is not always true has been demonstrated in Reference 40 where it is shown that a stationary stochastic process with a power-law spectrum, that is,

$$S_x(\omega) = \omega^{-\gamma},$$

possesses a low correlation dimension depending upon the value of the exponent γ. In fact its fractal dimension, D_f, and the self-similarity parameter, ν, are related to γ, $D_f = \frac{5-\gamma}{2}$, and $\nu = \frac{\gamma-1}{2}$.[33] Interestingly, such a stochastic process with a power-law spectrum is also found to be non-Gaussian. Therefore, the commonly believed criterion based on a nonsaturating correlation dimension cannot always be used to distinguish a stochastic process from a low-dimensional chaos.

The fact that the dimension of the deterministic chaos can be quite small, while its spectrum is broadband like that of a pure stochastic signal, is very remarkable and has many potential applications in the area of signal processing. For example, it has been demonstrated that the predictability of a time series modeled as a chaos rather than as a stochastic process can be improved considerably.[41,42] There is a detectable change in the fractal dimension of sea-scattered signals when a target is present.[34] A deconvolution or inverse filtering scheme has been proposed where the volume occupied by an attractor is minimized.[43]

2.8 Exercises

1. Let x(t), t=0,1, ... , T − 1 be a finite time series. Form a complex time series as follows:

$$y(t) = x(t) + jx(T - 1 - t),$$

(a) Obtain the DFT of y(t) in terms of the DFT of x(t).
(b) Show that $Y(T-k) = Y^{*}(k)$ where Y(k) is DFT of y(t).

2. The spectrum of a discrete process is related to that of the continuous process through (2.32) (see also Figure 2.8). Now answer the following questions: (1) A continuous stochastic process with a peak at 400 Hz is sampled at 500 Hz. Sketch the spectrum of the sampled stochastic process. (2) Is the variance of the discrete process equal to that of the continuous process? (3) Define a difference process, $x_1(t) = x(t) - x(t-1)$, where x(t) is the sampled stochastic process. Obtain the spectrum of $x_1(t)$.

3. A sum of p random sinusoids satisfies the following difference equation (2.88):

$$s(t) = -\sum_{i=1}^{2p} a_i s(t-i),$$

where parameters a_i, $i = 1,2,\ldots,2p$ satisfy (2.87). Show that

$$a_{2p} = 1$$

and

$$a_{2p-1} = a_1.$$

[Hint: Multiply with $e^{j2p\omega_k}$ on both sides of the second equation in (2.87) and then compare term by term with the first equation of (2.87).]

4. Consider the following AR process:

$$x(t) - 1.269\,x(t-1) + .81\,x(t-2) = \varepsilon(t),$$

(a) Obtain the transient solution.
(b) Compute the covariance function using reverse Yule-Walker equations (2.60).
(c) Express the above AR process as an MA process and compute the MA coefficients.

5. Let s(t) be a sum of p complex random sinusoids,

$$s(t) = \sum_{i=1}^{p} \alpha_i \exp(-j\omega_i t + \phi_i) \qquad t \geq 0$$

Show that

$$s(t) + \sum_{i=1}^{p} a_i s(t-i) = 0 \qquad t \geq 0,$$

analogous to (2.88) for real sinusoids. Define the following matrices:

$$\mathbf{A} = \begin{bmatrix} a_p \cdots & a_2 & a_1 & 1 & & & & \\ & a_p \cdots & a_2 & a_1 & 1 & & \mathbf{0} & \\ \mathbf{0} & & \ddots & & & & \ddots & \\ & & & a_p \cdots & a_2 & a_1 & & 1 \end{bmatrix}$$

$$p \times (2p+1)$$

$$\mathbf{H} = \begin{bmatrix} 1 & 1 & \cdots & 1 \\ e^{j\omega_1} & e^{j\omega_2} & \cdots & e^{j\omega_p} \\ e^{j2\omega_1} & e^{j2\omega_2} & \cdots & e^{j2\omega_p} \\ \vdots & & \vdots & \\ e^{j(2p+1)\omega_1} & e^{j(2p+1)\omega_2} & \cdots & e^{j(2p+1)\omega_p} \end{bmatrix}.$$

$$(2p+1) \times p.$$

Show that $\mathbf{AH} = \mathbf{0}$ where $\mathbf{0}$ is the $p \times p$ zero matrix.

6. Let h(t) be a minimum phase causal filter and $h_{INV}(t)$ be its inverse. Show that

$$h_{INV}(t) * C_h(t) = \Omega\delta(t) \quad t \geq 0.$$

where $C_h(t)$ is autocorrelation function of h(t) and W is a constant. Express the above equation for finite t, $0 \leq t \leq T_0$ in a matrix form. For example, for $T_0 = 3$

$$\begin{bmatrix} C_x(0) & C_x(1) & C_x(2) & C_x(3) \\ C_x(1) & C_x(0) & C_x(1) & C_x(2) \\ C_x(2) & C_x(1) & C_x(0) & C_x(1) \\ C_x(3) & C_x(2) & C_x(1) & C_x(0) \end{bmatrix} \begin{bmatrix} h_{INV}(0) \\ h_{INV}(1) \\ h_{INV}(2) \\ h_{INV}(3) \end{bmatrix} = \begin{bmatrix} \Omega \\ 0 \\ 0 \\ 0 \end{bmatrix}.$$

Show that the autocorrelation matrix is of full rank, that is, $\|C_h(t)\| > 0$ for all values of T_0 (see Reference 44).

7. Show that the extended covariance matrix of an AR (p) process is of rank p.[45] Note that the covariance matrix is of size $(p+1) \times (p+1)$. [Hint: Use the Yule-Walker equations to show that the 1st column is a linear combination of the remaining p columns.]

8. Consider a nonminimum phase ARMA(2,6) process. Enumerate all possible MA filters leading to same spectrum. Thus, verify the statement on page 115 that there are $2^{\frac{q}{2}}$ combinations. What happens when some of the zeros are on the real axis or on the unit circle?

9. On page 126 we have shown that an AR(p) process plus white noise is equal to an ARMA(p,p). The proof was given in the time domain. Now obtain the same result in the frequency domain with the help of the generalized Fourier transform (GFT) or the z-transform.

10. Show that the rank of the covariance matrix of a bandlimited process is equal to or less than time (lag) bandwidth product. [Hint: Let $\mathbf{C}_x = E\{\mathbf{x}\mathbf{x}^H\}$ where

$$\mathbf{x} = \text{col}\{x(0), x(1), \ldots, x(T-1)\}.$$

Express x(t) in terms of its spectral representation (1.57) and evaluate the expected operation.]

11. Show that the covariance function of an ARMA process is equal to the convolution of covariance functions of AR and MA components taken separately. This may be demonstrated by convolving the covariance function of an MA process mentioned in Example 2.13 with that of the AR process in Example 2.9. The result is the covariance function given in Example 2.13.

12. Let x(t) be q^{th}-order non-Gaussian MA process. Show that the covariance and bicovariance functions are related as follows:

$$\sum_{i=0}^{q} b_i b_{i+m} C_X(\tau - i) = \frac{\sigma_\varepsilon^2}{\beta_\varepsilon^3} \sum_{i=0}^{q} b_i Bc(\tau - i, \tau - i + m) \qquad m = 0, 1, 2, \ldots$$

[Hint: Evaluate the z-transform on both sides of (2.49) and (2.109) and find the ratio of the z-transform of the covariance sequence and that of bicovariance sequence on h=s+m, m=0, 1, 2, ... lines (see Reference 46).]

13. Let

$$x(t) = \sum_{i=1}^{p} A_i \cos(\omega_i t + \phi_i),$$

where ϕ_i, i=1, 2, ..., p are all independent random variables uniformly distributed over $\pm\pi$. Show that the fourth-order cumulant on the diagonal s=g=h= τ is given by[47]

$$T_0(\tau) = -\frac{3}{8}\sum_{i=1}^{p} A_i^4 \cos(\omega_i \tau).$$

14. Let x(t) be the output of a filter driven by a non-Gaussian process. Show that the magnitude of the bicoherence (defined in (1.78)) of x(t) is a constant over all frequencies. A non-Gaussian process is said to be linear (third order) if its bicoherence magnitude is constant.

15. Consider a time series,

$$x(t) = \sum_{i=1}^{p} a_i x(t-i) + \sum_{i=1}^{p} b_i x(t-i)\varepsilon(t-1) + \varepsilon(t).$$

known as a bilinear time series. Show that the covariance function satisfies an equation similar to that of an ARMA(p,1) process,

$$C_X(\tau) = \sum_{i=1}^{p} a_i C_X(\tau - i) \qquad \tau > 1$$

The nonlinear interaction term in the time series model does not appear in the equation.[48]

16. A stochastic process, x(t), where t is a continuous variable, is sampled at time instants $t_n = n\Delta t + \phi_n$, where ϕ_n is a random variable uniformly distributed over an interval, $\pm \frac{\Delta t}{2}$. Such a time series is subjected to spectrum analysis. Relate the spectrum of the randomly sampled time series to that of x(t).

17. Starting from (2.55) show that the group delay of an AR filter is inversely proportional to the AR spectrum; in particular,

$$\tau_p(\omega) = \frac{\frac{1}{\sigma_\varepsilon^2}\sum_{\tau=0}^{\infty}\sum_{\tau'=0}^{\infty}\tau C_x(\tau)C_x(\tau')\cos(\omega(\tau-\tau'))}{S_x(\omega)}.$$

18. Consider an AR(2) process with poles at $\rho\exp(\pm j\theta)$. Show that the condition number $k(\mathbf{C}) = \frac{\lambda_{max}}{\lambda_{min}}$ of the covariance matrix (3 × 3) is given by

$$k(\mathbf{C}) = \frac{1+|q|}{1-|q|},$$

where

$$q = \frac{2\rho\cos(\theta)}{1+\rho^2}.$$

Note that

$$k(\mathbf{C})\big|_{min} = 1 \quad \text{at} \quad \theta = \pm\frac{\pi}{2}$$

and

$$k(\mathbf{C})\big|_{max} = \infty \quad \text{at } \theta = 0.$$

It is well known that the condition number plays an important role in stable inversion of the covariance matrix and thus in the estimation of AR coefficients and the AR spectrum. For minimum error, one would like to have the poles close to the imaginary axis. The position of the poles can be manipulated by choosing the sampling interval in the case when the time series is obtained by sampling a continuous process or by resampling a discrete time series.[49]

19. Consider the following signal model:

$$x(t) = \sum_{i=1}^{P}\exp(j\omega_i t + \phi_i) + \eta(t),$$

where

$$\eta(t) + \sum_{i=1}^{P_{noise}} a_i\eta(t-i) = \varepsilon(t)$$

and $\varepsilon(t)$ is a white noise process. Show that x(t) can be modeled as an AR process of order $P + P_{Noise}$. Suggest a method to distinguish the poles of the sinusoids from those of $\eta(t)$.

20. Consider a spectral density function of the form

$$S_x(\omega) = \exp\left(\sum_{k=-p}^{p} \lambda_k e^{-jk\omega}\right) \qquad \lambda_k \neq 0$$

the so-called exponential model of order p. Show that the parameters of the exponential, $\lambda_k = \lambda_{-k}$, $k = 0,1,\ldots,p$, are related to the covariance function as[50]

$$\sum_{k=1}^{p} k\lambda_k (C_x(\tau - k) - C_x(\tau + k)) = \tau C_x(\tau), \quad -\infty < \tau < \infty.$$

21. A nonlinear time series in its simplest form is represented by

$$y(t) = \sum_{\tau_1} h_1(\tau_1)x(t - \tau_1) + \sum_{\tau_1}\sum_{\tau_2} h_2(\tau_1,\tau_2)x(t - \tau_1)x(t - \tau_1),$$

where x(t) is a stationary zero-mean Gaussian time series. $h_1(\tau)$ is the impulse-response function of the linear part, and $h_1(\tau_1, \tau_2)$ is the impulse response of the quadratic part. Show the following relationships:[51]

$$H_1(\omega) = \frac{S_{xy}(\omega)}{S_x(\omega)}$$

$$H_2(\omega_1, \omega_2) = \frac{Bs_{xxy}(\omega_1, \omega_2)}{2S_x(\omega_1)S_y(\omega_2)},$$

where $Bs_{xxy}(\omega)$ is a cross-bispectrum between the input and the output. We have assumed that we have access to both input and output.

References

1. P. R. Gutowski, E. A. Robinson, and S. Treitel, Spectral estimation: fact or fiction, *IEEE Trans. Geosci. Electron.*, GE-16, 80-84, 1978.

2. L. H. Koopmans, *The Spectral Analysis of Time Series*, Academic Press, New York, 242, 1974.

3. V. V. Solodovnikov, *Introduction to the Statistical Dynamics of Automatic Control Systems.* Translated from Russian, Dover Publication, New York, 48, 1960.

4. A. J. Berkhout, On the minimum length property of one sided signals, *Geophysics,* 38, 657-672, 1973.

5. D. G. Childers, D. P. Skinner, and R. C. Kemeriat, The cepstrum: a guide to processing, *Proc. IEEE,* 65, 1428-1443, 1977.

6. B. Yegnanarayana, D. K. Saikia, and T. R. Krishnan, Significance of group delay functions in signal reconstruction from spectral magnitude or phase, *IEEE Trans.*, SP-32, 610-623, 1984.

7. B. Yegnanarayana and H. A. Murthy, Significance of group delay functions in spectral estimation, *IEEE Trans.*, SP-40, 2281-2289, 1992.

8. F. M. Reza, *An Introduction to Information Theory*, McGraw-Hill Book Co, New York, 1961.

9. H. C. Andrews, Entropy consideration in the frequency domain, *Proc. IEEE*, 55, 113-114, 1968.

10. D. J. Thompson, Spectrum estimation and harmonic analysis, *Proc IEEE*, 70, 1055-1096, 1982.

11. Funbg-I. Tseng and T. K. Sarkar, Enhancement of poles in spectral analysis, *IEEE Trans. Geosci. and Remote Sensing,* GE-20, 161-168,1982.

12. H. Wold, *A Study in the Analysis of Stationary Time Series,* Stockholm University, Stockholm, Sweden, Almqvist and Wiksell, 1954.

13. J. W. Woods, Two dimensional discrete markovian fields, *IEEE Trans.*, IT-18, 232-240, 1972.

14. K. Konstantinides, Threshold bounds in SVD and a new iterative algorithm for order selection in AR models, *IEEE. Trans.*, SP-39, 1218-1221, 1991.

15. B. Cernuschi-Frias, A derivation of the Gohberg-Semencol relation, *IEEE. Trans.*, ASSP-39, 190-192, 1991.

16. D. Lysne and D. Tjostheim, Loss of spectral peaks in autoregressive spectral estimation, *Biometrica*, 74, 200-206, 1987.

17. S. Sivananda, M.E. dissertation, Dept. of ECE, IISc, Bangalore-560012, India, 1987.

18. H. W. Turnbull, *Theory of Equations*, Oliver and Boyd, Edinburgh, Scotland, 1952.

19. S. M. Kay, A new ARMA spectral estimator, *IEEE. Trans.,* ASSP-28, 585-588, 1980.

20. E. A. Talkhan, A. F. Hassan, and N. A. Amin, An ARMA model for power spectrum estimation, *IEEE. Proc.* ICASSP-83, 1080-1084, 1983.

21. Xian-Da Zhang and Yuan-Sheng Zhang, Determination of the MA order of an ARMA process using sample correlations, *IEEE. Trans.*, SP-41, 2277-2280, 1993.

22. Y. T. Chan, J. M. M. Lavoie, and, J. B. Plant, A parameter estimation approach to estimation of frequencies of sinusoids, *IEEE. Trans.*, ASSP-29, 214-219, 1981.

23. A. M. Walker, Some consequences of superimposed error in time series analysis, *Biometrika*, 47, 33-43, 1960.

24. M. Pagano, Estimation of models of autoregressive signal plus white noise. *Ann. Statist.,* 2, 99-108, 1974.

25. M. Kaveh and S. P. Bruzzone, Statistical efficiency of correlation based methods for ARMA spectral estimation, *Proc. IEEE*, 130, pt. F, 211-217, 1983.

26. J. V. Candy, *Signal Processing, The Model Based Approach*, McGraw-Hill Book Co, New York, 1986.

27. T. Matsuoka and T. J. Ulrych, Phase estimation using bispectrum, *Proc. IEEE*, 72, 1403-1411, 1984.

28. A. Swami and J. M. Mendel, ARMA parameter estimation using only output cumulants, *IEEE. Trans.*, ASSP-38, 1257-1265, 1990.

29. M. R. Raghuveer and C. L. Nikias, Bispectrum estimation via AR modeling, *Signal Process*, 10, 35-48, 1986.

30. T. S. Parker and L. O. Chua, Chaos: A tutorial for engineers, *Proc. IEEE*, 75, 982-1008, 1987.

31. P. R. Fenstermacher, H. L. Swinney, and J. P. Gollub, Dynamical instabilities and the transition to chaotic Taylor vortex flow, *J. Fluid Mech.*, 94, part 1, 103-128, 1979.

32. F. Takes, *Detecting strange attractors in turbulence*, *Lecture Notes in Math,* Vol. 898, Springer-Verlag, Berlin, Germany, 366-381, 1980.

33. G. W. Wornell and A. V. Oppenheim, Estimation of fractal signals from noisy measurements using wavelets, *IEEE. Trans.*, SP-40, 611-623, 1992.

34. T. Lo, H. Leung, J. Litva, and S. Haykin, Fractal characterisation of sea-scattered signals and detection of sea-surface targets, *Proc. IEEE, part-F*, 140, 243-250, 1993.

35. D. S. Broomhead and G. P. King, Extracting qualitative dynamics from experimental data, *Physica D*, 20, 217-236, 1986.

36. R. Vautard and M. Ghil, Singular spectrum analysis in nonlinear dynamics, with applications to paleoclimatic time series, *Physica D*, 35, 395-424, 1989.

37. R. Badii, G. Broggi, B. Derighetti, M. Ravani, S. Cliberto, A. Politi, and M. A. Rubio, Dimension increase in filtered chaotic signals, *Phys. Rev. Lett.*, 60, 976-982, 1988.

38. S. H. Isabelle, A. V. Oppenheim, and G. W. Wornell, Effects of convolution on chaotic signals, *IEEE, Proc.* ICASSP-92, IV133-136, 1992.

39. G. Mayer-Kress, Ed, Dimension and entropies in chaotic systems, *Quantification of Complex Behavior,* Springer-Verlag, Berlin, 180-256, 1986.

40. A. R. Osborne and A. Provenzale, Finite correlation dimension for stochastic systems with power-law spectra, *Physica D*, 35, 357-381, 1989.

41. J. D. Farmer and J. J. Sidorowich, Predicting chaotic time series, *Phys. Rev. Lett.,* 59(8), 845-848, 1987.

42. A. Kumar and S. Mullick, Application of deterministic chaos to signal modeling, Workshop on 'Signal Processing, Communications, & Networking', July 23-26, 1990, Indian Institute of Science, Bangalore, India.

43. J. D. Scargle, Studies in astronomical time series analysis. IV. Modeling chaotic and random processes with linear filters, *Astrophys, J.*, 359, 469-482, 1990.

44. A. J. Berkhout, On the minimum phase criterion of sampled signals, *IEEE Trans., Geosci. Electron.,* GE-11, 186-198, 1973.

45. J. A. Cadzow, Spectral estimation: an overdetermined rational model equation approach, *Proc. IEEE*, 70, 907-939, 1982.

46. J. K. Tugnait, Approaches to FIR system identification with noisy data using higher order statistics, *IEEE Trans,* ASSP-18, 1307-1317, 1990.

47. A. Swami and J. M. Mendel, Cumulant-based approach to the harmonic retrieval problem, *IEEE Proc., ICASSP-88*, 2264-2267, 1988.

48. V. J. Mathew and T. K. Moon, Parameter estimation for a bilinear time series model, *IEEE Proc., ICASSP-91*, 5, 3513-3516, 1991.

49. M. J. Villalba and B. K. Walker, Spectrum manipulation for improved resolution, *IEEE Trans.*, ASSP-37, 820-831, 1989.

50. Q. Cheng, Parameter estimation of exponential model for spectra, *IEEE Proc., ICASSP-90*, 2611-2613, 1990.

51. L. J. Tick, The estimation of transfer functions of quadratic systems, *Technometrics*, 3, 563-567, 1961.

Chapter Three

Spectrum Estimation: *Low-Resolution Methods*

3.1 An Overview

In this chapter we shall pursue the classical methods of estimating the power spectral density function or simply the spectrum of a stationary stochastic process. No further constraints on the nature of the time series will be imposed. In the next chapter we shall impose some specific model on the time series and evolve the so-called high resolution methods. There are two well-known approaches to spectrum estimation; namely, (1) estimate the covariance function and then take its Fourier transform, which is the desired spectrum (Wiener-Khintchin theorem, page 17) and (2) compute the Fourier transform of many uncorrelated segments and then average the magnitude square of the Fourier transforms. The most popular method of the first type is the Blackman-Tukey (BT)[1] method and of the second type is the Welch[2] method. The important issues in the estimation of the spectrum are those arising out of finite data length, for example, bias, variance, resolution, etc. Finite data automatically brings in the need for a spectral window which controls the bias, resolution, and to some extent the variance of the spectral estimate. All methods described in this chapter are characterized by low resolution which is controlled solely by the length of the time series and the type of window function used. Two sinusoids separated by $\Delta\omega < \frac{2\pi}{T}$ cannot be resolved. This resolution limit is often known as Rayleigh limit. The actual resolution achieved, however, depends upon the type of the window function employed in the spectrum estimation. Furthermore, the window also controls the bias caused by the leakage of power through the sidelobes, in particular when the dynamic range of the spectrum is very high, for example, when there are sharp peaks in the spectrum. The design of the window function has engaged the attention of many workers over several decades[3]. Scores of window functions have been proposed. Of these the most popular ones are the Hanning and Hamming windows. There are optimum windows where the energy, by design, is concentrated to a prescribed frequency bandwidth. When the prescribed bandwidth is an integer multiple of $\frac{2\pi}{T}$, there is more than one optimum window, all of which may be used to produce independent spectral estimates which may then be averaged to yield a stable estimate of the spectrum. Use of multiple windows results in a narrow mainlobe and low sidelobe level.

Given a finite length data there is a trade off between the number of uncorrelated segments or the number of lags at which the covariance function may be computed and the degree of resolution which one can achieve. Since the variance of the spectral estimate is inversely proportional to the number of segments, we aim at as many segments as possible consistent with the desired resolution. One ingenious method of increasing this number is to partly overlap, up to 50%, the adjacent data segments[4] or to overlap many short windows to synthesize a single long window[5]. Such an overlap and add method was suggested by Mathews and Youn,[6] who called it a STUSE (short-time unbiased spectrum estimation).

For non-Gaussian time series the bispectrum and trispectrum are important quantities. Their estimation, though a straightforward extension of the methods for estimation of spectrum, is difficult to statistically characterize. Analogous to the covariance matrix we define a bicovariance matrix by expressing the elements on the diagonal of the bicovariance function in a matrix form. Such a matrix will be useful in the estimation of AR and ARMA coefficients of non-Gaussian time series. The spectrum analysis of nonstationary time series has been a major area of research. We shall only briefly outline the computational issues in the estimation of spectrogram, multiscale spectrogram (scalogram), and Wigner-Ville distribution. The statistical characterization of these estimates, yet to be fully understood, is beyond the scope of this book.

The low-resolution spectrum estimation methods described here are linear in the sense that the law of superposition will hold good; that is, the spectrum of the sum of two *uncorrelated* time series is equal to the sum of the spectra of individual time series. On the other hand, the high-resolution methods to be described in the next chapter are nonlinear, for they do not satisfy the law of superposition. The linear methods, like linear systems, are easy to understand and easy to describe mathematically.

3.2 Covariance Function

The covariance function of a stationary stochastic time series (see (1.19)) involves the use of ensemble averaging, which requires many realizations of the stochastic process. In almost all practical cases just one realization is available from which we will have to estimate the covariance function. This is done by invoking the property of ergodicity (see page 23) which states that for certain stationary stochastic processes (known as ergodic processes) the ensemble average may be replaced by the time average. Since the ergodicity cannot be ascertained beforehand, it is preferable to call the estimate of the covariance function obtained by time averaging by another name, for example, autocorrelation function; this, however, is not a standard practice. An estimate of the covariance function of a finite stationary time series is given by

$$\hat{C}_x(\tau) = \frac{1}{T-\tau} \sum_{t=0}^{T-\tau-1} (x(t) - \hat{m}_x)(x(t+\tau) - \hat{m}_x), \tag{3.1}$$

where $\hat{m}_x$ is an estimate of the mean. From (3.1) we note that as $T \to \infty$. $\hat{C}_x(\tau) \to C_x(\tau)$ provided the time series is ergodic.

A. Estimation of Mean

The estimate of the mean of a time series is given by

$$\hat{m}_x = \frac{1}{T} \sum_{t=0}^{T-1} x(t). \tag{3.2}$$

It is easy to show that the above estimate is unbiased,

$$E\{\hat{m}_x\} = \frac{1}{T} \sum_{t=0}^{T-1} E\{x(t)\} = m_x \quad ,$$

and its variance is given by

$$\begin{aligned} \text{var}\{\hat{m}_x\} &= E\{(\hat{m}_x - m_x)^2\} \\ &= \frac{1}{T^2} \sum_{t=0}^{T-1} \sum_{t'=0}^{T-1} E\{(x(t) - m_x)(x(t') - m_x)\} \\ &= \frac{1}{T^2} \sum_{t=0}^{T-1} \sum_{t'=0}^{T-1} C_x(t-t') = \frac{1}{T} \sum_{\tau=-T+1}^{T-1} \left(1 - \frac{|\tau|}{T}\right) C_x(\tau). \end{aligned} \tag{3.3}$$

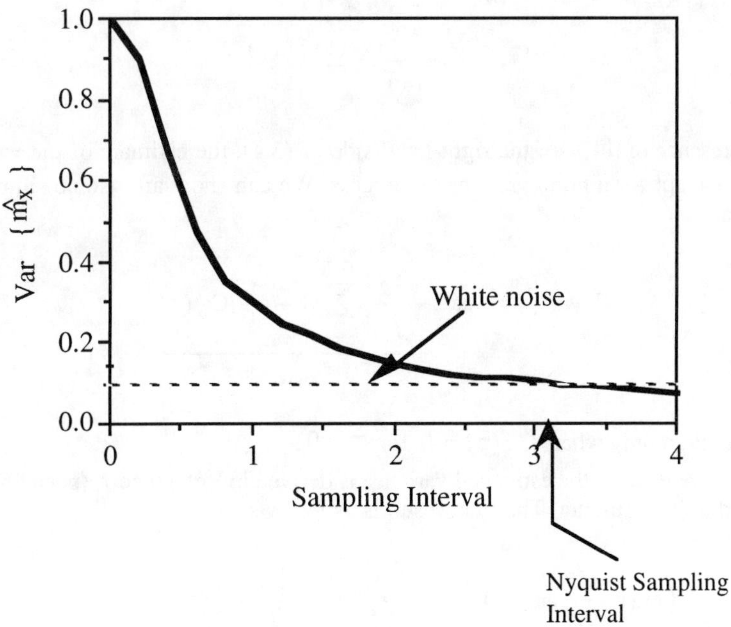

Figure 3.1: The effect of sample correlation on the variance of the estimated mean (3.3) for a band-limited white noise process with a covariance function ($C_x(\tau) = \sin(\tau)/\tau$). The dashed line is for completely uncorrelated samples. T=10.

For a white noise process, (3.3) reduces to

$$\text{var}\{\hat{m}_x\} = \frac{\text{Var}_x}{T}.$$

For a correlated process the variance of $\hat{m}_x$ would depend upon the sampling interval; for example, for a band-limited white noise process with a covariance function given by $\frac{\sin(\tau)}{\tau}$, $\text{var}\{\hat{m}_x\}$ as a function of sampling interval is shown in Figure 3.1.

From (3.3) we notice that for large T (>> correlation distance)

$$\text{var}\{\hat{m}_x\} \underset{T \to \infty}{\rightarrow} \frac{1}{T} \sum_{-T+1}^{T-1} C_x(\tau) = \frac{S_x(\omega)\big|_{\omega=0}}{T}.$$

B. **Estimation of Variance**

The variance of stationary stochastic time series is estimated as follows:

$$\hat{var}_x = \frac{1}{T-1}\sum_{t=0}^{T-1}(x(t) - \hat{m}_x)^2. \tag{3.4}$$

Because of the presence of $\hat{m}_x$ on the right-hand side of (3.4), the estimate of the variance turns out to be biased except for a pure white noise process. We can show after some simple algebraic manipulation that

$$E(\hat{var}_x\} = var_x - \underbrace{\frac{2}{T-1}\sum_{\tau=1}^{T-1}(1 - \frac{\tau}{T})C_x(\tau)}_{\text{bias}}. \tag{3.5}$$

The bias becomes zero only when $C_x(\tau) = 0$ for $\tau > 0$.

The mean square error in the estimated variance is derived in Reference 7, (page 452). The time series is assumed to be Gaussian. The final result is as follows:

$$E\{(\hat{var}_x - var_x)^2\} = \frac{2}{T}\sum_{\tau=-T+1}^{T-1}(1 - \frac{|\tau|}{T})C_x^2(\tau) - \frac{4}{T^3}\sum_{t}^{T-1}\sum_{s}^{T-1}\sum_{s'=0}^{T-1}C_x(t-s)C_x(t-s') + 3[\frac{1}{T}\sum_{\tau=-T+1}^{T-1}(1 - \frac{|\tau|}{T})C_x(\tau)]^2. \tag{3.6}$$

For a pure white noise process, (3.6) reduces to a simpler form as follows:

$$E\{(\hat{var}_x - var_x)^2\} = (\frac{2}{T} - \frac{1}{T^2})var_x^2. \tag{3.7}$$

For a correlated process the mean square error would depend upon the sampling interval; for example, for a band-limited white noise process the mean square error as a function of sampling interval is shown in Figure 3.2. Note that the mean square error is large when the sampling interval is less than the Nyquist interval when the samples are correlated. For large T the mean square error goes to zero; that is,

$$E\{(\hat{var}_x - var_x)^2\} = \frac{2}{T}\sum_{\tau=-T+1}^{T-1}C_x^2(\tau) \to 0, \quad T \to \infty \tag{3.8}$$

provided $\sum_{\tau=-T+1}^{T-1} C_x^2$ is finite; and then the stationary stochastic process is said to be ergodic with respect to variance.

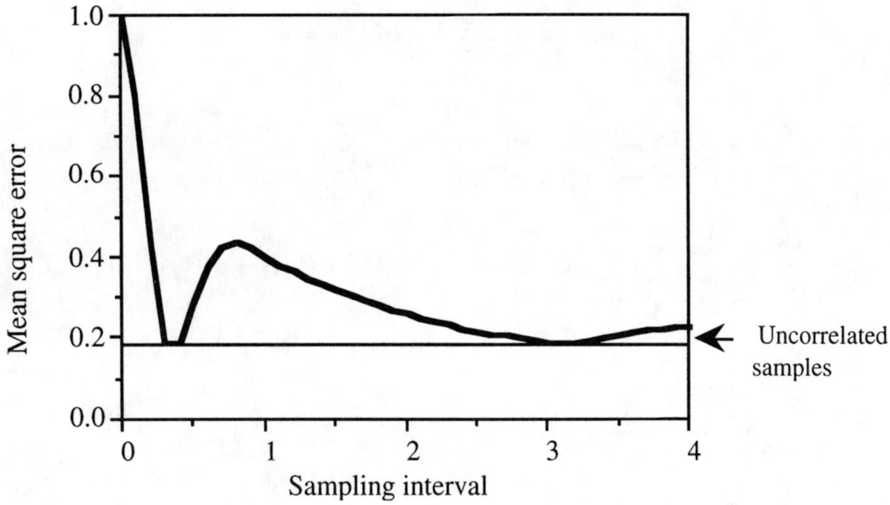

Figure 3.2: The mean square error in the estimation of variance given by (3.6) for a band-limited white noise process. The thin line refers to the completely uncorrelated samples. T=10.

C. Estimation of Covariance

We shall now study the properties of the estimate of covariance given by (3.1). The expected value may be shown to be

$$E\{\hat{C}_x(\tau)\} = C_x(\tau) - \frac{1}{T(T-\tau)}\sum_{t}^{T-1}\sum_{s=0}^{T-\tau-1} C_x(t-s-\tau)$$
$$- \frac{1}{T(T-\tau)}\sum_{t}^{T-1}\sum_{s=0}^{T-\tau-1} C_x(t-s) + \frac{1}{T^2}\sum_{t}^{T-1}\sum_{s=0}^{T-1} C_x(t-s) \qquad \tau \geq 0. \tag{3.9}$$

From (3.9) we get an estimate of the bias:

$$E\{\hat{C}_x(\tau)\} - C_x(\tau) = \frac{1}{T^2}\sum_{t}^{T-1}\sum_{s=0}^{T-1} C_x(t-s)$$
$$- \frac{1}{T(T-\tau)}\left[\sum_{t}^{T-1}\sum_{s=0}^{T-\tau-1} C_x(t-s-\tau) + \sum_{s}^{T-1}\sum_{t=0}^{T-\tau-1} C_x(t-s)\right].$$
$$\tau \geq 0 \tag{3.10}$$

For an uncorrelated process the bias turns out to be equal to $-\frac{\sigma_x^2}{T}$.

The expression for the variance of the estimate is far more complex, involving the expected value of the products of four random variables, which may be evaluated only when the stochastic process is Gaussian (see page.11). After some tedious algebra we obtain the following result:

$$E\{(\hat{C}_x(\tau) - C_x(\tau))^2\} =$$

$$\frac{4}{T^2(T-\tau)}\sum_{s}^{T-1}\sum_{s'}^{T-1}\sum_{t=0}^{T-1} C_x(s-t)C_x(s'-t-\tau)$$

$$-\frac{2}{T(T-\tau)^2}\sum_{s}^{T-1}\sum_{t'}^{T-1}\sum_{t=0}^{T-1}[C_x(s-t)C_x(t-t') + C_x(s-t')C_x(t-t'+\tau)$$

$$+ C_x(t-t')C_x(s-t'-\tau) + C_x(s-t-\tau)C_x(t-t'-\tau)]$$

$$+\frac{1}{(T-\tau)^2}\sum_{t}^{T-\tau-1}\sum_{t'=0}^{T-\tau-1}[C_x^2(t-t') + C_x(t-t'+\tau)C_x(t-t'-\tau)]$$

$$+[\frac{1}{T^2}\sum_{t}^{T-1}\sum_{t'=0}^{T-1} C_x(t-t')]^2. \qquad \tau \geq 0$$

(3.11)

For $\tau = 0$ (3.11) reduces to (3.6). The mean square error of the covariance function of a white noise process may be obtained from (3.11) by letting

$$C_x(\tau) = \sigma_x^2 \delta(\tau)$$

and dropping all terms of order $\frac{1}{T^2}$ and above.

$$E\{(\hat{C}_x(\tau) - C_x(\tau))^2\} \approx \frac{\sigma_\varepsilon^4}{T-|\tau|} \qquad |\tau| \neq 0 \quad \text{and} \quad |\tau| << T$$

$$\approx \frac{2\sigma_\varepsilon^4}{T} \qquad \tau = 0.$$

D. Covariance Matrix

Let x(t), t=0, 1, 2, ... $T-1$ be the time series from which we would like to estimate the covariance matrix. The time series is segmented into P segments where $P = \frac{T}{\Delta T}$. The segments are defined as follows:

$$\mathbf{x}_i = \text{col}\{x(i\Delta T) \;\; x(i\Delta T+1) \;\; x(i\Delta T+2) \;\; \ldots \;\; x(i\Delta T+\Delta T-1)\}.$$

$$i = 0, 1, \ldots P-1$$

An estimate of the covariance matrix is given by

$$\hat{\mathbf{C}}_X = \frac{1}{P}\sum_{i=0}^{P-1}\mathbf{x}_i\mathbf{x}_i^T. \tag{3.12}$$

As a new data segment arrives, the covariance matrix may be updated as follows:

$$\hat{\mathbf{C}}_x^{P+1} = \frac{P}{P+1}\hat{\mathbf{C}}_x^{P} + \frac{1}{P+1}\mathbf{x}_P\mathbf{x}_P^T. \tag{3.13}$$

Note that the covariance matrix as defined in (3.12) is always positive semidefinite. To show that, consider a quadratic:

$$\mathbf{A}^T\hat{\mathbf{C}}_x\mathbf{A} = \frac{1}{P}\sum_{i=0}^{P-1}\mathbf{A}^T\mathbf{x}_i\mathbf{x}_i^T\mathbf{A}$$

$$= \frac{1}{P}\sum_{i=0}^{P-1}\left|\mathbf{A}^T\mathbf{x}_i\right|^2 \geq 0.$$

Further, it will be positive definite only when $P \geq \Delta T$. As noted in Chapter One, for stationary time series the covariance matrix must always be Toeplitz. However, the estimate of the covariance matrix for finite P will rarely be Toeplitz. Assuming that the time series is stationary the Toeplitz character may be imposed on the covariance matrix. This is done by replacing the main diagonal and all subdiagonals by their respective averages. In doing so we will be averaging out the statistical errors arising on account of finite data.

E. **Structured Covariance Matrix**

The covariance matrix of a stationary stochastic process has a Toeplitz structure (see page 13). In addition, it may possess other types of structural features, for example, finite rank. A method for estimating a covariance matrix of specified structure that would maximize the probability of observing the data vector has been proposed by Burg.[8] In Chapter One we have noted the probability density function of a Gaussian random vector (1.11); for convenience we reproduce the same here:

$$f(\mathbf{x}) = \frac{1}{\sqrt{2^P\pi^P|\mathbf{Var}_x|}}\exp\left(-\frac{1}{2}\mathbf{x}^T\mathbf{Var}_x^{-1}\mathbf{x}\right).$$

Let the elements of random vector be derived from a time series; then $\mathbf{Var}_x$ is simply the covariance matrix, $\mathbf{C}_x$. Further, note that

$$\mathrm{tr}(\mathbf{AB}) = \mathrm{tr}(\mathbf{BA}),$$

where tr(.) stands for the trace of a matrix, that is, the sum of all elements on the main diagonal of a square matrix. We shall assume that both **AB** and **BA** are square matrices. Now, let $\mathbf{A}=\mathbf{x}^T$ and $\mathbf{B} = \mathbf{C}_x^{-1}\mathbf{x}$ and use the above matrix result to obtain

$$\mathbf{x}^T\mathbf{C}_x^{-1}\mathbf{x} = \mathrm{tr}(\mathbf{C}_x^{-1}\mathbf{x}\mathbf{x}^T). \tag{3.14}$$

Using (3.14) in (1.11) and remembering that the length of the vector is now ΔT, we obtain

$$f(\mathbf{x}) = \frac{1}{\sqrt{2^{\Delta T}\pi^{\Delta T}|\mathbf{C}_x|}}\exp\left(-\frac{1}{2}\mathrm{tr}(\mathbf{C}_x^{-1}\mathbf{x}\mathbf{x}^T)\right)$$

Given a set of P uncorrelated Gaussian vectors (real), their joint distribution function will be given by

$$f(\mathbf{x}_0, \mathbf{x}_1, \ldots \mathbf{x}_{P-1}) = \frac{1}{\sqrt{2^T \pi^T |\mathbf{C}_x|^P}} \exp\left(-\frac{P}{2} \operatorname{tr}\left(\mathbf{C}_x^{-1} \frac{1}{P} \sum_{m=0}^{P-1} \mathbf{x}_m \mathbf{x}_m^T\right)\right).$$

It may be noted from (3.12) that $\frac{1}{P}\sum_{p=0}^{P-1} \mathbf{x}_p \mathbf{x}_p^T$ is by definition an estimate of the covariance matrix. Replacing the sum by the estimate of the covariance matrix we obtain

$$f(\mathbf{x}_0, \mathbf{x}_1, \ldots \mathbf{x}_{P-1}) = \frac{1}{\sqrt{2^T \pi^T |\mathbf{C}_x|^P}} \exp\left(-\frac{P}{2} \operatorname{tr}(\mathbf{C}_x^{-1} \hat{\mathbf{C}}_x)\right). \tag{3.15}$$

Given a set of data vectors from which $\hat{\mathbf{C}}_x$ is computed, we choose a covariance matrix of a stationary time series satisfying all its usual properties listed in page 11 plus any other property imposed by the signal source and maximize the probability of observing the data vectors, that is, maximize (3.15). Such an estimate is known as *maximum likelihood estimate (MLE)* of the covariance matrix.

Burg[8] has shown that the maximum of (3.15) or of its logarithm is achieved under the condition that $\hat{\mathbf{C}}_x$ is positive definite when

$$\operatorname{tr}[(\mathbf{C}_x^{-1} \hat{\mathbf{C}}_x \mathbf{C}_x^{-1} - \mathbf{C}_x^{-1}) \delta \mathbf{C}_x] = 0, \tag{3.16}$$

where $\delta\mathbf{C}_x$ is any feasible variation of $\mathbf{C}_x$. It may be assumed that $\delta\mathbf{C}_x$ also satisfies all the properties of $\mathbf{C}_x$. We can the replace $\delta\mathbf{C}_x$ by $\mathbf{C}_x$ and obtain

$$\operatorname{tr}[\mathbf{C}_x^{-1} \hat{\mathbf{C}}_x] = \Delta T, \tag{3.17}$$

where ΔT stands for the size of the covariance matrix. That $\hat{\mathbf{C}}_x$ is positive definite implies that $P \geq \Delta T$. Using (3.17) in (3.15) we obtain the following for the probability density function:

$$f(\mathbf{x}_0, \mathbf{x}_1, \ldots \mathbf{x}_{P-1}) = \frac{1}{\sqrt{2^T \pi^T |\mathbf{C}_x|^P}} \exp\left(-\frac{T}{2}\right).$$

The principle of maximum likelihood estimate (MLE) may be restated as follows:

Minimize the determinant of the covariance matrix under the constraints that $\delta\mathbf{C}_x$ *satisfies all properties of a covariance matrix, and the trace of* $\mathbf{C}_x^{-1}\hat{\mathbf{C}}_x$ *equals* ΔT.

When there are no further constraints on $\mathbf{C}_X$, other than positive definiteness, the maximum likelihood solution of $\mathbf{C}_x$ is simply $\hat{\mathbf{C}}_x$. This follows from (3.16) where it is assumed that $\delta\mathbf{C}_x$ is arbitrary and hence,

$$(\mathbf{C}_x^{-1}\hat{\mathbf{C}}_x\mathbf{C}_x^{-1} - \mathbf{C}_x^{-1}) = 0.$$

Evidently this is not enough, as we know well that the covariance matrix of a stationary stochastic process has, in addition, Toeplitz symmetry. Furthermore, the eigenstructure of the covariance matrix may have a certain special form depending upon the model governing the time series.

F. MLE of Covariance Matrix

Burg[8] has proposed an algorithm known as Inverse Iteration Algorithm for minimizing the determinant of the covariance matrix under the constraint that (3.18) is satisfied. It is, however, not known whether the solution is unique. In References 9 and 10 it is shown that a positive definite solution exists with probability one if $P \geq \frac{\Delta T}{2}$ (for ΔT even). This condition can be further relaxed in some special cases, a few of which are outlined below:

(a) Let $\mathbf{C}_x = \sigma_x^2 \mathbf{I}$. From (3.17) the MLE of σ_x^2 is given by

$$\hat{\sigma}_x^2 = \frac{1}{T}\sum_{p=1}^{P}\sum_{t=1}^{\Delta T} x_p^2(t).$$

Note that $\hat{\sigma}_x^2 > 0$ for any value of P including P=1, that is, a single data vector.

(b) Another special case, where from a single data vector it is possible to obtain the MLE covariance matrix occurs when the true covariance matrix is a positive definite circulant matrix. If the stochastic process consists of random sinusoids whose frequencies are spaced at $\Delta\omega$ apart, the covariance matrix of size $\Delta T \times \Delta T$, where $\Delta T = \frac{2\pi}{\Delta\omega}$, will be truly circulant (see Example 1.2 for illustration). The circulant covariance matrix may be expressed as (see (1.22b))

$$\mathbf{C}_x = \sum_{r=-L}^{r=L} s_r \mathbf{a}(r)\mathbf{a}^H(r),$$

where

$$2L+1 = \Delta T$$

and

$$\mathbf{a}(r) = \frac{1}{\sqrt{\Delta T}}\text{col}\{1, e^{j\frac{2\pi}{\Delta T}r}, e^{j\frac{2\pi}{\Delta T}r2}, \ldots, e^{j\frac{2\pi}{\Delta T}r(\Delta T-1)}\},$$

which is also known as a DFT vector. The power in an rth sinusoid, s_r, is an unknown quantity. We shall now show how to get the MLE of s_r.

We shall apply the principle of MLE. For this, minimize the determinant of $\mathbf{C}_x$ $\left(= \prod_{r=-L}^{L} s_r\right)$ under the constraint given by (3.15). We minimize the following Lagrange expression:

$$\ln \left|\mathbf{C}_x\right| + \gamma[\operatorname{tr}\{\mathbf{C}_x^{-1}\hat{\mathbf{C}}_x\} - \Delta T] = \min, \tag{3.18}$$

where γ is a Lagrange parameter. The covariance matrix is estimated as given in (3.12) where we shall assume that different data vectors are obtained by segmenting the observed data. Using (3.12) and (3.14) in (3.18) we obtain

$$\ln \left|\mathbf{C}_x\right| + \gamma[\frac{1}{P}\sum_{p=0}^{P-1}\{\mathbf{x}_p^T\mathbf{C}_x^{-1}\mathbf{x}_p\} - \Delta T] = \min. \tag{3.19}$$

Now use (1.22b) to replace $\mathbf{C}_x$ in terms of unknowns, s_r, and then differentiate the expression with respect to s_r and set the derivative to zero. After some algebraic simplifications we obtain the following result[11]:

$$\begin{aligned} s_r &= \frac{1}{T}\sum_{p=0}^{P-1}\left|\mathbf{x}_p^T\mathbf{a}(r)\right|^2 \\ &= \frac{1}{T}\sum_{p=0}^{P-1}\left|\sum_{t=0}^{\Delta T-1} x_p(t)e^{-j\frac{2\pi}{T}rt}\right|^2 . \end{aligned} \tag{3.20}$$

This is indeed the popular spectrum estimation procedure known as the Welch[2] method, which will be described in the next section. To recapitulate, the MLE estimate of power given by (3.20) has been derived under the assumptions that random sinusoids occur at frequencies which are integer multiples of the fundamental frequency equal to $\frac{2\pi}{\Delta T}$ where ΔT is the size of the covariance matrix. Given a time series as a sum of a finite number of sinusoids, it is sometimes possible to choose a sampling interval and the size of the covariance matrix such that all sinusoids belong to a set of harmonics whose fundamental frequency is $\frac{1}{\Delta T}$.

Example 3.1: Let x(t) be a sum of five random real sinusoids, the frequencies being 9.0, 15.0, 27.0, 33.0, and 36.0 Hz. Observe that the lowest common factor (other than one) is 3 Hz. We shall sample the continuous process, x(t), at a rate of 96 samples per second. If we now select the covariance matrix size, $\Delta T = 32$, the fundamental frequency shall be 3 Hz. The MLE of the power at fundamental and its harmonics, which include the given frequencies, can be obtained from (3.20). The power at other harmonics is zero, though in practice some power may be present even at other harmonics (End of Example).

By imposing the periodic extension constraint, a nonsingular estimate of the covariance matrix is ensured.[10] Miller and Snyder[12] have proposed the Expectation and Maximization (EM) algorithm of Dempster, Laird, and Rubin[13] for ML estimation of the covariance matrix.

G. **Suboptimal Alternative**

The above-mentioned algorithms for ML estimation of the covariance matrix are computationally intensive. A simpler but sub-optimal alternative is to impose the Toeplitz property directly on $\hat{\mathbf{C}}_x$ by replacing the elements on the main diagonal and those on the subdiagonals by their respective averages. As shown in Reference 14 this is equivalent to finding a Toeplitz matrix closest to the estimated covariance matrix, closeness measured in terms of Hilbert-Schmidt norm (or Frobenius norm). Define a matrix $\bar{\mathbf{C}}_x$,

$$\bar{\mathbf{C}}_x = \sum_{i=1}^{T} \gamma_i \mathbf{Q}_i, \tag{3.21}$$

where γ_i , i=1, ..., T are unknown constants, and $\mathbf{Q}_i$ is a square $T \times T$ matrix with 1s in the ith upper and the ith lower subdiagonals and zeros elsewhere; for example, $\mathbf{Q}_2$ is defined as shown below:

$$\mathbf{Q}_2 = \begin{bmatrix} 0100 \\ 1010 \\ 0101 \\ 0010 \end{bmatrix},$$

and similarly are defined $\mathbf{Q}_3, \mathbf{Q}_4, \ldots$; however, $\mathbf{Q}_1 = \mathbf{I}$. We would like to minimize the difference between $\bar{\mathbf{C}}_x$ and the estimated $\hat{\mathbf{C}}_x$ in the sense of the Hilbert-Schmidt norm; that is,

$$\left\|\bar{\mathbf{C}}_x - \hat{\mathbf{C}}_x\right\|^2 = \mathrm{tr}\{(\bar{\mathbf{C}}_x - \hat{\mathbf{C}}_x)(\bar{\mathbf{C}}_x - \hat{\mathbf{C}}_x)^T\}. \tag{3.22}$$

Using (3.21) we substitute for $\bar{\mathbf{C}}_x$ in the above equation. After some simplification we obtain

$$\left\|\bar{\mathbf{C}}_x - \hat{\mathbf{C}}_x\right\|^2 = \gamma^T \mathbf{M} \gamma - 2\mathbf{b}^T \gamma + \mathrm{tr}\{\hat{\mathbf{C}}_x \hat{\mathbf{C}}_x^T\}, \tag{3.23}$$

where

$$\mathbf{b} = \mathrm{col}\{\mathrm{tr}\{\mathbf{Q}_1 \hat{\mathbf{C}}_x^T\}, \mathrm{tr}\{\mathbf{Q}_2 \hat{\mathbf{C}}_x^T\}, \ldots, \mathrm{tr}\{\mathbf{Q}_T \hat{\mathbf{C}}_x^T\},$$

$$\gamma = \mathrm{col}\{\gamma_1, \gamma_2, \ldots, \gamma_T\},$$

and

$$\mathbf{M} = \mathrm{diag}\{T, 2(T-1), 2(T-2), \ldots, 2\}.$$

Note that an element of vector **b** corresponds to the sum of elements on the diagonal or on one of the subdiagonals of $\hat{\mathbf{C}}_x$. On minimizing (3.23) with respect to γ we obtain

$$\gamma = \mathbf{M}^{-1}\mathbf{b}.$$

Theoretical	Raw	1st Iteration	5th Iteration	10th Iteration
499.33	541.93	481.82	481.69	481.59
464.87	408.03	442.72	444.65	444.74
337.12	327.85	327.16	327.79	327.94
302.66	296.19	295.09	291.33	291.20
1.0000	1.8615	17.25	3.94	3.34
1.0000	1.6406	10.47	3.57	3.32
1.0000	1.3185	4.72	3.33	3.32
1.0000	1.2242	2.47	3.31	3.31
1.0000	1.1581	2.38	3.31	3.31
0.99999	0.9075	1.87	3.31	3.31
0.99999	0.7504	1.23	3.31	3.31
0.99998	0.7003	1.22	3.31	3.31
0.99996	0.5922	1.07	3.29	3.31
0.99996	0.4580	0.95	3.28	3.31
0.99994	0.3781	0.16	3.10	3.30
0.99992	0.2675	-5.38	2.70	3.29

Table 3.1:. Eigenvalues of the covariance matrix whose diagonal elements are averaged and the noise eigenvalues are replaced by their average.

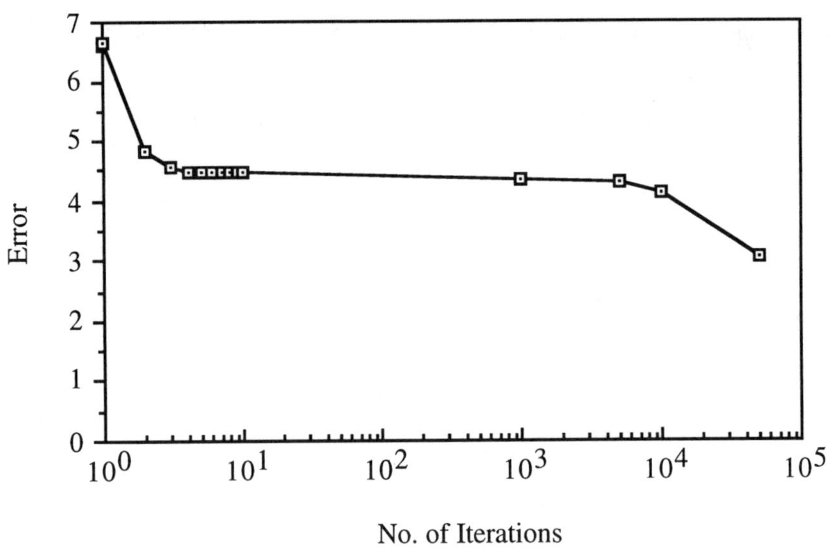

Figure 3.3: Error (Frobenius norm of the difference matrix) as a function of the number of iterations.

The minimum value of the norm is given by

$$\left\| \overline{\mathbf{C}}_x - \hat{\mathbf{C}}_x \right\|^2_{min} = \mathrm{tr}\{\hat{\mathbf{C}}_x \hat{\mathbf{C}}_x^T\} - \mathbf{b}^T \mathbf{M}^{-1} \mathbf{b}. \tag{3.24}$$

Additional constraints on the covariance matrix, arising when the time series is modeled as a sum of finite number of sinusoids and white noise or a sum of finite order AR process and white noise, can be imposed along with the above scheme of creating a Toeplitz matrix closest to the estimated matrix. Assume, for example, that the true covariance matrix is of rank p; then, the first p eigenvalues will be dominant, and the remaining $T-p$ eigenvalues will be very close to zero. However, the eigenvalues of the estimated Toeplitz matrix are more likely to be negative, thus violating the fundamental requirement that the covariance matrix be nonnegative definite. A way out is to replace all $T-p$ eigenvalues by a small positive number or by the average of $T-p$ eigenvalues when there is background white noise in the time series. This will, however, destroy the Toeplitz structure, which may be reimposed by resorting to diagonal averaging. The Toeplitz structure and the nonnegative definite character of a covariance matrix are alternatively imposed until a stable acceptable covariance matrix is found. However, it is not known whether such an iterative process will converge to a true covariance matrix.[14]

Example 3.2: A time series consisting of two equal powered (fifty units each) random sinuoids of 10 Hz and 20 Hz, sampled at 50 Hz, along with Gaussian white noise (unit variance and hence SNR=20 dB) was generated. Fifty independent data segments each of 16 samples were generated. The covariance matrix was estimated as per (3.13). Ideally, the eigenvalue structure of the covariance matrix consists of four nonequal eigenvalues (signal eigenvalues) and 12 equal eigenvalues (noise eigenvalues) as shown in the first column of Table 3.1. The suboptimal procedure for estimation of structured covariance matrix (SCM) as described on page 171 was used. The noise eigenvalues of the estimated covarinace matrix were replaced by their average. The results of successive application of the algorithm are shown in Table 3.1. The Schmidt-Hilbert norm (Frobenius norm) of the difference matrix between the structured covariance matrix and the theoretical matrix as a function of the number of iterations is shown in Figure 3.3. The error decreases in the begining quite rapidly, but after a few initial iterations the decrease slows down (End of Example).

3.3 Estimation of Spectrum and Cross-Spectrum

The classical approach to estimating spectrum and cross-spectrum (of a pair of time series) has been via Fourier transformation of autocorrelation and cross-correlation functions, respectively. This method is often called the indirect method. After the discovery of fast Fourier transform the direct method of first computing the magnitude square of the Fourier transform of the time series followed by averaging over different samples of the time series or over a band of frequncies has gained considerable popularity. There are different variations of the direct method. The spectral estimate is essentially a random variable as it is derived from a random process. It is, therefore, necessary to statistically characterize the spectral estimate, that is, obtain its mean, variance, and probability density. Analytical expressions for mean and variance are derived. But for the probability density function only an outline of the derivation is given.

A. Blackman-Tukey (BT)

The Fourier transform of the estimated covariance function straightaway gives an estimate of the spectrum, and similarly, the Fourier transform of the estimated cross-covariance function gives us an estimate of the cross-spectrum. Let $\hat{C}_x(\tau)$, $\tau = 0, \pm 1, \pm 2, \ldots, \pm \Delta T$ be the estimate of the covariance function and $\Delta T << T$ be the maximum lag. The estimated spectrum and cross-spectrum are given by

$$\hat{S}_x(\omega,\Delta T) = \sum_{\tau=-\Delta T}^{\tau=\Delta T} \hat{C}_x(\tau)\exp(-j\omega\tau),$$

$$\hat{S}_{xy}(\omega,\Delta T) = \sum_{\tau=-\Delta T}^{\tau=\Delta T} \hat{C}_{xy}(\tau)\exp(-j\omega\tau). \tag{3.25}$$

The dependence on the maximum lag for which the covariance function is available is explicitly shown. We can relate $\hat{S}_x(\omega,\Delta T)$ to $\hat{S}_x(\omega)$, that is, when the covariance function is available over infinte lags. For this, define a window function, $w_1(t)$, smoothly decreasing to zero for lags greater than ΔT. We shall now rewrite (3.25):

$$\begin{aligned}\hat{S}_x(\omega,\Delta T) &= \sum_{\tau=-\infty}^{\tau=\infty} w_1(\tau)\hat{C}_x(\tau)\exp(-j\omega\tau)\\ &= W_1(\omega) * \hat{S}_x(\omega),\\ \hat{S}_{xy}(\omega,\Delta T) &= \sum_{\tau=-\infty}^{\tau=\infty} w_1(\tau)\hat{C}_{xy}(\tau)\exp(-j\omega\tau)\\ &= W_1(\omega) * \hat{S}_{xy}(\omega),\end{aligned} \tag{3.26}$$

where $w_1(t)$ is the Fourier transform of the window. The window function has another important role; namely, it underweighs those covariance estimates at large lags where the estimation error is greater. The window function is a symmetric function with $w_1(0) = 1.0$ so that the total power practically remains unaltered. To show this we shall assume that the spectrum is a smooth function, practically constant over the range of rapid variation in $W_1(\omega)$. Then, we may approximate (3.26) as

$$\hat{S}_x(\omega,\Delta T) = W_1(\omega) * \hat{S}_x(\omega) \approx \hat{S}_x(\omega)\ \frac{1}{2\pi}\int_{-\pi}^{\pi} W_1(\omega)\,d\omega = \hat{S}_x(\omega),$$

where we have set,

$$\frac{1}{2\pi}\int_{-\pi}^{\pi} W_1(\omega)\,d\omega = 1\ ,$$

since $w_1(0) = 1.0$.

The spectral estimate in Blackman-Tukey (BT) can become negative on account of the Fourier transform of the window function, $W_1(\omega)$, not being nonnegative everywhere. The popular Hanning and Hamming windows suffer from this problem. The window function must then be slightly altered so that its Fourier transform becomes positive everywhere. A recursive procedure exists to generate such a closest positive sequence.[15] The window sequence is enlarged by inserting zeros in the middle without altering its symmetry,

$$[\Omega(t),\ t = 0,1,\ldots,N-1] =$$
$$[w(0),\ w(1),\ w(2),\ldots,w(\tfrac{\Delta T-1}{2}),\ 0,\ \ldots,0,\ w(\tfrac{\Delta T-1}{2}),\ \ldots,\ w(2),\ w(1)],$$

τ	Hanning	Closest positive Sequence
0	1.0	1.0329
1	0.9755	0.9938
2	0.9045	0.9016
3	0.7939	0.7760
4	0.6545	0.6317
5	0.5	0.4826
6	0.3455	0.3409
7	0.2061	0.2166
8	0.0955	0.1173
9	0.0245	0.0474
10	0.0	0.0085

Table 3.2: Closest positive sequence.

where $N >> \Delta T$ and ΔT is assumed to be odd. The iterative process consists of two steps:

(1) Compute DFT of $\Omega(t)$. The negative DFT coefficients, if any, are set to zero. If all DFT coefficients are positive the recursion is stopped. Compute inverse DFT of the modified sequence.

(2) In the window sequence thus obtained replace the central $N - \Delta T$ numbers with zeros and go back to step one. As an example, the approximate closest positive sequence to the Hanning window after ten iterations is shown in Table 3.2 (from Reference 15).

There is an alternate method of computing the spectrum from the covariance matrix, which is obtained from the data (see Section.3.2).

$$\hat{S}_x(\omega) = \mathbf{a}^H \hat{\mathbf{C}}_x \mathbf{a}, \tag{3.27}$$

where $\mathbf{a}$ is the frequency vector defined in (1.22a). The two estimates of the spectrum turn out to be same when the window is given by

$$w_1(\tau) = 1 - \frac{|\tau|}{\Delta T}.$$

In other words, there is an implicit window of the above type (often known as a triangular window or Bartlett window) in the spectral estimate given by (3.27). This can be verified by expanding the matrix product in (3.27) (see Exercise 3.1). To compute the power from the power spectral density given by (3.25) or (3.27) we need to divide the right-hand-side by ΔT.

B. Mean and Variance of BT Spectral Estimate

The BT spectral estimate given by (3.26) may be expressed in a different form as

$$\hat{S}_x(\omega) = \sum_{\tau=-T+1}^{\tau=T-1} w_1(\tau)\left(1-\frac{|\tau|}{T}\right)\hat{C}_x(\tau)\exp(-j\omega\tau)$$

$$= \sum_{\tau=-T+1}^{\tau=T-1} w_1(\tau)\left(1-\frac{|\tau|}{T}\right)\frac{1}{T-|\tau|}\sum_{t=0}^{T-|\tau|-1} x(t)x(t+\tau)\exp(-j\omega\tau), \tag{3.28}$$

where we have introduced an additional window, $(1-\frac{|\tau|}{T})$. Observe that the covariance function is estimated to a maximum possible lag of $T-1$. Equation (3.28) may be written as a convolution product in the frequency domain:

$$\hat{S}_x(\omega) = \frac{1}{T}\sum_{k=0}^{T-1} W_1\left(\frac{2\pi k}{T}\right)\left|X\left(\omega-\frac{2\pi k}{T}\right)\right|^2. \tag{3.29}$$

We shall now use (3.29) for evaluation of mean and variance of the estimate.

$$E\{\hat{S}_X(\omega)\} = \sum_{k=0}^{T-1} W_1\left(\frac{2\pi k}{T}\right)E\left\{\frac{1}{T}\left|X\left(\omega-\frac{2\pi k}{T}\right)\right|^2\right\}. \tag{3.30}$$

From the properties of the DFT coefficients of a finite stationary stochastic process (2.23) we note that

$$E\left\{\frac{1}{T}\left|X\left(\omega-\frac{2\pi k}{T}\right)\right|^2\right\} = S_x\left(\omega-\frac{2\pi k}{T}\right),$$

and hence (3.30) reduces to

$$E\{\hat{S}_x(\omega)\} = \sum_{k=0}^{T-1} W_1\left(\frac{2\pi k}{T}\right)S_x\left(\omega-\frac{2\pi k}{T}\right). \tag{3.31}$$

The BT spectrum will be unbiased only when the actual spectrum is a smoothly varying function and

$$\sum_{k=0}^{T-1} W_1\left(\frac{2\pi k}{T}\right) = Tw_1(0) = 1$$

or

$$w_1(0) = \frac{1}{T}.$$

The variance may be expressed as

$$\mathrm{Var}\{\hat{S}_x(\omega)\} = E\{\hat{S}_x(\omega)^2\} - [E\{\hat{S}_x(\omega)\}]^2,$$

where we need to evaluate $E\{\hat{S}_x(\omega)^2\}$. From (3.29) we get

$$E\{\hat{S}_x(\omega)^2\} = \sum_{k'=0}^{T-1}\sum_{k''=0}^{T-1} W_1(\tfrac{2\pi k'}{T})W_1(\tfrac{2\pi k''}{T})E\{\frac{1}{T^2}\left|X(\omega - \tfrac{2\pi k'}{T})\right|^2\left|X(\omega - \tfrac{2\pi k''}{T})\right|^2\}. \tag{3.32}$$

Here we shall assume that x(t) is a Gaussian process so that we can invoke (1.18) for the expected value of the product of four Gaussian random variables. We obtain

$$\begin{aligned} E\{\frac{1}{T^2}\left|X(\omega - \tfrac{2\pi k'}{T})\right|^2\left|X(\omega - \tfrac{2\pi k''}{T})\right|^2\} &= S_x(\omega - \tfrac{2\pi k'}{T})S_x(\omega - \tfrac{2\pi k''}{T}) \\ &+ S_x(\omega - \tfrac{2\pi k'}{T})S_x(\omega - \tfrac{2\pi k''}{T})\delta_{k'k''}. \end{aligned} \tag{3.33}$$

Using (3.33) in (3.32) and the final expression for variance we obtain

$$\mathrm{Var}\{\hat{S}_x(\omega)\} = \sum_{k=0}^{T-1} W_1^2(\tfrac{2\pi k}{T})S_x^2(\omega - \tfrac{2\pi k}{T}). \tag{3.34a}$$

Assuming that the spectrum is a smooth function such that it is almost a constant over the mainlobe of the window spectrum, we can express the variance as follows:

$$\begin{aligned} \mathrm{Var}\{\hat{S}_x(\omega)\} &\approx S_x^2(\omega)\sum_{k=0}^{T-1} W_1^2(\tfrac{2\pi k}{T}), \\ &= S_x^2(\omega)T\sum_{t=0}^{T-1} w_1^2(t). \end{aligned} \tag{3.34b}$$

Let

$$\begin{aligned} w_1(\tau) &= \frac{1}{T} \quad \tau = 0,1,2, \ldots \Delta T - 1, \\ &= 0 \quad \text{otherwise} \quad , \end{aligned}$$

where ΔT is the effective width of the window; then (3.34b) reduces to

$$\mathrm{Var}\{\hat{S}_x(\omega)\} \approx \frac{\Delta T}{T} S_x^2(\omega),$$

from which it follows that we must have $\Delta T << T$ so that the variance of the estimate is small compared to the spectrum. In general, however, when the condition of smoothness is not satisfied, the spectrum of the window plays a significant role in determining the variance; in particular, in the neighborhood of a large spectral peak the variance is likely to be affected by the sidelobes of the window spectrum.

C. Welch Method

When the length of the time series is large, it is computationally efficient to segment the time series and compute the square of the discrete Fourier transform coefficients and finally average them over all segments. The method of segmentation, first suggested by Welch,[2] exploits the basic statistical properties of DFT coefficients (see page 81),

$$S_x(k) = \frac{1}{\Delta T} E\{X(k)X^*(k)\}$$
$$S_{xy}(k) = \frac{1}{\Delta T} E\{X(k)Y^*(k)\}.$$
$$\Delta T \to \infty \tag{3.35}$$

There are basically two difficulties in the implementation of the procedure indicated in (3.35). First, given just a single time series, how does one carry out the expected operation, and second, since all practical data records are finite, how does one carry out the limiting operation, i.e., $\Delta T \to \infty$? By segmenting a time series into, say, P uncorrelated (independent when the process is Gaussian) segments, the expected operation is carried out by taking an average over all uncorrelated segments. Alternatively, since the DFT coefficients are uncorrelated, the expected operation may be replaced by an arithmetic average over a narrow band of frequencies. We shall consider both alternatives.

D. Segmentation

Let us segment a time series x(t), t=0, 1, 2, ... T − 1 into P equal length partially overlapping segments, (Figure 3.4).

$$x_i(t) = x(i(1-\alpha)\Delta T + t), \quad i = 0,1,2,\ldots, P-1$$
$$t = 0, 1, 2, \ldots, \Delta T - 1, \tag{3.36}$$

where a is the fraction of overlap and ΔT is the length of each segment.

Note that the number of segments, $P = \dfrac{T}{(1-\alpha)\Delta T}$, increases as the overlap increases; consequently, more computational effort is required for the same data length. Each segment is multiplied by a window function, and its DFT is computed. The direct estimation of spectrum involves evaluation of the following:

$$\hat{S}_x(k) = \frac{1}{P\Delta T} \sum_{i=0}^{P-1} X_i(k) X_i^*(k),$$
$$\hat{S}_{xy}(k) = \frac{1}{P\Delta T} \sum_{i=0}^{P-1} X_i(k) Y_i^*(k). \tag{3.37}$$

It was assumed that the segments are uncorrelated. This would require that the segment length, ΔT, be neither too small nor the overlap be too large. We shall provide a theoretical basis for selecting ΔT and α. Let us compute the correlation between two segments:

$$C_{m,n}(t,\tau) = E\{x_m(t)w_0(t)x_n(t+\tau)w_0(t+\tau)\}$$
$$= w_0(t)w_0(t+\tau)E\{x(t+m(1-\alpha)\Delta T)x(t+\tau+n(1-\alpha)\Delta T)\}$$
$$= w_0(t)w_0(t+\tau)C_x((m-n)(1-\alpha)\Delta T + \tau).$$

The average (time average) correlation is given by

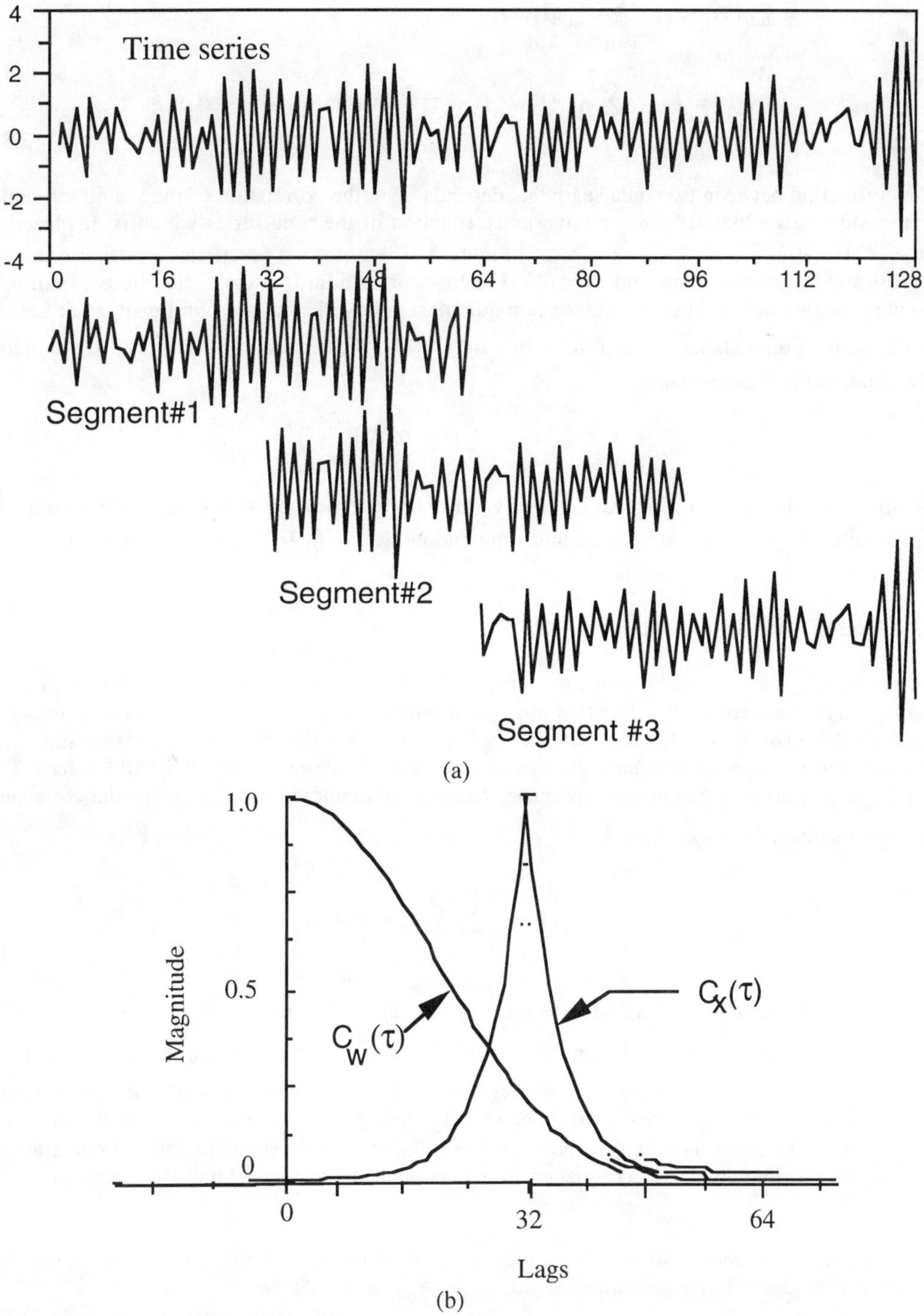

Figure 3.4: (a) Segmentation of a long time series (1st order AR process a_1=0.8) with 50% overlap. (b) The cross-correlation between two adjacent segments is equal to the product of $C_W(\tau)$ and $C_x(\tau)$.

$$C_{m,n}(\tau) = \frac{1}{\Delta T}\sum_{t=0}^{\Delta T-\tau} C_{m,n}(t,\tau)$$

$$= \{\frac{1}{\Delta T}\sum_{t=0}^{\Delta T-\tau} w_0(t)w_0(t+\tau)\}\, C_x((m-n)(1-\alpha)\,\Delta T+\tau). \tag{3.38}$$

The correlation between two data segments depends upon the correlation of the window function (expression inside braces) and the covariance function of the time series which is displaced by $(m-n)(1-\alpha)\Delta T$. We have sketched in Figure 3.4(b) these two functions for a first-order AR process and Hanning window and α=50%. For low correlation, it is clear that the correlation of the window function and the covariance function must not overlap to any significant level. Let τ_0 be the correlation distance of the time series and τ_w be the correlation distance of the window. The condition for nonoverlap is

$$\tau_0 + \tau_w \le (1-\alpha)\Delta T. \tag{3.39}$$

For the example shown in Figure 3.4 nonoverlap is possible when $\alpha \le 25\%$. Further, only for white noise ($\tau_0 = 0$), one may use a uniform window ($\tau_w = 0.9\Delta T$) with zero overlap.

E. **Circular Autocorrelation**

On performing the inverse Fourier transform of the discrete spectrum obtained from (3.37), we shall get a discrete covariance function but with a difference. It is a periodic function with period equal to ΔT. The reason for this periodicity lies in the fact that the DFT operation implicitly assumes that the time series is periodic (see Chapter Two for properties of DFT and Exercise 3.2). Indeed we can express the circular covariance function as an aliased form of the regular covariance function estimate (3.1) scaled by $(1-\frac{|\tau|}{\Delta T})$,

$$C_{cir}(\tau) = \sum_{i=-\infty}^{\infty} \hat{C}_x(\tau + i\Delta T). \tag{3.40}$$

The nature of aliasing is similar to the aliasing arising in the discrete spectrum (see (2.32)). However, since the time series is limited to 0 to $\Delta T - 1$, $\hat{C}_x(\tau)$ is confined to a lag interval of $\pm\Delta T$. The aliasing can be avoided if each segment of the time series is padded with as many zeros as its length. By doing so we would indeed be increasing the period of the circular covariance function to $2\Delta T$; then there will be no aliasing. In the example given below we demonstrate the role of zero padding on the autocorrelation function obtained via inverse DFT of the spectrum.

Example 3.3: To demonstrate the aliasing error in the estimation of covariance (autocorrelation) function, consider a simple random sinusoid,

$$x(t) = \cos(\omega t + \phi), \quad t = 0,1, \ldots \Delta T - 1,$$

where $\omega = 0.2\pi$, ϕ is the random phase, and ΔT=64. The data segment is padded with zeros, 64 and 32, and then its DFT is computed. The inverse DFT of the Fourier coefficient magnitude square is shown below (see Figure 3.5). There is no aliasing when the number of zeros padded is equal to the segment length. The role of the implicit triangular window is also seen.

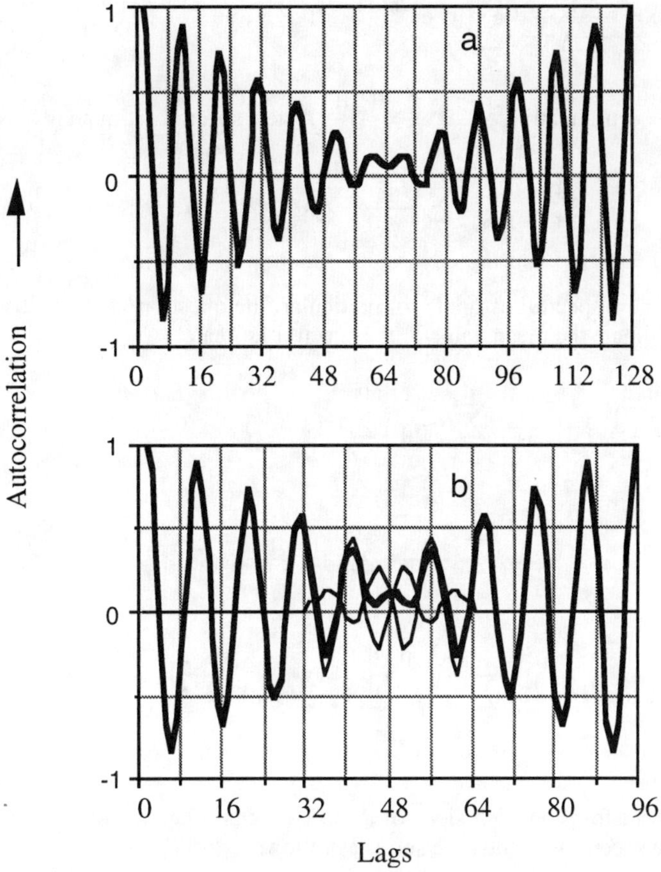

Figure 3.5: Aliasing arises in the estimation of covariance function. (a) 64 zeros were padded, doubling the length of each segment. There is no aliasing. (b) 32 zeros were padded. Aliasing error is now seen in the central part, on account of partial overlapping with the i = – 1 term in (3.40).

F. **Overlap and Add**

Consider the following generalization of the Welch method of spectrum estimation. A long time series is segmented as described earlier:

$$x_i(t) = x(i(1-\alpha)\Delta T + t), \quad i = 0, 1, 2, \ldots, P-1$$
$$t = 0, 1, 2, \ldots, \Delta T - 1 .$$

Each segment is first multiplied by a window, and then its DFT is evaluated. Next, we define the following vector:

$$\mathbf{X} = \text{col}\,\{X_0(k),\ X_1(k),\ X_2(k),\ \ldots,\ X_{P-1}(k)\}$$

and a matrix

$$\mathbf{D} = \frac{1}{\sqrt{\Delta T}} \mathrm{diag}\{1, e^{j2\pi k(1-\alpha)}, e^{j4\pi k(1-\alpha)} \ldots e^{j(P-1)2\pi k(1-\alpha)}\},$$

and form a new spectral matrix, $\mathbf{S} = \mathbf{D}\mathbf{X}\mathbf{X}^H\mathbf{D}^H$. A new spectral estimate is introduced:

$$\hat{S}_x = \frac{1}{2r+1} \sum_{i=-r}^{r} \frac{1}{P-|i|} \sum_{l=0}^{P-|i|-1} s_{l,l+i} \qquad 0 \le r \le P-1. \tag{3.41}$$

Let us study the new spectral estimate, in particular, the question of how close it is to the actual spectrum. We compute the mean value of the spectral estimate:

$$E\{\hat{S}_x\} = \frac{1}{2r+1} \sum_{i=-r}^{r} \frac{1}{P-|i|} \sum_{l=0}^{P-|i|-1} E\{s_{l,l+i}\} \qquad 0 \le r \le P-1$$

$$= \frac{1}{2r+1} \sum_{i=-r}^{r} \frac{1}{P-|i|} \sum_{l=0}^{P-|i|-1} E\{\frac{1}{\Delta T} X_l X^*_{l+i}\}\, e^{(\frac{2\pi}{\Delta T}k(1-\alpha)\Delta Ti)}$$

$$= \frac{1}{2r+1} \sum_{i=-r}^{r} \frac{1}{P-|i|} \sum_{l=0}^{P-|i|-1} S_{x_l x_{l+i}}(k)\, e^{(\frac{2\pi}{\Delta T}k(1-\alpha)\Delta Ti)}. \tag{3.42}$$

Taking Fourier transform on both sides of (3.38), we can relate the cross-spectrum between two segments and the spectrum of the time series as follows:

$$S_{x_l x_{l+i}}(k) = \frac{1}{2\pi} \int_{-\pi}^{\pi} S_x(\omega) e^{-j\omega(1-\alpha)\Delta Ti} \left| W_0(\frac{2\pi k}{\Delta T} - \omega) \right|^2 d\omega. \tag{3.43}$$

Using the result (3.43) in (3.42) we obtain

$$E\{\hat{S}_x(k)\} = \frac{1}{2\pi} \int_{-\pi}^{\pi} S_x(\omega) \left| W_0(\frac{2\pi k}{\Delta T} - \omega) \right|^2 \frac{1}{2r+1} \sum_{i=-r}^{r} e^{-j(\omega - \frac{2\pi}{\Delta T}k)(1-\alpha)\Delta Ti}\, d\omega. \tag{3.44}$$

Define an effective window by combining the last two terms of the integrand in (3.44):

$$w_{eff}(\tau) = \frac{1}{2r+1} \sum_{i=-r}^{r} w_1(\tau + i(1-\alpha)\Delta T),$$

or

$$w_{eff}(\tau) = \frac{1}{2r+1} \sum_{i=-r}^{r} w_1(\tau + i(1-\alpha)\Delta T), \tag{3.45}$$

where

$$w_1(\tau) = \sum_{t=0}^{\Delta T-\tau} w_0(t) w_0(t+\tau).$$

Given $w_0(t)$ it is possible to arrive at an optimum overlap which will result in an effective window having a flat top and sloping edges. The sidelobes of $W_{eff}(\omega)$ are largely controlled by those of $\left|W_0(\omega)\right|^2$, but its mainlobe is controlled by that of $\dfrac{\sin(\omega(1-\alpha)\frac{\Delta T}{2}(2r+1))}{(2r+1)\sin(\omega(1-\alpha)\frac{\Delta T}{2})}$, whose mainlobe width is on the order of $\dfrac{2}{(1-\alpha)\,\Delta T(2r+1)}$.

Example 3.4: First let us consider a uniform window,

$$\begin{aligned} w_0(t) &= 1 & & 0 \le t \le \Delta T - 1 \\ &= 0 & & \text{otherwise} \end{aligned},$$

whose autocorrelation function is

$$\begin{aligned} w_1(\tau) &= |\Delta T| - \tau & & 0 \le \tau \le \Delta T - 1 \\ &= 0 & & \text{otherwise} \end{aligned}.$$

Using the above function in (3.45) it is easy to show that the effective window will be flattopped if $\alpha = 0$ but for any $r\ (\neq 0)$. Next we shall consider the Hanning window whose autocorrelation is shown in Figure 3.6a. The effective window for 50% overlap and r=2 is shown in Figure 3.6b (End of Example).

The fundamental difference between the Welch method and the overlap and add method is that in the former the segments are assumed to be uncorrelated, but in the latter the segments may be correlated. The idea was first introduced by Rabiner and Allen[5] and applied to spectrum estimation by Mathew and Youn.[6,16]

Example 3.5: In this example we shall demonstrate the relative superiority of the overlap and add method over simple segment averaging as in the Welch method. The time series consisted of two equal power random sinusoids with white background noise whose power is one tenth of the combined signal power. The normalized frequencies of the sinusoids are 0.125 and 0.15625 Hz. A time series of length 1024 points was segmented into 32 overlapping segments each of length 64 points (overlap=50%). The Hanning window was applied to each segment. The estimated spectrum via the overlap and add method is shown in Figure 3.7a for two different values of r (defined in 3.41), namely, two and eight. The spectrum obtained via the Welch method is shown in Figure 3.7b. The overlap and add method certainly shows better resolution without any bias. Furthermore, the depth of the valley separating the two peaks increases for higher r. The sinusoids, however, remain unresolved in the Welch method (End of Example).

G. Frequency Averaging

The entire time series is considered as a single long segment. It is first windowed and then padded with zeros to double its length, and finally it is Fourier transformed. A running average, with or without overlap, of the squared Fourier coefficients (periodogram) is carried out. The method of

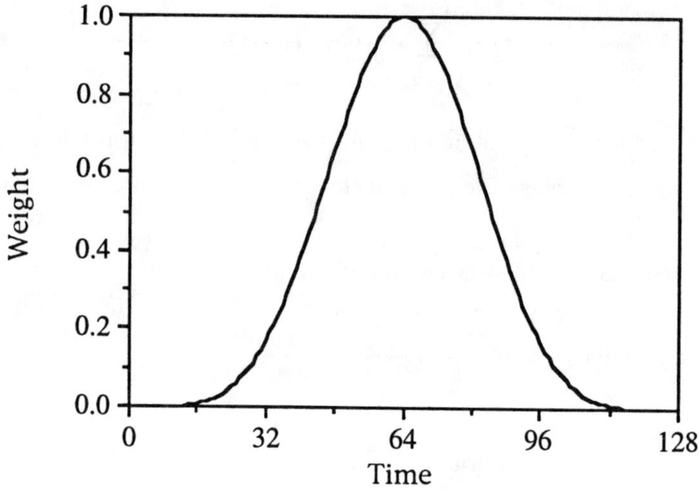

(a)

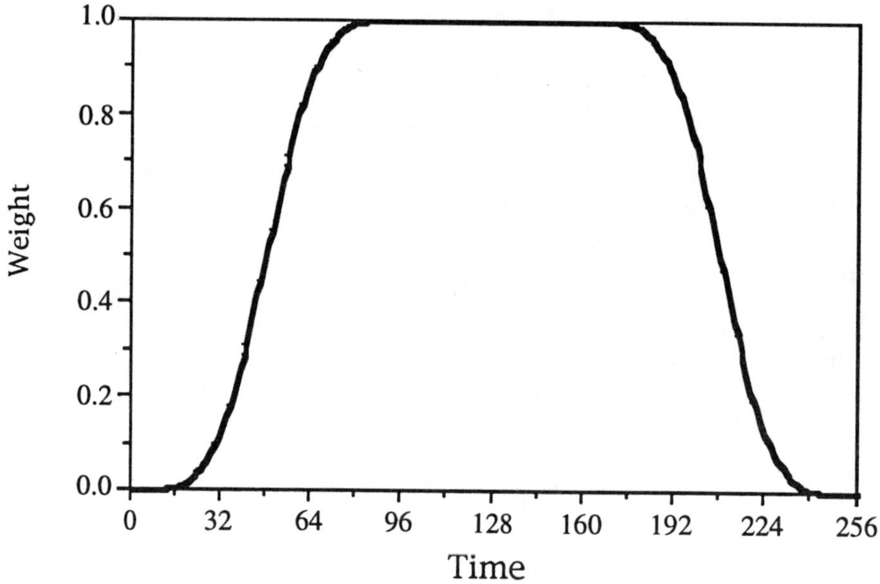

(b)

Figure 3.6: (a) Autocorrelation of the Hanning window. (b) The effective window obtained by the overlap and add process described in (3.45).

frequency averaging may be considered as a complement to the Welch method but in the frequency domain. The estimate of the spectrum is given as follows:

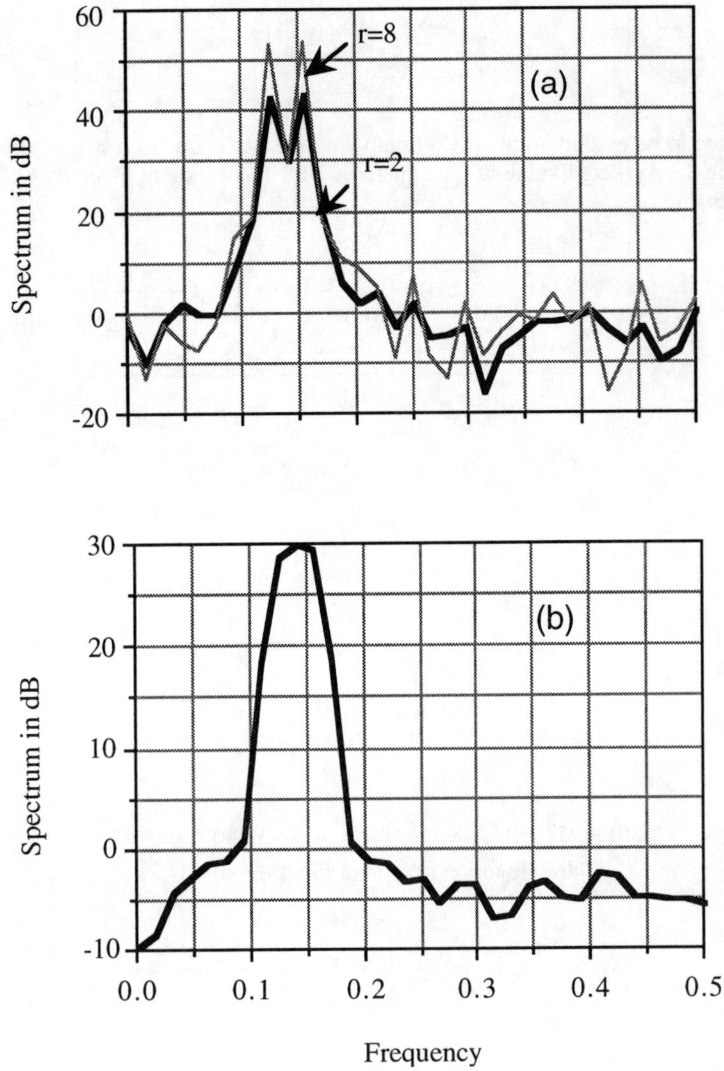

Figure 3.7: Spectrum obtained from (a) overlap and add method and (b) Welch method. The sinusoids are well resolved by the overlap and add method but not by the Welch method.

$$\hat{S}_x(k) = \frac{1}{TP} \sum_{i=-\frac{P-1}{2}}^{i=\frac{P-1}{2}} W_1(i)\left|X(k+i)\right|^2, \tag{3.46}$$

where $W_1(i)$, $|i| = 0, 1, \ldots \frac{P-1}{2}$ are weighting coefficients such that

$$\frac{1}{P}\sum_{i=-\frac{(P-1)}{2}}^{\frac{(P-1)}{2}} W_1(i) = 1,$$

and P is assumed to be an odd number. Compute the expected value of the quantities on both sides of (3.46). Using the expected value of the square of a DFT coefficient given by (2.34) for uniform window, we obtain

$$E\{\hat{S}_x(k)\} = \frac{1}{2\pi P}\sum_{i=-\frac{(P-1)}{2}}^{\frac{(P-1)}{2}} W_1(i)\int_{-\pi}^{\pi} S_x(\omega)\frac{\left|W_0(\frac{2\pi(k+i)}{T}-\omega)\right|^2}{T}d\omega$$

$$= FT\{C_x(\tau)w_1(\tau)\frac{1}{T}\sum_{t=1}^{T-\tau} w_0(t)w_0(t+\tau)\},$$

(3.47)

where

$$w_1(\tau) = \frac{2}{P}\sum_{i=0}^{\frac{(P-1)}{2}} W_1(i)\cos(\frac{2\pi i}{P}\tau),$$

Example 3.6: Let $S_x(\omega) = \sigma_x^2$ and assume that T is very large so that we can approximate the Fourier transform of the window function as a delta function; that is,

$$\left|W_0(\frac{2\pi(k+i)}{T} - \omega)\right|^2 \approx T\,\delta(\frac{2\pi(k+i)}{T} - \omega).$$

Then,

$$E\{\hat{S}_x(\frac{2\pi k}{T})\} = \frac{\sigma_x^2}{P}\sum_{i=-\frac{(p-1)}{2}}^{i=\frac{(p-1)}{2}} W_1(i) = \sigma_x^2,$$

if we choose the spectral weighting coefficients such that the average of the coefficients is equal to unity (End of Example).

The role of the spectral weighting coefficients may be examined in the lag domain. The covariance function is multiplied by a new window function, $w_1(\tau)$, which is the Fourier transform of the spectral averaging coefficients (see (3.47)). In fact there are two different windows, namely, $w_0(t)$ applied to the time series and $w_1(\tau)$ implicitly applied to the covariance function. The window applied to the time series is known as linear window (also known as data window) and that applied to the covariance function as quadratic window[17] (also known as lag window). We may combine the two windows into a single composite window,

$$w_{Comp}(\tau) = w_1(\tau)\frac{1}{T}\sum_{t=1}^{T-\tau} w_0(t) w_0(t+\tau). \tag{3.48}$$

Using (3.48) in (3.47) we obtain a simplified expression,

$$E\{\hat{S}_x(\frac{2\pi k}{T})\} = \frac{1}{2\pi}\int_{-\pi}^{\pi} S_x(\omega) W_{Comp}(\frac{2\pi k}{T} - \omega)\, d\omega. \tag{3.49}$$

The composite window has better sidelobe characteristics, for we are free to choose $w_1(\tau)$ so as to reduce the bad sidelobe structure of $w_0(t)$.[18]

Example 3.7: To demonstrate the sidelobe structure of a composite window consider the following combination:

Linear window: Uniform window
Quadratic window: Hanning window

The Fourier transform of the composite window is given by

$$W_{comp}(\omega) = \frac{1}{4}\left[\frac{\sin(\omega - \frac{2\pi}{T})\frac{T}{2}}{T\sin(\omega - \frac{2\pi}{T})\frac{1}{2}}\right]^2 + \frac{1}{2}\left[\frac{\sin\omega\frac{T}{2}}{T\sin\omega\frac{1}{2}}\right]^2 + \frac{1}{4}\left[\frac{\sin(\omega + \frac{2\pi}{T})\frac{T}{2}}{T\sin(\omega + \frac{2\pi}{T})\frac{1}{2}}\right]^2.$$

A plot of this function is shown in Figure 3.8 (End of Example).

The two methods of direct spectrum estimation, namely, the Welch method and its counterpart in the frequency domain, produce equivalent results insofar as their first-order properties is concerned; however, they are different in terms of their second-order properties. From the point of computational speed, the method of segmentation is more efficient; consequently, it has a possibility of real time implementation (with lag of one segment). In addition, the method of segmentation allows one to check for possible nonstationarity. The method of frequency averaging has a unique possibility that the resolution can be varied as a function of frequency, depending upon the spectral features. Moreover, it is possible to adaptively alter the quadratic window.

H. Mean and Variance of Spectral Estimate

We shall first obtain the mean of the spectral estimate. We shall give the detailed derivation for the Welch spectral estimate, but only the final results for the method of frequency averaging. The DFT of the ith segment may be expressed following (2.24) as follows:

$$X_i(k) = [dX(\omega) e^{j(1-\alpha)i\Delta T\omega}] * W_0(\omega). \tag{3.50}$$

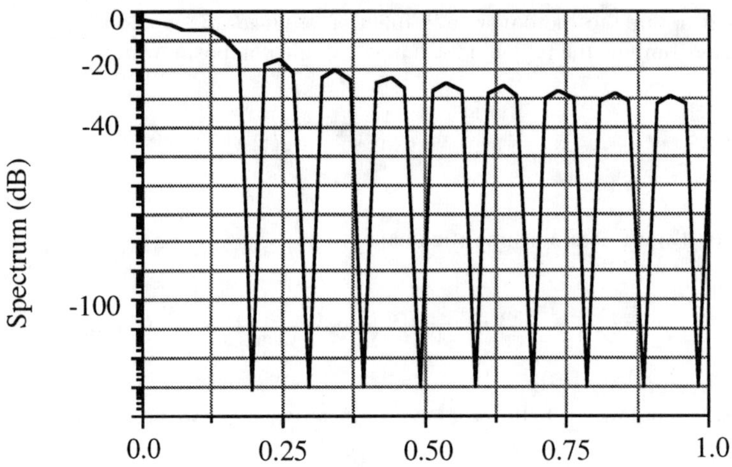

Figure 3.8: Spectrum of the composit window in frequency averaging method.

Using the above representation in (3.37) and computing the expected value, we obtain

$$E\{\hat{S}_x(\frac{2\pi k}{\Delta T})\} = \frac{1}{P\Delta T}E\{\sum_{i=0}^{P-1}\left|[dX(\omega)e^{j(1-\alpha)i\Delta T\omega}]*W_0(\omega)\right|^2\}$$

$$= \frac{1}{2\pi}\int_{-\pi}^{\pi} S_x(\frac{2\pi k}{\Delta T} - \omega)\frac{W_0^2(\omega)}{\Delta T}d\omega \quad , \qquad (3.51a)$$

and similarly for the cross-spectrum,

$$E\{\hat{S}_{xy}(\frac{2\pi k}{\Delta T})\} = \frac{1}{2\pi}\int_{-\pi}^{\pi} S_{xy}(\frac{2\pi k}{\Delta T} - \omega)\frac{W_0^2(\omega)}{\Delta T}d\omega. \qquad (3.51b)$$

The spectral estimate is unbiased only when the actual spectrum is a smoothly varying function (see Exercise 3.4 where the bias is related to the even-order derivatives of the actual spectrum) and

$$\frac{1}{2\pi}\int_{-\pi}^{\pi}\frac{W_0^2(\omega)}{\Delta T}d\omega = 1$$

or

$$\frac{1}{\Delta T}\sum_{t=0}^{\Delta T-1} w_0^2(t) = 1.0.$$

Next we shall compute the variance of the spectral estimate.

$$E\{(\hat{S}_x(k) - E\{\hat{S}_x(k)\})^2\} = E\{[\frac{1}{p\Delta T}\sum_{i=0}^{p-1}|X_i(k)|^2 - E\{\hat{S}_x(k)\}]^2\}$$

$$= (\frac{1}{p\Delta T})^2 \sum_{i}^{p-1}\sum_{i'=0}^{p-1} E\{|X_i(k)|^2 |X_{i'}(k)|^2\} - [E\{\hat{S}_x(k)\}]^2. \tag{3.52}$$

Using the result of (2.24) in the first expression on the right-hand side of (3.52) we get

$$E\{|X_i(k)|^2|X_{i'}(k)|^2\} = \frac{1}{(2\pi)^4}\int\int_{-\pi}^{+\pi}\int\int E\{dX(\omega)dX^*(\omega')dX(\omega'')dX^*(\omega''')\}$$
$$e^{j\Delta T\{i(1-\alpha)\omega - i(1-\alpha)\omega' + i'(1-\alpha)\omega'' - i'(1-\alpha)\omega'''\}}$$
$$W_0(k-\omega)W_0^*(k-\omega')W_0(k-\omega'')W_0^*(k-\omega'''). \tag{3.53}$$

Further, the expected value of the product of four Gaussian random variables may be derived following (1.18):

$$\frac{1}{(2\pi)^4}E\{dX(\omega)dX^*(\omega')dX(\omega'')dX^*(\omega''')\}$$
$$= \frac{1}{(2\pi)^2}[S_x(\omega)\delta(\omega-\omega')S_x(\omega'')\delta(\omega''-\omega''')d\omega d\omega''$$
$$+ S_x(\omega)\delta(\omega-\omega''')S_x(\omega')\delta(\omega'-\omega'')d\omega d\omega'], \tag{3.54}$$

which is valid except at $\omega, \omega', \omega'', \omega''' = (0,\pi)$, when there will be an additional term (third term in (1.18)), $S_x^2(\omega)\Big|_{\omega=(0,\pi)}$ (see Reference 16 where it is argued that the third term is practically zero except at $\omega = (0,\pi)$)). Using (3.54) in (3.53) we obtain

$$\mathrm{Var}\{\hat{S}_x(\frac{2\pi k}{\Delta T})\} =$$

$$\frac{1}{(2\pi)^2}\int_{-\pi}^{+\pi}\int \frac{S_x(\omega)S_x(\omega')}{P^2}\left|\frac{\sin(P\bar{\alpha}\Delta T\frac{\omega-\omega'}{2})}{\sin(\bar{\alpha}\Delta T\frac{\omega-\omega'}{2})}\right|^2 \frac{\left|W_0(\frac{2\pi k}{\Delta T}-\omega)W_0(\frac{2\pi k}{\Delta T}-\omega')\right|^2}{\Delta T^2} d\omega d\omega',$$
$$k \neq 0 \text{ or } \frac{\Delta T}{2} \tag{3.55}$$

where $\bar{\alpha} = 1 - \alpha$. For large $P\bar{\alpha}\Delta T$ we can approximate

$$\left|\frac{\sin(P\bar{\alpha}\Delta T\frac{\omega-\omega'}{2})}{\sin(\bar{\alpha}\Delta T\frac{\omega-\omega'}{2})}\right|^2 \approx P\delta(\omega-\omega'). \tag{3.56}$$

Using this approximation in (3.55) we get a simplified result,

$$\mathrm{Var}\{\hat{S}_x(\frac{2\pi k}{\Delta T})\} \approx \frac{1}{2\pi}\int_{\pi}^{\pi}\frac{S_x(\omega)^2}{P}\frac{\left|W_0(\frac{2\pi k}{\Delta T}-\omega)\right|^4}{\Delta T^2}d\omega.$$

$$k \neq 0 \quad \text{or} \quad \frac{\Delta T}{2} \tag{3.57a}$$

At $k = 0$ or $\frac{\Delta T}{2}$, on account of an additional term in (3.54), the variance of the spectral estimate will be twice that given in (3.57). Assume that $S_x(\omega)$ is approximately flat over the bandwidth of $W_0(\omega)$; then, we can reduce (3.57) to

$$\mathrm{Var}\{\hat{S}_x(k)\} \approx \frac{S_x(k)^2}{P}\frac{1}{2\pi}\int_{\pi}^{\pi}\frac{\left|W_0(\omega)\right|^4}{\Delta T^2}d\omega. \tag{3.57b}$$

The spectral estimate, $\hat{S}_x(k)$, may be considered as a χ^2_ν random variable with ν degrees of freedom, where

$$\nu = \frac{2}{\mathrm{var}\{\frac{\hat{S}_x(k)}{S_x(k)}\}} = 2P/\frac{1}{2\pi}\int_{\pi}^{\pi}\frac{\left|W_0(\omega)\right|^4}{\Delta T^2}d\omega.$$

The denominator on the right-hand side is a characteristic of the window function; for example, for the uniform window it is 1.0, for the Hanning window it is 1.944, and for the optimum window it is 2.802.[19]

In the above analysis we have assumed that the successive segments are uncorrelated. This assumption is now relaxed (as in overlap and add method). Assume that r (r<<P) successive segments are correlated. The analysis for correlated segments follows the approach used above. We shall give the final result only. (The detailed derivation is given by Mathews and Youn.[6])

$$\mathrm{Var}\{\hat{S}_x(k)\} =$$

$$\frac{1}{(2\pi)^2}\int_{-\pi}^{+\pi}\int_{-\pi}^{\pi}\frac{S_x(\omega)S_x(\omega')}{P^2}\left|\frac{\sin(P\bar{\alpha}\Delta T\frac{\omega-\omega'}{2})}{\sin(\bar{\alpha}\Delta T\frac{\omega-\omega'}{2})}\right|^2\left|\frac{\sin((\frac{2\pi}{\Delta T}k-\omega')\bar{\alpha}\frac{\Delta T}{2}(2r+1))}{(2r+1)\sin((\frac{2\pi}{\Delta T}k-\omega')\bar{\alpha}\frac{\Delta T}{2})}\right|^2$$

$$\frac{\left|W_0(\frac{2\pi}{\Delta T}k-\omega)W_0(\frac{2\pi}{\Delta T}k-\omega')\right|^2}{\Delta T^2}d\omega d\omega', \qquad k \neq 0 \quad \text{or} \quad \frac{\Delta T}{2} \tag{3.58}$$

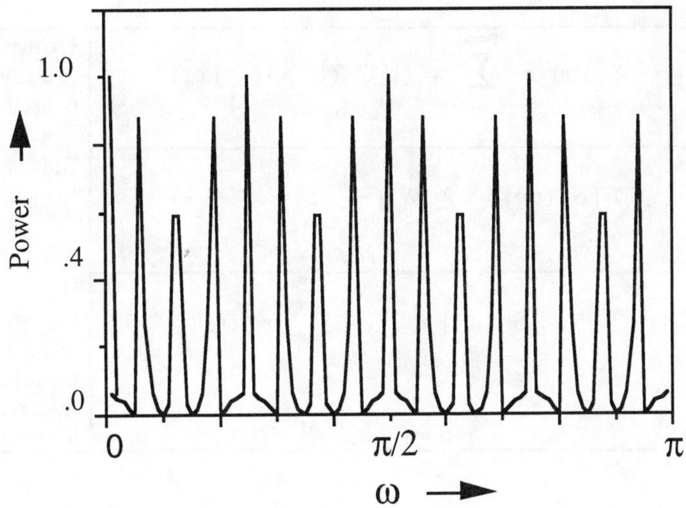

Figure 3.9: A plot of $\left|\dfrac{\sin(\omega\,\bar{\alpha}\frac{\Delta T}{2}(2r+1))}{(2r+1)\sin(\omega\,\bar{\alpha}\frac{\Delta T}{2})}\right|^2$ for $\bar{\alpha} = 0.5$, $\Delta T = 64$ and r=2.

Using the approximation (3.56) in (3.58) we obtain

$$\operatorname{Var}\{\hat{S}_x(k)\} \approx \frac{1}{2\pi}\int_{\pi}^{\pi} \frac{S_x(\omega)^2}{P}\left|\frac{\sin\left(\left(\frac{2\pi}{\Delta T}k-\omega\right)\bar{\alpha}\frac{\Delta T}{2}(2r+1)\right)}{(2r+1)\sin\left(\left(\frac{2\pi}{\Delta T}k-\omega\right)\bar{\alpha}\frac{\Delta T}{2}\right)}\right|^2 \frac{\left|W_0\left(\frac{2\pi}{\Delta T}k-\omega\right)\right|^4}{\Delta T^2}\,d\omega.$$

$$k \neq 0 \quad \text{or} \quad \frac{\Delta T}{2} \tag{3.59}$$

Note that when r=0, which corresponds to uncorrelated segments, (3.59) reduces to (3.57). Further, note that the expression for the variance when the segments are correlated has an additional term,

namely, $\left|\dfrac{\sin(\omega\,\bar{\alpha}\frac{\Delta T}{2}(2r+1))}{(2r+1)\sin(\omega\,\bar{\alpha}\frac{\Delta T}{2})}\right|^2$ which is sketched in Figure 3.9.

In the frequency averaging method the variance of the spectral estimate may be derived in an analogous fashion. The final result is as follows:

$$\operatorname{Var}\{\hat{S}_x(k)\} = \frac{1}{(2\pi TP)^2}\sum_{i,\,i'=-\frac{P-1}{2}}^{\frac{P-1}{2}}\sum W_1(i)W_1(i')$$

BT Method

Formula	$\hat{S}_x(\omega) = \sum_{\tau=-\infty}^{\tau=\infty} w_1(\tau)\hat{C}_x(\tau)\exp(-j\omega\tau)$	**Comments:** A quadratic window is used.
Mean, (3.31)	$E\{\hat{S}_x(\omega)\} = \sum_{k=0}^{T-1} W_1(\frac{2\pi k}{T})S_x(\omega - \frac{2\pi k}{T})$	Bias is zero only for white noise.
Variance, (3.34b)	$\mathrm{Var}\{\hat{S}_x(\omega)\} \approx S_x^2(\omega)T\sum_{t=0}^{T-1} w_1^2(t)$	For low variance, the effective width of the window must be a small fraction of the data length.

Welch Method

Formula	$\hat{S}_x(k) = \frac{1}{P\Delta T}\sum_{i=0}^{P-1} X_i(k)X_i^*(k)$	**Comments:** Uncorrelated segments are assumed.
Mean, (3.51a)	$E\{\hat{S}_x(\frac{2\pi k}{\Delta T})\} = \frac{1}{2\pi}\int_{-\pi}^{\pi} S_x(\frac{2\pi k}{\Delta T} - \omega)\frac{W_0^2(\omega)}{\Delta T}d\omega$	Bias is zero only for white noise.
Variance, (3.57a)	$\mathrm{Var}\{\hat{S}_x(\frac{2\pi k}{\Delta T})\} \approx \frac{1}{2\pi}\int_{\pi}^{\pi}\frac{S_x(\omega)^2}{p}\frac{\lvert W_0(\frac{2\pi k}{\Delta T} - \omega)\rvert^4}{\Delta T^2}d\omega$ $k \neq 0 \quad \text{or} \quad \frac{\Delta T}{2}$	Note the concept of degrees of freedom given on page 190.

Table 3.3(a): A summary of the spectrum estimation methods (low-resolution or conventional type).

$$\int_{-\pi}^{+\pi}\int^{\pi} S_x(\omega)S_x(\omega')W_0(k+i-\omega)W_0^*(k+i-\omega') W_0^*(k+i'-\omega)W_0(k+i'-\omega')d\omega d\omega'.$$

$$k \neq 0 \quad \text{or} \quad \frac{\Delta T}{2} \tag{3.60a}$$

For large T we can approximate

$$\lvert W_0(k+i-\omega)\rvert^2 \approx T\delta(k+i-\omega)$$

Overlap and Add

Formula, Equation (3.41)	$\hat{S}_x = \frac{1}{2r+1}\sum_{i=-r}^{r}\frac{1}{P-\lvert i\rvert}\sum_{l=0}^{P-\lvert i\rvert-1} s_{l,l+i} \qquad 0 \le r \le P-1$	**Comments:** Averaging of diagonal elements.
Mean, Equation (3.44)	$E\{\hat{S}_x(k)\} = \frac{1}{2\pi}\int_{-\pi}^{\pi} S_x(\omega)\left\lvert W_0\left(\frac{2\pi k}{\Delta T}-\omega\right)\right\rvert^2 \frac{1}{2r+1}\sum_{i=-r}^{r} e^{-j(\omega-\frac{2\pi}{\Delta T}k)(1-\alpha)\Delta T i}\, d\omega$	Estimate can be unbiased if the effective window is flat.
Variance, Equation (3.59)	$\mathrm{Var}\{\hat{S}_x(k)\} \approx$ $\frac{1}{2\pi}\int_{\pi}^{\pi} \frac{S_x(\omega)^2}{P}\left\lvert\frac{\sin\left(\left(\frac{2\pi}{\Delta T}k-\omega\right)\alpha\frac{\Delta T}{2}(2r+1)\right)}{(2r+1)\sin\left(\left(\frac{2\pi}{\Delta T}k-\omega\right)\alpha\frac{\Delta T}{2}\right)}\right\rvert^2 \frac{\left\lvert W_0\left(\frac{2\pi}{\Delta T}k-\omega\right)\right\rvert^4}{\Delta T^2}\, d\omega$ $k \neq 0$ or $\frac{\Delta T}{2}$	The variance is reduced on account of the middle term in the integrand. Its value lies between 0 and 1 (see Figure 3.9).

Frequency Averaging

Formula, Equation (3.46)	$\hat{S}_x(k) = \frac{1}{TP}\sum_{i=-\frac{P-1}{2}}^{\frac{P-1}{2}} W_1(i)\lvert X(k+i)\rvert^2$	**Comments:** Weighting coeff. can reduce the sidelobe height.
Mean, Equation (3.47)	$E\{\hat{S}_x(k)\} = \frac{1}{2\pi P}\sum_{i=-\frac{(P-1)}{2}}^{\frac{(P-1)}{2}} W_1(i)\int_{-\pi}^{\pi} S_x(\omega)\frac{\left\lvert W_0\left(\frac{2\pi(k+i)}{T}-\omega\right)\right\rvert^2}{T}\, d\omega$	Variable resolution is possible, but it is computationally inefficient.
Variance, Equation (3.60)	$\mathrm{Var}\{\hat{S}_x(k)\} = \frac{1}{P^2}\sum_{i=-\frac{P-1}{2}}^{\frac{P-1}{2}} \lvert W_1(i)\rvert^2 S_x^2(k+i)$ $k \neq 0$ or $\frac{\Delta T}{2}$	

Table 3.3(b): A summary of the spectrum estimation methods (continued from previous page).

Then, (3.60) may be simplified to

$$\mathrm{Var}\{\hat{S}_x(k)\} = \frac{1}{p^2}\sum_{i=-\frac{P-1}{2}}^{\frac{P-1}{2}} \left|W_1(i)\right|^2 S_x^2(k+i).$$

$$k \neq 0 \quad \text{or} \quad \frac{\Delta T}{2} \tag{3.60b}$$

We have looked at four different methods of spectrum estimation and their bias and variability. A quick comparison among the four methods is made in Table 3.3.

I. Probability Density Function of Spectral Estimate

Let x(t) and x(t), t=0, 1, 2, ... $T-1$ be a pair of Gaussian time series and $X(k)$ and $Y(k)$, k=0, 1, ..., $T-1$ be their discrete Fourier transform (DFT) coefficients which are also Gaussian. For a fixed k, $X(k)$ and $Y(k)$ constitute a pair of random variables whose joint probability density function is given by

$$f(X, Y) = \frac{1}{2\pi\sqrt{|S|}} \exp\left(\frac{-1}{2|S|}\begin{vmatrix} S_x & S_{xy} & X \\ S_{yx} & S_y & Y \\ X & Y & 0 \end{vmatrix}\right), \tag{3.61}$$

where

$$\mathbf{S} = \begin{bmatrix} S_x & S_{xy} \\ S_{yx} & S_y \end{bmatrix}.$$

is the spectral matrix of the pair of time series. The spectral matrix may be estimated using the method of segmentation where we split the entire length of the available time series into p uncorrelated segments, each of length ΔT. The estimate of the spectral matrix is given by

$$\hat{\mathbf{S}} = \frac{1}{P}\sum_{i=1}^{P}\begin{bmatrix} X_i \\ Y_i \end{bmatrix}\begin{bmatrix} X_i^* & Y_i^* \end{bmatrix}. \tag{3.62}$$

The estimate is known to be a maximum likelihood estimate of **S**, and its elements (multiplied by $\sqrt{P}$) are distributed according to complex Wishart distribution (adapted from Reference 20),

$$f(\sqrt{P}\hat{S}_x, \sqrt{P}\hat{S}_y, \sqrt{P}\hat{R}, \sqrt{P}\hat{Q}) = \frac{\left|P\hat{\mathbf{S}}\right|^2}{I(\mathbf{S})}\exp\left(-\frac{P}{2}\mathrm{tr}\{\mathbf{S}^{-1}\hat{\mathbf{S}}\}\right) \qquad \left|\hat{\mathbf{S}}\right| > 0$$

$$= \quad 0 \qquad \text{otherwise}, \tag{3.63}$$

where

$$\hat{R} + j\hat{Q} = \hat{S}_{xy},$$

$$I(\mathbf{S}) = \pi\Gamma(P)\Gamma(P-1)\left|\mathbf{S}\right|^2.$$

$\Gamma(.)$ is a gamma function, and tr{.} stands for the trace of a matrix. Writing the full form of the spectral martices in $\mathrm{tr}\{\mathbf{S}^{-1}\hat{\mathbf{S}}\}$, we get

$$\mathrm{tr}\{\mathbf{S}^{-1}\hat{\mathbf{S}}\} = \frac{1}{1-|\mathrm{coh}|^2}\left[\frac{\hat{S}_x}{S_x} + \frac{\hat{S}_y}{S_y} - \frac{2\alpha\hat{R}}{\sqrt{S_x S_y}} + \frac{2\beta\hat{Q}}{\sqrt{S_x S_y}}\right],$$

where

$$\alpha + j\beta = \frac{S_{xy}}{\sqrt{S_x S_y}}.$$

Using above result in (3.63) and writing $|\hat{\mathbf{S}}|$ in the explicit form, we obtain the final result,

$$f(\hat{S}_x, \hat{S}_y, \hat{R}, \hat{Q}) = \frac{1}{I(\mathbf{S})}\left[\frac{P}{1-|\mathrm{coh}|^2}\left\{\frac{\hat{S}_x\hat{S}_y - \hat{R}^2 - \hat{Q}^2}{S_x S_y}\right\}\right]^{p-2}$$

$$\exp\left(-\frac{1}{2}\frac{P}{1-|\mathrm{coh}|^2}\left[\frac{\hat{S}_x}{S_x} + \frac{\hat{S}_y}{S_y} - \frac{2\alpha\hat{R}}{\sqrt{S_x S_y}} + \frac{2\beta\hat{Q}}{\sqrt{S_x S_y}}\right]\right)$$

$$|\hat{\mathbf{S}}| > 0,$$

$$= \quad 0 \quad \text{otherwise} . \tag{3.64}$$

We can in principle, using (3.64), compute the confidence limits on the estimated spectrum and cross-spectrum. We shall in the next section show how to obtain the confidence limits on the coherence. It may be noted, the expression for the distribution function contains the true spectrum and cross-spectrum which are themselves being estimated. Thus, these results are of limited practical utility in actual computation of the confidence bounds. In Chapter Five we shall outline the method of bootstrap which uses the observed data for the computation of the confidence bounds.

3.4 Estimation of Coherence

The coherence between two time series is defined as a cross-spectrum that is normalized with respect to the geometric mean of the spectra of the two time series (see (1.38)),

$$\mathrm{Coh}_{xy}(\omega) = \frac{S_{xy}(\omega)}{\sqrt{S_x(\omega) S_y(\omega)}},$$

where for estimation of coherence the exact spectra and cross-spectrum are to be replaced by their estimates, respectively,

$$\hat{Coh}_{xy}(\omega) = \frac{\hat{S}_{xy}(\omega)}{\sqrt{\hat{S}_x(\omega)\hat{S}_y(\omega)}}. \tag{3.65}$$

The statistical characterization of the estimate of coherence is a highly complex topic, but of great significance. Some of the important analytical results on this topic are summarized here.

A. **Probability Density Function of Coherence Estimate**

The estimate of the coherence function, being a ratio of two random variables, namely, the estimated cross-spectrum and square root of the product of estimated spectra, is also a random variable evidently having a more complex statistical structure. For example, it is highly biased, and it has a large variability, particularly, when the true coherence is low. In contrast, when the true coherence is high both bias and variability of the estimated coherence are low. Accurate coherence estimates require a large number of independent data segments (at least a few tens). We shall now briefly outline the probability structure of the estimated coherence function.

In (3.64) express $\hat{R}$ and $\hat{Q}$ in terms of $|\hat{Coh}|$ and ϕ:

$$\hat{R} = \sqrt{\hat{S}_x\hat{S}_y}\,|\hat{Coh}|\cos(\phi),$$

$$\hat{Q} = \sqrt{\hat{S}_x\hat{S}_y}\,|\hat{Coh}|\sin(\phi).$$

We obtain

$$f(\hat{S}_x, \hat{S}_y, \hat{Coh}, \phi) = \frac{1}{I(S)}\left[P\left\{\frac{\hat{S}_x\hat{S}_y(1-|\hat{Coh}|^2)}{S_xS_y(1-|Coh|^2)}\right\}\right]^{P-2}$$

$$\exp\left(-\frac{1}{2}\frac{P}{1-|Coh|^2}\left[\frac{\hat{S}_x}{S_x}+\frac{\hat{S}_y}{S_y}-2\sqrt{\frac{\hat{S}_x\hat{S}_y}{S_xS_y}}\,|\hat{Coh}|(\alpha\cos(\phi)-\beta\sin(\phi))\right]\right),$$

$$|\hat{S}| > 0$$

$$= 0 \qquad \text{otherwise}\,. \tag{3.66}$$

As we are interested in the probability of $|\hat{Coh}|$ and ϕ, we shall integrate (3.66) with respect to $\hat{S}_x$ and $\hat{S}_y$. Without going into the details of integration we give the final result taken from Reference 20,

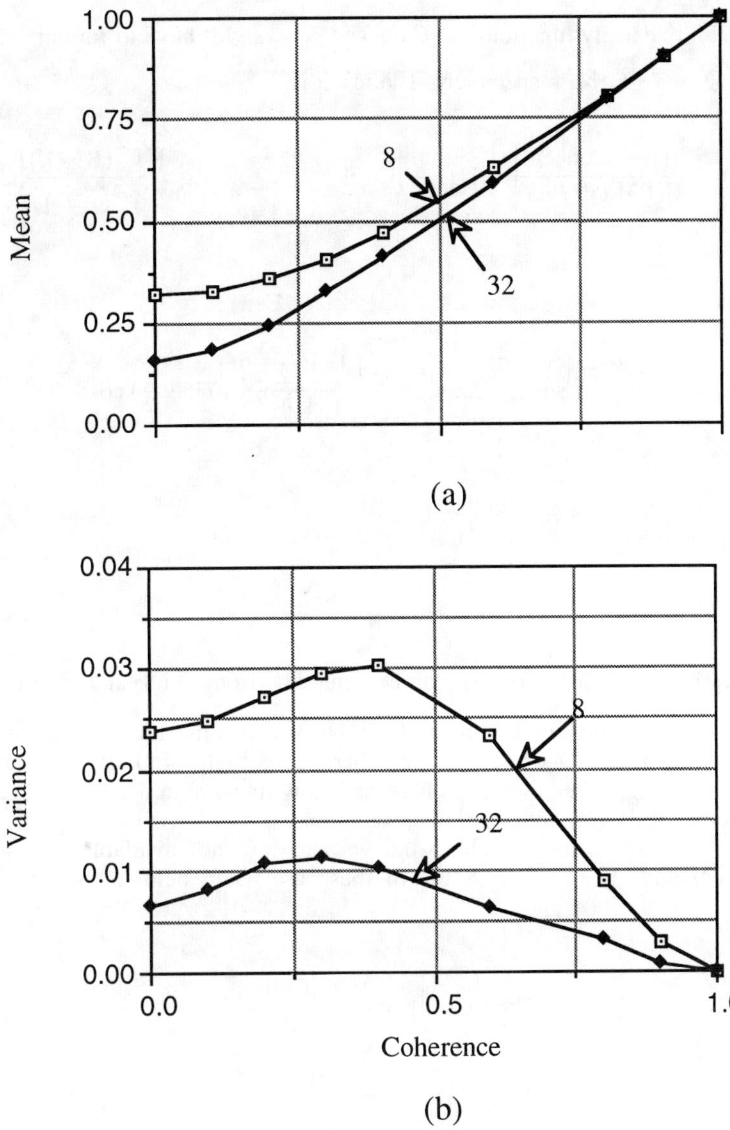

Figure 3.10: Theoretical (a) mean and (b) variance of the estimated coherence function using eight and thirty-two data segments.

$$f(|\hat{Coh}|, \hat{\phi}) = \frac{(1 - |Coh|^2)^P}{\pi\Gamma(P)\Gamma(P-1)} |\hat{Coh}| (1 - |\hat{Coh}|^2)^{P-2}$$

$$\sum_{k=0}^{\infty} 2^k |Coh|^k \frac{\Gamma^2(P + \frac{k}{2})}{\Gamma(k+1)} |\hat{Coh}|^k \cos^k(\hat{\phi} - \varphi). \tag{3.67}$$

To get the probability density functions of $|\hat{Coh}|$ and $\hat{\phi}$ we will have to further integrate (3.67) with respect to $\hat{\phi}$ and $|\hat{Coh}|$, respectively. The result is,

$$f(|\hat{Coh}|) = \frac{2(1-|Coh|^2)^P}{\Gamma(P)\Gamma(P-1)}|\hat{Coh}|(1-|\hat{Coh}|^2)^{P-2}\sum_{k=0}^{\infty}|Coh|^{2k}\frac{\Gamma^2(P+k)}{\Gamma^2(k+1)}|\hat{Coh}|^{2k} \tag{3.68}$$

and

$$f(\hat{\phi}) = \frac{(1-|Coh|^2)^P}{\pi\Gamma(P)}\sum_{k=0}^{\infty}|Coh|^k\, 2^{k-1}\frac{\Gamma(P+\frac{k}{2})}{\Gamma(k+1)}\Gamma(1+\frac{k}{2})\cos^k(\varphi-\hat{\phi}), \tag{3.69}$$

which converges for $|Coh||\cos(\varphi-\hat{\phi})| < 1$.

B. Bias and Variance

The mean and variance of $|\hat{Coh}|$ and $\hat{\phi}$ can be computed from (3.68) and (3.69). The bias and variance of $|\hat{Coh}|$ are shown in Figure 3.10.[21] Notice that the bias is high when the true coherence is low, but it is low when the true coherence is high. Similarly, the variance of the coherence estimate is high when the true coherence is low (maximum between 0.25 and 0.5). The variance goes to zero as the true coherence approaches one.

As closed form expressions for bias and variance are not available, many attempts to approximate them have been made. Some of these are listed below. For large P and high coherence,[22]

$$\begin{aligned} \mathrm{Var}\{|\hat{Coh}|\} &\approx \frac{(1-|Coh|^2)^2}{2P} \\ \mathrm{Var}\{\hat{\phi}\} &\approx \frac{(|Coh|^{-2}-1)}{2P}. \end{aligned} \tag{3.70}$$

$|\hat{Coh}|$ may be mapped into a Gaussian random variable using a mapping function[23] valid for $P > 20$ and $0.4 < |\hat{Coh}| < 0.95$,

$$z = \tanh^{-1}(|\hat{Coh}|). \tag{3.71}$$

The mean and the variance of of z are given by

$$\begin{aligned} E\{z\} &= \tanh^{-1}(|Coh|) + \frac{1}{2(P-1)}, \\ \mathrm{Var}\{z\} &= \frac{1}{2(P-1)}. \end{aligned} \tag{3.72}$$

Both bias and variance of the coherence estimate can be reduced by overlapping the adjacent segments, within the constraints of (3.39). A reduction by a factor of two or three was reported by

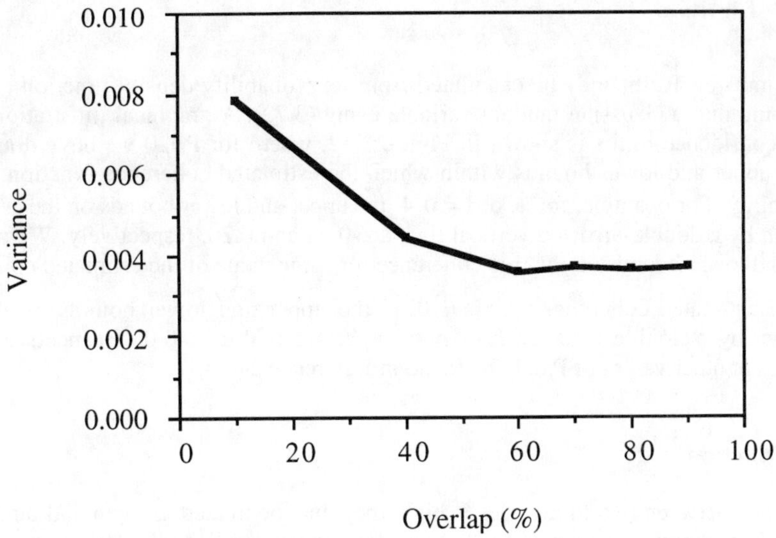

Figure 3.11: Variance of the coherence estimate as a function of the overlap.

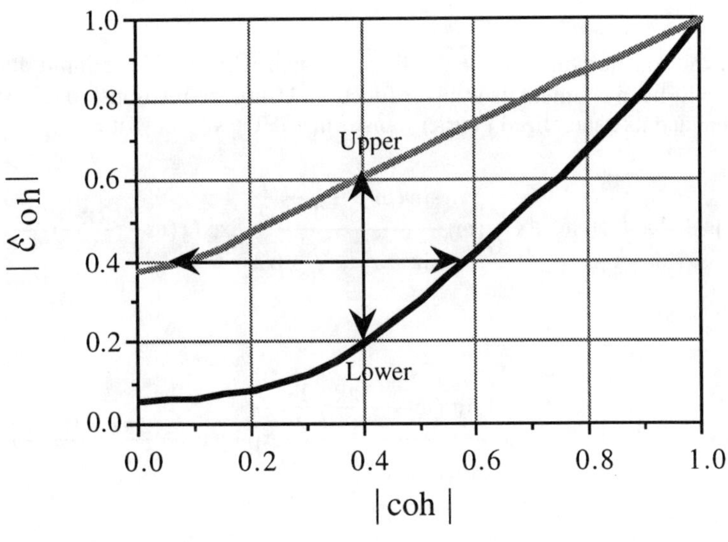

Figure 3.12: 90% confidence bounds for P=20 (forty degrees of freedom).

Carter et al[4] who have applied 50% overlap on synthetic white noise and have used a Hanning window (correlation distance=0.58ΔT). A similar experiment was repeated, but with an optimum window (correlation distance = 0.63ΔT). Two 1024-point uncorrelated white noise sequences were divided into overlapped segments each of length 64 points. The estimated coherence (magnitude square) was averaged over all frequency bins. The experiment was then repeated one hundred times for the purpose of computing the bias and variance. The outcome (variance only) of this experiment for different overlaps is shown in Figure 3.11.

C. Confidence Limits

The confidence limits on $|\hat{Coh}|$ may be calculated from its probability density function (3.68)[24] or after its mapping into a Gaussian random variable using (3.72). A graphical illustration of the meaning of the confidence limits is shown in Figure 3.12, where for P=20 we have drawn two curves denoting upper and lower bounds within which the estimated coherence function will lie with 90% probability. For example, for $|Coh| = 0.4$ the upper and lower bounds on the estimated coherence, shown by a double-arrowed vertical line, are 0.62 and 0.20, respectively. We can also infer the upper and lower bounds on the true coherence for some value of the estimated coherence. For example, for estimated coherence, $|\hat{Coh}| = 0.4$, the upper and lower bounds on the true coherence, shown by a double arrowed horizontal line, are 0.58 and 0.06, respectively. The confidence bounds at other values of P may be found in Reference 24.

D. Transfer Function

When the coherence between two time series is high, they may be treated as input and output of a linear system, and an attempt to estimate the transfer function of the linear system is worthwhile. Let x(t) be the input and y(t) be the output. Analogous to (2.11), the input and output of a linear system may be expressed as

$$y(t) = \frac{1}{2\pi}\int_{-\pi}^{\pi} H(\omega)\, dX(\omega) \exp(j\omega t), \tag{3.73}$$

where $H(\omega)$ is the transfer function of the linear system. Given a finite input and output of a system, our aim will be to estimate its transfer function. Using the relationship (2.24) between the Fourier coefficient and its generalized Fourier transform (GFT), we can write

$$Y(k) = \frac{1}{2\pi}\int_{-\pi}^{\pi} H(\omega)\, dX_G(\omega) \frac{\sin\left(\omega - \frac{2\pi k}{T}\right)\frac{T}{2}}{\sin\left(\omega - \frac{2\pi k}{T}\right)/2} \exp\left(j\left(\omega - \frac{2\pi k}{T}\right)\frac{T-1}{2}\right)$$

and

$$X(k) = \frac{1}{2\pi}\int_{-\pi}^{\pi} dX_G(\omega) \frac{\sin\left(\omega - \frac{2\pi k}{T}\right)\frac{T}{2}}{\sin\left(\omega - \frac{2\pi k}{T}\right)/2} \exp\left(j\left(\omega - \frac{2\pi k}{T}\right)\frac{T-1}{2}\right). \tag{3.74}$$

From (3.74) we obtain

$$E\{Y(k)X^*(k)\} = \frac{1}{2\pi}\int_{-\pi}^{\pi} H(\omega) S_x(\omega) \left\{\frac{\sin\left(\omega - \frac{2\pi k}{T}\right)\frac{T}{2}}{\sin\left(\omega - \frac{2\pi k}{T}\right)/2}\right\}^2 d\omega. \tag{3.75}$$

When T is large, the Dirichlet kernel may be approximated (see (3.56)) by $T\delta(\omega - \frac{2\pi k}{T})$, which simplifies (3.75) to

$$\frac{E\{Y(k)X^*(k)\}}{T} = H(\frac{2\pi k}{T})S_x(\frac{2\pi k}{T}). \tag{3.76}$$

The above simplification is appropximately valid when the transfer function is constant over the bandwidth of the window spectrum.

Next, let us assume that the transfer function is a smoothly varying function, such that in its Taylor series expansion we need only consider the first two terms. Equation (3.75) becomes

$$E\{Y(k)X^*(k)\} = H(\frac{2\pi k}{T})\frac{1}{2\pi}\int_{-\pi}^{\pi} S_x(\omega)\left\{\frac{\sin(\omega - \frac{2\pi k}{T})\frac{T}{2}}{\sin(\omega - \frac{2\pi k}{T})/2}\right\}^2 d\omega +$$

$$+ H'(\frac{2\pi k}{T})\frac{1}{2\pi}\int_{-\pi}^{\pi}(\omega - \frac{2\pi k}{T})S_x(\omega)\left\{\frac{\sin(\omega - \frac{2\pi k}{T})\frac{T}{2}}{\sin(\omega - \frac{2\pi k}{T})/2}\right\}^2 d\omega$$

$$= H(\frac{2\pi k}{T})S_x(\frac{2\pi k}{T}) + H'(\frac{2\pi k}{T})\tilde{S}_x(\frac{2\pi k}{T}), \tag{3.77}$$

where

$$\tilde{S}_x(\frac{2\pi k}{T}) = \frac{1}{2\pi}\int_{-\pi}^{\pi}(\omega - \frac{2\pi k}{T})S_x(\omega)\left\{\frac{\sin(\omega - \frac{2\pi k}{T})\frac{T}{2}}{\sin(\omega - \frac{2\pi k}{T})/2}\right\}^2 d\omega,$$

and $H'(\frac{2\pi k}{T})$ is the derivative of the transfer function evaluated at $\frac{2\pi k}{T}$. Equation (3.77) may be solved for $H(\frac{2\pi k}{T})$ and $H'(\frac{2\pi k}{T})$ if we use two (or more) different types of spectrum estimates using different types of windows, for example, eigenspectra obtained by using eigenfunctions as windows (see next section on windows).[25] It is easy to see how the above idea can be generalized to handle higher order derivatives in the Taylor series expansion of the transfer function. The transfer function is now defined within the window bandwidth by its value and one or more derivatives, which should enable us to faithfully reconstruct the entire transfer function.[25]

E. **Effect of Misalignment**

When one of the time series is delayed with respect to the second time series, the estimation of coherence or the transfer function can be in serious error. To demonstrate this situation, consider a pair of time series, x(t) and y(t), where $y(t) = x(t - \tau_0)$. The transfer function of such a simple delay operator is $H(\omega) = \exp(-j\tau_0\omega)$, which we substitute in (3.75),

$$E\{Y(k)X^*(k)\} = \frac{1}{2\pi}\int_{-\pi}^{\pi} e^{-j\tau_0\omega} S_x(\omega)\left\{\frac{\sin\left(\omega - \frac{2\pi k}{T}\right)\frac{T}{2}}{\sin\left(\omega - \frac{2\pi k}{T}\right)/2}\right\}^2 d\omega. \tag{3.78}$$

The right-hand side of (3.78) is a convolution of $e^{-j\tau_0\omega} S_x(\omega)$ and the Dirichlet kernel; hence it may be written as a Fourier transform of the product of $C_x(\tau - \tau_0)$ and $(1 - \frac{|\tau|}{T})$. The transfer function estimated from the misaligned time series will be given by

$$\frac{E\{Y(k)X^*(k)\}}{E\{|X(k)|^2\}} = \frac{\int_{-\pi}^{\pi} e^{-j\tau_0\omega} S_x(\omega)\left\{\frac{\sin\left(\omega - \frac{2\pi k}{T}\right)\frac{T}{2}}{\sin\left(\omega - \frac{2\pi k}{T}\right)/2}\right\}^2 d\omega}{\int_{-\pi}^{\pi} S_x(\omega)\left\{\frac{\sin\left(\omega - \frac{2\pi k}{T}\right)\frac{T}{2}}{\sin\left(\omega - \frac{2\pi k}{T}\right)/2}\right\}^2 d\omega}. \tag{3.79}$$

To get some concrete results on the effect of misalignment, we shall further simplify the time series model as in the example given below.

Example 3.8: For simplicity, assume that $S_x(\omega) = \sigma_x^2$; then, an estimate of the transfer function will be given by

$$\frac{E\{Y(k)X^*(k)\}}{E\{|X(k)|^2\}} = \int_{-\pi}^{\pi} e^{-j\tau_0\omega}\left\{\frac{\sin\left(\omega - \frac{2\pi k}{T}\right)\frac{T}{2}}{\sin\left(\omega - \frac{2\pi k}{T}\right)/2}\right\}^2 d\omega.$$

The right-hand side of the above equation, being a convolution of $e^{-j\tau_0\omega}$ and $\left\{\frac{\sin\left(\frac{\omega T}{2}\right)}{\sin(\omega/2)}\right\}^2$, may be written as a Fourier transform of the product of $\delta(\tau - \tau_0)$ and $(1 - \frac{|\tau|}{T})$. Hence, (3.79) may be reduced to

$$\frac{E\{Y(k)X^*(k)\}}{E\{|X(k)|^2\}} = \left(1 - \frac{|\tau_0|}{T}\right) e^{j\tau_0\frac{2\pi k}{T}}.$$

From the above it is clear that the transfer function magnitude is considerably reduced when the misalignment is comparable to the segment length. A similar effect of misalignment on the estimation of coherence is discussed in Reference 26 (End of Example).

3.5 Spectrum of Window Function

A study of window functions for spectrum estimation had been a topic of abiding interest until the high-resolution methods, which do not demand a window, were introduced. It is not possible to describe the large number of spectral windows which are available in the literature. Instead, some of the recent developments in this area are covered. These include optimum windows and discrete prolate spheroidal sequences used as a collection of windows.

A. Role of Spectral Window

From (3.31) we note that the expected value of the spectral estimate is equal to the convolution of the actual spectrum of the time series and the spectrum of the linear window or the Fourier transform of the quadratic window. It is therefore important to study the spectrum of different windows and how it affects the quality of the resulting estimate of the spectrum. To get a qualitative picture of how the spectrum of the window function affects the computed spectrum, consider a random sinusoid, $x(t) = A_0 \sin(\omega_0 t + \varphi)$, whose spectrum is to be estimated. The actual spectrum of a random sinusoid is a delta function, which we shall use in (3.31) and obtain the computed spectrum as equal to the spectrum of the window function,

$$E\{\hat{S}_x(\omega)\} = \frac{A_0^2}{2} W_1(\omega - \omega_0) + \frac{A_0^2}{2} W_1(\omega + \omega_0).$$

The computed spectrum will show a peak at the frequency corresponding to the sinusoid. Moreover, there will be many secondary peaks on account of the sidelobes of $W_1(\omega)$. Thus, a true line spectrum appears as a broad peak flanked by many sidelobes spread over a wide frequency bandwidth. Further, some of the sidelobes of $W_1(\omega)$ may be negative; for example, if $w_1(t)$ is a uniform window, the alternate sidelobes of $W_1(\omega)$ will be negative. Since the spectrum cannot be negative, such a window is not acceptable as a quadratic window.

Example 3.9: As an example let us consider a uniform window used as a linear window. The spectrum of the uniform window is plotted in Figure 3.13 for two different lengths. As noted earlier, the width of the peak or mainlobe is inversely proportional to the data length, but the sidelobe level remains independent of the data length. However, as data length increases, more and more sidelines are packed into a given frequency band. The 3 dB width is equal to $\frac{1.56}{\Delta T}$ (in hertz if time is measured in seconds), and the maximum sidelobe level is -13.5 dB. The first null occurs at $\pm\frac{1}{\Delta T}$, and the successive nulls are spaced $\frac{1}{\Delta T}$ apart (End of Example).

A spectral window is characterized by two attributes, namely, a mainlobe of finite width and a series of sidelines. While the former is responsible for blurring sharp spectral features, if any, the latter is responsible for leakage of power from a strong spectral peak into a weak neighboring peak unless it happens to lie in one of the nulls of the spectrum of the window.

Example 3.10: In this example we would like to demonstrate the leakage of power. Consider a time series as a sum of two random sinusoids with unit variance background noise,

$$x(t) = 2.0 \sin(\omega_1 t + \varphi_1) + 8.0 \sin(\omega_2 t + \varphi_2) + \eta(t),$$

where φ_1 and φ_2 are random phases. The frequencies were so selected that for minimum leakage the weaker sinusoid is placed at the null of the spectrum of the window and for maximum leakage

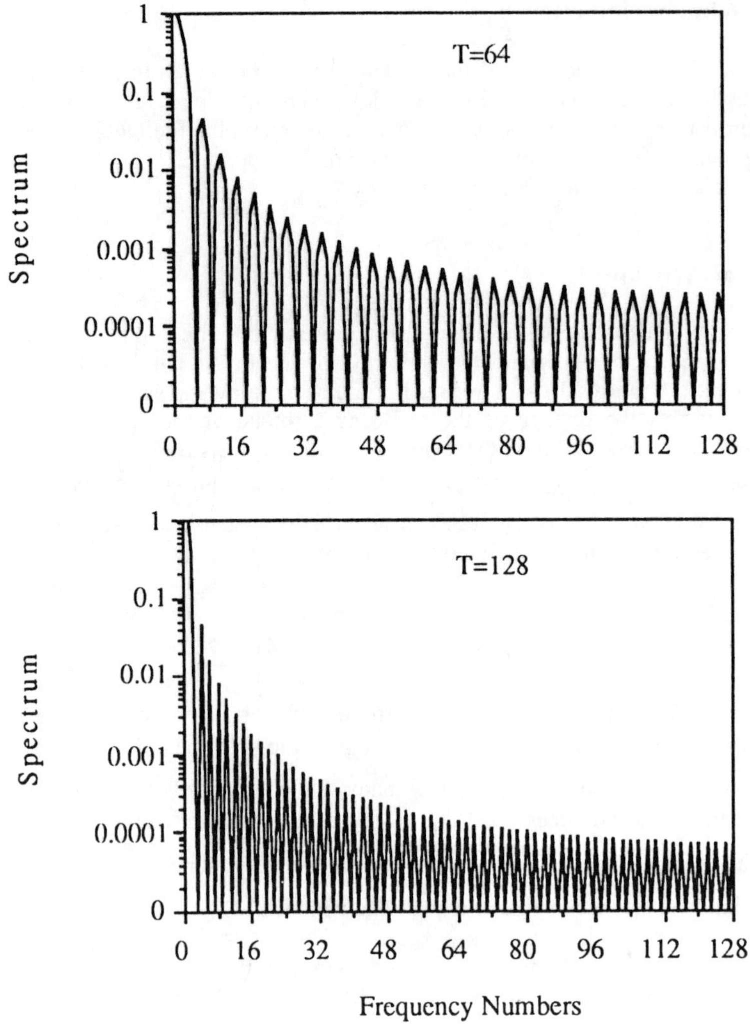

Figure 3.13: Spectrum of uniform window. Each window was padded with zeros to increase its length to 256.

at its sidelobe. Twenty uncorrelated segments each of length 16 points were generated and used to compute the covariance matrix. Using this covariance matrix, the power spectrum was computed following (3.26a). The first frequency was held fixed at 0.2π. The second frequency, ω_2, was so selected that the second null or the sidelobe of the spectrum of the window falls on the weaker sinusoid. The computed power spectrum for both cases is shown in Figure 3.14.

The weak peak is at 0.1 (normalized frequency) and the large spectral peak at 0.225 corresponding to the second null of the spectral window. The spectrum is shown in (a). The measured power at 0.1 is 1.0108 against the theoretical value of 1.0. Next, the large spectral peak is placed at 0.2875 corresponding to the second sidelobe of the spectral window. The spectrum is shown in (b) where the measured power at 0.1 is 1.316. The extra power is evidently due to the leakage of power from the large spectral peak via the second sidelobe. The power leakage has additionally distorted the weak spectral peak (End of Example).

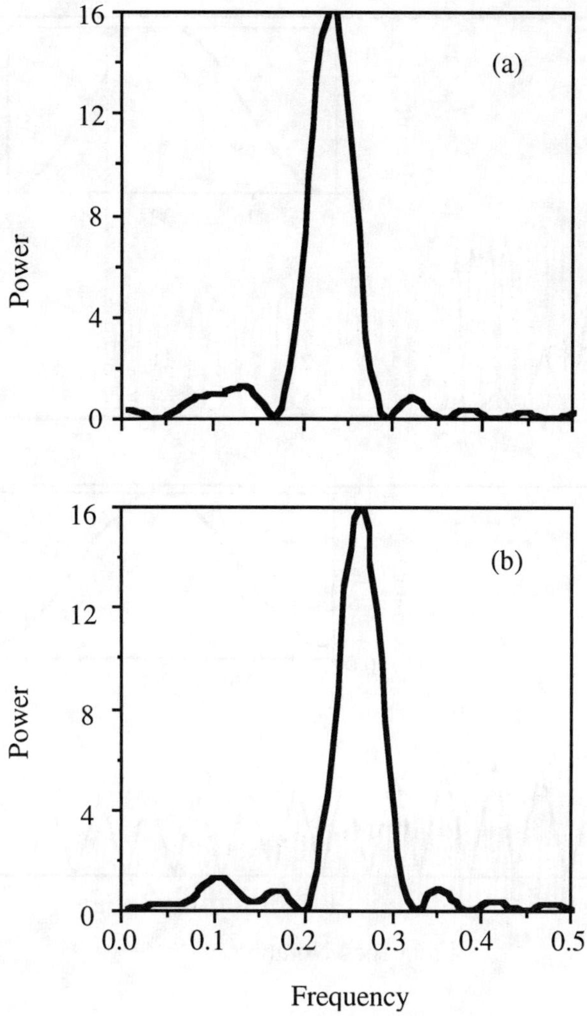

Figure 3.14: The effect of the presence of a large spectral peak close to a weak peak is shown above.

The leakage of power through the sidelobes is perhaps more serious than the broadening of a line spectrum, as the former results in a high bias error, while the latter reduces the resolution capability. A lot of effort has been spent in the design of better windows, having as low a level as possible, but at the cost of a broader mainlobe. A popular spectral window is the cosine window or the Hanning window,

$$w(t) = 0.5(1 - \cos(\frac{2\pi t}{\Delta T})), \quad t = 0,1,2, \ldots, \Delta T - 1,$$

which is sketched in the inset of Figure 3.15 for ΔT=64. The Fourier transform of the Hanning window is given by

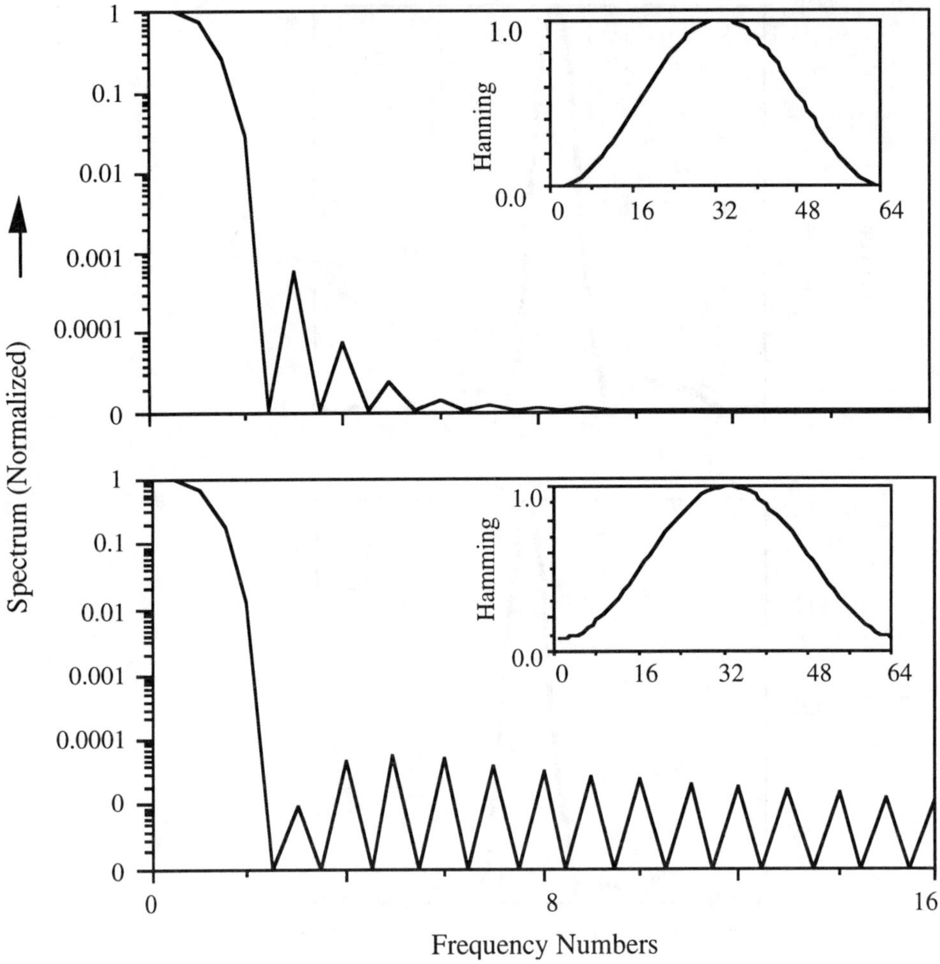

Figure 3.15: Hanning and Hamming windows (shown in the inset) and their spectra.

$$W(\omega) = 0.5D(\omega) + 0.25[D(\omega - \frac{2\pi}{\Delta T}) + D(\omega + \frac{2\pi}{\Delta T})],$$

where

$$D(\omega) = \frac{\sin(\omega\frac{\Delta T}{2})}{\sin(\frac{\omega}{2})}\exp(-j\frac{\omega}{2}(\Delta T - 1)).$$

The spectrum of the Hanning window is shown in Figure 3.15 for ΔT=64. The mainlobe width is approximately equal to $\frac{2.88}{\Delta T}$, and the largest sidelobe is -32 dB below the mainlobe height (compare this with that for uniform window). The correlation distance (a lag at which the correlation drops down to one tenth of its maximum value at zero lag) is equal to 0.58ΔT. The sidelobe level has been reduced considerably, though at the cost of the broad mainlobe. Closely related to the Hanning window is the Hamming window given by

$$w(t) = 0.54 - 0.46 \cos\left(\frac{2\pi}{\Delta T} t\right), \qquad t = 0, 1, 2, \ldots, \Delta T - 1,$$

whose Fourier transform is given by

$$W(\omega) = 0.54 D(\omega) + 0.23\left[D\left(\omega - \frac{2\pi}{\Delta T}\right) + D\left(\omega + \frac{2\pi}{\Delta T}\right)\right],$$

which has a new null at $\omega = 2.6\frac{2\pi}{\Delta T}$, not present in the Hanning window. The spectrum of the Hamming window is shown in Figure 3.15. An important difference between the Hanning and Hamming windows is in their sidelobe structure. Notice that on account of theoretical null at $\omega = 2.6\frac{2\pi}{\Delta T}$, the first sidelobe of the Hamming window spectrum is considerably smaller than that of the Hanning window. A good review of different types of windows is given in Reference 3.

B. Optimum Windows

A window is optimized in the sense that the area under the mainlobe of its spectrum is maximum in relation to the total area.[27] Let $w_0, w_1, \ldots, w_{\Delta T-1}$ be the window coefficients. The area under the mainlobe is given by

$$\frac{1}{2\pi} \int_{-\frac{2\pi}{\Delta T}}^{\frac{2\pi}{\Delta T}} |W(\omega)|^2 d\omega,$$

and the total area is given by

$$\frac{1}{2\pi} \int_{-\pi}^{\pi} |W(\omega)|^2 d\omega,$$

where W(w) is the discrete Fourier transform of the window sequence. Substituting for W(w) in terms of the window coefficients, we obtain the area in the mainlobe and the total area as follows:

$$\frac{1}{2\pi} \int_{-\frac{2\pi}{\Delta T}}^{\frac{2\pi}{\Delta T}} |W(\omega)|^2 d\omega = \sum_{k}^{\Delta T-1} \sum_{l=0}^{\Delta T-1} w_k w_l \frac{\sin\left(\frac{2\pi}{\Delta T}(k-l)\right)}{\pi(k-l)}$$

and

$$\frac{1}{2\pi} \int_{-\pi}^{\pi} |W(\omega)|^2 d\omega = \sum_{k=0}^{\Delta T-1} w_k^2.$$

The ratio of the area under the mainlobe to the total area is given by

$$\beta = \frac{\sum_{k}^{\Delta T-1} \sum_{l=0}^{\Delta T-1} w_k w_l \frac{\sin\left(\frac{2\pi}{\Delta T}(k-l)\right)}{\pi(k-l)}}{\sum_{k=0}^{T-1} w_k^2}. \tag{3.80a}$$

Next, we define the following quantities:

$$\mathbf{w} = \text{col}\{w_0, w_1, \ldots, w_{T-1}\}$$

$$\mathbf{M} = \begin{bmatrix} \frac{2}{\Delta T} & \frac{\sin\frac{2\pi}{\Delta T}}{\pi} & \cdots & \frac{\sin\frac{2\pi}{\Delta T}(\Delta T-1)}{(\Delta T-1)\pi} \\ \frac{\sin\frac{2\pi}{\Delta T}}{\pi} & \frac{2}{\Delta T} & \cdots & \frac{\sin\frac{2\pi}{\Delta T}(\Delta T-2)}{(\Delta T-2)\pi} \\ & & & \\ \frac{\sin\frac{2\pi}{T}(\Delta T-1)}{(\Delta T-1)\pi} & \frac{\sin\frac{2\pi}{\Delta T}(\Delta T-2)}{(\Delta T-2)\pi} & \cdots & \frac{2}{\Delta T} \end{bmatrix}.$$

In terms of **w** and **M** defined above we can rewrite the power ratio defined in (3.80a) as follows:

$$\beta = \frac{\mathbf{w}^T\mathbf{M}\mathbf{w}}{\mathbf{w}^T\mathbf{w}}. \tag{3.80b}$$

We would like to maximize the power ratio under the constraint that

$$\sum_{k=0}^{T-1} w_k^2 = 1 \quad .$$

The solution to the above constrained maximization problem is that the window sequence must be equal to the eigenvector corresponding to the largest eigenvalue of matrix **M**, and b is the corresponding largest eigenvalue. The optimum window has the property of concentrating energy in a narrow band, $\pm\frac{2\pi}{\Delta T}$, just like the well-known prolate spheroidal functions, introduced by Slepian, Pollack, and Landau.[28] The energy of the prolate spheroidal functions is concentrated in a specific band, $\pm\Delta\omega$. In fact, if we let $\Delta\omega = \frac{2\pi}{\Delta T}$, the optimum window turns out to be the discrete version of the prolate spheroidal function. The prolate spheroidal function has been approximated by more familiar functions, like the Bessel function of the second kind and the hyperbolic sine.[29]

Example 3.11: The maximum eigenvector of matix **M** was obtained for three different window lengths, namely, 16, 32, and 64. The results of numerical computation are shown in Table 3.4. The spectrum of the window is shown in Figure 3.16. The 3 dB width of the main lobe is $\frac{2.12}{\Delta T}$. (Compare this with the mainlobe width of the uniform window, $\frac{1.56}{\Delta T}$ and that of Hanning

T=16, β=0.98155523

1.0,	0.9647692	0.8969838
0.8017354	0.6860384	0.5581594
0.4268421	0.3005089	

T=32, β=0.9811725

1.0	0.9911078	0.9734944
0.9474967	0.9136096	0.8724740
0.8248617	0.7716553	0.7138285
0.6524216	0.5885176	0.5232161
0.4576095	0.3927591	0.3296615
0.2694880		

T=64, β=0.9811725

1.0	0.9977716	0.9933259
0.9866837	0.9778773	0.9669485
0.9539502	0.9389438	0.9220010
0.9032019	0.8826353	0.8603978
0.8365931	0.8113318	0.7847307
0.7569119	0.7280011	0.6981291
0.6674292	0.6390368	0.6040891
0.5717239	0.5390787	0.5062901
0.4734933	0.4408209	0.4084023
0.3763631	0.3448242	0.3139016
0.2837055	0.2543395	

Table 3.4: Optimum window coefficients taken from Reference 24. Note that only the right half of the window is shown above.

window, $\frac{2.88}{\Delta T}$.) The height of the first sidelobe is -22 dB. The correlation distance of the optimum window is $0.63\Delta T$ (End of Example).

C. **Multiple Window**

The optimum window described above may be further improved by increasing the width of the mainlobe of the window spectrum.[30] As we increase the width of the mainlobe, the number of significant eigenvalues of **M** matrix will also increase. In fact, the number of significant eigenvalues of magnitude close to unity are approximately equal to 2DfT, where $2\Delta f$ is the bandwidth (in hertz) of energy concentration. These eigenvectors are known as discrete prolate spheroidal sequences (DPSS) whose energy is concentrated within a band of $\pm\Delta f$ Hz. The fraction of energy concentrated within this bandwidth is equal to the corresponding eigenvalue.

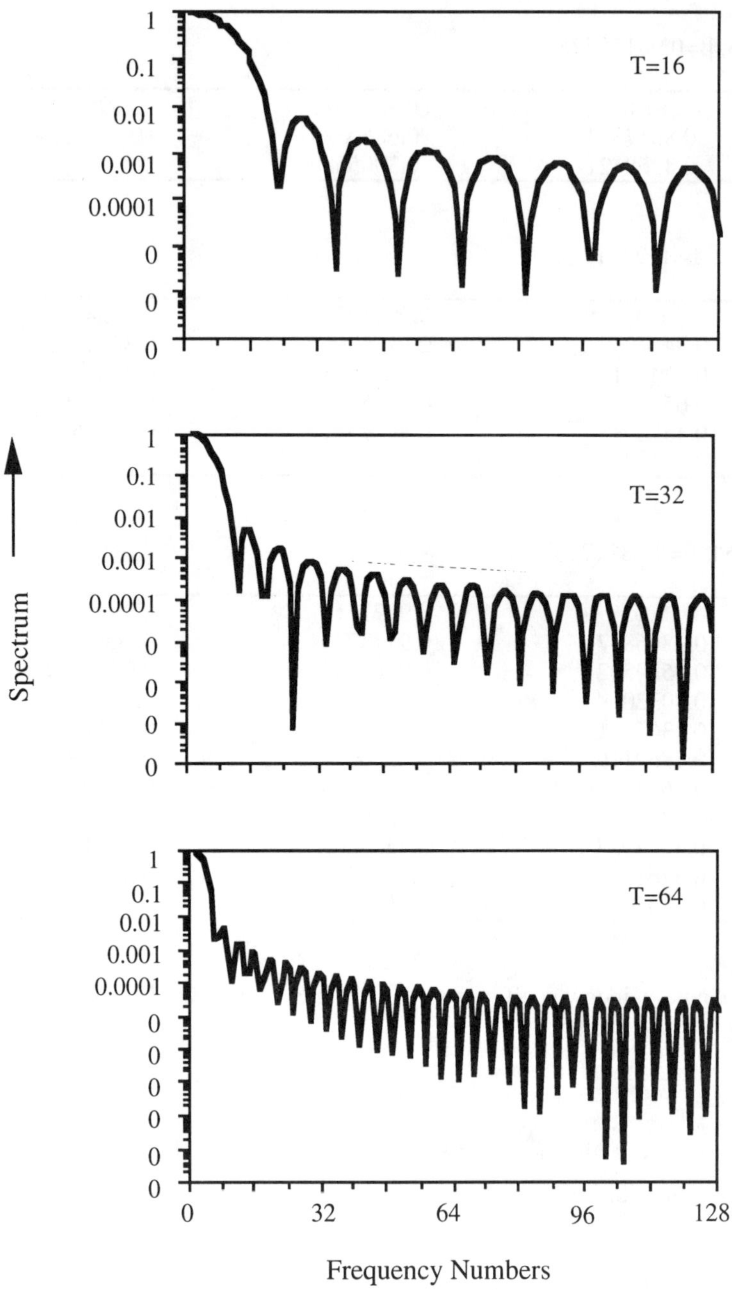

Figure 3.16: Spectrum of optimum window. Each window function given in Table 3.4 was first mirror reflected and then padded with zeros to make each of length 256.

The DFT coefficients of DPSS are known as the DPS wave function (DPSW). Both DPSS and DPSW are doubly orthogonal, DPSS over 0 to $T-1$ and $\pm\infty$ and DPSW over $\pm\Delta f$ and $\pm\frac{1}{2}$.[31] We may consider all or some of these significant eigenvectors as possible candidates for the spectral window.

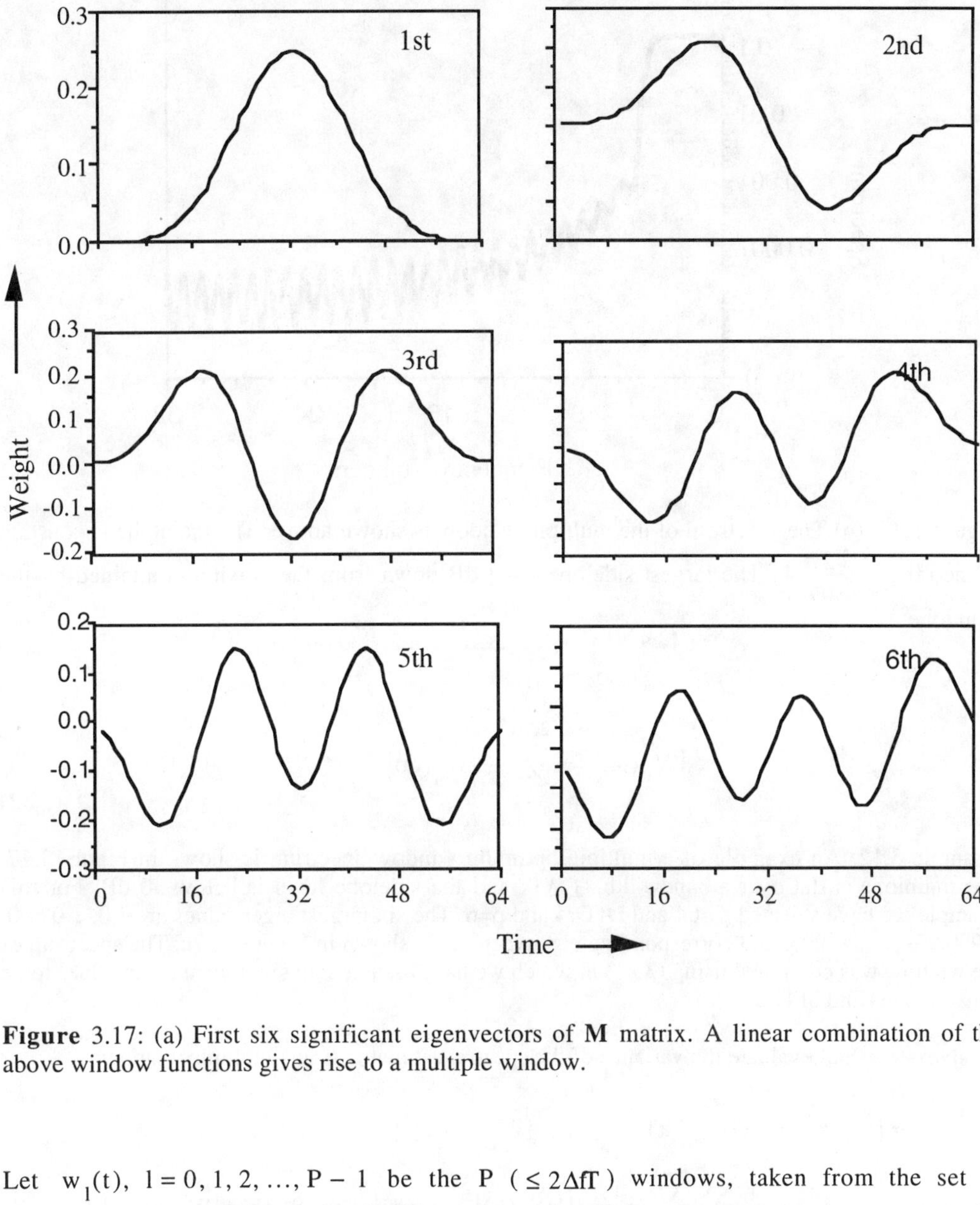

Figure 3.17: (a) First six significant eigenvectors of **M** matrix. A linear combination of the above window functions gives rise to a multiple window.

Let $w_l(t),\ l = 0, 1, 2, \ldots, P-1$ be the $P\ (\le 2\Delta fT)$ windows, taken from the set of eigenvectors corresponding to the significant eigenvalues. It is no longer necessary to segment the time series. The spectral estimate is given by

$$\hat{S}_x(k) = \frac{1}{P}\sum_{i=0}^{P-1}\frac{1}{\lambda_i}\left|\sum_{t=0}^{T-1} x(t)w_i(t)\exp\left(-j\frac{2\pi k}{T}t\right)\right|^2, \tag{3.81}$$

where $\lambda_i,\ i = 0, 1, \ldots, P-1$ are significant eigenvalues. The spectrum of the multiple window is equal to the sum of the spectra of the significant eigenvectors,[31]

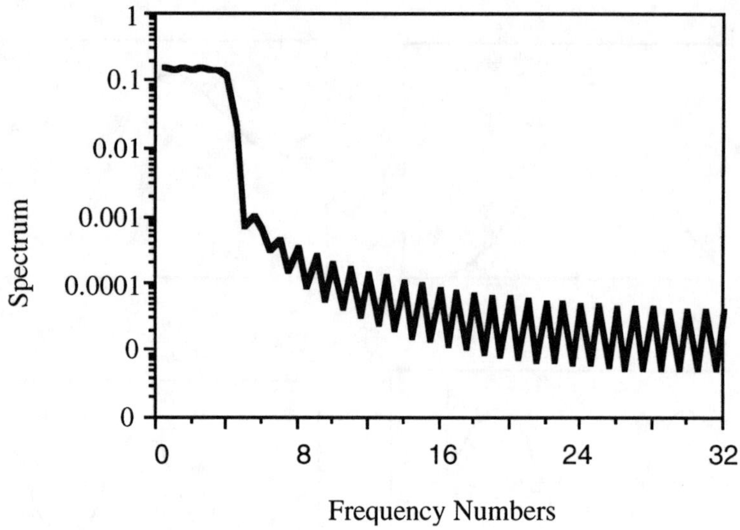

Figure 3.17: (b) The spectrum of the multiple window is shown above. The mainlobe is constant in the range $\pm\frac{2\pi}{64}4$. The largest sidelobe is 23 dB down from the maximum attained by the mainlobe.

$$\left|W_c(\omega)\right|^2 = \sum_{i=1}^{2\Delta fT} \frac{1}{\lambda_i}\left|W_i(\omega)\right|^2 \quad . \tag{3.82}$$

Example 3.12: An example of a multiple optimum window spectrum is shown in Figure 3.17. The mainlobe is flat over a bandwidth, $\pm\Delta f$, and the sidelobe level is below 30 dB. For this example we have selected T=64 and DfT=4 and p=6. The six largest eigenvalues are 1.0, 1.0, 1.0, 1.0, 0.999, and 0.993, and corresponding eigenvectors are shown in Figure 3.17a. The spectrum of the window was computed using (3.81) in which we have assumed all significant eigenvalues to be equal to one (End of Example).

Next we shall evaluate the variance of the spectral estimate using a multiple window.

$$\begin{aligned}
\text{Var}\{\hat{S}_x(k)\} &= E\{[\hat{S}_x(k) - S_x(k)*\left|W_C(\omega)\right|^2]^2\} \\
&= \frac{1}{P^2}\sum_{1}^{P-1}\sum_{m=0}^{P-1}\frac{1}{\lambda_1\lambda_m}E\{\left|X_1(k)\right|^2\left|X_m(k)\right|^2\} - [S_x(k)*\left|W_C(\omega)\right|^2]^2 \\
&= \frac{1}{P^2}\sum_{1}^{P-1}\sum_{m=0}^{P-1}\frac{1}{\lambda_1\lambda_m}\frac{1}{(2\pi)^2}\int\int_{-\pi}^{+\pi} S_x(\omega)S_x(\omega')W_1(k-\omega)W_m^*(k-\omega) \\
&\qquad W_m(k-\omega')W_1^*(k-\omega')\,d\omega\,d\omega'
\end{aligned} \tag{3.83}$$

We shall assume that $S_X(\omega)$ is approximately flat over a bandwidth of $\pm\Delta\omega$ and hence (3.83) may be reduced to

$$\mathrm{Var}\{\hat{S}_x(k)\} \approx \frac{1}{P}|S_x(k)|^2 \frac{1}{P}\sum_l \sum_m \frac{1}{\lambda_l \lambda_m} \frac{1}{4\pi^2} \left| \int_{-\Delta\omega}^{\Delta\omega} W_l(k-\omega) W_m^*(k-\omega)\, d\omega \right|^2 . \tag{3.84}$$

Since the DPSW are orthogonal on $\pm\Delta\omega$, the integral (3.84) may be written as

$$\delta_{ml} \int_{-\Delta\omega}^{\Delta\omega} |W_l(\omega)|^2 d\omega$$

Hence (3.84) reduces to

$$\mathrm{Var}\{\hat{S}_x(k)\} \approx \frac{1}{P}|S_x(k)|^2 \frac{1}{P}\sum_l \frac{1}{\lambda_l^2}\Big[\frac{1}{2\pi}\int_{-\Delta\omega}^{\Delta\omega} |W_l(k-\omega)|^2 d\omega\Big]^2 . \tag{3.85}$$

We can further simplify (3.85) by virtue of the fact that the window functions used are normalized eigenvectors whose spectral power is concentrated within a bandwidth of $\pm\Delta\omega$,

$$\mathrm{Var}\{\hat{S}_x(k)\} \approx \frac{|S_x(k)|^2}{P} . \tag{3.86}$$

Comparing (3.86) with (3.57) we find that if we equate the number of significant eigenvalues to the number of uncorrelated segments, the two results are the same except for a numerical constant. However, for the same resolution (see Exercise 3.12), the sidelobes of the multiple window are about 10 dB lower. This is the main advantage of the multiple-window method over the traditional single-window method.

D. Multiple Window and MLE of Spectrum

The eigenvectors of the **M** matrix possess several interesting properties, in particular, double orthogonality, concentration of energy, etc. The eigenvectors corresponding to the significant eigenvalues (close to unity) have their energy concentrated to a band, $\pm\frac{1}{2}\Delta\omega$. We shall call these signal eigenvectors and the others noise eigenvectors. The role of the noise eigenvectors is to transfer some energy from outside the band into the band of interest, $\pm\frac{1}{2}\Delta\omega$. Since this is undesirable we shall treat this as noise. The signal eigenvectors, on the other hand, keep the power practically undisturbed. Let $\mathbf{E}_s$ and $\mathbf{E}_\eta$ represent the subspaces spanned by signal eigenvectors and noise eigenvectors, respectively. The signal space has a dimension of $T \times P$, and the noise space has a dimension of $T \times (T-P)$.

Let us now model a stochastic process in the frequency range, $\pm\frac{1}{2}\Delta\omega$, as a sum of two components, $\mathbf{X} = \mathbf{E}_s\alpha + \mathbf{E}_\eta\beta$, where α and β are uncorrelated zero mean Gaussian random vectors of size $P \times 1$ and $(T-P) \times 1$), respectively. The first component represents a low-rank approximation of signal and the second component represents the model error, caused by the power which has leaked into the band from outside.[31] The covariance will be given by

$$\mathbf{C}_x = \mathbf{E}_s\mathbf{C}_\alpha\mathbf{E}_s^T + \mathbf{E}_\eta\mathbf{C}_\beta\mathbf{E}_\eta^T , \tag{3.87}$$

where $\mathbf{C}_\alpha$ and $\mathbf{C}_\beta$ are covariance matrices of α and β, respectively. Note that $\mathbf{E}_s$ and $\mathbf{E}_\eta$, being the eigenvectors of $\mathbf{M}$ matrix, are known. Let us now attempt MLE of the unknown quantities, $\mathbf{C}_\alpha$ and $\mathbf{C}_\beta$. The orthogonality of $\mathbf{E}_s$ and $\mathbf{E}_\eta$ implies that

$$|\mathbf{C}_X| = |\mathbf{C}_\alpha||\mathbf{C}_\beta|$$

and

$$\mathbf{C}_x^{-1} = \mathbf{E}_s\mathbf{C}_\alpha^{-1}\mathbf{E}_s^H + \mathbf{E}_\eta\mathbf{C}_\beta^{-1}\mathbf{E}_\eta^H. \tag{3.88}$$

Substituting (3.87) and (3.88) in (3.18) we obtain

$$\ln|\mathbf{C}_\alpha| + \gamma[\operatorname{tr}\{\mathbf{C}_\alpha^{-1}\mathbf{E}_s^H\hat{\mathbf{C}}_x\mathbf{E}_s\} - p] + \ln|\mathbf{C}_\beta| + \gamma[\operatorname{tr}\{\mathbf{C}_\beta^{-1}\mathbf{E}_\eta^H\hat{\mathbf{C}}_x\mathbf{E}_\eta\} - T + p] = \min, \tag{3.89}$$

which represents two separate unconstrained maximizations whose solution yields MLE of $\mathbf{C}_\alpha$ and $\mathbf{C}_\beta$,[32]

$$\hat{\mathbf{C}}_\alpha = \mathbf{E}_s^H\hat{\mathbf{C}}_x\mathbf{E}_s,$$
$$\hat{\mathbf{C}}_\beta = \mathbf{E}_\eta^H\hat{\mathbf{C}}_x\mathbf{E}_\eta. \tag{3.90}$$

The signal power in frequency range $\pm\frac{1}{2}\Delta\omega$ is given by

$$\operatorname{tr}\{\hat{\mathbf{C}}_\alpha\} = \sum_{i=1}^{P}\mathbf{e}_i^H\hat{\mathbf{C}}_x\mathbf{e}_i, \tag{3.91}$$

and the power that has leaked into the same range is given by

$$\operatorname{tr}\{\hat{\mathbf{C}}_\beta\} = \sum_{i=p+1}^{T}\mathbf{e}_i^H\hat{\mathbf{C}}_x\mathbf{e}_i. \tag{3.92}$$

Now consider a frequency band centered at w, that is, in the range $\omega \pm \frac{1}{2}\Delta\omega$. A stochastic process in this range may be modeled as

$$\mathbf{X} = \mathbf{h}(\omega)\mathbf{E}_s\alpha + \mathbf{h}(\omega)\mathbf{E}_\eta\beta,$$

where

$$\mathbf{h}(\omega) = \frac{1}{\sqrt{T}}\operatorname{diag}\{1, e^{-j\omega}, e^{-j2\omega}, e^{-j3\omega}, \ldots, e^{-j(T-1)\omega}\}.$$

Going through all steps shown above, we obtain the MLE of power in the frequency range $\omega \pm \frac{1}{2}\Delta\omega$,

$$\mathrm{tr}\{\hat{\mathbf{C}}_{\alpha}(\omega)\} = \sum_{i=1}^{P} \mathbf{e}_i^H \mathbf{h}^H(\omega)\hat{\mathbf{C}}_x \mathbf{h}(\omega)\mathbf{e}_i .$$

The average power in the frequency band, that is the power spectrum, is given by

$$S_x(\omega) = \sum_{i=1}^{P} \mathbf{e}_i^H \mathbf{h}^H(\omega)\hat{\mathbf{C}}_x \mathbf{h}(\omega)\mathbf{e}_i . \tag{3.93}$$

Substituting for the estimate of covariance matrix (3.12 with $\Delta T = T$) in (3.93), we can express (3.93) in a more familiar form,

$$\begin{aligned} S_x(\omega) &= \sum_{i=1}^{P} \left| \mathbf{X}^H \mathbf{h}(\omega)\mathbf{e}_i \right|^2 \\ &= \frac{1}{T}\sum_{i=1}^{P}\sum_{t=0}^{T-1} \left| x(t)e_i(t)e^{-jt\omega} \right|^2 . \end{aligned} \tag{3.94}$$

The above equation is similar to (3.81) where we have introduced the idea of multiple window. Thus, the spectral estimate obtained by using a multiple window is closely related to the MLE of the average power in a frequency band, $\omega \pm \frac{1}{2}\Delta\omega$.

3.6 Estimation of Bicovariance and Bispectrum

The bicovariance function on certain slices including the diagonal slice is of practical interest. It may be expressed in a matrix form (bicovariance matrix) and may be estimated just like a covariance matrix. The estimation of bispectrum and trispectrum follows essentially the same procedure as in the Welch spectrum estimation. However, their statistical characterization is far more involved and therefore not included. These topics and other issues are extensively covered in a recent book by C. L. Nikias and A. P. Petropulu.[50]

A. Bicovariance Matrix

The bicovariance function on the right diagonal, a line passing through the center with slope = +1.0 (see Figure 3.18), may be arranged in a matrix form. Such a matrix will be called here a bicovariance matrix, which is useful in the estimation of non-Gaussian AR and ARMA models of time series (see Section 2.6).

$$\mathbf{Bc} = \begin{bmatrix} Bc(0,0) & Bc(1,1) & \cdots & & Bc(\Delta T-1,\Delta T-1) \\ Bc(-1,-1) & Bc(0,0) & \cdots & & Bc(\Delta T-2,\Delta T-2) \\ & & \cdots & & \\ & & \cdots & & \\ & & \cdots & Bc(0,0) & Bc(1,1) \\ Bc(-\Delta T+1,-\Delta T+1) & & \cdots & Bc(-1,-1) & Bc(0,0) \end{bmatrix} . \tag{3.95}$$

We define the following data vectors:

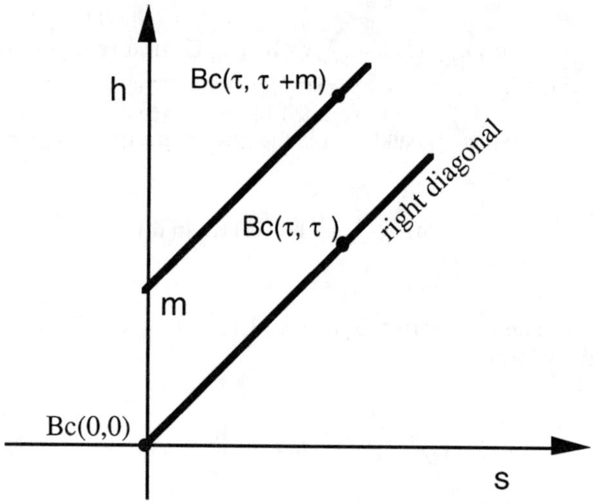

Figure 3.18: The bicovariance matrix is formed out of elements lying on the right diagonal starting at zero or any point on the diagonal. A similar bicovariance matrix on any subdiagonal can be formed.

$$\mathbf{x}_t = \mathrm{col}\{x(t),\ x(t+1),\ \ldots,\ x(t+\Delta T-1)\},$$
$$\mathbf{x}_t^2 = \mathrm{col}\{x^2(t),\ x^2(t+1),\ \ldots,\ x^2(t+\Delta T-1)\},$$

and form an outer product,

$$E\{\mathbf{x}_t \mathbf{x}_t^{2^T}\} = \begin{bmatrix} E\{x(t)x^2(t)\} & \ldots & E\{x(t)x^2(t+\Delta T-1)\} \\ & \vdots & \\ E\{x(t+\Delta T-1)x(t)^2\} & \ldots & E\{x(t+\Delta T-1)x^2(t+\Delta T-1)\} \end{bmatrix}$$

$$= \mathbf{Bc}.$$

The bicovariance matrix, like the covariance matrix, may be computed from uncorrelated data segments, each of length ΔT. Using each data segment we form the vectors $\mathbf{x}_t$ and $\mathbf{x}_t^2$ and then compute an outer product, namely, $\mathbf{x}_t \mathbf{x}_t^{2^T}$. Finally, we average the outer product matrices over all data segments,

$$\hat{\mathbf{B}}\mathrm{c} = \frac{1}{P}\sum_{i=0}^{P-1} \mathbf{x}_{i\Delta T}\mathbf{x}_{i\Delta T}^{2^T} \quad . \tag{3.96a}$$

The bicovariance matrix is Toeplitz but not symmetric. This structure of the bicovariance matrix may be used to obtain a better estimate, just as in the case of covariance matrix (Section 3.1). This has not been attempted so far.

Sometimes we would like to compute a bicovariance matrix whose first element is at a point s=h=t >0. For example, we would need a bicovariance matrix for the estimation of AR coefficients in ARMA model starting from s=h=t >q. Such a bicovariance matrix is called an extended bicovariance matrix (see page 138). The extended bicovariance matrix may be computed just as in (3.96a), but now the vectors $\mathbf{x}_t$ and $\mathbf{x}_t^2$ are defined differently, as follows:

$$\mathbf{x}_{t-\tau} = \mathrm{col}\,\{\, x(t-\tau),\, x(t-\tau+1),\, x(t-\tau+2),\, \ldots,\, x(t-\tau+\Delta T-1)\}$$
$$\mathbf{x}_t^2 = \mathrm{col}\,\{x^2(t),\;\; x^2(t+1),\;\; x^2(t+2),\;\; \ldots,\; x^2(t+\Delta T-1)\}$$

$$E\{\mathbf{x}_{t-\tau}\mathbf{x}_t^{2\,T}\} = \begin{bmatrix} E\{x(t-\tau)\,x^2(t)\} & \cdots & E\{x(t-\tau)\,x^2(t+\Delta T-1)\} \\ \vdots & & \vdots \\ E\{x(t-\tau+\Delta T-1)x^2(t)\} & \cdots & E\{x(t-\tau+\Delta T-1)\,x^2(t+\Delta T-1)\} \end{bmatrix}$$

$$= \begin{bmatrix} Bc\,(\tau,\tau) & \cdots & Bc\,(\tau+\Delta T-1,\tau+\Delta T-1) \\ & \cdots & \\ Bc\,(\tau-\Delta T+1,\tau-\Delta T+1) & \cdots & Bc\,(\tau,\tau) \end{bmatrix}$$

$$= \mathbf{B}c_\tau^e \qquad ,$$

where $\mathbf{B}c_\tau^e$ stands for the extended bicovariance matrix. With finite data the extended bicovariance matrix is estimated by computing an average of the outer product of the vectors defined above:

$$\hat{\mathbf{B}}c_\tau^e = \frac{1}{P}\sum_{i=0}^{P-1}\mathbf{x}_{i\Delta T-\tau}\,\mathbf{x}_{i\Delta T}^{2\;T}\,. \tag{3.96b}$$

Similarly, for the estimation of a bicovariance matrix on, say, the mth subdiagonal (Figure 3.18), we redefine the two data vectors as

$$\mathbf{x}_{t-\tau} = \mathrm{col}\,\{\, x(t-\tau)\quad x(t-\tau+1)\quad \ldots\; x(t-\tau+\Delta T-1)\}$$
$$\mathbf{x}_{t,m}^2 = \mathrm{col}\,\{x(t)\;\; x(t+m)\;\;\; x(t+1)\;\; x(t+m+1)\;\ldots$$
$$x(t+\Delta T-1)\;\; x(t+m+\Delta T-1)\}$$

and form an average of their outer product as in (3.96a),

$$\hat{\mathbf{B}}c_{\tau,m}^e = \frac{1}{p}\sum_{i=0}^{p-1}\left[\mathbf{x}_{i\,\Delta T-\tau}\right]\left[\mathbf{x}_{i\,\Delta T,m}^2\right]^T. \tag{3.96c}$$

Note that the actual number of data segments, each of length ΔT, which may be formed from the given time series of length T, would depend upon the parameters τ and m.

B. Bispectrum

The bispectrum may be computed either directly from the data or from the bicovariance function which is estimated from the data. First, we shall show how to directly compute the bispectrum from the data. As in the Welch method for spectrum estimation, the available data record is divided into partially overlapping segments. Each segment is multiplied with a window function, and the DFT of the product is computed. An estimate of the bispectrum is given by an average of the triple product of the DFT coefficients,

$$\hat{B}_x(k,l) = \frac{1}{p}\sum_{i=1}^{p} X_i(k)X_i(l)X_i(-k-l).$$

We shall briefly examine the bias in the estimate. The variance of the estimate, however, is far more involved; hence, it will not be considered. Consider the following triple product:

$$\begin{aligned}E\{X(k)X(l)X(-k-l)\} = \frac{1}{8\pi^3}\int_{-\pi}^{+\pi}\int\int E\{dX_G(\omega')dX_G(\omega'')dX_G(\omega''')\}\\ H(\omega'-\frac{2\pi k}{T})H(\omega''-\frac{2\pi l}{T})H(\omega'''+\frac{2\pi(k+l)}{T})\\ \exp(j(\omega'+\omega''+\omega''')\frac{T-1}{2}.\end{aligned} \tag{3.97}$$

where $H(\omega)$ is the Fourier transform of the window function; for example, for uniform linear window (see (2.24)),

$$H(\omega'-\frac{2\pi k}{T}) = \frac{\sin\left(\omega'-\frac{2\pi k}{T}\right)\frac{T}{2}}{\sin\frac{(\omega'-\frac{2\pi k}{T})}{2}}.$$

We have noted in Chapter One (1.75) that

$$\begin{aligned}&E\{\frac{1}{2\pi}dX_G(\omega_1)\frac{1}{2\pi}dX_G(\omega_2)\frac{1}{2\pi}dX_G(\omega_3)\}\\ &= \frac{1}{8\pi^3}Bs(\omega_1,\omega_2)d\omega_1 d\omega_2 \delta(\omega_1+\omega_2+\omega_3)d\omega_3.\end{aligned}$$

Using (1.75) in (3.97) we obtain

$$\begin{aligned}&E\{X(k)X(l)X(-k-l)\}\\ &= \frac{1}{4\pi^2}\int_{-\pi}^{+\pi}\int B_x(\omega',\omega'')H(\omega'-\frac{2\pi k}{T})H(\omega''-\frac{2\pi l}{T})H(-(\omega'+\omega'')+\frac{2\pi(k+l)}{T})d\omega' d\omega''.\end{aligned}$$

We define discrete bispectrum as follows:

$$B_x(k,l) = \frac{1}{4\pi^2}\int_{-\pi}^{+\pi}\int B_x(\omega',\omega'')H_b(\omega'-\frac{2\pi k}{T},\omega''-\frac{2\pi l}{T})d\omega' d\omega'', \tag{3.98}$$

where $H_b(.)$ stands for bispectral window defined as

$$H_b(\omega', \omega'') = H(\omega')H(\omega'')H^*(\omega' + \omega''). \tag{3.99}$$

It may be shown that $H_b(\omega', \omega'')$ satisfies the symmetry properties of the bispectrum (see (1.72)). A window may also be used in the lag domain (quadratic window); that is, a window is applied to the bicovariance function. A discussion on quadratic window for bispectrum estimation may be found in Reference 33.

Example 3.13: For uniform data window the bispectral window is given by

$$H_b(\omega', \omega'') = \frac{\sin\left(\omega'\frac{T}{2}\right)\sin\left(\omega''\frac{T}{2}\right)\sin\left((\omega'+\omega'')\frac{T}{2}\right)}{\sin\left(\frac{\omega'}{2}\right)\sin\left(\frac{\omega''}{2}\right)\sin\left(\frac{\omega'+\omega''}{2}\right)}.$$

Note that on the main axes and on the left diagonal, that is, a line through the center with slope = − 1, the bispectral window is equal to the spectral window,

$$H_b(\omega, 0) = H_b(0, \omega) = H_b(\omega, -\omega) = T\left[\frac{\sin\left(\omega\frac{T}{2}\right)}{\sin\left(\frac{\omega}{2}\right)}\right]^2.$$

However, on the right diagonal,

$$H_b(\omega, \omega) = \left[\frac{\sin\left(\omega\frac{T}{2}\right)}{\sin\left(\frac{\omega}{2}\right)}\right]^2 \frac{\sin(\omega T)}{\sin(\omega)}.$$

The main lobe of the bispectral window will be elongated in the direction of the left diagonal but compressed in the direction of right diagonal, which results into a cross section similar to the hexagonal passband of a triple product of three ideal bandpass filters (see Figure 1.15).

The bispectral estimate on l=0 and k=0 axes is highly biased, as it is always zero when the mean is subtracted from the time series. To see this, we note that when the mean is subtracted from the time series, the first DFT coefficient is always zero; that is, $X(0) = 0$. Hence,

$$\begin{aligned} B_x(k,0) &= E\{X(k)X(0)X(-k)\} = 0 \\ B_x(0,l) &= E\{X(0)X(l)X(-l)\} = 0. \end{aligned} \tag{3.100}$$

The bispectrum may also be computed as a discrete Fourier transform of the bicovariance function which is computed from the data,

$$\hat{B}c(s,h) = \frac{1}{T-\delta}\sum_{t=0}^{T-\delta} x(t)x(t+s)x(t+h),$$

where

$$\delta = \max\{|s|, |h|\}$$

and $-\Delta T \le s \le \Delta T$ and $-\Delta T \le h \le \Delta T$, where $\Delta T << T$. The bispectrum is given by

$$\hat{B}s(\omega_1,\omega_2) = \sum_{h}^{\Delta T} \sum_{s=-\Delta T}^{\Delta T} \hat{B}c(s,h)\exp(-j(\omega_1 s + \omega_2 h)).$$

The bispectrum on the right diagonal assumes a particularly simple form. It is shown as the Fourier transform of a (left) diagonally summed bicovariance function.

$$\begin{aligned} Bs(\omega,\omega) &= \sum_{s} \sum_{h=-\Delta T}^{+\Delta T} Bc(s,h)\exp(-j(s+h)\omega) \\ &= \sum_{p=-2\Delta T}^{2\Delta T} \sum_{s=N_1}^{N_2} Bc(s,p-s)\exp(-jp\omega), \end{aligned} \tag{3.101a}$$

where $p=s+h$, $N_1=-\Delta T+p$ and $N_2=\Delta T$ for $p>0$ and $N_1=-\Delta T$ and $N_2=\Delta T+p$ for $p<0$. The inner sum in (3.101a) implies summation over all elements on a pth left diagonal of the bicovariance function; that is,

$$D_\Sigma(p) = \sum_{s=N_1}^{N_2} Bc(s,p-s).$$

We can then express the bispectrum on the right diagonal as a Fourier transform of $D_\Sigma(p)$:

$$Bs(\omega,\omega) = \sum_{p=-2\Delta T}^{2\Delta T} D_\Sigma(p)\exp(-j\omega p). \tag{3.101b}$$

C. **Aliasing Effect in Bispectrum**

When a bispectrum or trispectrum is computed from the samples of an analog signal, the question of sampling rate is important to avoid aliasing errors. The Nyquist sampling rate required for an alias-free spectrum is two times the highest frequency. For the bispectrum the required sampling rate is three times, and for the trispectrum it is four times the highest frequency.[51,52] Consider the bispectral estimate using the DFT coefficients (3.97). Assume that $x(t)$, $t=0, 1, \ldots$ is a *deterministic* digital signal obtained by sampling an analog band-limited (highest frequency is b Hz, $b<\pi$ Hz) signal at a rate 1 Hz. Using the relation between the continuous and the discrete Fourier transform (2.31) in (3.97) (without ensemble averaging),

$$\begin{aligned} \tilde{B}s(\omega_1,\omega_2) &= X(\omega_1)X(\omega_2)X^*(\omega_1+\omega_2) \\ &= \sum_{n=-\infty}^{\infty} X^c(\omega_1 - 2\pi n_1) \sum_{n_2=-\infty}^{\infty} X^c(\omega_2 - 2\pi n_2) \sum_{n_3=-\infty}^{\infty} X^{c^*}(\omega_1+\omega_2 - 2\pi n_3), \end{aligned} \tag{3.102}$$

where superscript c stands for the Fourier transform of continuous function. Observe that the first two terms in (3.102) are periodic, and they give rise to a periodic pattern of pass and stop regions,

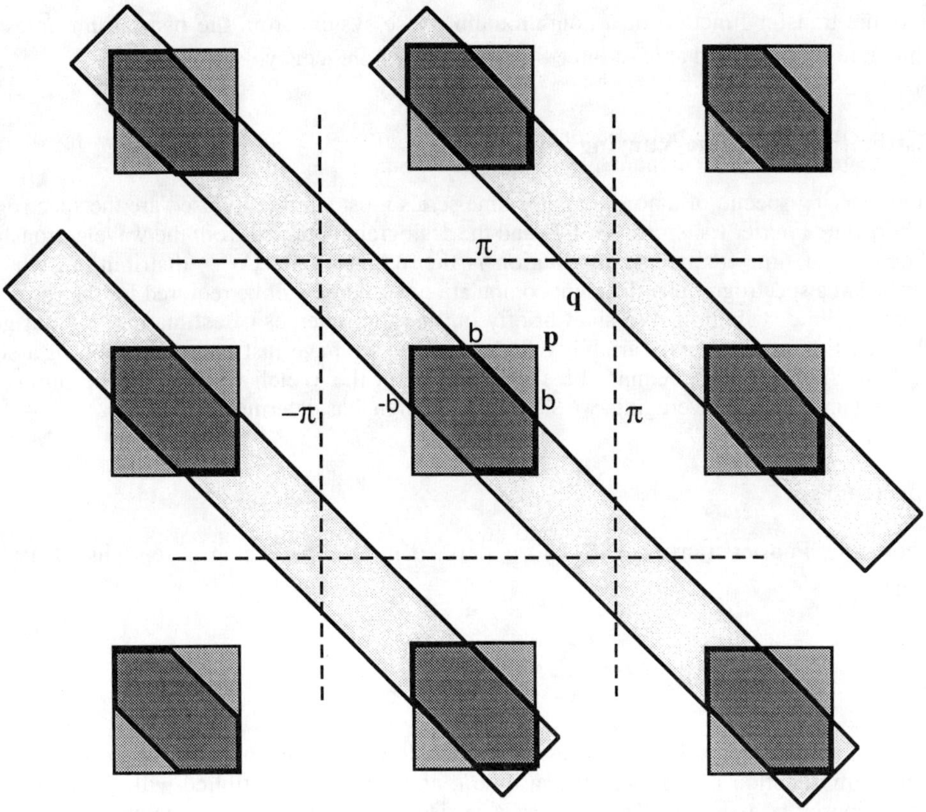

Figure 3.19: The checkerboard pattern due to the first two terms in (3.102) and the oblique stripes produced by the third term. The product of these two patterns results in replicated hexagons.

like a checkerboard (see Figure 3.19). The third term produces an oblique striped pattern also shown in Figure 3.19. A product of these two patterns results in repeated hexagons, similar to the one shown in Figure 1.13. One of the neighboring oblique stripes may overlap with the principal square, giving rise to a nonzero term in the triple product in (3.102). This will happen when points p and q in Figure 3.19 come together, that is, when

$$\pi - \frac{b}{2} = b, \quad \text{or}$$

$$b = \frac{2}{3}\pi.$$

Thus, aliasing error will be present whenever $b \geq \frac{2}{3}\pi$; that is, the sampling rate must be at least three times the highest frequency. Using a similar approach it is shown in References 51 and 52 that the aliasing error in the trispectrum will be zero if the sampling rate is at least four times the highest frequency.

The aliasing error described above will not be present when the signal is a stationary stochastic process. The triple-product term, when statistically averaged (ensemble averaging), vanishes. Sampling at a rate at least twice the highest frequency is adequate to yield alias-free estimates of the bispectrum and trispectrum. However, there will be some complication when trying to estimate the transfer function of a nonminimum phase system from the bispectrum unless the sampling rate is maintained at least three times the highest frequency.[52]

3.7 Estimation of Time-Varying Spectrum:

The time-varying spectra of a nonstationary time series most commonly used are the spectrogram from short-time Fourier transform (STFT) and the scalogram obtained from the wavelet transform. Another type of time-frequency distribution is the Wigner-Ville (WV) distribution, which is different from a spectrogram. A different computational scheme will be required for the estimation of Wigner-Ville distribution. We shall briefly outline the schemes of estimation without going into the question of statistical variability. Indeed, since we have just one realization of a signal which, being nonstationary, cannot be segmented as in the Welch method for the purpose of carrying out an ensemble averaging, we shall treat the signal as deterministic.

A. Spectrogram

The short-time Fourier transform (STFT) is defined in (1.115), which is reproduced here for convenience:

$$X_1(t,\omega) = \sum_{\tau=-\infty}^{\infty} x(\tau)w(t-\tau)\exp(-j\omega\tau).$$

Note that w(t) is a lowpass filter; any one of the window functions discussed in Section 3.3 will do. A window function is first centered at time t, and it is then multiplied with the signal. The product is Fourier transformed. This process is carried on at a sequence of time instants which are separated by the effective duration of the window. For finite data and finite window, a computational scheme based on fast Fourier transform has been suggested.[34] The scheme involves the following steps:

(1) Divide the total signal into overlapping segments each of length ΔT points, equal to the length of the window function, and the nonoverlapping part is equal to the effective length of the window, ΔT_{eff}. For example, for a uniform window, $\Delta T_{eff}=\Delta T$, and for a Hamming window, $\Delta T_{eff} = \frac{1}{4}\Delta T$.

(2) Each segment is now multiplied with the window, and its DFT is computed.

(3) The Fourier transform is next squared to obtain the spectrogram at time equal to $(n+\frac{1}{2})\Delta T_{eff}$, where n=0, 1, 2, ... stands for the segment number. Steps (b) and (c) are repeated until all segments are exhausted.

Remember that there will be $P = \frac{T}{\Delta T_{eff}}$ segments in all. The width of each frequency bin will be proportional to $\frac{\lambda}{\Delta T}$, where λ is a constant depending upon the type of window used. Thus, the time-frequency plane of the spectrogram is sampled at ΔT_{eff} in time and $\frac{\lambda}{\Delta T}$ in frequency. Evidently, it is not possible to simultaneously achieve both high time and frequency resolution. The smallest resolution cell is a rectangle of size $\Delta T_{eff} \times \frac{\lambda}{\Delta T}$ (see Figure 3.20). The discrete spectrogram is given by

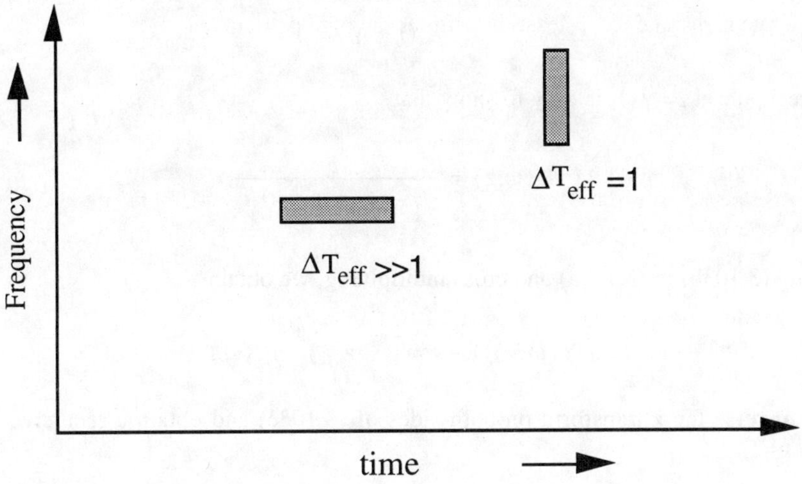

Figure 3.20: Resolution cell in the time-frequency plane is a rectangle. Two possible types of resolution cells corresponding to high frequency resolution and high time resolution are shown.

$$S_{x_1}(n,m) = \left|X_1(n\Delta T_{eff}, m\Delta\omega)\right|^2, \quad n = 0, 1, \ldots P$$
$$m = 0, 1, \ldots \frac{\pi}{\lambda}\Delta T\,.$$

In (1.115) if we choose a window that is a causal filter, it is possible to express the STFT in a recursive form reference.[53] To show this, let us rewrite (1.115):

$$\begin{aligned} X_1(t,\omega) &= e^{j\omega t}\sum_{\tau=-\infty}^{t} x(\tau)w(t-\tau)\exp(-j\omega(t-\tau)) \\ &= e^{j\omega t}\sum_{\tau'=0}^{\infty} x(t-\tau')w(\tau')\exp(-j\omega\tau') \\ &= e^{j\omega t}\sum_{\tau'=0}^{\infty} x(t-\tau')h(\omega,\tau'), \end{aligned}$$

where w(τ) is a causal filter and $h(\omega,\tau') = w(\tau')\exp(-j\omega\tau')$. Let

$$y(\omega,t) = \sum_{\tau'=0}^{\infty} x(t-\tau')h(\omega,\tau')$$

or its z-transform,

$$Y(\omega,z) = X(z)H(\omega,z)\,. \tag{3.103a}$$

For the window function we shall assume a particular filter whose z-transform is

$$W(z) = \frac{1}{(1 - \gamma z^{-1})^{p}},$$

having a real pole at z= γ. It is easy to show that

$$H(\omega,z) = \frac{1}{(1 - (\gamma e^{-j\omega})z^{-1})^{p}}. \tag{3.103b}$$

Substituting (3.103b) in (3.103a) and cross multiplying, we obtain

$$Y(\omega, z)(1 - (\gamma e^{-j\omega})z^{-1})^{p} = X(z). \tag{3.103c}$$

Now take inverse the z-transform on both sides of (3.103c) and obtain a recursive relation for y(w,t) (for p=3):

$$y(\omega,t) = x(t) + 3\gamma e^{-j\omega t}y(\omega,t-1) - 3\gamma^{2}e^{-j2\omega t}y(\omega,t-2) + \gamma^{3}e^{-j3\omega t}y(\omega,t-3),$$

which may be solved with initial condition,

$$y(\omega, t) = 0 \text{ for } t < 0.$$

Note that the window function used here is asymmetric with its peak shifted away (delayed) from the point where the time-varying Fourier transform is sought. The output is therefore correspondingly delayed, depending upon the pole position.[53]

B. Multiscale Spectrogram

The square of the wavelet transform, which was introduced in Chapter One, is known as a scalogram, since a scale factor is used in defining the wavelet. Here we shall use relative frequency as the scale factor. The wavelets can be looked upon as bandpass filters of constant Q. The square of the wavelet transform would represent power in a cell in the time-frequency plane. The dimension of the cell is controlled by the bandwidth and time duration of the wavelet used for analysis (see Figure 3.20). In the low-frequency region the time resolution is coarse. The reverse is true in the high-frequency region. The scalogram has a similar physical meaning as the spectrogram except for the varying size of the resolution cell. Hence, it is suggested that the scalogram may be aptly called a multiscale spectrogram. In the usual spectrogram the cell size is fixed. The multiscale spectrogram is defined as a square of the wavelet transform (see page 52), that is, convolution of the signal with a wavelet,

$$X_{2}(t,\omega) = \sqrt{\frac{\omega_{0}}{\omega}} \sum_{\tau=-\infty}^{\infty} x(\tau)\, w\!\left(\frac{\omega}{\omega_{0}}(t-\tau)\right). \tag{3.104a}$$

Hence, the multiscale spectrogram is given by

$$S_{X_{2}}(t,\omega) = \left|X_{2}(t,\omega)\right|^{2}. \tag{3.104b}$$

For evaluation of the multiscale spectrogram it is not necessary to require that the wavelets derived from the mother wavelet by varying the scale factor be orthogonal as in other applications; for

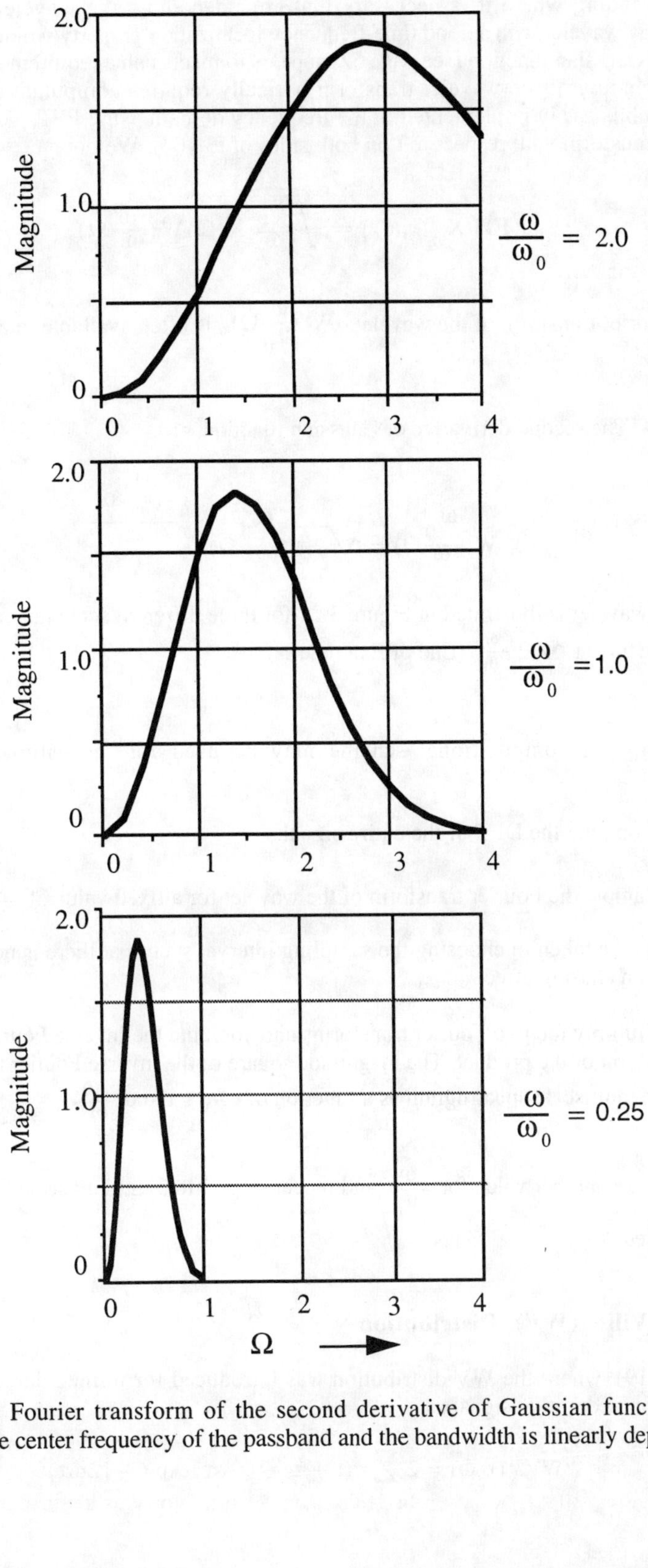

Figure 3.21: The Fourier transform of the second derivative of Gaussian function for three different scales. The center frequency of the passband and the bandwidth is linearly dependent on the scale.

example, in coding where it is necessary that after decoding, the wavelet transform must be invertible. Any wavelet with a good time-frequency localization property would suffice. The second derivative of Gaussian function (see page 52) appears to meet such a requirement.

The estimation of the wavelet transform basically requires computation of the convolution product, which is easily implemented in the frequency domain using FFT. To show this, compute the Fourier transform with respect to t on both sides of (3.104). We obtain

$$\mathrm{FT}\{X_2(t,\omega)\} = \sqrt{\frac{\omega_0}{\omega}}\, X(\Omega)\, W\left(\frac{\omega}{\omega_0}\Omega\right), \tag{3.105}$$

where the Fourier transform of the wavelet, $W(\frac{\omega}{\omega_0}\Omega)$, is often available in a closed form.

Example 3.14: The second derivative of Gaussian function,

$$W\left(\frac{\omega_0}{\omega}\Omega\right) = \sqrt{2\pi}\left(\frac{\omega_0}{\omega}\Omega\right)^2 e^{\frac{-\left(\frac{\omega_0}{\omega}\Omega\right)^2}{2}},$$

as a possible wavelet is illustrated in Figure 3.21 for three different scale factors. The maximum of the wavelet occurs at $\sqrt{2}\,\frac{\omega}{\omega_0}$ (End of Example).

The following computational scheme may be used for the estimation of multiscale spectrogram:

(1) Compute the DFT of the entire signal.

(2) Sample the Fourier transform of the wavelet for a fixed value of $\frac{\omega}{\omega_0}$. Care should be taken in choosing the sampling interval such that there is no time domain aliasing effect.

(3) Multiply the two Fourier transforms and compute the inverse Fourier transform of the product. The magnitude square of the inverse Fourier transform is the multiscale spectrogram as a function of t for a fixed $\frac{\omega}{\omega_0}$.

(4) Take another value for $\frac{\omega}{\omega_0}$ and repeat steps 1 to 3 until all scale factors have been covered.

C. Wigner-Ville (WV) Distribution

Equation (1.119), where the WV distribution was introduced for infinite duration signal, may be rewritten for finite duration signal as

$$WV_x(t,\omega) = 2\sum_{\tau\in I} x(t+\tau)\,x(t-\tau)\exp(-j2\omega\tau), \tag{3.106}$$

where I is a set of integers, and the product, $x(t+\tau)x(t-\tau)$, is nonzero. For example, if the time series is given over an interval, 0 to T – 1, τ lies in the set shown in Figure 3.21. For a fixed t, the range of τ is

$$|\tau| \le t \qquad \text{for} \quad 0 \le t \le \frac{T-1}{2}$$

$$|\tau| \le (T-1-t) \quad \text{for} \quad \frac{T-1}{2} < t \le T-1, \tag{3.107}$$

as shown in Figure 3.22 by a double arrowed line.

It may be observed from (3.106) that the discrete WV distribution will be periodic in ω with the period being π or $\pm\frac{\pi}{2}$. To avoid aliasing, there should be no power outside the frequency band, $\pm\frac{\pi}{2}$. This may be achieved by first sampling the signal at Nyquist rate followed by midpoint interpolation. As an example, let the original signal be band limited to 100 Hz (that is, spectral power is zero outside the band $\pm$ 100 Hz). The signal is sampled at 200 Hz and then interpolated at midpoints. In principle, the interpolation can be carried out without any error as the signal is assumed to be band limited.

The following computational scheme may be used for numerical evaluation of the discrete WV distribution.

(1) For fixed t, form a product time series, $y(t,\tau) = x(t+\tau)x(t-\tau)$, over the range of τ as determined in (3.107).

(2) Multiply the product time series, $y(t,\tau)$, with a spectral window such as the Hanning window.

(3) Compute the discrete Fourier transform. Since the length of the product series, $Y(t,\tau)$, depends on t, we may have to pad zeros to make all product series of equal length, so that the frequency interval is the same for all t.

Equation (3.106) can be expressed in a standard DFT form as shown below:

$$\begin{aligned} WV_x(t,\omega) &= 2\sum_{-T+1}^{T-1} y(t,\tau)w(\tau)\exp(-j2\omega\tau) \\ &= 4\,\mathrm{Re}\left\{\sum_{0}^{T-1} y(t,\tau)w(\tau)\exp(-j2\omega\tau)\right\} \\ &\quad - 2x^2(t)w(0) \end{aligned} \tag{3.108}$$

where w(t) is a symmetric spectral window. Let us assume $\omega = \frac{\pi k}{T}$ in (3.108) and rewrite as

$$\begin{aligned} WV_x(t,\omega) &= 4\,\mathrm{Re}\left\{\sum_{0}^{T-1} y(t,\tau)\,w(\tau)\exp\left(-j2\frac{\pi k}{T}\tau\right)\right\} \\ &\quad - 2x^2(t)w(0). \end{aligned} \tag{3.109}$$

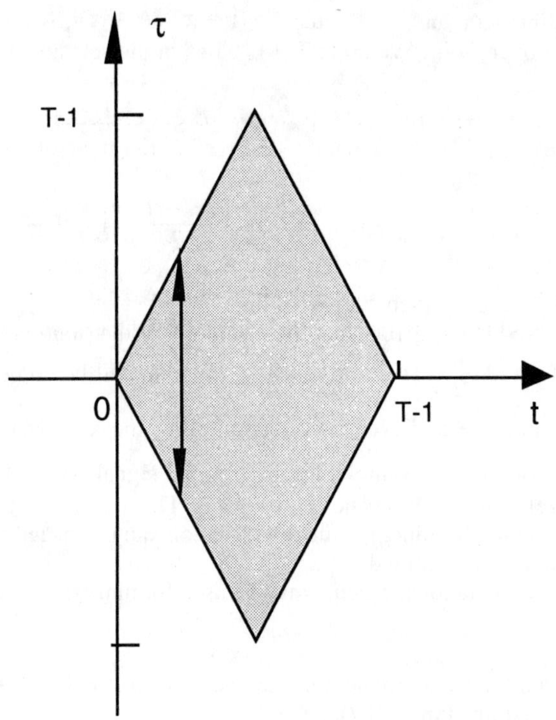

Figure 3.22: The product, $x(t+\tau)x(t-\tau)$, in (3.106) for the evaluation of the discrete WV distribution has a diamond-shaped support region as shown above.[35] (T is odd.)

which is in a standard DFT form. An efficient algorithm for the computation of discrete WV is given in Reference 36.

D. Ambiguity Function

The estimation of the ambiguity function is along the same line as the estimation of WV distribution. The discrete ambiguity function is given by (1.120), which is reproduced here for convenience:

$$A(\Omega,\tau) = \sum_{t=-\infty}^{\infty} x(t+\tfrac{\tau}{2})x(t-\tfrac{\tau}{2})e^{-j\Omega t}.$$

We shall remove the fractional lag, $\frac{\tau}{2}$, by introducing a new variable, $\tau' = \frac{\tau}{2}$. When the time series is of finite duration, (1.120) may be expressed as

$$A(\Omega,2\tau') = \sum_{t\in I} x(t+\tau')x(t-\tau')e^{-j\Omega t}, \tag{3.110}$$

where I is a set of integers where the product, $x(t+\tau')x(t-\tau')$, is nonzero. The set is a diamond-shaped figure as in Figure 3.22. For a fixed τ, it is easily shown that t is bounded as

$$|\tau| \le t \le T - |\tau|.$$

As for the computation scheme, we follow the procedure outlined for the estimation of WV distribution with the only difference that the discrete Fourier transform is now over t for a fixed τ, which must be an even integer.

We noted in Chapter One that the ambiguity function and the WV distribution form a Fourier transform pair. The cross terms in WV distribution are often reduced by means of suitable filtering, which in the frequency domain is a weighting function or a frequency-dependent window. Hence, weighting is first applied to the ambiguity function, and then it is Fourier transformed. Both operations can be combined into one,[37] giving the following expression for WV distribution:

$$WV_x(t,\omega) = \frac{1}{2\pi}\int_{-\pi}^{\pi} d\Omega e^{j\Omega t} \sum_{\tau' \in I} \sum_{t' \in I} x(t' + \tau')x(t' - \tau')\phi(\Omega,\tau')e^{-j\Omega t'}e^{-j2\omega\tau'}, \tag{3.111}$$

where $\phi(\Omega,\tau')$ is the weighting function. Let us take a close look at the integrand in (3.111). The inner sum represents the windowed (frequency-dependent window) WV distribution, and the outer sum represents the one-dimensional Fourier transform of the WV distribution.

Equation (3.111) may also be expressed in a convolution form. For this, we first carry out the integration with respect to Ω and rewrite (3.111) as

$$WV_x(t,\omega) = \sum_{\tau' \in I} \sum_{t' \in I} x(t' + \tau')x(t' - \tau')\psi(t - t', \tau')e^{-j2\omega\tau'}, \tag{3.112}$$

where

$$\psi(t - t', \tau') = \frac{1}{2\pi}\int_{-\pi}^{\pi} \phi(\Omega,\tau')\exp(j\Omega(t - t'))d\Omega.$$

You may note that

$$\sum_{t' \in I} x(t' + \tau')x(t' - \tau')\psi(t - t', \tau')$$

is a convolution between $x(t' + \tau')x(t' - \tau')$ and $\psi(t', \tau')$ with respect to variable t'. The smoothed WV distribution as given in (3.111) is indeed equal to the spectrogram, provided

$$\psi(t,\tau) = w(t + \tau)w(t - \tau).$$

E. **Relation between Spectrogram and WV Distribution**

The spectrogram and the multiscale spectrogram (i.e., scalogram) are indeed related to the WV distribution through a convolution-type integral. To show the relationship between the spectrogram and the distribution, take the square on both sides of (1.115),

$$S_{x_1}(t,\omega) = |X_1(t,\omega)|^2 = \sum_{\tau=-\infty}^{\infty} \sum_{-\infty}^{\infty} x(\tau)x(\tau')w(t - \tau)w(t - \tau')\exp(-j\omega(\tau - \tau')). \tag{3.113}$$

Setting $\tau = n + \frac{k}{2}$ and $\tau' = n - \frac{k}{2}$ in (3.113), we obtain

$$S_{x_1}(t,\omega) = \sum_{n=-\infty}^{\infty} \sum_{k=-\infty}^{\infty} x(n+\frac{k}{2})x(n-\frac{k}{2})w(t-n-\frac{k}{2})w(t-n+\frac{k}{2})\exp(-j\omega k), \tag{3.114}$$

which, if we set

$$\psi(t-n,k) = w(t-n+\frac{k}{2})\,w(t-n-\frac{k}{2}),$$

may be reduced to a smoothed WV distribution as in (3.112). Further, notice that in (3.114) the inner sum may be looked upon as a DFT of the product of two functions, namely, $x(n+k)x(n-k)$ and $\psi(t-n,k)$. Hence, it may be expressed as a convolution of their Fourier transforms, that is, a convolution of WV distributions of signal and window,

$$S_{x_1}(t,\omega) = \frac{1}{2\pi}\sum_{n=-\infty}^{\infty}\int_{-\pi}^{\pi} WV_x(n,\nu)\,WV_w(t-n,\omega-\nu)\,d\nu. \tag{3.115}$$

Similarly, to obtain a relation between the multiscale spectrogram (scalogram) and the WV distribution, square both sides of (3.104a). We have

$$S_{X_2}(t,\omega) = \frac{\omega_0}{\omega}\sum_{\tau''=-\infty}^{\infty}\sum_{\tau'=-\infty}^{\infty} x(\tau')x(\tau'')\,w(\frac{\omega}{\omega_0}(t-\tau'))w(\frac{\omega}{\omega_0}(t-\tau'')). \tag{3.116}$$

In (3.116), as before, we introduce a change of variables, $\tau = n + \frac{k}{2}$ and $\tau' = n - \frac{k}{2}$, giving

$$S_{X_2}(t,\omega) = \frac{\omega_0}{\omega}\sum_{n=-\infty}^{\infty}\sum_{k=-\infty}^{\infty} x(n+\frac{k}{2})x(n-\frac{k}{2})\,w(\frac{\omega}{\omega_0}(t-n-\frac{k}{2}))w(\frac{\omega}{\omega_0}(t-n+\frac{k}{2})). \tag{3.117}$$

Further, the product, $w(\frac{\omega}{\omega_0}(t-n-\frac{k}{2}))w(\frac{\omega}{\omega_0}(t-n+\frac{k}{2}))$, in (3.117) may be expressed as inverse Fourier transform of the WV distribution (see (1.119)),

$$w(\frac{\omega}{\omega_0}(t-n-\frac{k}{2}))w(\frac{\omega}{\omega_0}(t-n+\frac{k}{2})) =$$

$$\frac{1}{2\pi}\int_{-\pi}^{\pi} WV_w(\frac{\omega}{\omega_0}(t-n,\Omega)\exp(j\Omega\frac{\omega}{\omega_0}k))d\Omega. \tag{3.118}$$

Using (3.118) in (3.117) we obtain after simplification,

$$S_{X_2}(t,\omega) = \frac{1}{2\pi}\sum_{n=-\infty}^{\infty}\int_{-\pi}^{\pi} WV_x(n,\Omega)\,WV_w(\frac{\omega_0}{\omega}(t-n),\frac{\omega}{\omega_0}\Omega)\,d\Omega. \tag{3.119}$$

Equations (3.115) and (3.119) establish an analytical relationship between the WV distribution and spectrograms. It is common knowledge that the WV distribution is troubled by the presence of cross terms, but the spectrograms are generally free of the cross terms. That this is not always true may be suspected from the above result connecting the WV distribution and spectrograms. Indeed, the existence of cross terms in spectrograms is clearly documented in Reference 55.

3.8 Exercises

1. In computing the covariance function via (3.1) the finite data is assumed to be either zero outside the data interval or periodic. These two assumptions lead to different estimates. The first assumption leads to the regular autocorrelation function and the second assumption to the circular autocorrelation (page 180). Both estimates can be expressed in matrix notations. Show

$$C_{cir}(\tau) = \mathbf{x}^T \mathbf{J}_\tau \mathbf{x} \quad \text{and}$$

$$C_{reg}(\tau) = \mathbf{x}^T \mathbf{L}_\tau \mathbf{x},$$

where

$$\mathbf{x} = \{x(0), x(1), \ldots x(T-1)\}^T$$

is a data vector, $\mathbf{J}_\tau$ is a permutation matrix obtained by applying circular shift of t units to each row of an identity matrix, and $\mathbf{L}_\tau$ is similar to $\mathbf{J}_\tau$, but in place of the circular shift, the regular shift is applied.

2. In the estimation of the BT spectrum via the covariance matrix, show that there is an implicit triangular window applied to the covariance function. [Hint: Expand the matrix product given in (3.27) and collect all terms from a diagonal of the covariance matrix.]

3. Let $X(k)$, k=0, 1, ..., $T-1$ be the DFT of a sequence, x(t), t=0,1, ..., $T-1$. Show that the inverse DFT of $|X(k)|^2$ is equal to the autocorrelation function of x(t). It is periodic with period T. The periodicity arises on account of an implicit assumption in the DFT that x(t) is periodic with a period, T. Let $C_{cir}(\tau)$ be the autocorrelation function of periodic sequence (e.g., obtained from inverse DFT of $|X(k)|^2$), and $\hat{C}_x(\tau)$ is the autocorrelation function of nonperiodic finite sequence computed via (3.1) scaled by $(1 - \frac{|\tau|}{T})$. Show that

$$C_{cir}(\tau) = \sum_{i=-\infty}^{\infty} \hat{C}_x(\tau + i\Delta T).$$

4. The cross-spectrum between two stationary time series, x(t) and y(t), is estimated as in (3.25). Show the following properties of the cross-spectral estimate (Welch method):

$$E\{\hat{S}_{xy}(\omega)\} = S_{xy}(\omega) * \frac{|\bar{W}_0(\omega)|^2}{\Delta T}$$

and

$$\operatorname{var}\{\hat{S}_{xy}(\omega)\} = \frac{1}{P}\left[\{S_x(\omega)S_y(\omega)\} * \frac{\left|\overline{W}_0(\omega)\right|^4}{\Delta T^2}\right] \qquad \text{for } \omega \notin 0,\pi,$$

$$= \frac{1}{P}\left[\left\{S_x(\omega)S_y(\omega) + \left|S_{xy}(\omega)\right|^2\right\} * \frac{\left|\overline{W}_0(\omega)\right|^4}{\Delta T^2}\right] \qquad \text{for } \omega = 0,\pi .$$

5. The bias in the estimated spectrum depends upon the window function. Let $W_0(\omega)$ be the Fourier transform of a linear window (used in time domain). Show that

$$E\{\hat{S}_x(\omega)\} \approx S_x(\omega) + \frac{S''_x(\omega)}{2!}\frac{1}{2\pi}\int_{-\pi}^{\pi} \xi^2 W_0^2(\xi)d\xi + \ldots ,$$

where $S''_x(\omega)$ is the second derivative. Papoulis.[38] has derived a window which minimizes the integral on the right-hand side, and hence is a minimum bias window. More recently, multiple windows that minimize both bias as well as the variance have been introduced.[56]

6. In (3.48) a composite window was introduced as a product of autocorrelation of a linear window and a quadratic window. Let the linear window be the default constant window. Design a quadratic window such that the composite window is equal to some prescribed function (see Reference 39).

7. In (3.80b) let the **M** matrix be defined as follows:

$$\mathbf{M} = \left\{\frac{\sin(n-k)2\pi\alpha}{(n-k)\pi}\right\},$$

where α is a variable (in (3.80b) $\alpha = \frac{1}{\Delta T}$). Show that as $\alpha \to 0$ the optimum window is simply the uniform window. The properties of the optimum window also tend to those of the uniform window. The optimum windows as a function of a are considered in Reference 40.

8. Recall that the mean of the spectral estimate obtained by the method of overlap and add is given by (3.44):

$$E\{\hat{S}_x(k)\} = \frac{1}{2\pi}\int_{-\pi}^{\pi} S_x(\omega)\left|W_0\left(\frac{2\pi k}{\Delta T} - \omega\right)\right|^2 \frac{1}{2r+1}\sum_{i=-r}^{r} e^{-j(\omega - \frac{2\pi}{\Delta T}k)(1-\alpha)\Delta T i}\, d\omega$$

Now let $1-\alpha = \frac{1}{\Delta T}$ and $w_0(0) = 1.0$ and $w_0(t) = 0$ for $t > 0$. Compute the effective window and show that the estimate is unbiased provided the stochastic process is an MA process of order r.

9. We often subtract the mean from the time series before computing its spectrum. As a result, the estimated spectrum (using, for example, the Welch method) at ω=0 is always zero, even though the true spectrum may have a finite value. How do you then estimate the spectrum at zero frequency[41]?

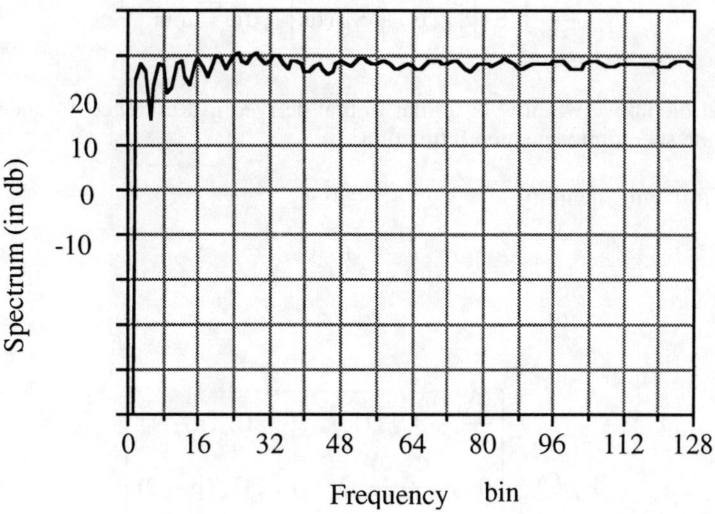

Figure 3.23: Spectrum of a chirp window of length equal to 64. The width of a frequency bin is $\frac{\pi}{128}$.

10. A linear FM (chirp signal), when used as a window, has some interesting properties as outlined in Reference 42. In particular the spectrum has a deep null at the correct frequency only if there is one complex sinusoid. On the other hand, no such null is observed when there are two or more sinusoids. Explain this phenomenon in the light of the actual spectrum of a chirp window shown in Figure 3.23. The discrete chirp window is given by

$$h(t) = \exp\left(j\frac{\pi}{T}(t(t+1))\right), \quad t = 0, 1, 2, \ldots, T-1.$$

11. In the overlap and add method let $\alpha = 0.5$, that is, 50% overlap; then the matrix **D** becomes

$$\mathbf{D} = \text{diag}\ \{1, (-1)^k, 1, (-1)^k, \ldots\}.$$

We notice that the spectral matrix as defined in (3.41) turns out to be the same as Rader's[43] segmentation procedure, which possesses certain computational advantages arising out of the fact that the DFT of the overlapped portion of the segment need be evaluated only once. Verify the above claim.

12. Show that the resolution of the spectral estimate obtained from a multiple window method is equal to the resolution of the spectrum obtained via the Welch method when the time series is segmented with 50% overlap into as many segments as the number of significant eigenvalues and a uniform window is employed.

13. Since the error in the estimate of covariance function increases with lag, a tapered window would help to reduce the effect of this error on the spectrum. This is another reason for using a window with a finite covariance function. Find a window $w_1(\tau)$ such that the total mean square error in the estimate of the spectrum is minimum; that is,

$$\frac{1}{2\pi}\int_{-\pi}^{\pi} E\{\left|\hat{S}_x(\omega) - S_x(\omega)\right|^2\} d\omega = \min$$

It turns out that the above window is similar to that derived in Reference 44 under an additional condition that the spectrum is a smooth function.

14. Define the following quantity:

$$S_A(\omega) = \mathbf{A}^H \mathbf{C}_x^{Ca} \mathbf{A} \,,$$

where

$$\mathbf{C}_x^{Ca} = \begin{bmatrix} \frac{C_x(0)}{2} & C_x(1) & \ldots & C_x(p-1) \\ 0 & \frac{C_x(0)}{2} & C_x(1) \ldots & C_x(p-2) \\ \cdots & \cdots & \cdots & \cdots \\ 0 & 0 & 0 & \ldots \frac{C_x(0)}{2} \end{bmatrix}$$

and

$$\mathbf{A} = \text{col}\,\{1, e^{j\omega}, e^{j2\omega}, \ldots, e^{j(P-1)\omega}\}.$$

Show that Re{ $S_A(\omega)$ } is equal to the spectrum of the time series. What about the imaginary part? $S_A(\omega)$ is known as analytic spectrum and its magnitude as spectral envelope.[45] It is claimed to be smooth and stable.

15. Define

$$\hat{S}_x(\omega) = \frac{\mathbf{A}^H \mathbf{w}^H \hat{\mathbf{C}}_x \mathbf{w} \mathbf{A}}{\|\mathbf{w}\,\mathbf{A}\|} \,,$$

where

$$\mathbf{A} = \text{col}\,\{1, e^{j\omega}, e^{j2\omega}, \ldots, e^{j(\Delta T - 1)\omega}\},$$
$$\mathbf{w} = \text{diag}\,\{w(0), w(1), \ldots, w(\Delta T - 1)\},$$
$$\hat{\mathbf{C}}_x = \frac{1}{P}\sum_{i=1}^{P} \mathbf{x}_i \mathbf{x}_i^H, \text{ and}$$
$$\mathbf{x}_i = \text{col}\,\{x(i\bar{\alpha}\Delta T), x(i\bar{\alpha}\Delta T + 1), \ldots, x(i\bar{\alpha}\Delta T + \Delta T - 1)\}.$$

Show that $\hat{S}_x(\omega)$ is identical to the Welch spectrum. Is $\hat{S}_x(\omega)$ always >0?

16. Consider the following feedback system shown in Figure 3.24. Let x(t) and y(t) be input and output time series tapped at points shown in Figure 3.24. Show that the transfer function between x(t) and y(t) is given by

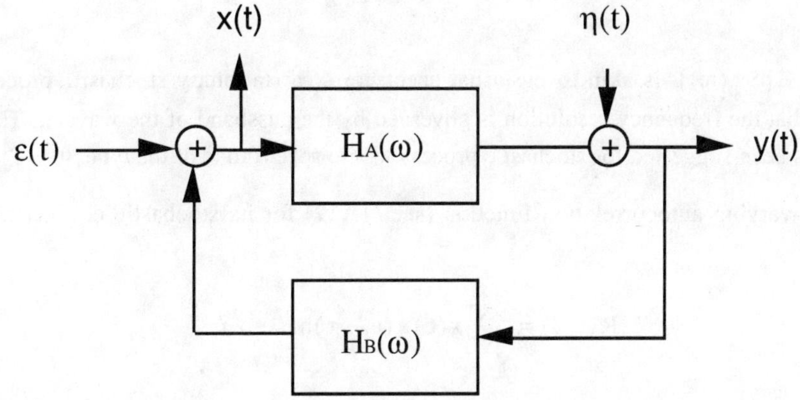

Figure 3.24: Feedback system. We like to find the transfer function between x(t) and y(t).

$$H_{xy}(\omega) = \frac{H_A(\omega) S_\varepsilon(\omega) + H_B^*(\omega) S_\eta(\omega)}{S_\varepsilon(\omega) + \left|H_B(\omega)\right|^2 S_\eta(\omega)},$$

where $H_A(\omega)$ is the transfer function of the system in the forward path and $H_B(\omega)$ is the transfer function of the system in the feedback path. You may assume that $\varepsilon(t)$ and $\eta(t)$ are uncorrelated stationary stochastic processes. $S_\varepsilon(\omega)$ and $S_\eta(\omega)$ are spectra of $\varepsilon(t)$ and $\eta(t)$, respectively.[46]

17. Let $x(k\Delta t)$ be a band-limited time series obtained by sampling a band-limited continuous process (highest frequency= ω_0) at a rate greater than two times the Nyquist rate $(\Delta t \le \frac{\pi}{2\omega_0})$. Show that the WV distribution in continuous time and that in discrete time are related as[47]

$$WV_x(t,\omega) = \sum_{k=-\infty}^{\infty} WV_x(k\Delta t,\omega) \frac{\mathrm{Sin}\frac{\pi}{\Delta t}(t - k\Delta t)}{\frac{\pi}{\Delta t}(t - k\Delta t)}.$$

18. A B-spline window of order m is defined as follows:[48]

$$w_m(t) = \frac{1}{2\pi}\int_{-\pi}^{\pi} \left\{ \frac{\sin(\omega\frac{T}{2})}{(\omega\frac{T}{2})} \right\}^m \exp(j\omega t)\, d\omega.$$

Show that the local support of the window, that is, where $w_m(t) \neq 0$, is equal to $(-m\frac{T}{2}, m\frac{T}{2})$.

19. A stationary stochastic process x(t) is subjected to wavelet transform. Let $\tilde{S}_x(t,\omega)$ be the multiscale spectrogram (scalogram) defined in (3.104). We now define another quantity,

$$\tilde{S}_x(\omega) = \sum_t \tilde{S}_x(t,\omega).$$

Show that $E\{\tilde{S}_x(\omega)\}$ is akin to the usual spectrum of a stationary stochastic process with a difference that the frequency resolution is governed by the passband of the wavelet. This type of estimate has been suggested for stochastic process whose spectrum is of the type $\omega^{-\nu}$.[49]

20. A time-varying autocorrelation function (see (1.112) for its stochastic equivalent) may be defined as

$$R(t,\tau) = \sum_{t'=-\infty}^{\infty} x(t')x(t'-\tau)h(t-t'),$$

where h(t) is a window, which when selected as a causal filter function, enables $R(t,\tau)$ to be evaluated recursively.[54] For example, let

$$h(t) = \frac{(1-\gamma)^p (t+p-1)!}{(t)!(p-1)!}\gamma^t,$$

whose z-transform is given by

$$H(z) = \frac{(1-\gamma)^p}{(1-\gamma z^{-1})^p},$$

where $|\gamma| < 1$ and p is an integer. Show that the time-varying autocorrelation function satisfies the following recursive relation:

$$R_{p+1}(t,\tau) = \gamma R_{p+1}(t-1,\tau) + (1-\gamma)R_p(t,\tau).$$

References

1. R. B. Blackman and J. W. Tukey, *The Measurement of Power Spectra from the Point of View of Communications Engineering*, Dover Publication, New York, 1958.

2. P. D. Welch, The use of fast Fourier transform for estimation of power spectra: A method based on time averaging over short, modified periodograms, *IEEE. Trans.*, AU-15, 70-73, 1967.

3. F. J. Harris, On the use of windows for harmonic analysis with the discrete Fourier transform, *Proc. IEEE*, 66, 51-83,1978.

4. G. C. Carter, C. H. Knapp, and A. H. Nuttal, Estimation of the magnitude squared coherence function via overlapped fast Fourier transform processing, *IEEE Trans.*, AU-21, 337-344, 1973.

5. L. R. Rabiner and J. B. Allen, Short time Fourier analysis technique for FIR system identification and power estimation, *IEEE Trans.*, ASSP-27, 182-192,1979.

6. V. J. Mathews and D. H. Youn, Analysis of the short-time unbiased spectrum estimation algorithm, *IEEE Trans.*, ASSP-33, 136-142, 1985.

7. T. W. Anderson, *The Statistical Analysis of Time Series*, John Wiley & Sons, Inc., New York, 452,1958.

8. J. P. Burg, D. G. Luenberger, and D. L. Wenger, Estimation of structured covariance matrices, *Proc. IEEE*, 70, 963-974, 1982.

9. D. R. Fuhrmann, Progress in structured covariance estimation, 4th ASSP Workshop on Spectrum Estimation and Modelling, Minneapolis, MIN, 158-161, 1988.

10. D. R. Fuhrmann and M. I. Miller, On the existence of positive-definite maximum-likelihood estimate of structured covariance matrices, *IEEE Trans.*, IT-34, 722-729, 1988

11. A. Dembo, The relation between maximum likelihood estimation of structured covariance matrices and periodograms, *IEEE Trans.*, ASSP-34, 1661-1662, 1986.

12. M. I. Miller and D. L. Snyder, The role of likelihood and entropy in incomplete-data problems: application to estimating point process intensities and Toeplitz constrained covariances, *Proc. IEEE*, 75, 892-907,1987.

13. A. Dempster, N. Laird, and D. Rubin, Maximum likelihood from incomplete data via the EM algorithm, *J. R. Stat. Soc.*, B-39, 1-37, 1977.

14. H. Wang and G. H. Wakefield, Signal subspace approximation for line spectrum estimation, *IEEE Proc. ICASSP-87*, 2054-2057, 1987.

15. J. A. Cadzow, Signal enhancement: A useful signal processing tool, IEEE ASSP Workshop on Spectrum Estimation, 162-167, 1988.

16. V. J. Mathews, D. H. Youn, and N. Ahmed, A unified approach to non-parametric spectrum estimation algorithms, *IEEE Trans.*, ASSP-35, 338-349, 1987.

17. C. K. Yuen, Quadratic windowing in the segment averaging method for power spectrum computation, *Technometrics*, 20, 195-200, 1978.

18. A. H. Nuttall and G. C. Carter, Spectrum estimation using combined time and lag weighting, *Proc. IEEE*, 70, 1115-1125, 1982.

19. A. T. Walden, Variance and degrees of freedom of a spectral estimator following data tapering and spectral smoothing, *Signal Process*, 20, 67-75, 1990.

20. N. R. Goodman, On the joint estimation of the spectra, cospectra and quadrature spectrum of a two dimensional spectrum of stationary process, New York University, New York, Sci. paper 10 (AD 134919), 1957.

21. G. C. Carter, C. H. Knapp, and A. H. Nuttal, Statistics of the estimate of the magnitude-coherence function, *IEEE Trans.*, AU-21, 388-389, 1973.

22. M. J. Hinich and C. S. Clay, The application of the discrete Fourier transform in the estimation of power spectra, coherence, and bispectra of geophysical data, *Rev. of Geophys.*, 6, 347-363, 1968.

23. L. D. Enochson and N. R. Goodman, Gaussian approximations to the distribution of sample coherence, AFFDL-TR-65-67 (AD 620987), June 1965.

24. D. E. Amos and L. H. Koopman, Tables of the distribution of the coefficient of coherence for stationary bivariate Gaussian process, Sandia Corp., Mono. SCR-483, 1963.
25. D. J. Thomson, Estimation of coherence between complicated processes, Fifth ASSP Workshop on Spectrum Estimation and Modeling, 256-259, 1990.

26. L. J. Tick, Estimation of coherency, in *Spectral Analysis of Time Series*, B. Harris (Ed), John Wiley & Sons, Inc., New York,1967.

27. A. Eberhard, An optimal discrete window for calculation of power spectra, *IEEE., Trans.*, AU-21, 37-43, 1973.

28. D. Slepian, H. O. Pollack, and H. J. Landau, Prolate spheroidal wave functions, Fourier analysis and uncertainty, *Bell Syst. Tech. J.*, 40, 43-84, 1961.

29. J. F. Kaiser and R. W. Schafer, On the use of the I0-Sinh Window for spectrum analysis, *IEEE Trans.*, ASSP-28, 105-107, 1980.

30. L. C. Barbosa, A maximum energy concentration window, *IBM J. Res. Dev.,* 30, 321-325, 1986.

31. D. J. Thomson, Spectrum estimation and harmonic analysis, *Proc. IEEE*, vol 70, pp. 1055-1096, 1982.

32. B. D. Van Veen and L. L. Scharf, Estimation of structured covariance matrices and multiple window spectrum analysis, *IEEE Trans.*, ASSP-38, 1467-1471, 1990.

33. I. Konvalinko, A new class of two dimensional windows for bispectrum estimation, *Signal Process*, 37, 147-156, 1994.

34. J. B. Allen, Short term spectral analysis, synthesis, and modification by discrete Fourier transform by discrete Fourier transform, *IEEE Trans.*, ASSP-25, 235-238, 1977.

35. M. A. Poletti, The development of a discrete transform for the Wigner distribution and ambiguity function, *J. Acoust. Soc. Am.*, 84, 238-252, 1988.

36. Mingui Sun, Ching-Chung Li, L. N. Sekhar, and R. J. Sclabassi, Efficient computation of the discrete Pseudo-Wigner distribution, *IEEE Trans.*, ASSP-37, 1735-1741, 1989.

37. W. Martin and P. Flandrin, Wigner-Ville spectral analysis of non-stationary processes, *IEEE Trans.*, ASSP-33, 1460-1470, 1985.

38. A. Papoulis, Minimum-bias windows for high resolution spectral estimates, *IEEE Trans.*, IT-19, pp. 9, 1973.

39. A. H. Nuttal and G. C. Carter, Spectral estimation using combined time and lag weighting, *Proc. IEEE*, 1115-1125, 1982.

40. H. Rohling and J. Schuermann, Discrete time window functions with arbitrarily low sidelobe level, *Signal Process*, 5, 127-138, 1983.

41. P. Heidelberger and P. D. Welch, A spectral method for confidence interval generation and run length control in simulation, *Commun. ACM*, 24, 233-245, 1981.

42. L. J. Griffiths, High resolution spectral estimation: Rethinking the Fourier transform, *IEEE Proc. ICASSP-86,* 169-172, 1986.

43. C. M. Rader, An improved algorithm for high speed autocorrelation with application to spectrum estimation. *IEEE Trans.,* AU-18, 439-442, 1970.

44. M. Bertran-Salvans, A generalized window approach to spectral estimation, *IEEE Trans.*, ASSP-32, 7-19, 1984.

45. M. A. Lagunas, M. Amengual, and M. E. Forcada, The periodogram envelope in parametric and nonparametric spectrum estimation, *IEEE Proc. ICASSP-89*, 2174-2177, 1989.

46. H. Akaike, Some problems in the application of the cross-spectral method, in *Spectral Analysis of Time Series*, B. Harris (Ed), John Wiley & Sons, 1967.

47. T. A. C. M. Claasen and W. F. G. Mecklenbrauker, The Wigner Distribution — A tool for time frequency signal analysis, *Philips J. Res.*, 35, part I, 217-250, part II, 276-300; and part III, 372-389, 1980.

48. K. Toraichi, M. Kamada, and R. Mori, A new series of digital filter windows B-spline windows, *IEEE Proc. ICASSP-86*, 1349-1352,1986.

49. P. Abry, P. Goncalves, and P. Flandrin, Wavelet-based spectral analysis of 1/f processes, *IEEE Proc. ICASSP-93*, III, 237-240, 1993.

50. C. L. Nikias and A. P. Petropulu, *Higher-Order Spectra Analysis, A Nonlinear Signal Processing Framework*, PTR Prentice Hall Inc, New York, 1993.

51. L. A. Pflug, Sampling requirements and aliasing for higher-order correlations, *J. Acoust. Soc. Am.*, 94, 2159-2172, 1993.

52. J. Le Roux, C. Coroyer, and D. Rossille, Illustration of the effects of sampling on higher-order spectra, *Signal Process*, 36, 375-390, 1994.

53. W. Chen, N. Kehtarnavaz, and T. W. Spencer, An efficient algorithm for time varying Fourier transform, *Proc. IEEE*, 41, 2488-2490, 1993.

54. M. G. Amin and K. D. Feng, Order recursive spectrum estimation, *Proc. IEEE*, 76, 289-290, 1988.

55. S. Kadambe and G. F. Boudreaux-Bartels, A comparison of the existance of "cross terms" in the Wigner-Ville distribution and the squared magnitude of the wavelet transform and the short time Fourier transform, *IEEE, Trans.*, SP-40, 2498-2517, 1992.

56. K. S. Riedel and A. Sidorenko, Minimum bias multiple taper spectral estimation, *IEEE Trans.*, SP-43, 188-195, 1995.

Chapter Four

Spectrum Estimation: *High-Resolution Methods*

4.1 An Overview

The linear methods of spectrum estimation, both the Blackman-Tukey (BT) and the direct method based on FFT suffer from a major problem of resolution. In both methods some invalid assumption, namely, zero data or repetitive data outside the duration of observations, is made. The assumption becomes implicit when one uses a window function and applies a finite Fourier transform as in FFT. We have shown in the last chapter that the estimated spectrum is a convolution between the true spectrum and the spectrum of the window function (see (3.31)). The estimated spectrum will be a smeared version of the true spectrum. In the process of smearing, closely spaced spectral peaks get merged into a broader peak and the spectral power from a strong peak leaks into its neighborhood, often leading to false estimates of spectral power or false spectral peaks. A good deal of effort has gone into the design of windows with "good" properties (see Section 3.5 for details on different types windows). Windows based on the discrete prolate spheroidal sequence appear to come closest to ideal windows, namely, in respect of the narrow mainlobe and low sidelobes.

A. Data Adaptive Window

When the dynamic range of the spectrum is very high, the best available windows become ineffective. Such a situation arises, for example, when there are closely spaced sinusoids in the presence of noise. To partially alleviate the problem of resolution Capon[1] came up with a data adaptive window where the spectrum of the window adapts itself to the true spectrum such that the bias or leakage of power is minimum. This is achieved through a continuous adjustment of the spectrum of the window, keeping in view the nature of the true spectrum. Let us treat a stochastic process as a sum of a sinusoid of a given frequency and the "rest" of the process as unwanted interference and apply the maximum likelihood detection principle. We seek a filter that extracts the sinusoid without any distortion and at the same time minimizes the influence of the "rest" of the process. The output power of the filter will be equal to the power of the sinusoid plus the power contribution from the interference, which may be looked upon as a bias or power leakage. The filter is designed to automatically minimize the bias. A bank of such filters, one at each frequency, can be thought of for estimation of power as a function of frequency, that is, the power spectrum. Such an estimate was called a maximum likelihood (ML) spectral estimate by Capon[1] because for a Gaussian process this is equivalent to maximizing the likelihood of detecting a sinusoid in the presence of Gaussian noise. But the time series model described above is strictly not equal to the usual model of a sinusoid plus Gaussian noise; hence, the filter defined by Capon cannot be treated as an ML estimate. For this reason Capon's ML spectral estimate is sometimes known as a minimum variance (MV) estimate. We shall, however, continue to use the original name. The ML spectrum delivers high resolution at moderate SNR (e.g., SNR > 0 dB). It is generally robust as precise modeling of the time series is not required, and hence the modeling errors, if any, do not ruin the results.

B. Extrapolation

In the next approach, to overcome the deleterious effect of the use of a window in linear spectral estimation, we completely eliminate the use of the window itself. This can be achieved if we are able to extrapolate the available data outside the duration of the data. Such a prediction is possible if we are able to fit the data with a finite rational model. From the data we must be able to estimate the finite number of coefficients of a rational model. This would naturally be more reliable than estimating the power at the large number of frequencies required to accurately define the power spectrum. Once the model parameters are estimated, the spectrum may be derived from the model parameters themselves or from the extrapolated data using the derived rational model. Such a spectrum is often known as the parametric spectrum. In Chapter Two we studied some of the most important time series models, namely, moving average (MA), autoregressive (AR), and autoregressive moving average (ARMA) models.

Instead of extrapolating the observed time series we may attempt to extrapolate the autocorrelation function, either computed from the observed time series or directly measured.[2] The extrapolation of the autocorrelation function is done under the constraint that the resulting spectrum leads to maximum entropy for the underlying stochastic process. The Fourier transform of the autocorrelation function including the extrapolated component gives us the so-called maximum entropy (ME) spectrum. It turns out that the maximum entropy extrapolation of the autocorrelation function is equivalent to fitting an AR model to the time series.[3] Indeed, the ME spectrum can directly be related to the AR model under an additional constraint that the AR filter is a minimum phase filter. It is interesting to note that the classical Blackman-Tukey(BT) spectrum can be compared with the MA spectrum. In both cases the autocorrelation function is of finite duration. The ARMA spectrum has been shown to be a maximum entropy spectrum with additional constraints derived from the ouput of certain linear filters acting on the time series.[4] We shall later show that the maximum entropy (ME) method has a better resolution capability than the maximum likelihood (ML) method. The number of autocorrelation coefficients used in the estimation of the ME spectrum (or the order of the AR model) plays a major role in determining the resolution capability of the maximum entropy method.

C. Subspace Approach

An alternate model of a time series is a sum of r random sinusoids in the background of noise of a known covariance function, for example, white noise whose covariance function is a delta function but of unknown variance. Given a covariance function at a finite number of lags, $C_x(0), C_x(1), C_x(2), \ldots, C_x(p)$, where p=r, the number of sinusoids, a theorem due to Caratheodory,[5] states that the frequencies and the amplitudes of all sinusoids can be determined uniquely. Pisarenko[6] gave a method for calculating the parameters of the sinusoids. The method involves finding the zeros of a polynomial whose coefficients are the elements of the null eigenvector of the covariance matrix. The background noise is assumed to be white with known variance which is then subtracted from the diagonal entries of the covariance function. If we enlarge the size of the covariance matrix, that is, for $p > r$, the minimum eigenvalue repeats $(p-r+1)$ times, and it is equal to the noise power. The eigenvectors corresponding to the repeated eigenvalues span a subspace which is orthogonal to the space spanned by the vectors representing r sinusoids. A vector $\mathbf{h}(\omega) = \mathrm{col}\{1, e^{j\omega}, e^{j2\omega}, e^{j3\omega}, \ldots, e^{j(p-1)\omega}\}$ is said to represent a sinusoid of frequency ω in p-dimensional space. We shall refer to it as a *frequency vector*. This property of the covariance matrix has been used not only in the estimation of frequencies of the sinusoids but also the number of sinusoids and the noise power. The most important point to keep in mind about the subspace method is that its resolution property is independent of the signal-to-noise ratio (SNR). This is only a theoretical property, and in practice the errors in the estimation of the covariance function, modeling errors, and unknown background noise can seriously limit its resolution capability.

It is clear from the above overview that the key to high resolution lies in precise modeling of the time series. In the BT/FFT method the time series is simply modeled as a stationary stochastic process, the lowest level of modeling; on the other hand, in the subspace method the time series is modeled as a sum of random sinusoids and white noise with all poles and zeros confined to the unit circle, the highest level of modeling. In Figure 4.1 the different spectrum estimation methods are arranged in increasing order of model refinement.

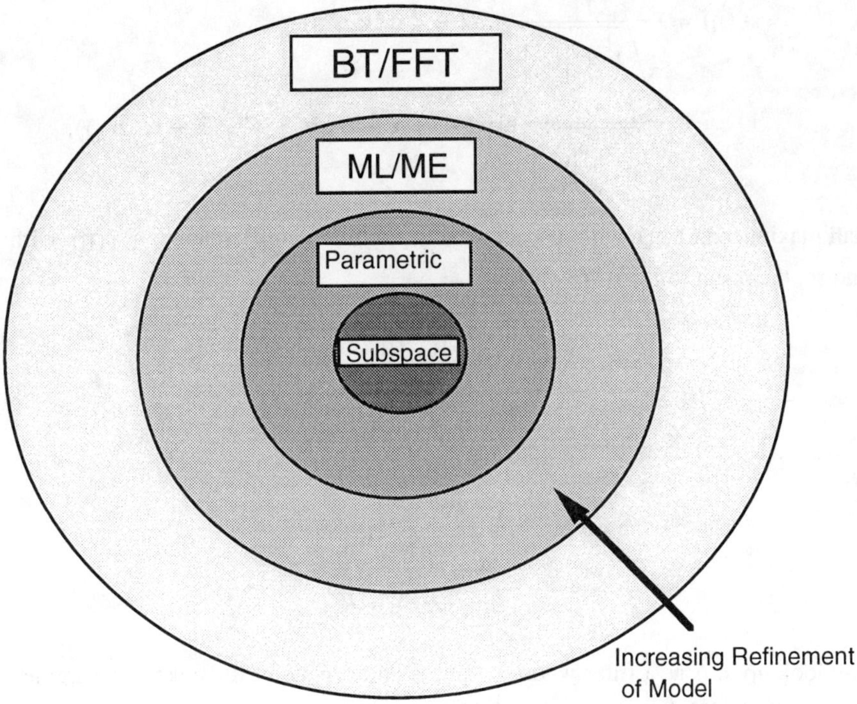

Figure 4.1: Domain of modern spectrum analysis. Higher resolution is achieved by means of refined modeling.

4.2 **Maximum Likelihood (ML) Spectrum**

Let x(t), t=0, 1, 2, ..., T − 1 be a time series, that is, the sum of sinusoids of known frequency in the background of Gaussian noise.

$$x(t) = \alpha_0 \exp(j\omega_0 t) + \eta(t). \tag{4.1}$$

As an introduction to Capon's ML method, an outline of maximum likelihood estimation of α_0 is first described. Define the following vectors:

$$\mathbf{x} = \text{col}\{x(0), x(2), x(2), \ldots, x(T-1)\},$$

$$\mathbf{h}_0 = \text{col}\{1, e^{j\omega_0}, e^{j2\omega_0}, e^{j\omega_0}, \ldots, e^{j(T-1)\omega_0}\},$$

$$\eta = \text{col}\{\eta(0), \eta(2), \eta(2), \ldots, \eta(T-1)\}.$$

In Section 1.2 we introduced the vector random variable and its Gaussian probability density function (pdf). In ML estimation the unknown parameter is so chosen that the probability of observing the stochastic process is maximized (see Reference 7 for more details on ML estimation). The pdf of **h** is given by

$$f(\eta) = \frac{1}{\sqrt{2^p \pi^p |\mathbf{C}_\eta|}} \exp\left(-\tfrac{1}{2}\eta^T \mathbf{C}_\eta^{-1} \eta\right)$$

$$= \frac{1}{\sqrt{2^p \pi^p |\mathbf{C}_\eta|}} \exp\left(-\tfrac{1}{2}(\mathbf{x} - \alpha_0 \mathbf{h}_0)^T \mathbf{C}_\eta^{-1} (\mathbf{x} - \alpha_0 \mathbf{h}_0)\right),$$

We shall maximize ln $f(\eta)$ with respect to α_0. For this we differentiate ln $f(\eta)$ with respect to α_0 and set the result to zero. We obtain

$$\hat{\alpha}_0 = \frac{\mathbf{h}_0^H \mathbf{C}_\eta^{-H} \mathbf{x}}{\mathbf{h}_0^H \mathbf{C}_\eta^{-H} \mathbf{h}_0} = \mathbf{A}^H \mathbf{x},$$

where

$$\mathbf{A} = \frac{\mathbf{C}_\eta^{-1} \mathbf{h}_0}{\mathbf{h}_0^H \mathbf{C}_\eta^{-H} \mathbf{h}_0}. \tag{4.2}$$

We may look upon **A** as a filter whose output is an estimate of the unknown parameter. Further, it is easy to show that $\hat{\alpha}_0$ is unbiased; that is,

$$E\{\hat{\alpha}_0\} = \alpha_0$$

and

$$E\{|\hat{\alpha}_0|^2\} = |\alpha_0|^2 + \frac{1}{\mathbf{h}_0^H \mathbf{C}_\eta^{-H} \mathbf{h}_0}. \tag{4.3}$$

A. Capon's Method

We now introduce Capon's approach to estimate the amplitude of a complex sinusoid. Let us define a filter vector, $\mathbf{A}=\text{col}\{a_0, a_1, a_2, \ldots, a_{T-1}\}$, having properties that (i) the power of the sinusoid remains undisturbed, and (ii) the variance of the filtered noise is minimum. The first condition requires that

$$|\mathbf{A}^H \mathbf{h}_0|^2 = 1, \tag{4.4}$$

and the second condition can be realized by minimizing

$$\mathbf{A}^H \mathbf{C}_\eta \mathbf{A} = \min. \tag{4.5}$$

Equations (4.4) and (4.5) constitute a variational problem which may be solved using the Lagrange multiplier technique where we minimize a quantity,

$$J = \mathbf{A}^H \mathbf{C}_\eta \mathbf{A} + \lambda(1 - \left|\mathbf{A}^H \mathbf{h}_0\right|^2) = \min,$$

with respect to **A**. We obtain

$$\mathbf{C}_\eta \mathbf{A} = \lambda \mathbf{h}_0 \mathbf{h}_0^H \mathbf{A}, \tag{4.6}$$

which may be written as a standard eigenvalue equation,

$$\frac{1}{\lambda}\mathbf{A} = \mathbf{C}_\eta^{-1} \mathbf{h}_0 \mathbf{h}_0^H \mathbf{A}. \tag{4.7}$$

Observe that **A** is an eigenvector of $\mathbf{C}_\eta^{-1} \mathbf{h}_0 \mathbf{h}_0^H$ and the corresponding eigenvalue is $\frac{1}{\lambda}$. Premultiplying on both sides of (4.7) with $\mathbf{A}^H \mathbf{C}_\eta$ and using (4.2) we can show that

$$\mathbf{A}^H \mathbf{C}_\eta \mathbf{A} = \lambda, \tag{4.8}$$

that is, the inverse of the eigenvalue which, for minimum noise power, should be the one having the largest value. Multiplying on both sides of (4.7) by $\mathbf{h}_0^H$ we can show that the noise power output is given by

$$\lambda = \frac{1}{\mathbf{h}_0^H \mathbf{C}_\eta^{-1} \mathbf{h}_0}. \tag{4.9}$$

To solve for the corresponding eigenvector we substitute the value of λ in (4.7), and by inspection we note that the following satisfies the equation:

$$\mathbf{A} = \frac{\mathbf{C}_\eta^{-1} \mathbf{h}_0}{\mathbf{h}_0^H \mathbf{C}_\eta^{-1} \mathbf{h}_0}. \tag{4.10a}$$

and the output power will be

$$\alpha_0^2 + \frac{1}{\mathbf{h}_0^H \mathbf{C}_\eta^{-1} \mathbf{h}_0}. \tag{4.10b}$$

Note that (4.10a) and (4.10b), whose derivation was based on constrained minimization, are identical to those obtained using the principle of maximum likelihood (ML), namely, (4.2) and (4.3).

Example 4.1: Let $\mathbf{C}_\eta = \sigma_\eta^2 \mathbf{I}$; then $\mathbf{h}_0^H \mathbf{C}_\eta^{-1} \mathbf{h}_0 = \sigma_\eta^{-2} T$. Hence, the power at the filter output is given by $\alpha_0^2 + \frac{\sigma_\eta^2}{T}$. The filtered noise power is reduced by a factor equal to the size of the filter (End of Example).

The noise covariance is rarely known because a sample of pure noise process, from which one could estimate its covariance function, is not available. Normally, what we observe is the signal plus noise; hence we can only estimate the covariance of the sum process. Let us consider a stochastic process as a sum of a large number of random sinusoids (see Section 1.4). Pick any one sinusoid as the desired signal and the rest as the unwanted noise. Clearly, this model of the stochastic process is akin to the problem of maximum likelihood estimation discussed a while ago. However, the Gaussian stochastic process without the selected sinusoid, that is, "the rest;" may not remain Gaussian. Naturally, the method of estimating the power of the buried sinusoid will no longer be a maximum likelihood method. It is more appropriate to describe it as a minimum variance method. The problem is reformulated as follows: A filter is sought such that its response (magnitude square) to a sinusoidal input of known frequency is unity; but its response to the entire process (consisting of all random sinusoids in a stochastic process) is minimum. Equations (4.4) and (4.5) may be rewritten as

$$\left|\mathbf{A}^H \mathbf{h}_0\right|^2 = 1, \tag{4.11a}$$

$$\mathbf{A}^H \mathbf{C}_x \mathbf{A} = \min. \tag{4.11b}$$

Following the same procedure as given above we obtain the desired filter and its output power

$$\mathbf{A} = \frac{\mathbf{C}_x^{-1}\mathbf{h}_0}{\mathbf{h}_0^H \mathbf{C}_x^{-1}\mathbf{h}_0},$$

$$\text{Output power} = \frac{1}{\mathbf{h}_0^H \mathbf{C}_x^{-1}\mathbf{h}_0}. \tag{4.11c}$$

It is instructive to note that the output power is a sum of the power in the selected sinusoid and the power that has leaked into it from the neighboring sinusoids. This point will be made clear in an example given below.

Example 4.2: Consider a time series as a sum of a random (complex) sinusoid and background noise,

$$X(t) = \alpha_0 e^{j\omega_0 t} + \eta(t),$$

whose covariance matrix is given by

$$\mathbf{C}_x = \left|\alpha_0\right|^2 \mathbf{h}_0 \mathbf{h}_0^H + \mathbf{C}_\eta .$$

The inverse of the covariance matrix can be obtained by applying Woodbury's identity[8]:

$$\mathbf{C}_x^{-1} = \mathbf{C}_\eta^{-1} - \frac{\alpha_0^2 \mathbf{C}_\eta^{-1}\mathbf{h}_0 \mathbf{h}_0^H \mathbf{C}_\eta^{-1}}{1 + \alpha_0^2 \mathbf{h}_0^H \mathbf{C}_\eta^{-1} \mathbf{h}_0}.$$

Substituting the above result in (4.11c) we obtain for the output power,

$$\text{Output power} = |\alpha_0|^2 + \frac{1}{\mathbf{h}_0^H \mathbf{C}_\eta \mathbf{h}_0},$$

which is the same as given in (4.11) (End of Example).

By definition, the maximum likelihood (ML) spectrum, or more appropriately Capon's spectrum of a stochastic process, is given by

$$S_{ML}(\omega) = \frac{1}{\mathbf{h}^H(\omega)\mathbf{C}_x^{-1}\mathbf{h}(\omega)}, \tag{4.12}$$

where we have explicitly shown the dependence of power on the frequency of the chosen sinusoid. By varying ω continuously over $\pm\pi$ we obtain the spectrum. Note that even though the covariance function is available only at discrete lags over a finite range, $S_{ML}(\omega)$ is defined as a continuous function of ω. This is not possible in the BT- or FFT-based method. The ML spectrum given by (4.12a) yields an estimate of the power and not power spectral density (PSD), that is, power per hertz. (See Exercise 4.8 where we introduce a normalized ML spectrum.)

B. Fast ML Method

To evaluate (4.12) we will have to compute the Hermitian form in the denominator for every frequency; however, the covariance matrix need be inverted only once. A considerable saving of the computation time can be achieved by expressing the Hermitian form as a DFT.[9] Express the denominator in (4.12) in full form and carry out the diagonal summation of

$$\begin{aligned}
\mathbf{h}^H(\omega)\mathbf{C}_x^{-1}\mathbf{h}(\omega) &= \sum_{m=0}^{T-1}\sum_{n=0}^{T-1} e^{-jm\omega}\,\bar{C}_x(m,n)\,e^{jn\omega} \\
&= \sum_{\tau=-T+1}^{T-1} e^{-j\tau\omega} \sum_{n=\tau}^{T-1-\tau} \bar{C}_x(\tau,\tau+n),
\end{aligned}$$

where $\bar{C}_x(m,n)$ is the $(m,n)^{th}$ term of $\mathbf{C}_x^{-1}$ and $\tau = (m-n)$. The second sum in the above equation is the diagonal summation as explained in Chapter Two (page 83). The first sum is simply the DFT of a sequence obtained by diagonally summing the elements of the inverse covariance matrix..

C. Adaptive Window

Equation (4.12) can be expressed in a fashion such that the hidden window in the ML method is made more apparent. For this let

$$\{\mathbf{C}_X^{-1}\}_{k,l^{th}\ \text{entry}} = q_{k,l}$$

and

$$S_{ML}(\omega) = \frac{1}{\sum_k \sum_l q_{k,l} e^{j((k-l)\omega)}}. \tag{4.13}$$

Now define

$$a_l(\omega) = \frac{\sum_k q_{k,l} e^{j(k-l)\omega}}{\sum_k \sum_l q_{k,l} e^{j((k-l)\omega)}}. \tag{4.14}$$

We shall show that $a_l(\omega)$, l=0, 1, 2, ..., $p-1$ acts as a frequency-dependent window. Let us evaluate the expected value of the square of the Fourier coefficient of the windowed time series.

$$E\left\{\left|\sum_l a_l(\omega) X(l) \exp(-j\omega l)\right|^2\right\} = \sum_k \sum_l a_k(\omega) a_l^*(\omega) C_x(k-l) \exp(-j\omega(k-l))$$

$$= \frac{\sum_{k'} \sum_{l'} \sum_{l}^{T-1} q_{k',l} \sum_k q_{l',k} C_x(k-l) e^{j\omega(k'-l')}}{\left|\sum_{k=0}^{T-1} \sum_{l=0}^{T-1} q_{k,l} e^{j\omega(k-l)}\right|^2}$$

$$= \frac{1}{\sum_{k=0}^{T-1} \sum_{l=0}^{T-1} q_{k,l} e^{j\omega(k-l)}} = S_{ML}(\omega).$$

From the above it is clear that $a_l(\omega)$ may be considered as a window. Since it is controlled by the covariance function of the data, it may be treated as a data-dependent window.

Example 4.3: Here we shall demonstrate the role of the data adaptive window. Consider two random sinusoids in the presence of unit variance white noise,

$$x(t) = \sqrt{8}\cos(\omega_1 t + \phi_1) + \sqrt{2}\cos(\omega_2 t + \phi_2) + \eta(t),$$

where $\omega_1 = 10$ Hz and $\omega_2 = 13$ Hz. A time series is obtained by sampling the signal at 100 samples per second. A covariance matrix of size 16×16 was formed using 400 independent segments each of length 16 samples. The ML spectrum of the time series is shown in Figure 4.2a. The spectrum of the window (4.14) tuned to frequency $\omega_2 = 10$ Hz is shown in Figure 4.2b, and the one tuned to 13 Hz is shown in Figure 4.2c. Notice a peak at the tuned frequency and a deep null at the second frequency where there is significant power. Thus, the leakage of power from the neighboring strong spectral line is stopped. This is the main feature of the ML method, and this is responsible for its high-resolution property and its ability to estimate true power (End of Example).

D. **Broadband ML**

In Capon's method we sought to preserve power at a particular frequency, or over very limited bandwidth around the frequency where it is desired to estimate the power. A modification suggested

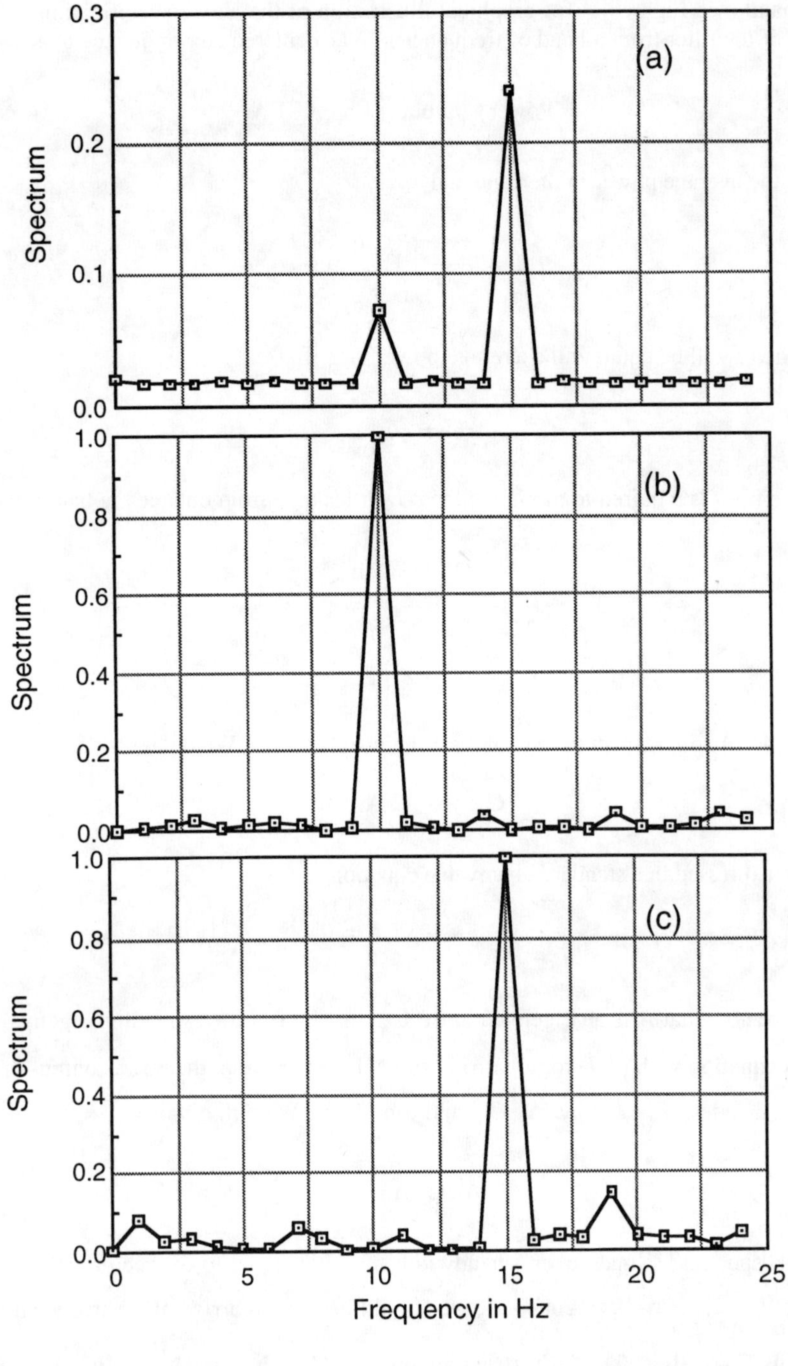

Figure 4.2: The spectrum of the adaptive window centered at frequency 10 Hz (b) and centered at 15 Hz (c). The ML spectrum is shown in (a).

in Reference 10 attempts to widen the bandwidth. We shall call this a broadband ML (Capon's original method as a narrowband ML). We seek a filter vector **A** which will preserve the power within a specified band of frequencies and at the same time minimize the power leakage from outside the band (see Figure 4.3 for graphical illustration of the above specifications). The power at the output of the filter from a band of frequencies $\Delta\omega$ centered at a frequency ω is given by

$$\text{Power output} = S_0 \mathbf{A}^H \Gamma \mathbf{A},$$

where S_0 is the average power in the band and

$$\Gamma = \frac{1}{\Delta\omega} \int_{\Delta\omega} \mathbf{E}\mathbf{E}^H d\omega.$$

The output power will be equal to the average power S_0 if

$$\mathbf{A}^H \Gamma \mathbf{A} = 1. \tag{4.15a}$$

In addition the filter is required to minimize the power leakage from outside the band; that is,

$$\mathbf{A}^H \mathbf{C}_x \mathbf{A} = \min. \tag{4.15b}$$

$$\mathbf{A}^H \mathbf{C}_x \mathbf{A} + \lambda(1 - \mathbf{A}^H \Gamma \mathbf{A}) = \min. \tag{4.16}$$

Differentiate (4.16) with respect to $\mathbf{A}^H$ and set the result to zero. We obtain

$$\mathbf{C}_x \mathbf{A} - \lambda \Gamma \mathbf{A} = 0,$$

which may be expressed in a standard eigenvalue equation

$$\mathbf{C}_x \mathbf{A} - \lambda \Gamma \mathbf{A} = 0. \tag{4.17}$$

From (4.17) we note that $\mathbf{A}$ is an eigenvector of $\mathbf{C}_x^{-1}\Gamma$, and the corresponding eigenvalue is $\frac{1}{\lambda}$. Compare this equation with (4.7) for a narrowband ML. To arrive at the power output of the filter, premultiply both sides of (4.17) by $\mathbf{A}^H \mathbf{C}_x$ and use (4.15). We will obtain

$$\mathbf{A}^H \mathbf{C}_x \mathbf{A} = \lambda. \tag{4.18}$$

Thus, the ML spectral estimate over a bandwidth of Dw is equal to the reciprocal of the largest eigenvalue of $\mathbf{C}_X^{-1}\Gamma$. If we let $\Delta\omega \to 0$, we should be able to arrive at a narrowband ML. For this, note that $\Gamma \underset{\Delta\omega \to 0}{\to} \mathbf{h}\mathbf{h}^H$. The largest eigenvalue of $\mathbf{C}_x^{-1}\mathbf{h}\mathbf{h}^H$ is $\mathbf{h}^H\mathbf{C}_x^{-1}\mathbf{h}$. As a matter of interest, let us consider the reverse case, namely, $\Delta\omega \to 2\pi$, that is a full band. It turns out that $\Gamma \underset{\Delta\omega \to 2\pi}{\to} \mathbf{I}$, and hence (4.15), reduces to

$$\mathbf{A}^H \mathbf{A} = 1. \tag{4.19}$$

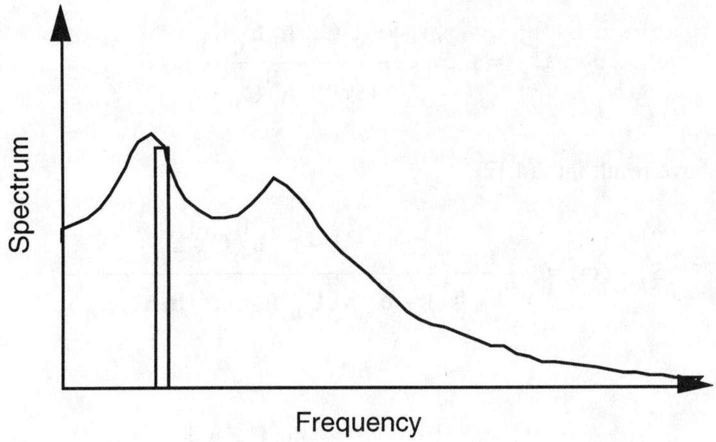

Figure 4.3: Wideband ML spectrum. A filter is sought such that the power inside a band of specified width is preserved and out-of-band leakage is minimized.

The solution of (4.15b) along with the constraint (4.19) is simply the minimum eigenvector of $\mathbf{C}_x$.

The power estimate derived as a reciprocal of the largest eigenvalue of $\mathbf{C}_x^{-1}\Gamma$ is related to the actual average power S_0 within the band. To show this let us decompose the covariance matrix into two parts, the first part corresponding to the average power within the band, and the second part corresponding to the power lying outside the band.

$$\mathbf{C}_x = S_0\Gamma + \mathbf{Q}. \tag{4.20}$$

Since $\mathbf{Q}$ is a valid covariance matrix, it must be positive definite. Since

$$\mathbf{Q} = \mathbf{C}_x[\mathbf{I} - S_0\mathbf{C}_x^{-1}\Gamma],$$

it will be positive semidefinite only if the bracketed quantity is also positive semidefinite. This is made possible if $S_0 \le$ reciprocal of the largest eigenvalue of $\mathbf{C}_x^{-1}\Gamma$. Thus, the ML spectrum provides *an upper bound* on the average power within the chosen band.

E. **Resolution of ML Method**

To understand the resolution properties of the ML method and compare them with those of the BT- and the direct FFT-based method, the simplest approach is to compute the 3-dB width of the ML spectral peak. Assume the signal model as in (4.1). The covariance matrix of the stochastic process is given by

$$\mathbf{C}_x = |\alpha_0|^2\mathbf{h}_0\mathbf{h}_0^H + \mathbf{C}_\eta. \tag{4.21}$$

The inverse of the covariance matrix is easily obtained by applying Woodbury's identity,[8]

$$\mathbf{C}_x^{-1} = \mathbf{C}_\eta^{-1} - \frac{|\alpha_0|^2 \mathbf{C}_\eta^{-1} \mathbf{h}_0 \mathbf{h}_0^H \mathbf{C}_\eta^{-1}}{1 + |\alpha_0|^2 \mathbf{h}_0^H \mathbf{C}_\eta^{-1} \mathbf{h}_0}. \tag{4.22}$$

Substitute the above result into (4.12):

$$S_{ML}(\omega) = \frac{1}{\mathbf{h}^H \mathbf{C}_\eta^{-1} \mathbf{h}} \left[\frac{1 + \alpha_0^2 \mathbf{h}_0^H \mathbf{C}_\eta^{-1} \mathbf{h}_0}{1 + \alpha_0^2 \mathbf{h}_0^H \mathbf{C}_\eta^{-1} \mathbf{h}_0 \sin^2(\mathbf{h}, \mathbf{h}_0, \mathbf{C}_\eta^{-1})} \right], \tag{4.23}$$

where

$$\sin^2(\mathbf{h}, \mathbf{h}_0, \mathbf{C}_\eta^{-1}) = 1 - \frac{\left| \mathbf{h}^H \mathbf{C}_\eta^{-1} \mathbf{h}_0 \right|^2}{\mathbf{h}^H \mathbf{C}_\eta^{-1} \mathbf{h} \mathbf{h}_0^H \mathbf{C}_\eta^{-1} \mathbf{h}_0}$$

$$= 1 - \cos^2(\mathbf{h}, \mathbf{h}_0, \mathbf{C}_\eta^{-1}) .$$

The angle between two vectors **h** and $\mathbf{h}_0$ in a space spanned by columns of a matrix, $\mathbf{C}_\eta^{-1}$ is, by definition,

$$\cos^2(\mathbf{h}, \mathbf{h}_0, \mathbf{C}_\eta^{-1}) = \frac{\left| \mathbf{h}^H \mathbf{C}_\eta^{-1} \mathbf{h}_0 \right|^2}{\mathbf{h}^H \mathbf{C}_\eta^{-1} \mathbf{h} \mathbf{h}_0^H \mathbf{C}_\eta^{-1} \mathbf{h}_0}.$$

Note that

$$\cos^2(\mathbf{h}, \mathbf{h}_0, \mathbf{C}_\eta^{-1}) = 1 \text{ when } \mathbf{h} = \mathbf{h}_0$$

and

$$\cos^2(\mathbf{h}, \mathbf{h}_0, \mathbf{C}_\eta^{-1}) = 0$$

when $\mathbf{h} \perp \mathbf{h}_0$.

Let us compute the 3-dB width of the ML spectrum and compare it with that of the BT method. The peak power is at $\mathbf{h} = \mathbf{h}_0$ and is equal to

$$S_{ML}(\omega = \omega_0) = |\alpha_0|^2 + \frac{1}{\mathbf{h}_0^H \mathbf{C}_\eta^{-1} \mathbf{h}_0}.$$

Let the 3-dB point be at ω_1; then

$$\frac{S_{ML}(\omega_1)}{S_{ML}(\omega_0)} = \frac{\mathbf{h}_0^H \mathbf{C}_\eta^{-1} \mathbf{h}_0}{\mathbf{h}_1^H \mathbf{C}_\eta^{-1} \mathbf{h}_1} \frac{1}{\{1 + \alpha_0^2 \mathbf{h}_0^H \mathbf{C}_\eta^{-1} \mathbf{h}_0 \sin^2(\mathbf{h}_1, \mathbf{h}_0, \mathbf{C}_\eta^{-1})\}} = \frac{1}{2}, \tag{4.24}$$

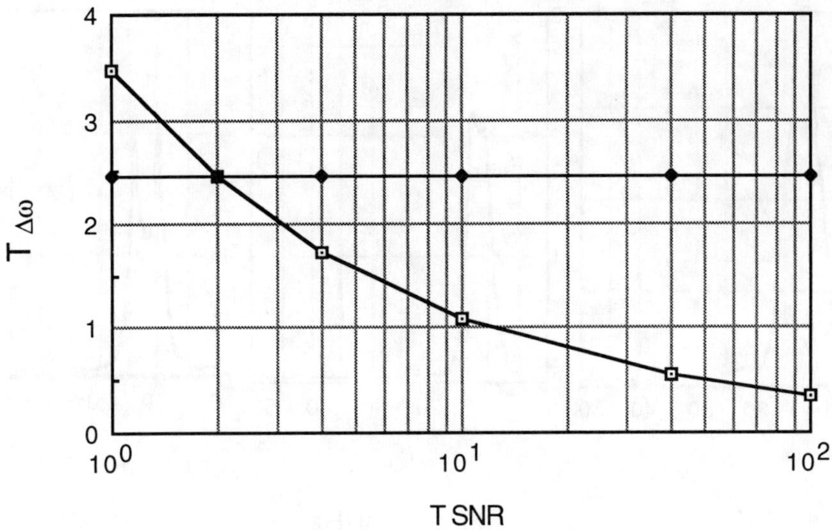

Figure 4.4: 3dB width of a spectral peak as function of SNR.

which may be simplified if we assume $\mathbf{C}_\eta = \sigma^2_\eta \mathbf{I}$, that is, a white noise background,

$$\frac{T\alpha_0^2}{\sigma_\eta^2}\left[1 - \frac{\sin^2(\omega_1 - \omega_0)\frac{T}{2}}{T^2 \sin^2\frac{(\omega_1 - \omega_0)}{2}}\right] = 1. \tag{4.25}$$

Let $\Delta\omega = \omega_1 - \omega_0 << \frac{2\pi}{T}$; then

$$\frac{\sin^2\frac{\Delta\omega T}{2}}{T^2 \sin^2\frac{\Delta\omega}{2}} \approx 1.0 - 0.3321\left(\frac{\Delta\omega T}{2}\right)^2. \tag{4.26}$$

Using the above approximation, (4.25) may be approximately expressed as

$$(\Delta\omega T)^2 \approx 12.04\frac{\sigma_\eta^2}{T\alpha_0^2}. \tag{4.27}$$

It may be noted that $\Delta\omega$ actually refers to half the width of the peak, halfway down from its maximum value. For comparison with the Blackman-Tukey (BT) method we will first express (3.25) in a matrix form,

$$S_{BT}(\omega) = \frac{1}{T}\mathbf{h}^H \mathbf{C}_x \mathbf{h}. \tag{4.28a}$$

Using (4.21) in (4.28a) and assuming that the background noise is white, we get

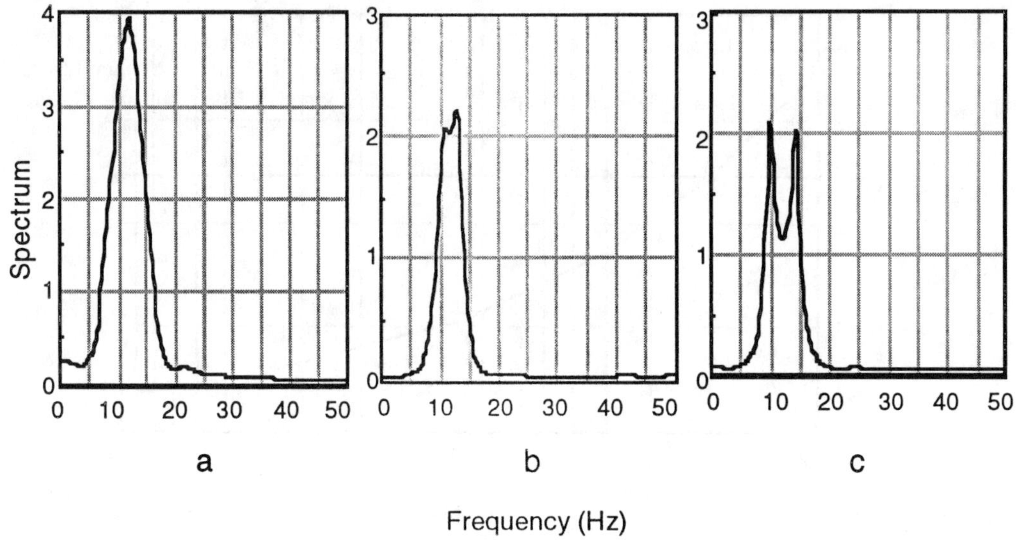

Figure 4.5: (a) BT spectrum (structured covariance matrix). (b) ML spectrum before and (c) after structuring the covariance matrix.

$$S_{BT}(\omega) = \frac{|\alpha_0|^2}{T}\left|\mathbf{h}^H \mathbf{h}_0\right| + \frac{1}{T}\mathbf{h}^H \mathbf{C}_\eta \mathbf{h}$$

$$-\left|\alpha_0\right|^2 T \frac{\sin^2(\omega_1 - \omega_0)\frac{T}{2}}{T^2 \sin^2\frac{(\omega_1 - \omega_0)}{2}} + \sigma_\eta^2.$$

Using approximation (4.26) we obtain the 3-dB width (half-width) of the spectral peak as being given by

$$\Delta\omega \approx \frac{\sqrt{6}}{T}. \tag{4.28b}$$

where we have assumed that the spectral peak stands well above the background noise. $\Delta\omega T$ as a function T SNR is shown in Figure 4.4.[11] For T SNR > 2 the 3-dB width of the ML spectral peak is less than that of the BT spectrum. Hence, the expectation is that the ML spectrum should yield higher resolution than the BT spectrum. This will be illustrated in the next example (Example 4.4).

Example 4.4: In this example we shall demonstrate the resolution capability of the ML spectrum, in particular, when the covariance matrix has been properly structured (see Chapter Three, p. 8). A time series consisting of two equal power sinusoids and unit variance white noise (SNR=10 dB) given by

$$x(t) = 3.16\cos(20\pi t + \phi_1) + 3.16\cos(26\pi t + \phi_2) + \eta(t)$$

was sampled at 100 Hz. A covariance matrix of size 16×16 was computed using 50 independent data segments each of length 16. Further, the estimated covariance matrix was structured as described in Chapter Three (page 171). The BT spectrum was computed using (3.26a), and the ML spectrum was computed before and after structuring the covariance matrix. The results are displayed in Figure 4.5. The ML spectrum obtained after structuring the covariance matrix is superior to the BT spectrum. The structured covariance matrix has yielded better resolution, but there is no improvement in the estimation of frequency (End of Example).

F. High-Resolution ML

It is important to note that the resolution in the ML method, unlike in the BT or the subspace method, is dependent on SNR. One way to increase the resolution is to increase the SNR by removing the noise power from the covariance matrix. This simple approach can be shown to enhance the resolution power of the ML method,[12,13] in particular, when the time series is a sum of sinusoids and white noise. For this we shall use some results which will be proved in Section 4.5. The covariance matrix of the sum of p real sinusoids and white noise is given by

$$\mathbf{C}_X = \mathbf{H}\mathbf{C}_0\mathbf{H}^H + \sigma_\eta^2\mathbf{I},$$

where $\mathbf{H}$ is a matrix whose columns are frequency vectors (2p frequency vectors), and $\mathbf{C}_0$ is a diagonal matrix whose elements represent the power of each sinusoid and σ_η^2 is noise variance. Let the eigen decomposition of the covariance matrix be

$$\mathbf{C}_x = \mathbf{E}\Gamma\mathbf{E}^H,$$

where (see Section 4.5)

$$\Gamma = \text{diag}\ \{\gamma_1 + \sigma_\eta^2, \gamma_2 + \sigma_\eta^2, \ldots, \gamma_{2p} + \sigma_\eta^2, \sigma_\eta^2, \ldots, \sigma_\eta^2\}$$

and $\mathbf{E}$ is a matrix whose columns are the eigenvectors. From the covariance matrix let us subtract $\sigma_0^2\mathbf{I}$, where σ_0^2 $(<\sigma_\eta^2)$ is an arbitrary constant. The eigen decomposition of the difference may be expressed as

$$\mathbf{C}_x - \sigma_0^2\mathbf{I} = \mathbf{E}\Gamma_0\mathbf{E}^H,$$

where

$$\Gamma_0 = \text{diag}\ \{\gamma_1 + \sigma_\eta^2 - \sigma_0^2\ \gamma_2 + \sigma_\eta^2 - \sigma_0^2 \ldots \gamma_{2p} + \sigma_\eta^2 - \sigma_0^2\ \sigma_\eta^2 - \sigma_0^2 \ldots \sigma_\eta^2 - \sigma_0^2\}.$$

Similarly, the eigen decomposition of $\mathbf{C}_X - \sigma_0^2$ may be expressed as

$$[\mathbf{C}_x - \sigma_0^2\mathbf{I}]^{-1} = \mathbf{E}\Gamma_0^{-1}\mathbf{E}^H. \tag{4.29a}$$

Substitute (4.29a) in place of $\mathbf{C}_x^{-1}$ in (4.12) defining the ML spectrum

$$S_{IML}(\omega) = \frac{1}{\mathbf{h}^H(\omega)\mathbf{E}\Gamma_0^{-1}\mathbf{E}^H\mathbf{h}(\omega)}$$

$$= \frac{1}{\mathbf{h}^H(\omega)\mathbf{E}_s\Gamma_s^{-1}\mathbf{E}_s^H\mathbf{h}(\omega) + \mathbf{h}^H(\omega)\mathbf{E}_\eta\Gamma_\eta^{-1}\mathbf{E}_\eta^H\mathbf{h}(\omega)}, \tag{4.29b}$$

where

$$\Gamma_s = \text{diag}\ \{\gamma_1 + \sigma_\eta^2 - \sigma_0^2, \gamma_2 + \sigma_\eta^2 - \sigma_0^2, \ldots, \gamma_{2p} + \sigma_\eta^2 - \sigma_0^2\},$$

$$\Gamma_\eta = \text{diag}\ \{\sigma_\eta^2 - \sigma_0^2, \ldots, \sigma_\eta^2 - \sigma_0^2\}.$$

Assuming that $\sigma_0^2 \approx \sigma_\eta^2$, the second term in the denominator of (4.29b) will be much greater than the first term except when $\omega = \omega_i$ $i = 1,2,\ldots,2p$. At these frequencies it turns out that $\mathbf{h}(\omega_i) \perp \mathbf{E}_\eta$ (see Section 4.5); therefore, the second term will be zero.

$$S_{IML}(\omega) = \frac{1}{\mathbf{h}^H(\omega)\mathbf{E}_s\Gamma_s^{-1}\mathbf{E}_s^H\mathbf{h}(\omega)} \tag{4.29c}$$

and at other frequencies

$$S_{IML}(\omega) = \frac{\sigma_\eta^2 - \sigma_0^2}{\left|\mathbf{h}^H(\omega)\mathbf{E}_\eta\right|^2}. \tag{4.29d}$$

Note that (4.29d) is similar to an eigenvector spectrum (see (4.107)) and (4.29c) is an ML spectrum without noise. Thus, the improved ML spectrum $S_{IML}(\omega)$ behaves like a normal ML spectrum at the unknown frequencies of the sinusoids but like an eigenvector spectrum at other frequencies. In practice, it is enough if we set σ_0^2 close to the minimum eigenvalue of the covariance matrix. In the example given below we demonstrate the resolution capability of $S_{IML}(\omega)$ in comparison with the ML spectrum.

Example 4.5: A time series consisting of two narrowband stochastic signals centered at 100 Hz and 115 Hz with 10 Hz bandwidth was generated by packing one hundred random sinusoids within a 10 Hz bandwidth. The sampling rate was 500 samples per second. A covariance matrix of size 8×8 was computed using 32 snapshots. The maximum and minimum eigenvalues were 334.68 and 5.52, respectively, giving an eigenvalue spread of 65. From the diagonal elements of the covariance matrix a small quantity was subtracted so as to make the eigenvalue spread equal to 3346, 33468, and 33468000. The maximum likelihood spectrums for the above cases are shown in Figure 4.6 (End of Example).

G. ML Cross-Spectrum

We shall now obtain the ML cross-spectrum and coherence.[14] Let x(t) and y(t) be two time series which we pass through two filters steered to the same frequency. The filter vectors are given by

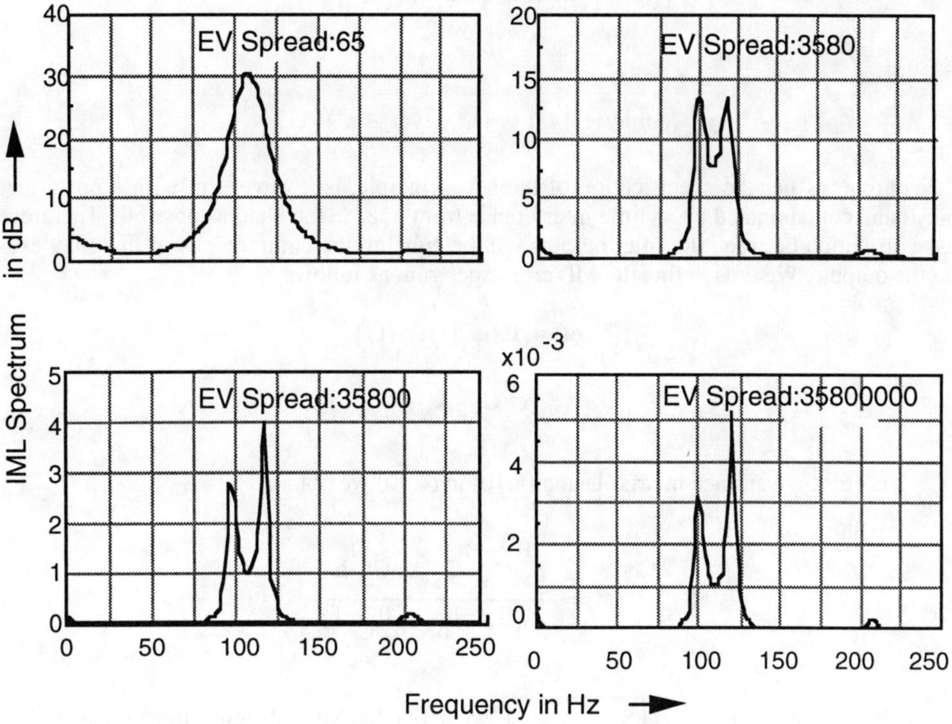

Figure 4.6: The improved ML spectrum obtained by modifying the eigenvalue spread of the covariance matrix. See Example 4.5 for more details.

$$\mathbf{A}_x = \frac{\mathbf{C}_x^{-1}\mathbf{h}}{\mathbf{h}^H \mathbf{C}_x^{-1}\mathbf{h}}$$

and

$$\mathbf{A}_y = \frac{\mathbf{C}_y^{-1}\mathbf{h}}{\mathbf{h}^H \mathbf{C}_y^{-1}\mathbf{h}},$$

respectively. The filter outputs, namely, $\varepsilon_x(t)$ and $\varepsilon_y(t)$, are given by

$$\varepsilon_x(t) = \mathbf{A}_x^H \mathbf{x}$$

and

$$\varepsilon_y(t) = \mathbf{A}_y^H \mathbf{y}.$$

where

$$\mathbf{x} = \text{col}\{x(t), x(t+1), \ldots x(t-p+1)\}$$

and

$$\mathbf{y} = \text{col}\{y(t), y(t+1), \ldots y(t-p+1)\}.$$

The filter output, as per the specification of the ML principle, is a waveform with a prescribed frequency band contaminated by as little interference from outside the band as possible. Therefore, the cross-correlation between the filter outputs will be equal to the common power that may exist between the outputs. We thus define the ML cross-spectrum as follows:

$$S_{xy_{ML}}(\omega) = E\{\varepsilon_x(t)\varepsilon_x(t)\}$$
$$= \mathbf{A}_x^H \mathbf{C}_{xy} \mathbf{A}_y, \tag{4.30}$$

where $\mathbf{C}_{xy}$ is cross-covariance matrix. Using (4.10) in (4.30) we obtain

$$S_{xy_{ML}}(\omega) = \frac{\mathbf{h}^H \mathbf{C}_x^{-H} \mathbf{C}_{xy} \qquad \mathbf{C}_y^{-1} \mathbf{h}}{\mathbf{h}^H \mathbf{C}_x^{-1} \mathbf{h} \quad \mathbf{h}^H \mathbf{C}_y^{-1} \mathbf{h}}. \tag{4.31}$$

Dividing both sides of (4.31) by $\sqrt{[S_{x_{ML}} \quad S_{y_{ML}}]}$ we define the ML coherence function as

$$\text{Coh}_{xy_{ML}}(\omega) = \mathbf{h}^H \mathbf{C}_x^{-H} \mathbf{C}_{xy} \mathbf{C}_y^{-1} \mathbf{h}[S_{x_{ML}} \quad S_{y_{ML}}]^{\frac{1}{2}}. \tag{4.32}$$

Example 4.6: Let

$$x(t) = \cos(\omega_0 t + \phi_1) + \eta_1(t),$$
$$y(t) = \cos(\omega_0 t + \phi_2) + \eta_2(t),$$

where $\eta_1(t)$ and $\eta_2(t)$ are uncorrelated noise and ϕ_1 and ϕ_2 are uncorrelated random phases, uniformly distributed over $\pm\pi$. It is easy to show that $\mathbf{C}_{xy} = \mathbf{0}$ and, hence, $\text{Coh}_{xy_{ML}} = 0$. Next, consider $\phi_1 = \phi_2 = \phi_0$. The covariance and cross-covariance matrices are given by

$$\mathbf{C}_x = |\alpha_0|^2 \mathbf{h}_0 \mathbf{h}_0^H + \sigma_\eta^2 \mathbf{I},$$
$$\mathbf{C}_y = |\alpha_0|^2 \mathbf{h}_0 \mathbf{h}_0^H + \sigma_\eta^2 \mathbf{I},$$
$$\mathbf{C}_{xy} = |\alpha_0|^2 \mathbf{h}_0 \mathbf{h}_0^H,$$

where we have assumed that $\sigma_{\eta_1}^2 = \sigma_{\eta_2}^2 = \sigma_\eta^2$. Using the above result along with Woodbury's formula for inversion, we obtain

$$\mathrm{Coh}_{xy_{ML}}(\omega) = \frac{|\alpha_0|^2}{\sigma_\eta^4}\mathbf{h}^H\left\{\mathbf{h}_0\mathbf{h}_0^H[\mathbf{I} - \frac{\frac{|\alpha_0|^2}{\sigma_\eta^2}\mathbf{h}_0\mathbf{h}_0^H}{1+\frac{|\alpha_0|^2}{\sigma_\eta^2}T}]^2\right\}\mathbf{h}\,[S_{x_{ML}} S_{y_{ML}}]^{-\frac{1}{2}},$$

which simplifies greatly when $\omega = \omega_0$:

$$\mathrm{Coh}_{xy_{ML}}(\omega)\Big|_{\omega=\omega_0} = \frac{\frac{|\alpha_0|^2}{\sigma_\eta^2}T}{1+\frac{|\alpha_0|^2}{\sigma_\eta^2}T} \approx 1, \quad \text{for } \frac{|\alpha_0|^2}{\sigma_\eta^2}T >> 1.$$

(End of Example).

H. Error Analysis

The statistical properties of the ML spectrum of a time series are not known. However, the statistical properties of the ML spectrum of array data (frequency-wavenumber spectrum) are known.[15,16] Computer simulation[11] appears to support the view that the variability of the ML spectrum is less than that of the AR spectrum.

4.3 Maximum Entropy (ME) Spectrum

The spectrum of a stationary stochastic discrete process is equal to the discrete Fourier transform of the covariance sequence (Wiener-Khintchin theorem) computed at discrete lags $\tau = 0, \pm 1, \pm 2, \ldots$

$$S_x(\omega) = \sum_{\tau=-\infty}^{\infty} C_x(\tau)\exp(-j\omega\tau). \tag{4.33}$$

When the covariance function is not known everywhere, there is no single unique spectrum but many spectrums, depending upon what value is assigned to the covariance sequence in the domain where it is not known. In the Blackman-Tukey method the covariance sequence is set to zero (or repeated) if the discrete Fourier transform is used to evaluate (4.33). This leads to the serious problem of bias. To minimize this effect a window function is introduced to scale down the covariance function near the edges so that there is a smooth transition to zero. Nevertheless, perfectly normal covariance values are undervalued.

Burg[2] founded a new principle of data reduction; namely, *the result of any transformation imposed on the experimental data shall incorporate and be consistent with all relevant data and be maximally noncommittal with regard to unavailable data.*[17] The last part of the statement may be expressed in terms of entropy, as a measure of uncertainty; that is, the entropy of the process must be maximum. We shall show next that a process whose entropy is maximum is one of the typical outcomes.

A. Principle of Entropy

The entropy of a random variable is defined in terms of its probablity of occurrence. Let x be a discrete random variable with possible outcomes as $x_1, x_2, \ldots, x_p$ and corresponding probabilities $f_1, f_2, \ldots, f_p$. The entropy of the random variable is given by

$$H = -\sum_{i=1}^{p} f_i \log_2 f_i \quad \text{bits}.$$

The entropy becomes maximum when $f_1, f_2, \ldots, f_p = \frac{1}{p}$; that is, all outcomes are equiprobable. When the random variable is continuous, the entropy is defined relative to some constant, C,

$$H = -\int_{-\infty}^{\infty} f(x)\log_2 f(x)dx + C$$

The entropy of a stationary stochastic process may be likewise written in terms of its multidimensional probability density function. But such an expression is hardly of any use if the aim is to estimate the spectral density function. Only in case of a Gaussian stationary stochastic process can one express the entropy in terms of its spectral density function. In other cases it is possible to relate the spectrum to entropy gain. We shall now show this result. Consider a filter with unit variance white noise input, and let its output be x(t). The input and output of the filter would be related to each other in the frequency domain as follows:

$$dX(\omega) = A(\omega)d\bar{\varepsilon}(\omega). \tag{4.34}$$

Let $f\{d\bar{\varepsilon}(\omega_1), d\bar{\varepsilon}(\omega_2), \ldots, d\bar{\varepsilon}(\omega_N)\}$ and $F\{dX(\omega_1), dX(\omega_2), \ldots, dX(\omega_N)\}$ be joint multidimensional probability density functions of N generalized Fourier components. On account of (4.34) the two probability density functions are related as follows:

$$F\{dX(\omega_1), dX(\omega_2), \ldots, dX(\omega_N)\} = \frac{f\left\{\frac{dX(\omega_1)}{A(\omega_1)}, \frac{dX(\omega_2)}{A(\omega_2)}, \ldots, \frac{dX(\omega_N)}{A(\omega_N)}\right\}}{\left|A(\omega_1)A(\omega_2)\ldots A(\omega_N)\right|}. \tag{4.35}$$

The entropy of N Fourier components turns out to be

$$H_X = H_\varepsilon + \sum_{i=1}^{N} \log_2\left|A(\omega_i)\right|,$$

where we have used the minimum-phase property of the filter (specifically, see Exercise 4.13). The entropy gain is given by

$$\Delta H_N = H_x - H_\varepsilon = \sum_{i=1}^{N} \log_2\left|A(\omega_i)\right|. \tag{4.36}$$

For large N the rate of entropy gain may be obtained from (4.36),[18]

$$\frac{\Delta H_N}{N} = \frac{1}{2\pi}\sum_{i=1}^{N}\log_2\left|A(\omega_i)\right|\frac{2\pi}{N} = \frac{1}{2\pi}\int_{-\pi}^{\pi}\log_2\left|A(\omega)\right|d\omega.$$
$$\lim N \to \infty \tag{4.37}$$

It is assumed that $|A(\omega)| > 0$. The entropy of the output of a filter may be increased by choosing a filter with a large entropy gain.

B. Concentration Theorem

Nature generates N realizations of a time series x: $\{x(t), t=0, 1, 2, \ldots, T-1\}$ of which n different types of sequences $x^0, x^1, x^2, \ldots, x^{n-1}$ are possible. In N realizations (N is large) the ith type of sequence occurs $N_i = f_i N$ times. Then the number of ways in which N_i, i=0, 1, ..., n − 1 can be selected out of N realizations is given by

$$W(f_0, f_1, \ldots f_{n-1}) = \frac{N!}{(f_0N)!(f_1N)!(f_2N)!\ldots(f_{n-1}N)!}, \tag{4.38}$$

where $W(f_0, f_1, \ldots, f_{n-1})$ is known as the multiplicity factor (introduced by Bernoulli and Laplace). For $N \to \infty$ we may use Stirling's approximation, $N! \approx \sqrt{2\pi}\,N^{N+\frac{1}{2}}e^{-N}$, for N! and simplify (4.38) as follows:

$$\lim_{N\to\infty} N^{-1}\log W = -\sum_{i=0}^{n-1} f_i \log f_i = H(f_0, f_1, \ldots f_{n-1}).$$

or

$$W = \exp(HN), \quad N \to \infty. \tag{4.39}$$

Thus, we note that *the distributions of higher entropy possess higher multiplicity*; that is, they can be realized by nature in more ways, and hence are more likely to occur.

The frequencies $f_0, f_1, \ldots, f_{n-1}$ may be considered as cartesian coordinates of a point p in n-dimensional space. Define in this space a hyperplane, S, where $f_i \geq 0$ and $\sum f_i = 1$. Let the point p lie on the hyper-plane, S. As point p moves on the hyperplane from any one of the vertices to the center, the entropy takes on all values ranging from 0 to log(n), the maximum being at the center. We shall now impose m independent linear constraints on f_i in the form,

$$\sum_{i=0}^{n-1} A_{ij} f_i = d_j, \quad j = 1, 2, \ldots m. \tag{4.40}$$

Let the hyperplane defined by (4.40) be denoted by M. The point p will now be confined to the intersection of two hyper-planes, namely, S and M. The intersection will be denoted by S', which is a hyperplane of dimension $n-m-1$. On S' the entropy will attain a maximum H_{max} ($\leq \log n$) at a unique point. Around this point of maximum entropy we expand the entropy in Taylor's series,

$$H(p) = H_{max} - ar^2 + \dots, \qquad a > 0,$$

and r is the radial distance from the point of maximum entropy. The domain of all possible frequency distributions that lie on S' and whose entropy is in the range $H_{max} - \Delta H \leq H \leq H_{max}$ is a sphere of radius R ($= \sqrt{\frac{\Delta H}{a}}$) in $n-m-1$ dimensions. In N trials the fraction of all possible outcomes which lie inside the sphere of radius of R is given by

$$F = \frac{\int_0^R \exp(-ar^2N)\, r^{n-m-2} dr}{\int_0^\infty \exp(-ar^2N)\, r^{n-m-2} dr}. \tag{4.41}$$

To evaluate the integrals in (4.41) we make the following substitution:

$$t = ar^2N \quad \text{or} \quad r = \sqrt{\frac{t}{Na}}$$

$$dt = 2aNrdr \quad \text{and} \quad dr = \sqrt{\frac{Na}{t}}\frac{1}{2aN}dt.$$

The denominator turns out to be

$$\int_0^\infty \exp(-ar^2N)\, r^{n-m-2} dr = \frac{\left(\frac{n-m-3}{2}\right)!}{2(aN)^{\frac{n-m-1}{2}}},$$

and the fraction,

$$F = \frac{1}{\left(\frac{n-m-3}{2}\right)!}\int_0^{aNR^2} e^{-t}\, t^{\frac{n-m-3}{2}}\, dt. \tag{4.42}$$

For large aNR^2 the integral in (4.42) may be expressed in a series form. We obtain an approximate expression for F as follows:[19]

$$F = 1 - \frac{(\Delta HN)^s}{s!}e^{-\Delta HN}\left[1 + \frac{s}{\Delta HN} + \frac{s(s-1)}{(\Delta HN)^2} + \dots\right], \tag{4.43}$$

where $s = \frac{(n-m-3)}{2}$. When s is an integer, the series in (4.43) terminates, and then it becomes exact. In Table 4.1 we give a sample calculation of F for different ΔHN.

F	ΔHN
0.99999	15
0.99995	10
0.95957	5
0.90842	4
0.80085	3
0.59399	2

Table 4.1: The fraction of all possible outcomes in N trials which lie inside a sphere of radius $=\sqrt{\frac{\Delta H}{a}}$ centered at the point of maximum entropy.

C. **Maximization of Entropy**

We maximize the entropy under constraints

$$\text{Entropy:} \qquad H(f) = -\sum_{i=1}^{N} f_i \log f_i,$$

$$\text{Constraints:} \qquad \sum_{i=1}^{N} A_{i,j} f_i = d_j \quad j = 1, 2, \ldots, M. \tag{4.44}$$

where $A_{i,j}$, i=1,2, ..., N and j=1,2, ..., M and d_j, j=1,2, ..., M are constants. Further, f_i, being a variable denoting probability, must satisfy

$$f_i \geq 0$$

and

$$\sum_{i}^{N} f_i = 1.$$

We define a Lagrange function,

$$J = H(f) - \sum_{j=1}^{M} \lambda_j \left(\sum_{i=1}^{N} A_{i,j} f_i - d_j\right) - \lambda_0 \left(\sum_{i=1}^{N} f_i - 1\right),$$

and maximize it with respect to f_i. We obtain

$$f_i = \exp(-1 - \lambda_0) \exp\left(-\sum_{j=1}^{M} \lambda_j A_{i,j}\right). \tag{4.45}$$

Since

$$\sum_{i}^{N} f_i = 1,$$

from (4.45) we obtain

$$z(\lambda_1,\lambda_2,\dots,\lambda_M) = \sum_{i=1}^{N} \exp\left(-\sum_{j=1}^{M} \lambda_j A_{i,j}\right) = \exp(1+\lambda_0), \tag{4.46}$$

where $z(\lambda_1,\lambda_2,\dots,\lambda_M)$ is known as the partition function. To solve for other Lagrange coefficients we make use of the constraint (4.44). The result is

$$\frac{\partial}{\partial\lambda_j}\log\left(z(\lambda_1,\lambda_1,\dots,\lambda_M)\right) + d_j = 0, \quad j = 1,2,\dots;M. \tag{4.47}$$

Using (4.45), (4.46), and (4.47) in (4.44) we obtain an expression for the maximum entropy

$$H_{max} = \log(z) + \sum_{j=1}^{M} \lambda_j d_j. \tag{4.48}$$

From random variable we go on to time series which may be looked upon as a vector random variable. Let x(t), t=0,1, ..., $T-1$ be a T-component vector random variable with a joint pdf $f(x_0, x_1, \dots, x_{T-1})$. The relative entropy is given by

$$H = -\int dx_0 \dots \int dx_{T-1} f(x_0, x_1, \dots, x_{T-1}) \log f(x_0, x_1, \dots, x_{T-1}). \tag{4.49}$$

The constraints are p autocorrelation coefficients defined as follows:

$$C_x(\tau) = \frac{1}{T-1-\tau} \sum_{t=0}^{T-1-\tau} x_t x_{t+\tau}, \qquad 0 \le \tau \le p-1 \tag{4.50}$$

The maximization of (4.49) under the autocorrelation constraints (4.50) leads to the following joint pdf:[19]

$$f(x_0, x_1, \dots x_{T-1}) = z^{-1} \exp\left(-\tfrac{1}{2}\mathbf{x}^T \Lambda\, \mathbf{x}\right), \tag{4.51}$$

where

$$\underset{(T\times 1)}{\mathbf{x}} = \text{col}\{x_0, x_1, \dots, x_{T-1}\},$$

$$\Lambda = \begin{bmatrix} \lambda_0, \lambda_1, \lambda_2, 0, & 0, & 0, & 0, & 0 \\ \lambda_1, \lambda_0, \lambda_1, \lambda_2, 0, & 0, & 0, & 0 \\ \lambda_2, \lambda_1, \lambda_0, \lambda_1, \lambda_2, 0, & 0, & 0 \\ 0, \lambda_2, \lambda_1, \lambda_0, \lambda_1, \lambda_2, 0, & 0 \\ 0, 0, \lambda_2, \lambda_1, \lambda_0, \lambda_1, \lambda_2, 0 \\ 0, 0, 0, \lambda_2, \lambda_1, \lambda_0, \lambda_1, \lambda_2 \\ 0, 0, 0, 0, \lambda_2, \lambda_1, \lambda_0, \lambda_1 \\ 0, 0, 0, 0, 0, \lambda_2, \lambda_1, \lambda_0 \end{bmatrix},$$

$$(T \times T, T=8)$$

and the partition function is given by

$$z = \int \ldots \int \exp\left(-\frac{1}{2}\mathbf{x}^T \Lambda\, \mathbf{x}\right) dx_0\, dx_1 \ldots dx_{T-1}. \tag{4.52}$$

The Lagrange coefficients are arranged in the form of a matrix Λ, which is a symmetric banded Toeplitz matrix. In the above example three Lagrange coefficients are arranged in a matrix form. From (4.51) we notice that the joint pdf that maximizes the entropy under the covariance constraints turns out to be a Gaussian function.

Let us express $\Lambda = \Omega^T \mathbf{g}\, \Omega$ where Ω is a matrix whose columns are eigenvectors of Λ and $\mathbf{g}$ is a diagonal matrix with eigenvalues of Λ on the diagonal. Then,

$$\mathbf{x}^T \Lambda\, \mathbf{x} = \sum_i g_i \left|\mathbf{x}^T \Omega_i\right|^2. \tag{4.53}$$

Using (4.53) in (4.52) and carrying out the integration we obtain

$$\log(z) = -\frac{1}{2}\sum_{i=0}^{T-1} \log(g_i) + \text{const}. \tag{4.54}$$

Define a polynomial in z,

$$g(z) = \sum_{i=-p+1}^{p-1} \lambda_i z^{-i},$$

where we have assumed $\lambda_{-i} = \lambda_i$. From the properties of a Toeplitz matrix for T>>p the eigenvalues of Λ are equal to g(z) evaluated over the unit circle at $z = \exp(j\frac{2\pi}{T}i)$. For large T the sum in (4.54) may be replaced by the integral

$$\log(z) = -\frac{T}{4\pi}\int_{-\pi}^{\pi} \log\left(g(e^{j\omega})\right) d\omega. \tag{4.55}$$

Differentiate (4.55) with respect to λ_i and use the following result:

$$\frac{\partial}{\partial \lambda_i}\log(z) = -\frac{T-\tau-1}{2}C_x(\tau)$$

,

which may be derived from (4.50) and (4.51). We finally obtain

$$C_x(\tau) = \frac{1}{2\pi}\int_{-\pi}^{\pi} \frac{\exp(j\omega\tau)}{\sum_{\tau}\lambda_\tau \exp(j\omega\tau)} d\omega. \tag{4.56}$$

Equation (4.56) is the basic result. First, given p covariance estimates we compute p Lagrange coefficients, and later, using the estimated Lagrange coefficients we extrapolate the covariance function outside the range, $\tau > p-1$. The extrapolated covariance function is maximally noncommittal with regard to the unknown covariance function.

D. **Filter Model**

Earlier in this section we have shown that the entropy gain of the filter output is related to the transfer function of the filter (4.37),

$$\Delta H = \frac{1}{4\pi}\int_{-\pi}^{\pi} \log\{S_x(\omega)\} d\omega, \tag{4.57}$$

where we have used the spectrum of the filter output in place of the filter transfer function (use the relation

$$S_x(\omega) = \sigma_\varepsilon^2 |A(\omega)|^2$$

with $\sigma_\varepsilon^2 = 1$). The observed data is in the form of a covariance function for p lags,

$$C_x(\tau) = \frac{1}{2\pi}\int_{-\pi}^{\pi} S_x(\omega)\exp(j\tau\omega) d\omega \qquad 0 \le \tau \le p-1. \tag{4.58}$$

It is desired to find the $S_x(\omega)$ that maximizes the entropy gain given by (4.57) and also satisfies the observed data, namely, (4.58). For this we form a Lagrange function and maximize it with respect to $S_x(\omega)$.

$$J = \frac{1}{4\pi}\int_{-\pi}^{\pi} \log\{S_x(\omega)\} d\omega + \sum_{\tau}\lambda_\tau [C_x(\tau) - \frac{1}{2\pi}\int_{-\pi}^{\pi} S_x(\omega)\exp(j\tau\omega) d\omega]$$

,

where λ_τ, $\tau = 0, +1, \ldots, +(p-1)$ are Lagrange coefficients. They are symmetric, $\lambda_\tau = \lambda_{-\tau}$ and $\lambda_\tau = 0 \quad \tau \ge p$. This leads to the following result:

$$C_x(\tau) = \frac{1}{2\pi}\int_{-\pi}^{\pi} \frac{\exp(j\tau\omega)}{\sum_{\tau=-p+1}^{p-1}\lambda_\tau \exp(-j\omega\tau)} d\omega \qquad 0 \le \tau \le p-1, \tag{4.59}$$

τ	0	1	2	3	4	5
$C_x(\tau)$	5.399	-1.2612	-4.242	3.259	2.1988	-4.041

Table 4.2: Covariance coefficients of ARMA(2,1) process.

τ	0	1	2	3	4	5
λ_τ	1.122	0.0813	0.3712	-0.1187	-0.0301	0.1179

Table 4.3: Estimated λ_τ for τ=0, 1, ..., 5.

τ	Extrapolated	True
6	0.0512	0.036
7	3.8045	3.6289
8	-1.9515	-1.847
9	-2.5393	-2.3516
10	3.1082	2.8427
11	0.7913	0.7010
12	-3.2736	-2.9160

Table 4.4: Extrapolated covariance function of ARMA(2,1) using λ_τ given in Table 4.3.

which is the same as (4.56), which was derived under a more general framework. In the range $0 \le \tau \le p-1$, (4.59) may be considered as a system of nonlinear equations in unknown coefficients, λ_τ, $\tau = 0, \pm 1, \ldots, \pm (p-1)$, which may be solved using any one of the standard optimization routines, for example, the quasi-Newton method (see Example 4.7). After λ_τ, $\tau = 0, \pm 1, \ldots, \pm (p-1)$ have been determined, (4.59) may be used to extrapolate the covariance function beyond the (p-1)th lag.

Example 4.7: In this example we shall show how to compute λ_τ from the given covariance coefficients (error free) and then to extrapolate the covariance function. We consider an ARMA(2,1) process with coefficients

$$a_1 = 0.5,$$
$$a_2 = 0.9025$$

and

$$b_1 = 0.3 .$$

Six error-free covariance coefficients are given as in Table 4.2. To determine λ_τ we have minimized the following cost function:

$$\sum_{\tau=0}^{p-1}\left\{C_x(\tau)-\frac{1}{2\pi}\int_{-\pi}^{\pi}\frac{\exp(j\tau\omega)}{\sum_{\tau=-p+1}^{p-1}\lambda_\tau\exp(-j\omega\tau)}d\omega\right\}^2$$

using Matlab routine FMINU. The iterative minimization process was terminated when the cost function was ≤ 0.005. The estimated λ_τ are as in Table 4.3. Using these values of λ_τ we have extrapolated the given covariance coefficients for $\tau > 5$. The extrapolated coefficients are in good agreement with the true coefficients as demonstrated in Table 4.4 (End of Example).

Finally, $\lambda_\tau, \tau = 0, \pm 1, \ldots, \pm(p-1)$ may be expressed as a Fourier transform of the reciprocal of the spectrum of x(t). From (4.59) we can write

$$\frac{1}{\sum_{\tau=-p+1}^{p-1}\lambda_\tau\exp(-j\omega\tau)} = \sum_{-\infty}^{\infty}C_x(\tau)\exp(-j\omega\tau),$$

or after further simplification we get

$$\lambda_\tau = \frac{1}{2\pi}\int_{-\pi}^{\pi}\frac{\exp(j\omega\tau)}{\sum_{\tau=-\infty}^{\infty}C_x(\tau)\exp(-j\omega\tau)}d\omega.$$

Note that coefficients $\lambda_\tau, \tau = 0, \pm 1, \ldots, \pm(p-1)$, sometimes known as *inverse covariance* coefficients, like the covariance coefficients, are symmetric and positive definite.

E. Minimum Phase Filter

Using the Fejer-Riesz theorem[20] we can express the denominator of the integrand in (4.59) as a product of a minimum phase filter and its conjugate,

$$\sum_{\tau=-p+1}^{p-1}\lambda_\tau\exp(j\omega\tau) = A(\omega)A^H(\omega), \tag{4.60}$$

where

$$A(\omega) = \sum_{m=0}^{p-1}a_m\exp(-j\omega m). \tag{4.61}$$

Note, by substituting (4.61) in (4.60) we can show that

$$\lambda_\tau = \sum_{m=0}^{p-\tau-1}a_m a_{m+\tau}.$$

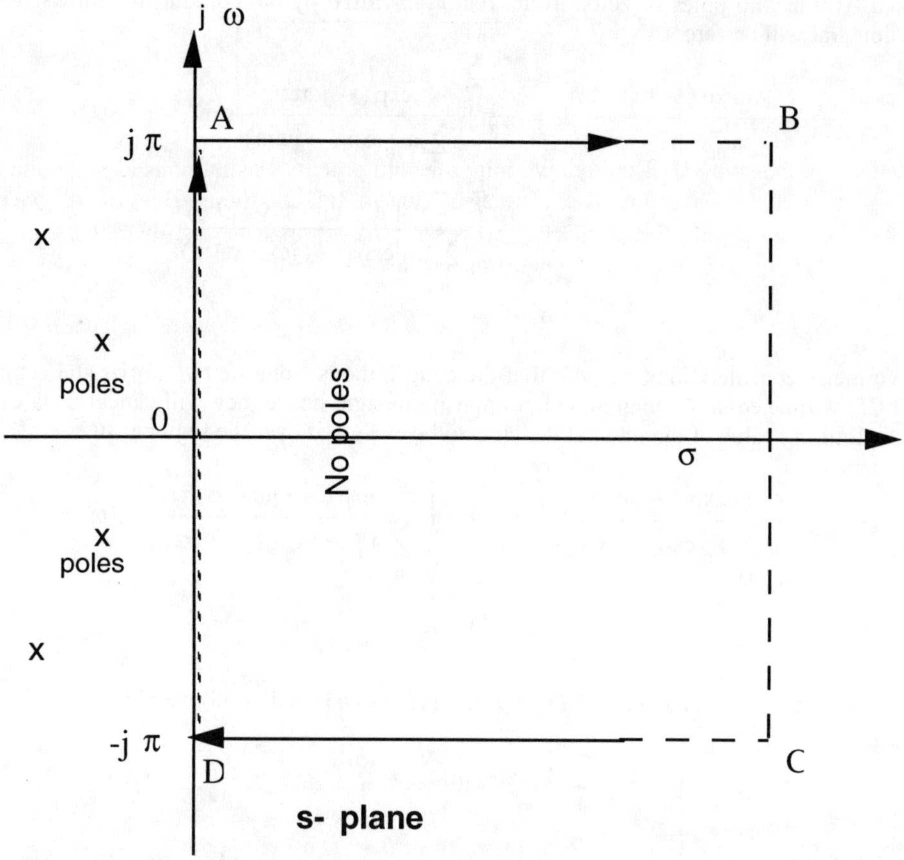

Figure 4.7: The integral is evaluated over a closed contour enclosing the right half of s-plane having no poles.

All poles and zeros of A(s), where $s = \sigma + j\omega$, are to the left of the imaginary axis in the s-plane. To determine the coefficients a_m we shall follow the approach described in Reference 21. From (4.59) it follows that

$$S_x(\omega) = \frac{1}{A(\omega)A^H(\omega)}$$

or

$$S_x(\omega)A^H(\omega) = A(\omega)^{-1}. \tag{4.62}$$

Taking the inverse Fourier transform on both sides of (4.62) we get

$$\sum_m C_x(\tau - m)a_m = \frac{1}{2\pi}\int_{-\pi}^{\pi} \frac{\exp(-j\omega\tau)}{\sum_m a_m \exp(-j\omega m)} d\omega, \quad \tau \geq 0. \tag{4.63}$$

We shall evaluate the integral on the right-hand side of (4.63) over a closed contour ABCD in the s-plane as shown in Figure 4.7.

Since A(s) has no poles or zeros in the region enclosed by the contour, it follows that the contour integral will be zero.

$$\frac{1}{2\pi}\oint \frac{\exp(-s\tau)}{\sum_m a_m \exp(-sm)} ds = \frac{1}{2\pi}\int_{-\pi}^{\pi} \frac{\exp(-j\omega\tau)}{\sum_m a_m \exp(-j\omega m)} d\omega$$
$$+ \frac{1}{2\pi}\int_{\pi}^{-\pi} \frac{\exp(-(j\omega+\sigma)\tau)}{\sum_m a_m \exp(-(j\omega+\sigma)m)} d\omega = 0.$$
$$\sigma \rightarrow \infty \tag{4.64}$$

From symmetry considerations we note that the contributions from the two horizontal segments AB and CD will be equal in magnitude but opposite in sign; hence they will cancel. It is easy to obtain the limiting value of the integral which is to be evaluated over the segment BC.

$$\frac{1}{2\pi}\int_{-\pi}^{\pi} \frac{\exp(-j\omega\tau)}{\sum_m a_m \exp(-j\omega m)} d\omega = \frac{1}{2\pi}\int_{-\pi}^{\pi} \frac{\exp(-(j\omega+\sigma)\tau)}{\sum_m a_m \exp(-(j\omega+\sigma)m)} d\omega$$
$$= \frac{1}{a_0}\delta_\tau, \tag{4.65}$$

where $\delta_\tau = 1$ for $\tau = 0$ and $\delta_\tau = 0$ for $\tau > 0$. From (4.63) and (4.65) we obtain

$$\sum_m c_x(\tau - m) a_m = \frac{1}{a_0}\delta_\tau.$$

or in matrix form

$$\mathbf{C}_x \mathbf{A} = \frac{1}{a_0}\delta, \tag{4.66a}$$

where $\mathbf{C}_X$ is the covariance matrix, **A** is the filter coefficient vector and $\delta = \mathrm{col}\{1, 0, 0, ..., 0\}$, and $a_0^2 = (\mathbf{C}_x^{-1})_{11}$, that is, the first element of $\mathbf{C}_x^{-1}$. A formal solution of (4.66) is given by

$$\mathbf{A} = \frac{1}{a_0}\mathbf{C}_x^{-1}\delta,$$

that is, the first column of $\mathbf{C}_x^{-1}$ normalized by the first element of the column. The maximum entropy spectrum may be expressed in terms of $\mathbf{C}_x^{-1}$ as follows:

$$S_{ME}(\omega) = \frac{(\mathbf{C}_X^{-1})_{11}}{\left|\mathbf{h}^H(\omega)\mathbf{C}_X^{-1}\delta\right|^2}, \tag{4.66b}$$

where $\mathbf{h}(\omega)$ is a frequency vector defined in (4.1). Note the similarity between the ME spectrum and the ML spectrum (4.12).

F. **Resolution of ME Spectrum**

As before, we shall study the resolution of the ME spectrum through an example. Consider the simple signal model of a complex sinusoid and white noise, whose covariance function is given by (4.21) and its inverse by (4.21a). Using Woodbury's identity in (4.66) we obtain

$$S_{ME}(\omega) = \frac{\left(1 - \frac{SNR}{1 + SNR\,T}\right)}{\left|1 - \frac{SNR\,\mathbf{h}^H \mathbf{h}_0}{1 + SNR\,T}\right|^2}, \tag{4.67}$$

where

$$SNR = \frac{\alpha_0^2}{\sigma_0^2}.$$

The peak value is equal to $(1 + SNR(T - 1))(1 + SNR\,T)$, or approximately,

$$S_{ME}(\omega)\big|_{max} \approx (SNR\,T)^2,$$

and 3 dB width (half-width) is found to be

$$\Delta\omega T = \frac{2}{SNR\,T}. \tag{4.68}$$

Compare (4.68) with that for the ML spectrum (4.27). The 3 dB width of the ME spectrum peak is narrower than that of the ML spectrum. Another interesting observation is that the ME spectrum peak height depends upon the size of the covariance matrix. However, we find that the product of peak height and peak width is proportional to the signal power,

$$2\Delta\omega * \text{Peak height} = 4\,SNR \tag{4.69}$$

The resolution of the ME spectrum is in general higher than that of the ML spectrum. To relate the ME spectrum to the ML spectrum we use the Cholesky decomposition of a covariance matrix (see Exercise 4.7),[22] $\mathbf{C}_x = \mathbf{A}^H \sigma \mathbf{A}$. Its inverse is $\mathbf{C}_x^{-1} = \mathbf{A}^{-1}\sigma^{-1}\mathbf{A}^{-H}$, which we shall use in (4.12) and obtain

$$\begin{aligned} S_{ML}(\omega) &= \frac{1}{[\mathbf{A}^{-H}\mathbf{h}(\omega)]^H \sigma^{-1} [\mathbf{A}^{-H}\mathbf{h}(\omega)]} \\ &= \frac{1}{\sum_{i=1}^{T} \sigma_i^{-2} \left|\mathbf{A}_i^{-H}\mathbf{h}(\omega)\right|^2}, \end{aligned} \tag{4.70}$$

where the subscript i on **A** denotes the ith column of **A**. From (4.66) we note that

$$\sigma_i^{-2}\left|\mathbf{A}_i^{-H}\mathbf{h}(\omega)\right|^2 = \frac{1}{S^i_{ME}},$$

where the superscript on S denotes the size of the covariance matrix used in the estimation of ME spectrum. Thus, we obtain from (4.70),

$$\frac{1}{S_{ML}(\omega)} = \sum_{i=1}^{T}\frac{1}{S^i_{ME}(\omega)}. \tag{4.71}$$

G. **Maximum Entropy**

To evaluate the maximum entropy that can be attained we substitute the spectrum in (4.57) in terms of the filter transfer function

$$\Delta H = \frac{1}{4\pi}\int_{-\pi}^{\pi}\log\sigma_\varepsilon^2|A(\omega)|^2 d\omega = \frac{1}{2}[\log\sigma_\varepsilon^2 - \log a_0^2]. \tag{4.72}$$

In deriving (4.72) we have used the fact that A(ω) is a minimum-phase filter having all poles and zeros to the left-hand side of the imaginary axis in the s-plane. This makes log(A(ω)) pole free on the right-hand side of the imaginary axis. Then, using the approach in deriving (4.65) we obtain (4.72).

H. **Minimum Cross Entropy** (MCE)

Cross entropy between two random variables x and y is defined as

$$H(q,p) = \int_{-\infty}^{\infty} dYq(y)\log\frac{q(y)}{p(y)}, \tag{4.73}$$

where p(x) is the pdf of x and q(y) is that of y. Cross entropy is a measure of information dissimilarity betwcen the two pdfs, p and q.[23] It is equal to zero for identical pdfs. When y=ax, the pdfs p and q are simply related to each other, $q(y) = \frac{p(\frac{y}{a})}{a}$. Substituting in (4.73) the cross entropy reduces to

$$\begin{aligned} H(q,p) &= \int_{-\infty}^{\infty} dyq(y)\log\frac{p(\frac{y}{a})}{p(y)} - \log(a) \\ &= \int_{-\infty}^{\infty} dxp(x)\log\frac{p(x)}{p(ax)} - \log(a). \end{aligned} \tag{4.74}$$

We shall now define cross entropy between two vector random variables,

$$\mathbf{dx} = \text{col}\,\{|dx(\omega_1)|^2, |dx(\omega_2)|^2, \ldots, |dx(\omega_N)|^2\},$$

$$d\varepsilon = \text{col}\,\{|d\bar{\varepsilon}(\omega_1)|^2, |d\bar{\varepsilon}(\omega_2)|^2, \ldots, |d\bar{\varepsilon}(\omega_N)|^2\},$$

whose components are linearly related as given by (4.34). It may be noted that $d\bar{\varepsilon}(\omega)$ is the differential of the generalized Fourier transform (GFT) of $\varepsilon(t)$, which is the input to a filter, and $dX(\omega)$ is the differential of the GFT of $x(t)$, it being the output of the filter. Generalizing (4.74) for vector random variables, we obtain the following expression for the cross entropy[24]:

$$H(q,p) = \int_{-\alpha}^{\alpha} d(d\varepsilon)p(d\varepsilon)\log \frac{p(d\varepsilon)}{p(|A(\omega_1)d\bar{\varepsilon}(\omega_1)|^2, \ldots, |A(\omega_N)d\bar{\varepsilon}(\omega_N)|^2)} - \sum_{n=1}^{N} \log|A(\omega_n)|^2 \tag{4.75}$$

where

$$d(d\varepsilon) = d(|d\bar{\varepsilon}(\omega_1)|^2)d(|d\bar{\varepsilon}(\omega_2)|^2) \ldots d(|d\bar{\varepsilon}(\omega_N)|^2),$$

and

$$p(d\varepsilon) = p(|d\bar{\varepsilon}(\omega_1)|^2, |d\bar{\varepsilon}(\omega_2)|^2, \ldots, |d\bar{\varepsilon}(\omega_N)|^2).$$

To make further progress we must assume that the input (therefore the output also) is Gaussian. The pdf of the vector random variable $d\varepsilon$ is given by

$$p(d\varepsilon) = \prod_{n=1}^{N} \frac{1}{S_\varepsilon(\omega_n)} \exp\left(-\frac{|d\bar{\varepsilon}(\omega_n)|^2}{S_\varepsilon(\omega_n)}\right). \tag{4.76}$$

We have assumed in obtaining (4.76) that the components of $d\varepsilon$ are independent. Use (4.76) in (4.75) and evaluate the integral. Note that

$$\int p(d\varepsilon)\frac{|d\bar{\varepsilon}(\omega_n)|^2}{S_\varepsilon(\omega_n)} d(d\varepsilon) = 1.$$

Dividing on both sides of (4.75) by N and letting $N \to \infty$, we obtain the normalized cross entropy,

$$H(q,p) = \frac{1}{2\pi}\int_{-\pi}^{\pi} |A(\omega)|^2 d\omega - 1 - \frac{1}{2\pi}\int_{-\pi}^{\pi} \log|A(\omega)|^2 d\omega. \tag{4.77a}$$

The cross entropy as given by (4.77a) is independent of the spectrum of the input process. The last term in (4.77a) without the minus sign is equal to twice the entropy gain of the filter as defined in (4.37). If the power gain of the filter is unity, that is,

$$\frac{1}{2\pi}\int_{-\pi}^{\pi} |A(\omega)|^2 d\omega = 1,$$

then,

$$H(q,p) = -\frac{1}{2\pi}\int_{-\pi}^{\pi} \log |A(\omega)|^2 d\omega$$
$$= -2\,\{\text{entropy gain of filter}\}.$$

The above expression may be written in a slightly different form as

$$H(q,p) + 2\,\{\text{entropy gain of filter}\} = \text{const}\ . \tag{4.77b}$$

from which it may be concluded that minimizing cross entropy is the same as maximizing the entropy gain of filter. For a pair of random variables, from (4.73) we can show that when one of the random variables is uniformly distributed, the sum of the cross entropy and the entropy of the second variable is constant. Hence, in selecting a distribution function consistent with the given constraints, one may either maximize the entropy or minimize the cross entropy.

The available data about the output process are the covariance estimates, $C_x(\tau)$, τ=0. 1, 2, ..., $p-1$, which we shall assume to be of measurement errors.

$$C_x(\tau) = \frac{1}{2\pi}\int_{-\pi}^{\pi} S_\varepsilon(\omega)|A(\omega)|^2 \exp(j\omega\tau)\, d\omega$$
$$\tau = 0, 1, \ldots, p-1 \tag{4.78}$$

The problem of spectrum estimation may be phrased as choosing a filter response function which will minimize the cross entropy and at the same time be consistent with the available data, namely, the covariance estimates. By minimizing the cross entropy we ensure that the pdf of the output process is as close as possible to that of the input. To minimize (4.77) under the constraints given by (4.78) we follow the same approach as in the case of maximum entropy (see Equations 4.57 and 4.58). We give the final result:

$$\hat{S}_x(\omega) = \frac{S_\varepsilon(\omega)}{1 + S_\varepsilon(\omega)\displaystyle\sum_{\tau=-p+1}^{p-1} \lambda_\tau \exp(-j\omega\tau)}. \tag{4.79}$$

Alternatively,

$$\frac{1}{\hat{S}_x(\omega)} = \frac{1}{S_\varepsilon(\omega)} + \sum_{\tau=-p+1}^{p-1} \lambda_\tau \exp(-j\omega\tau), \tag{4.80}$$

where $\lambda_\tau = \lambda_{-\tau}$, τ =0,1, ..., $p-1$ are to be determined from p covariance estimates. $\hat{S}_x(\omega)$ is the minimum cross entropy spectral estimate of the output process, and it is as close as possible to $S_\varepsilon(\omega)$ and consistent with the covariance constraints. $S_\varepsilon(\omega)$ is the *a priori* estimate of the

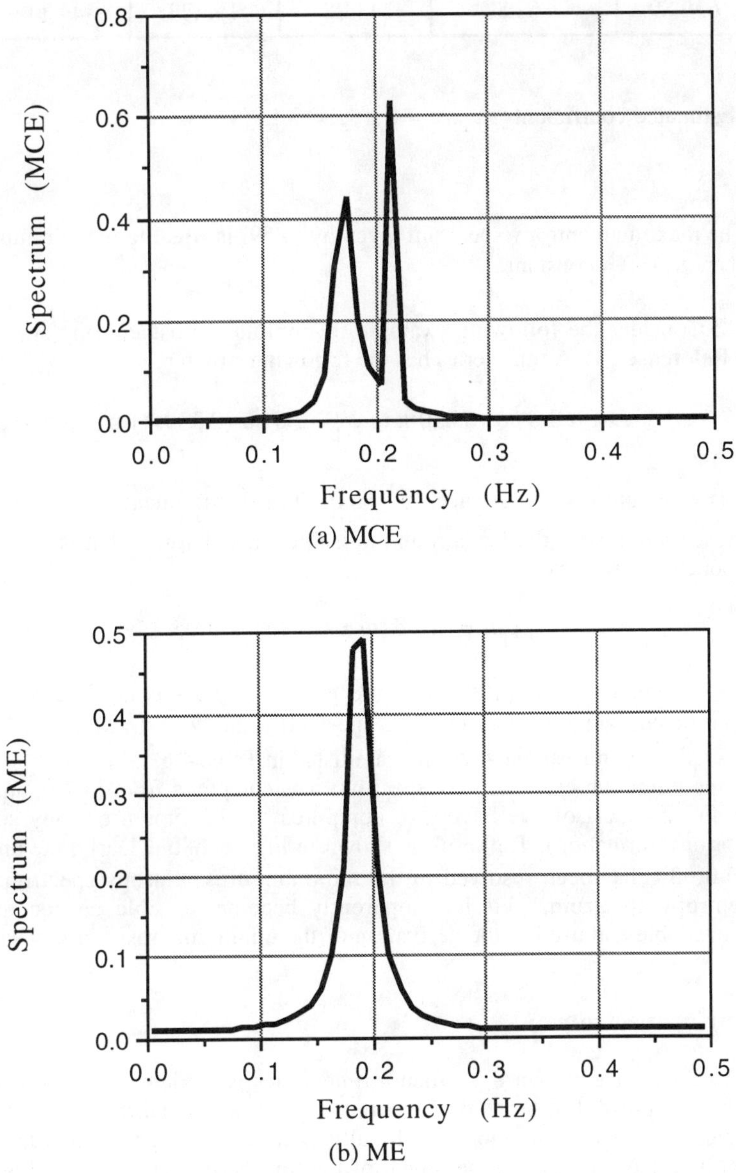

Figure 4.8: MCE spectrum of one of the sinusoid where the white noise power is assumed to be known. This *a priori* information helps resolve the unknown sinusoid (a), which was not possible in ME spectrum (b).

t	0	1	2	3	4	5
$C_x(t)$	4.05	1.4544	-2.7732	-3.2248	.2032	2.6900

Table 4.5: Error-free covariance coefficients.

t	0	1	2	3	4	5
λ_τ	636.89	-162.87	244.28	193.23	-44.11	95.38

Table 4.6: Estimated coefficients λ_τ, $\tau = 0,1, \ldots, 5$ in (4.79).

spectrum. The maximum entropy spectrum given by (4.59) is equal to the minimum cross entropy spectrum when $S_\varepsilon(\omega) = \text{constant}$.

Example 4.8: Consider the following example of minimum cross entropy spectrum estimation (taken from Reference 24). A time series has the following structure:

$$x(t) = 2.0 \cos(0.430\,\pi\, t + \phi_1) + 2.0 \cos(0.33\,\pi\, t + \phi_2) + \eta(t),$$

where $\eta(t)$ is white noise with variance, $\sigma_\eta^2 = 0.05$. The first frequency ($f_1 = 0.215$) and the power at this frequency along with the background noise power is known; that is, the spectrum of the known component is given by

$$S_\varepsilon(\omega) = 1.0\,\delta(\omega \pm 0.43\pi) + 0.05.$$

The aim is to estimate the frequency and the power of the second sinusoid. The error-free covariance coefficients are given in Table 4.5. The coefficients λ_τ, $\tau = 0,1, \ldots, 5$ in (4.79) were computed as explained in Example 4.7. They are listed in Table 4.6.

Finally, the minimum cross entropy spectrum was computed from (4.79), and it is shown in Figure 4.8a. For comparison we have also computed the maximum entropy spectrum (which requires no prior information). The result is shown in Figure 4.8b. Clearly, the unknown spectral peak at $\omega = 0.33\pi$ has been resolved in the minimum cross entropy spectrum but not in the maximum entropy spectrum. This has apparently become possible on account of the prior information available and used in the derivation of the minimum cross entropy spectrum (End of Example).

4.4 Parametric Spectrum

In Chapter Two we have shown that when a time series is modeled as an AR/MA process, its spectrum can be expressed in terms of the parameters of the model. A spectrum thus defined is known as a parametric spectrum, more specifically as an AR spectrum, or an MA spectrum, or an ARMA spectrum, depending upon the model of the time series. The AR coefficients play a key role in the estimation of the parametric spectrum, including the ARMA spectrum, without having to determine the MA coefficients (see Section 2.5). We shall naturally emphasize the estimation of AR coefficients given covariance estimates (Levinson-Durbin algorithm). Later, in the next chapter, we shall look into different methods of estimation of AR coefficients directly from the data. Such data-dependent methods, which are also known as adaptive methods, are ideal for real-time applications.

A. AR/MA/ARMA Spectrum

Here we shall recapitulate some of the important formulas of the parametric spectrum. The MA spectrum is given by (2.47),

$$S_x(\omega) = \sigma_\varepsilon^2 \left| \sum_n b_n \exp(-j\omega n) \right|^2,$$

which may be expressed as a conventional Blackman-Tukey (BT) spectrum.

$$\begin{aligned} S_x(\omega) &= \sigma_\varepsilon^2 \sum_n \sum_m b_m b_n \exp(-j\omega(m-n)) \\ &= \sigma_\varepsilon^2 \sum_\tau \{\sum_n b_{n+\tau} b_n\} \exp(-j\omega\tau). \end{aligned}$$

But, from (2.48) $C_x(\tau) = \sigma_\varepsilon^2 \sum_n b_{n+\tau} b_n$. Therefore,

$$S_x(\omega) = \sum_\tau C_x(\tau) \exp(-j\omega\tau).$$

Thus, the MA spectrum is the same as the conventional BT spectrum.

The AR spectrum is given by (2.65):

$$S_{AR}(\omega) = \frac{\sigma_\varepsilon^2}{[1 + \sum_{i=1}^{p} a_i \exp(-j\omega i)]^2}.$$

The spectrum of the ARMA time series can be expressed purely in terms of the AR coefficients and the covariance function in the range 0 to p+q lags without having to estimate the MA coefficients. The spectrum is given by (2.81):

$$S_{ARMA}(\omega) = \frac{\sum_{\tau=-q}^{q} \sum_{k=0}^{p} \sum_{l=0}^{p} a_k a_l C_x(k-l+\tau) \exp(-j\omega\tau)}{\left| \sum_{k=0}^{p} a_k \exp(j\omega k) \right|^2}.$$

The AR coefficients will have to be computed from the extended covariance matrix (2.82).

B. Implied Window

The autoregressive filter can be considered as a whitening filter or an inverse filter but of finite duration; that is,

$$a(t) = h_{INV}(t)\, w(t), \quad t = 0,1,2,\dots,p-1, \tag{4.81}$$

where

$$\begin{aligned} w(t) &= 1 \quad \text{for } 0 \le t \le p-1, \\ &= 0 \quad \text{otherwise} \end{aligned},$$

and $h_{INV}(t)$ is the inverse filter which we introduced in Chapter Two (Section 2.2). Using (4.81) in the spectral representation of the AR process, (2.65), we obtain

$$S_{AR}(\omega) = \frac{\sigma_\varepsilon^2}{\left|H_{INV}(\omega)\right|^2 * \left|\dfrac{\sin\left(\frac{1}{2}p\omega\right)}{\sin\frac{1}{2}\omega}\right|^2}. \tag{4.82}$$

Noting

$$S_x(\omega) = \frac{\sigma_\varepsilon^2}{\left|H_{INV}(\omega)\right|^2}$$

we can rewrite (4.82) in terms of the actual spectrum of the time series

$$S_{AR}^{-1}(\omega) = S_x^{-1}(\omega) * \left|\frac{\sin\left(\frac{1}{2}p\omega\right)}{\sin\frac{1}{2}\omega}\right|^2. \tag{4.83}$$

The spectrum of the implied window is the well-known Dirichlet function illustrated in Figure 2.7. The implied window in the parametric spectrum was pointed out by Durrani.[25]

C. Levinson-Durbin Algorithm

We shall now describe an iterative method of computing the AR coefficients. The algorithm exploits the Toeplitz structure of the covariance matrix to compute the AR coefficients of mth order given the coefficients of the m-1st order. To achieve this let us write the Yule-Walker equations for an AR process of order $m-1$ from (2.60).

$$\begin{bmatrix} C_x(0) & C_x(1) & C_x(2) & \cdots & C_x(m-1) \\ C_x(1) & C_x(0) & C_x(1) & \cdots & C_x(m-2) \\ C_x(2) & C_x(1) & C_x(0) & \cdots & C_x(m-3) \\ \vdots & & & & \vdots \\ C_x(m-1) & C_x(m-2) & \cdots & & C_x(0) \end{bmatrix} \begin{bmatrix} 1 \\ a_{1,m-1} \\ a_{2,m-1} \\ \vdots \\ a_{m-1,m-1} \end{bmatrix} = \begin{bmatrix} \sigma_{m-1}^2 \\ 0 \\ \vdots \\ 0 \\ 0 \end{bmatrix}, \tag{4.84}$$

where $1, a_{1,m-1}, \ldots, a_{m-1,m-1}$ are the AR coefficients of the $m-1^{st}$ order AR process. The actual order, p, of the AR process is greater than the assumed order, $m-1$. We now increase the dimension of the covariance matrix in (4.76) by padding a row and a column. The enlarged system of equations is given by

$$\begin{bmatrix} C_x(0) & C_x(1) & C_x(2) & \cdots & C_x(m) \\ C_x(1) & C_x(0) & C_x(1) & \cdots & C_x(m-1) \\ C_x(2) & C_x(1) & C_x(0) & \cdots & C_x(m-2) \\ \vdots & & & & \vdots \\ C_x(m) & C_x(m-1) & & \cdots & C_x(0) \end{bmatrix} \begin{bmatrix} 1 \\ a_{1,m-1} \\ \vdots \\ a_{m-1,m-1} \\ 0 \end{bmatrix} = \begin{bmatrix} \sigma^2_{m-1} \\ 0 \\ \vdots \\ 0 \\ \Delta_{m-1} \end{bmatrix}, \tag{4.85}$$

where

$$\Delta_{m-1} = C_x(m) + \sum_i C_x(m-i)\, a_{i,m-1} .$$

Now express (4.85) in a reverse order, and then linearly combine it with itself:

$$\begin{bmatrix} C_x(0) & C_x(1) & \cdots & C_x(m) \\ C_x(1) & C_x(0) & \cdots & C_x(m-1) \\ C_x(2) & C_x(1) & \cdots & C_x(m-2) \\ \vdots & & \vdots & \\ C_x(m) & C_x(m-1) & \cdots & C_x(0) \end{bmatrix} \left\{ \begin{bmatrix} 1 \\ a_{1,m-1} \\ \vdots \\ a_{m-1,m-1} \\ 0 \end{bmatrix} + \alpha_m \begin{bmatrix} 0 \\ a_{m-1,m-1} \\ \vdots \\ a_{1,m-1} \\ 1 \end{bmatrix} \right\} = \begin{bmatrix} \sigma^2_{m-1} \\ 0 \\ \vdots \\ 0 \\ \Delta_{m-1} \end{bmatrix} + \alpha_m \begin{bmatrix} \Delta_{m-1} \\ 0 \\ \vdots \\ 0 \\ \sigma^2_{m-1} \end{bmatrix}$$

Since we are free to choose α_m, we do so that $\Delta_{m-1} + \alpha_m \sigma^2_{m-1} = 0$. This leads us to the Yule-Walker equations of the mth order:

$$\begin{bmatrix} C_x(0) & C_x(1) & \cdots & C_x(m) \\ C_x(1) & C_x(0) & \cdots & C_x(m-1) \\ C_x(2) & C_x(1) & \cdots & C_x(m-2) \\ \vdots & & \vdots & \\ C_x(m) & C_x(m-1) & \cdots & C_x(0) \end{bmatrix} \begin{bmatrix} 1 \\ a_{1,m-1} + \alpha_m a_{m-1,m-1} \\ \vdots \\ a_{m-1,m-1} + \alpha_m a_{1,m-1} \\ \alpha_m \end{bmatrix} = \begin{bmatrix} \sigma^2_{m-1} + \alpha_m \Delta_{m-1} \\ 0 \\ \vdots \\ 0 \end{bmatrix}. \tag{4.86}$$

The coefficients of the mth order can be obtained from (4.86). They are

$$\begin{aligned} a_{0,m} &= 1.0 \\ a_{1,m} &= a_{1,m-1} + \alpha_m a_{m-1,m-1} \\ a_{2,m} &= a_{2,m-1} + \alpha_m a_{m-2,m-1} \\ &\cdots \\ a_{m,m} &= \alpha_m \end{aligned} \tag{4.87}$$

and the noise power

$$\sigma_m^2 = \sigma_{m-1}^2(1 - \alpha_m^2).$$

α_m is known as a reflection coefficient, which may be obtained from the first equation of the system of equations represented by (4.86),

$$\alpha_m = -\frac{C_x(m) + \sum_{i=1}^{m-1} a_{i,m-1} C_x(m-i)}{\sigma_{m-1}^2}. \tag{4.88}$$

The recursive system of equations (4.87) need an initial condition to get started; that is, $\sigma_0^2 = C_x(0)$. The recursion is stopped on reaching the stage when m=p, the order of the AR process which may be determined from the eigenvalue distribution of its covariance matrix (see Example 2.8). Other methods of order estimation from the finite data will be discussed later in Chapter Five.

Example 4.9: Given an autocorrelation function, the problem of computing the reflection coefficients is a relatively involved process. An expression can be derived only for a few coefficients. In this example we show how to compute the reflection coefficients given an autocorrelation function. From (4.80) it is easy to obtain an expression for

$$\alpha_1 = -\frac{C_x(1)}{C_x(0)}.$$

For other reflection coefficients we will need to use (4.79) and (4.80). A few of these coefficients are listed below:

$$\alpha_2 = -\frac{C_x(2) - \dfrac{C_x(1)^2}{C_x(0)}}{C_x(0)(1-\alpha_1^2)}$$

$$\alpha_3 = -\frac{C_x(3) + C_x(2)(\alpha_1 + \alpha_1\alpha_2) + C_x(1)\alpha_2}{C_x(0)(1-\alpha_1^2)(1-\alpha_2^2)}.$$

The reflection coefficients can be computed directly from the data without having to compute the autocorrelation function. This will be dealt with in Chapter Five (End of Example).

D. **Extended Covariance Matrix**

In Chapter Two we have shown that the AR coefficients of an ARMA model lie in the null space of the extended covariance matrix, in particular in the null space of $\mathbf{C}_x^{eH}\mathbf{C}_x^e$, where $\mathbf{C}_x^e$ is the extended matrix defined in Chapter Two (see (2.82)). In practice, because of estimation errors, the minimum eigenvalue of $\mathbf{C}_x^e$ will not be zero but a finite, small positive quantity. The AR coefficient vector will be the one which minimizes

$$\mathbf{A}^H \mathbf{C}_x^{e^H} \mathbf{C}_x^e \mathbf{A} = \min , \tag{4.89}$$

and its first coefficient is unity; that is,

$$\mathbf{A}^H \delta = 1,$$

where $\delta = \mathrm{col}\{1, 0, 0, \ldots , 0\}$. The coefficient vector is a solution of the following system of equations:

$$\mathbf{C}_x^{e^H} \mathbf{C}_x^e \mathbf{A} = \lambda_{\min} \delta , \tag{4.90}$$

where $\lambda_{\min}$ is the minimum eigenvalue of $\mathbf{C}_x^{e^H} \mathbf{C}_x^e$. Hence,

$$\mathbf{A} = \lambda_{\min} \{\mathbf{C}_x^{e^H} \mathbf{C}_x^e\}^{-1} \delta ,$$

that is, the first column of $\{\mathbf{C}_x^{e^H} \mathbf{C}_x^e\}^{-1}$ multiplied by $\lambda_{\min}$.

In the formation of $\mathbf{C}_x^{e^H} \mathbf{C}_x^e$ we assumed that the order of the AR component of the ARMA process, namely, p, is known. The dimension of $\mathbf{C}_x^{e^H} \mathbf{C}_x^e$ is (p+1); therefore, the dimension of null space will be one, in the ideal case, giving a unique eigenvector spanning the null space. In practice p is not known *a priori.* It will have to be estimated from the computed autocorrelation function. The approach is similar to that used for the order estimation of the AR model in Chapter Two (page 103) and is based on the approach suggested in Reference 26. Briefly, we form an extended covariance matrix so that the dimension of $\mathbf{C}_x^{e^H} \mathbf{C}_x^e$ is greater than (p+1), about two times, and then we examine its eigenvalue spectrum for significant eigenvalues. The order of the AR component is equal to the number of significant eigenvalues. An example is given below to illustrate the method. An alternate approach, described in Reference 27, is to look for stable (i.e., with significant power) poles of the z-transform of the AR coefficients obtained from an oversized extended covariance matrix.

Example 4.10: An ARMA(4,2) process was generated with the following coefficients:

$a_1 = -2.7607$	$b_0 = 1.0$
$a_2 = 3.8106$	$b_1 = 0.81$
$a_3 = -2.6555$	$b_2 = -1.2728$
$a_4 = 0.9238$	

and $\sigma_\varepsilon^2 = 1.0$. An extended covariance matrix of size 16×16 was formed from an autocorrelation function, which was computed using 650 samples after dropping the first 2000 samples so that the transients are practically dead. We assumed that the order of the MA component is unknown, but less than a preselected number; in the present experiment this number is three. Next, the

1142 E05	708 E05	5.55 E05	1.39 E05	0.22 E01	0.06 E00
882 E05	548 E05	4.68 E05	1.17 E05	2.28 E01	0.25 E00
613 E05	378 E05	2.54 E05	0.45 E05	3.06 E01	0.89 E00
611 E05	373 E05	3.56 E05	0.73 E05	0.90 E00	0.10 E00
461 E05	282 E05	3.33 E05	0.70 E05	0.34 E00	0.02 E00
559 E05	342 E05	3.04 E05	0.65 E05	3.75 E00	0.01 E00
950 E05	591 E05	7.87 E05	1.82 E05	3.88 E01	1.10 E00
476 E05	292 E05	3.77 E05	0.71 E05	1.30 E00	0.02 E00

Significant Eigenvalues

Figure 4.9: Eigenvalues of an extended covariance matrix of ARMA(4,2) process. The number of significant eigenvalues is equal to the order of the AR component, namely, 4.

1.677E08	1.088E08	5.091E05	9.187E04	2.661E02	1.306E02
1.391E08	1.078E08	3.074E05	7.663E04	2.372E02	5.872E01
1.247E08	9.643E07	1.538E05	7.351E04	1.029E02	4.507E00
1.234E08	7.502E07	1.414E05	3.568E04	8.577E00	2.438E00

Figure 4.10: Eigenvalue distribution of extended covariance matrix of ARMA(4,2) process described in Example 4.10 for different assumed MA order, namely, 2 (top row), 3 (second row), 4 (third row), and 5 (bottom row). 160 data points were used to compute the 16×16 covariance matrix.

λ_1	λ_2	λ_3	λ_4	λ_5
97066352	52519200	419520	26204	0.09

Table 4.7: Eigenvalues of the extended covariance matrix.

Computed values	-2.7772	3.8364	-2.6757	0.9258
True AR Coefficients	-2.7607	3.8106	-2.6555	0.9238

Table 4.8: Computed eigenvector normalized with respect to the first coefficient and the true AR coefficients are compared.

eigenvalues of $\mathbf{C}_x^{e\,H}\,\mathbf{C}_x^e$ were computed. The above experiment was repeated eight times with independent sets of 650 samples of time series. The computed eigenvalues are listed in Figure 4.9. The choice of the unknown MA order is not very important in the method for estimation of the AR order from the eigenvalue distribution. Indeed, the greater the selected value for the unknown MA order, the sharper is the transition between the signal eigenvalues and noise eigenvalues, as may be seen from Figure 4.10 (End of Example).

Having estimated the order of the AR component of the right size so that there is a unique null space, that is, spanned by a single eigenvector, the eigenvector corresponding to the smallest eigenvalue is the desired AR coefficient vector, provided the first element of the eigenvector is normalized. This is illustrated in the example given below.

Example 4.11: An extended covariance matrix of size 5×5 was formed from the autocorrelation function computed in the previous example. The eigenvalues are listed in Table 4.7. The eigenvector corresponding to the minimum eigenvalue was found after normalization (a_0=1.0). The result is compared with the true AR coefficients in Table 4.8. Finally, we shall show the spectrum of the ARMA process (Example 4.10) computed from the covariance function and the AR coefficients. The theoretical basis for such an estimation was given in Chapter Two (see (2.81)). The order of the AR process was determined as illustrated in Example 4.9, and the coefficients were determined as shown in Example 4.10. The covariance function was computed up to 16 lags. The ARMA spectrum was computed by evaluating the numerator and the denominator of (2.81), and the result is shown in Figure 4.11 where we also show the exact ARMA spectrum obtained by using the actual AR and MA coefficients in (2.75) (End of Example).

From the results of the current numerical example it is clear that the AR coefficients derived from the eigenvector corresponding to the minimum eigenvalue of the extended covariance matrix are very close to the true values (They differ only in the second decimal place). Naturally the ARMA spectrum obtained from the estimated AR coefficients is also close to the true spectrum, except for the noise at high frequencies, probably caused by estimation errors. It also demonstrates that the ARMA spectrum can be estimated without estimating the MA coefficients, which is a definite advantage in practical terms.

E. **Singular Value Decomposition** (SVD)

An extended covariance matrix can be expressed as

$$\mathbf{C}_x^e = \sum_{k=1}^{r} \lambda_k \mathbf{U}_k \mathbf{V}_k^H, \tag{4.91}$$

where vectors $\mathbf{U}_k$ and $\mathbf{V}_k$ are solutions of the following eigen decomposition equations:

$$\mathbf{C}_x^{e\,H}\,\mathbf{C}_x^e \mathbf{U}_k = \lambda_k^2 \mathbf{U}_k,$$

$$\mathbf{C}_x^e \mathbf{C}_x^{e\,H}\,\mathbf{V}_k = \lambda_k^2 \mathbf{V}_k,$$

where λ_k, k = 1, 2,...,r are singular values of $\mathbf{C}_x^e$ and r is the rank of $\mathbf{C}_x^e$ ($r \le \min\{p+1, N\}$). Note that $\mathbf{U}_k$ is a $(p+1) \times 1$ vector, and $\mathbf{V}_k$ is a $N \times 1$ vector. Vectors $\mathbf{U}_k$ and $\mathbf{V}_k$ satisfy the property of orthogonality; namely,

$$\mathbf{u}_k^H \mathbf{u}_{k'} = \delta_{kk'},$$
$$\mathbf{v}_k^H \mathbf{v}_{k'} = \delta_{kk'}.$$

Using this orthogonality property in (4.91) we can show that

$$\mathbf{C}_x^e \mathbf{V}_k = \lambda_k \mathbf{U}_k,$$
$$\mathbf{C}_x^{e\,H} \mathbf{U}_k = \lambda_k \mathbf{V}_k.$$

For more information on singular value decomposition the reader is referred to Reference 28. There has been a growing interest in the SVD decomposition for signal processing applications, in particular in the estimation of parameters of a linear system from a reduced rank covariance matrix which is derived from the estimated covariance matrix by nulling all insignificant singular values.[29] Through this process we seem to remove the estimation errors as well as noise, which results in a significant reduction of variance of the estimated parameters. A small amount of signal power is also lost, which introduces a small degree of bias in the estimated parameters.

As described in the previous subsection the effective rank (r_{eff}) of an extended covariance function can be estimated from the number of significant singular values.

The singular vectors corresponding to the insignificant singular values will span the null space which also contains the AR coefficient vector, **A**. We shall express **A** as a linear combination of singular vectors spanning the null space,

$$\mathbf{A} = \sum_{k=r_{eff}+1}^{N} \alpha_k \mathbf{U}_k$$

and select the coefficients such that

$$\sum_{k=r_{eff}+1}^{N} \alpha_k u_k(0) = 1$$

and

$$\sum_{k=r_{eff}+1}^{N} \alpha_k^2 = \sum_{i=1}^{p} a_i^2 = \min,$$

where $u_k(0)$ is the first coefficient of the kth singular vector. The constrained minimization problem leads to the following solution:[30]

$$\alpha_k = \frac{u_k(0)}{\sum_{k=r_{eff}+1}^{N} u_k^2(0)}.$$

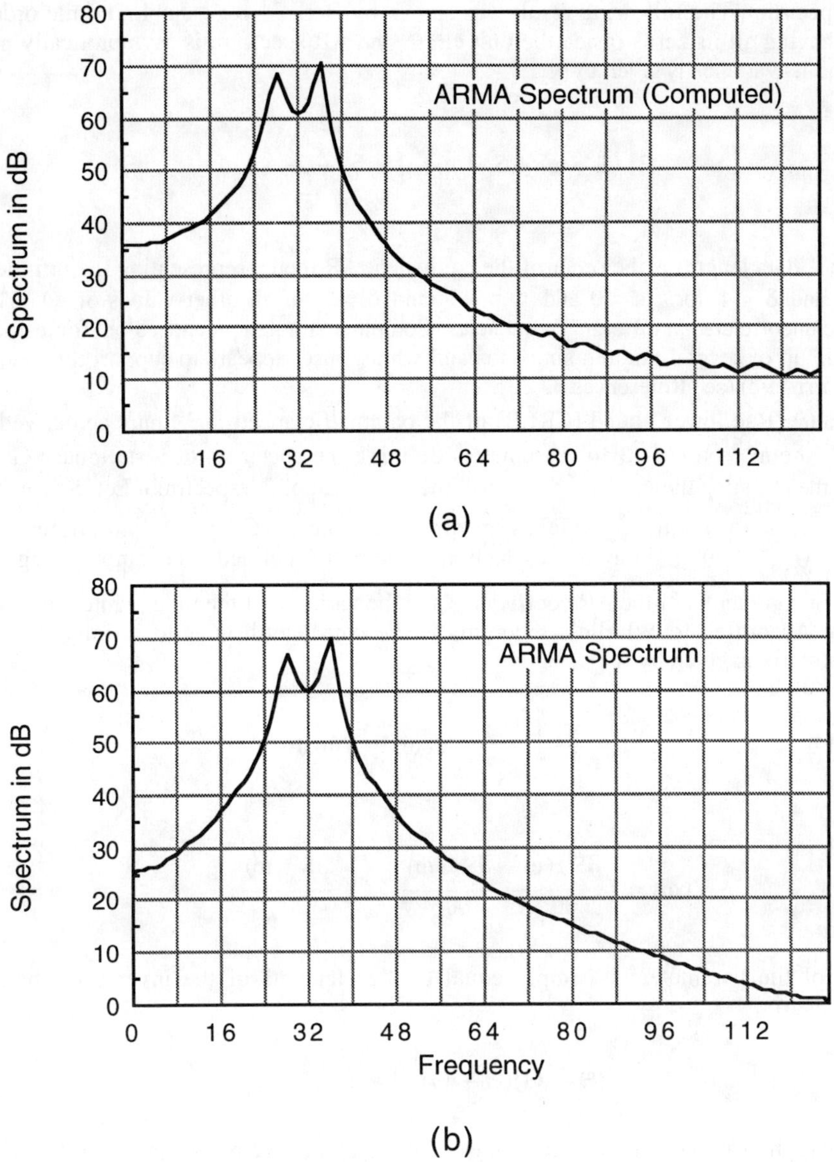

Figure 4.11: ARMA spectrum using the AR coefficients and covariancematrix but without MA coefficients is shown in (a), and the actual spectrum of the ARMA process in Example 4.10 is shown in (b). To convert the frequency number into Hz, multiply by $\frac{1}{256}$.

F. **Error Analysis**

We have assumed that an exact covariance matrix is available for the purpose of parameter estimation. This in practice is not true. We have only a finite length of data from which one can make an approximate estimate of the autocorrelation function. It is of great significance to study the effects of the errors in the autocorrelation function on parameter estimation, hence on the parameter spectrum. The following result was shown by Berk:[31] for a stable infinite order MA time series having all its zeros inside the unit circle, the AR spectrum is asymptotically normal, unbiased, and its variance is given by

$$E\{(\hat{S}_x(\omega) - S_x(\omega))^2\} \approx \delta \frac{S_x^2(\omega)}{\frac{T}{p}}, \tag{4.92}$$

where p (<<T, data length) is the order of the equivalent AR model representing the infinite order MA process and $\delta = 4$ for $\omega = 0$ and $\omega = \pm\pi$ and $\delta = 2$ at all other values of ω. The AR estimates are uncorrelated at different frequencies. Computer simulation, carried out for a stochastic process made up of several random sinusoids and white noise, appears to support the theoretical result given in (4.92) (see References 32,33).

The Cramer-Rao lower bound (CRLB) of the parametric spectrum[34] may be derived using Whittle's theorem[35] on CRLB for parameters defining the spectrum of a stationary Gaussian process and then relating the variance of the parameters to that of the spectrum. Let $S_x(\omega, \Theta)$ be the spectrum of a stationary Gaussian process where Θ is a parameter vector $\Theta = \text{col}\{\theta_1, \theta_2, \ldots, \theta_p\}$ in terms of which the spectrum is defined, an example being the AR spectrum defined in terms of the AR coefficients and the variance of the input white noise process (see (2.65)). According to Whittle's theorem, as the data length goes to infinity, the Fisher information matrix is given by

$$\mathbf{J} = \frac{T}{4\pi} \int_{-\pi}^{\pi} \frac{1}{S_x^2(\omega)} \mathbf{V}(\omega) \mathbf{V}^H(\omega) \, d\omega, \tag{4.93}$$

where

$$\mathbf{V}(\omega) = \left\{ \frac{\partial S_x(\omega)}{\partial \theta_1}, \frac{\partial S_x(\omega)}{\partial \theta_2}, \ldots, \frac{\partial S_x(\omega)}{\partial \theta_p} \right\}^T.$$

The CRLB of the parameters is simply equal to the elements of the inverse of the Fisher information matrix,

$$E\{(\hat{\Theta} - \Theta)(\hat{\Theta} - \Theta)^H\} \geq \mathbf{J}^{-1}. \tag{4.94}$$

Next, we relate the variance of the spectrum estimate to that of the parameter estimates. For this we expand the estimate of the spectrum in a Taylor series around the true spectrum,

$$\hat{S}_x(\omega, \hat{\Theta}) \approx S_x(\omega, \Theta) + \frac{\partial S_x(\omega)}{\partial \theta_1}(\hat{\theta}_1 - \theta_1) + \frac{\partial S_x(\omega)}{\partial \theta_2}(\hat{\theta}_2 - \theta_2) + \ldots + \frac{\partial S_x(\omega)}{\partial \theta_p}(\hat{\theta}_p - \theta_p)$$

which may be expressed in a compact vector notation:

$$(\hat{S}_x(\omega, \hat{\Theta}) - S_x(\omega, \Theta)) \approx \mathbf{V}^H(\omega)(\hat{\Theta} - \Theta).$$

Squaring on both sides and evaluating the expected value, we obtain

$$E\{(\hat{S}_x(\omega,\hat{\Theta}) - S_x(\omega,\Theta))(\hat{S}_x(\omega,\hat{\Theta}) - S_x(\omega,\Theta))\} \approx \mathbf{V}^H(\omega) E\{(\hat{\Theta} - \Theta)(\hat{\Theta} - \Theta)^H\} \mathbf{V}(\omega),$$

where we shall use (4.94) to obtain the CRLB of the spectral estimate,

$$E\{(\hat{S}_x(\omega,\hat{\Theta}) - S_x(\omega,\Theta))(\hat{S}_x(\omega,\hat{\Theta}) - S_x(\omega,\Theta))\} \geq \mathbf{V}^H(\omega)\mathbf{J}^{-1}\mathbf{V}(\omega). \tag{4.95}$$

Example 4.12: The calculation of CRLB is generally an involved process. We shall illustrate by means of an example of a second-order AR process with three unknown parameters, namely, $\Theta = \{a_1, a_2 \text{ and } \sigma_\varepsilon^2\}$. First by differentiating the expression for the spectrum (2.65a) we get

$$\mathbf{V}(\omega) = -S(\omega,\Theta)\left\{\frac{e^{-j\omega}}{A(\omega)} + \frac{e^{+j\omega}}{A^*(\omega)}, \frac{e^{-j2\omega}}{A(\omega)} + \frac{e^{+j2\omega}}{A^*(\omega)}, -\frac{1}{\sigma_\varepsilon^2}\right\}^T,$$

where

$$A(\omega) = (1 + a_1 e^{-j\omega} + a_2 e^{-j2\omega}).$$

Now substituting for $\mathbf{V}(w)$ in (4.93) Fisher Information matrix turns out to be

$$\mathbf{J} = \frac{T}{4\pi}\int_{-\pi}^{\pi}\begin{bmatrix} \Omega_1\Omega_1^* & \Omega_1\Omega_2^* & -\frac{\Omega_1}{\sigma_\varepsilon^2} \\ \Omega_2\Omega_1^* & \Omega_2\Omega_2^* & -\frac{\Omega_2}{\sigma_\varepsilon^2} \\ -\frac{\Omega_1^*}{\sigma_\varepsilon^2} & -\frac{\Omega_2^*}{\sigma_\varepsilon^2} & \frac{1}{\sigma_\varepsilon^4} \end{bmatrix} d\omega,$$

where

$$\Omega_1 = \frac{e^{-j\omega}}{A(\omega)} + \frac{e^{+j\omega}}{A^*(\omega)}$$

and

$$\Omega_2 = \frac{e^{-j2\omega}}{A(\omega)} + \frac{e^{+j2\omega}}{A^*(\omega)}.$$

Note that because A(w) is a transfer function of a causal system,

$$\frac{1}{2\pi}\int_{-\pi}^{\pi}\left(\frac{e^{-j\omega}}{A(\omega)} + \frac{e^{+j\omega}}{A^*(\omega)}\right)d\omega = 2h_{-1} = 0.$$

Similarly, we can show that

$$\frac{1}{2\pi}\int_{-\pi}^{\pi}\Omega_1\Omega_2 d\omega = \frac{1}{2\pi}\int_{-\pi}^{\pi}\left(\frac{e^{-j\omega}}{A(\omega)}+\frac{e^{+j\omega}}{A^*(\omega)}\right)\left(\frac{e^{-j2\omega}}{A(\omega)}+\frac{e^{+j2\omega}}{A^*(\omega)}\right)d\omega$$
$$= \frac{2C_x(1)}{\sigma_\varepsilon^2}.$$

Using these results the information matrix becomes[8]

$$\mathbf{J} = \frac{T}{\sigma_\varepsilon^2}\begin{bmatrix} C_x(0) & C_x(1) & 0 \\ C_x(-1) & C_x(0) & 0 \\ 0 & 0 & \frac{1}{2\sigma_\varepsilon^2} \end{bmatrix}.$$

Finally, from (4.86b) we get the following result for CRLB on the AR spectrum,

$$\mathrm{Var}\{\hat{S}_x(\omega)\} \geq \frac{S_x^2(\omega)}{T}[(|\Omega_1|^2+|\Omega_2|^2)\frac{\sigma_\varepsilon^2 C_x(0)}{\Delta} - (\Omega_1^H\Omega_2+\Omega_1\Omega_2^H)\frac{\sigma_\varepsilon^2 C_x(1)}{\Delta} + 2],$$

where $\Delta = C_x(0)^2 - C_x(1)^2$ (End of Example).

4.5 Subspace Methods

The poles of an AR or ARMA time series play a dominating role in shaping the spectrum, in particular, the spectral peaks. As a pole approaches the unit circle, the corresponding spectral peak becomes sharper, and its location correctly points to the pole position (see page 107 for more discussion on this issue). Therefore, an AR process with poles on the unit circle, which corresponds to a sum of sinusoids, is of great theoretical and practical significance. Additionally, background noise, if present, introduces zeros on the unit circle, leading to a special ARMA model which characterizes a sum of sinusoids and white noise (see page 124).

Pisarenko[6] first observed that the zeros of the z-transform of the eigenvector corresponding to the minimum eigenvalue of the covariance matrix lie on the unit circle, and their angular positions correspond to the frequencies of the sinusoids. It is assumed that the size of the covariance matrix is equal to 2p+1, where p is the number of sinusoids. In a later development, a larger covariance matrix was considered, and it was shown that the eigenvectors may be be divided into two groups, namely, the eigenvectors spanning the signal subspace and eigenvectors spanning the orthogonal noise space. The eigenvectors spanning the noise space are the ones whose eigenvalues are the smallest (repeated) eigenvalues equal to the noise power. The frequency vector comprising the discrete samples of a complex sinusoid of the unknown frequency lies in the signal subspace. These features of the above model form the basis of MUSIC (Multiple Signal Classification) algorithm discovered by Schmidt[36] and at about the same time by Bienvenu and Kopp.[37] We shall examine in some detail the basic concepts involved in the Pisarenko decomposition and Schmidt's subspace approach and many related ideas.

A. Pisarenko Decomposition

Let us go back to (2.95) and express it in a matrix notation,

$$\mathbf{x}_t^T\mathbf{A} = \varepsilon_t^T\mathbf{A}, \tag{4.96}$$

where

$$\mathbf{x}_t = \{x(t)\ x(t-1)\ \ldots\ x(t-2p)\}^T,$$

$$\mathbf{A} = \{1\ a_1\ a_2\ \ldots\ a_{2p}\}^T,$$

$$\varepsilon_t = \{\varepsilon(t)\ \varepsilon(t-1)\ \ldots\ \varepsilon(t-2p)\}^T.$$

Premultiply on both sides of (4.96) with $\mathbf{x}_t$ and evaluate the expected value. We obtain

$$\mathbf{C}_x\mathbf{A} = \sigma_\varepsilon^2\mathbf{A}, \tag{4.97}$$

where $\mathbf{C}_x$ is a covariance matrix of size $(2p+1)\times(2p+1)$. Note that

$$E\{\mathbf{X}_t\varepsilon_t^T\} = E\{(\mathbf{S}_t + \varepsilon_t)\varepsilon_t^T\} = \sigma_\varepsilon^2\mathbf{I},$$

since signal and noise are uncorrelated. From (4.97) we notice that **A** is an eigenvector of $\mathbf{C}_x$, and its eigenvalue is σ_ε^2. Let us now rewrite (4.97) as

$$(\mathbf{C}_x - \sigma_\varepsilon^2\mathbf{I})\mathbf{A} = \mathbf{0}.$$

Therefore, $(\mathbf{C}_x - \sigma_\varepsilon^2\mathbf{I})$ is a singular matrix with one of its eigenvalues being zero and the remaining, positive. We shall show in the next subsection that $(\mathbf{C}_x - \sigma_\varepsilon^2\mathbf{I})$ is of rank 2p. Having obtained the **A** vector the frequencies are estimated from the roots of a polynomial (see page 123 for more details),

$$1 + \sum_{k=1}^{2p} a_k z^{-k} = 0.$$

It is possible to reduce the size of the covariance matrix by exploiting the symmetry properties of the **A** vector (see Example 2.16). Let us split **A** into two halves; namely,

$$\mathbf{A}_1 = \text{col}\,\{1, a_1\, a_2, \ldots, \frac{a_p}{\sqrt{2}}\}$$

and

$$\mathbf{A}_2 = \text{col}\,\{\frac{a_p}{\sqrt{2}}, a_{p+1}, \ldots, a_{2p-1}, 1\}$$

$$\mathbf{A} = [\mathbf{A}_1 | \mathbf{A}_2].$$

Because of the symmetry shown in Example (2.16) we can relate $\mathbf{A}_1$ to $\mathbf{A}_2$,

$$\mathbf{A}_2 = \mathbf{J}\mathbf{A}_1, \tag{4.98}$$

where

$$\mathbf{J} = \begin{bmatrix} 0 & 0 & \dots & 0 & 1 \\ 0 & 0 & \dots & 1 & 0 \\ & & & & \\ 0 & 1 & \dots & 0 & 0 \\ 1 & 0 & \dots & 0 & 0 \end{bmatrix}$$

is an exchange matrix. We shall split in the same fashion the vectors $\mathbf{x}_t$ and ε_t,

$$\mathbf{x}_t = [\mathbf{x}_t^1 | \mathbf{x}_t^2]$$

and

$$\varepsilon_t = [\varepsilon_t^1 | \varepsilon_t^2]$$

(for example, $\mathbf{x}_t^1 = \text{col}\{x(t), x(t-1), \dots, \frac{x(t-p)}{\sqrt{2}}\}$). We will now rewrite (4.96) in terms of the above subvectors as follows:

$$(\mathbf{x}_t^1 + \mathbf{J}^T \mathbf{x}_t^2)^T \mathbf{A}_1 = (\varepsilon_t^1 + \mathbf{J}^T \varepsilon_t^2)^T \mathbf{A}_1, \tag{4.99}$$

where we used (4.98). Premultiply both sides of (4.99) with $(\mathbf{x}_t^1 + \mathbf{J}^T \mathbf{x}_t^2)$ and evaluate the expected value. We obtain the following result:

$$\tilde{\mathbf{C}}_x \mathbf{A}_1 = 2\sigma_\varepsilon^2 \mathbf{A}_1, \tag{4.100}$$

where

$$\begin{aligned} \tilde{\mathbf{C}}_x &= E\{(\mathbf{x}_t^1 + \mathbf{J}^T \mathbf{x}_t^2)(\mathbf{x}_t^1 + \mathbf{J}^T \mathbf{x}_t^2)^T\} \\ &= E\{\mathbf{x}_t^1 \mathbf{x}_t^{1^T}\} + E\{\mathbf{x}_t^1 \mathbf{x}_t^{2^T}\}\mathbf{J} + \mathbf{J}^T E\{\mathbf{x}_t^2 \mathbf{x}_t^{1^T}\} + \mathbf{J}^T E\{\mathbf{x}_t^2 \mathbf{x}_t^{2^T}\}\mathbf{J}. \end{aligned}$$

Note that $\tilde{\mathbf{C}}_x$ is now a $(p+1) \times (p+1)$ matrix, and $\mathbf{A}_1$ is eigenvector of $\tilde{\mathbf{C}}_x$. Thus, we have been able to exploit the symmetry present in the AR coefficients to reduce the dimension of the eigen decomposition problem.[33]

B. Subspace Approach

In the approach described above it is assumed that the number of sinusoids is known and the matrix size is $(2p+1) \times (2p+1)$ (or $(p+1) \times (p+1)$) if we were to exploit the symmetry properties of the AR coefficients). In the following approach some of these drawbacks will be removed. We shall model the observed time series as a sum of real random sinusoids and white noise,

$$x(t) = \sum_{i=1}^{p} \frac{A_i}{2} \{e^{j(\omega_i t + \phi_i)} + e^{-j(\omega_i t + \phi_i)}\} + \eta(t). \tag{4.101}$$

Let us now define the following vectors and matrices:

$$\mathbf{x}_t = \text{col}\{x(t) \quad x(t-1) \quad \dots \quad x(t-N+1)\},$$
$$\eta_t = \text{col}\{\eta(t) \quad \eta(t-1) \quad \dots \quad \eta(t-N+1)\},$$

$$\mathbf{H} = \begin{bmatrix} 1 & 1 & \dots & 1 & 1 \\ e^{-j\omega_1} & e^{+j\omega_1} & \dots & e^{-j\omega_p} & e^{+j\omega_p} \\ e^{-j2\omega_1} & e^{+j2\omega_1} & \dots & e^{-j2\omega_p} & e^{+j2\omega_p} \\ \vdots & & & \vdots & \\ e^{-j(N-1)\omega_1} & e^{+j(N-1)\omega_1} & \dots & e^{-j(N-1)\omega_p} & e^{+j(N-1)\omega_p} \end{bmatrix},$$

$$\mathbf{G}_t = \text{diag}\,\{e^{j\omega_1 t} \; e^{-j\omega_1 t} \; e^{j\omega_2 t} \; e^{-j\omega_2 t} \; \dots \; e^{j\omega_{2p} t} \; e^{-j\omega_{2p} t}\},$$

and

$$\mathbf{A} = \text{col}\,\{\frac{A_1}{2}e^{j\phi_1} \; \frac{A_1}{2}e^{-j\phi_1} \; \dots . \; \frac{A_p}{2}e^{j\phi_1} \; \frac{A_p}{2}e^{-j\phi_1}\}.$$

Using the above quantities we shall express (4.101) in a matrix form as shown below:

$$\mathbf{x}_t = \mathbf{H}\mathbf{G}_t\mathbf{A} + \eta_t. \tag{4.102}$$

From (4.102) we can evaluate the covariance matrix,

$$\mathbf{C}_x = \mathbf{H}\mathbf{C}_0\mathbf{H}^H + \sigma_\eta^2\mathbf{I}, \tag{4.103}$$

where

$$\mathbf{C}_0 = \text{diag}\,\{\frac{A_1^2}{4} \; \frac{A_1^2}{4} \; \frac{A_2^2}{4} \; \frac{A_2^2}{4} \; \dots \; \frac{A_{2p}^2}{4} \; \frac{A_{2p}^2}{4}\}.$$

In (4.103) we note that **H** is a Vandermonde matrix; hence it has full column rank, namely, 2p. Further, $\mathbf{C}_0$ is a full rank (2p) diagonal matrix. Consider the matrix difference,

$$[\mathbf{C}_x - \sigma_\eta^2\mathbf{I}] = \mathbf{H}\mathbf{C}_0\mathbf{H}^H. \tag{4.104}$$

Since the rank of the right-hand side of (4.104) is 2p, the rank of the left-hand side is also 2p, but the matrix size is $N \times N$. Assume that N>2p; then $\mathbf{C}_x - \sigma_\eta^2\mathbf{I}$ will have N – 2p zero eigenvalues. Now consider the collection of eigenvectors corresponding to the zero eigenvalues represented by $\mathbf{E}_\eta$. Consider the following product obtained from (4.104) by pre- and postmultiplying both sides of (4.104) by $\mathbf{E}_\eta$. Clearly,

$$\mathbf{E}_\eta^H[\mathbf{C}_x - \sigma_\eta^2\mathbf{I}]\mathbf{E}_\eta = \mathbf{0},$$

and hence,

$$\mathbf{E}_\eta^H \mathbf{H} \mathbf{C}_0 \mathbf{H}^H \mathbf{E}_\eta = \mathbf{0}. \tag{4.105}$$

Since $\mathbf{C}_0$ is full rank, (4.105) implies that

$$\mathbf{E}_\eta^H \mathbf{H} = \mathbf{0}. \tag{4.106}$$

This is an important result. It states that the space spanned by the columns of the **H** matrix is orthogonal to the space spanned by the eigenvectors corresponding to the zero eigenvalues of $\mathbf{C}_x - \sigma_\eta^2 \mathbf{I}$. We use this result to estimate the unknown frequencies of the sinusoids. Consider a vector defined as (frequency vector)

$$\mathbf{h}(\omega) = \mathrm{col}\{1 \;\; e^{-j\omega} \;\; e^{-j2\omega} \;\; \ldots \;\; e^{-j(N-1)\omega}\}$$

and the product $\mathbf{h}^H(\omega)\mathbf{E}_\eta \mathbf{E}_\eta^H \mathbf{h}(\omega)$, which will be equal to zero whenever ω is equal to the frequency of one of the sinusoids. In fact, it is more dramatic if we were to consider the inverse of the product. Consider the following quantity as a function of ω:

$$S_{EV}(\omega) = \frac{1}{\mathbf{h}^H(\omega)\mathbf{E}_\eta \mathbf{E}_\eta^H \mathbf{h}(\omega)}. \tag{4.107}$$

$S_{EV}(\omega)$ will be infinite whenever $\omega = \omega_1$ or ω_{-1}, ..., ω_p or ω_{-p}. It is often called a spectrum (eigenvector (EV) spectrum), even though it does not possess units of power. The method of frequency estimation described here is an adaptation of the well-known algorithm MUSIC for estimation of direction of arrival[36] to the time series problem.[38]

Having determined the frequencies of the sinusoids let us look into the question of power estimation.[39] Pre- and postmultiply on both sides of (4.94) by $\mathbf{H}^H$ and $\mathbf{H}$, respectively. We obtain

$$\mathbf{H}^H[\mathbf{C}_x - \sigma_\eta^2 \mathbf{I}]\mathbf{H} = \mathbf{H}^H \mathbf{H} \mathbf{C}_0 \mathbf{H}^H \mathbf{H}. \tag{4.108}$$

Note that $\mathbf{H}^H \mathbf{H}$ is of full rank; hence its inverse is defined. We can then simplify (4.108),

$$\mathbf{C}_0 = [\mathbf{H}^H \mathbf{H}]^{-1} \mathbf{H}^H [\mathbf{C}_x - \sigma_\eta^2 \mathbf{I}]\mathbf{H}[\mathbf{H}^H \mathbf{H}]^{-1}.$$

C. Colored Noise

When the background noise is nonwhite the Pisarenko decomposition needs to be modified. We have shown in Chapter Two that a sum of sinusoids and a MA process of order q as a background noise can be represented as an ARMA(2p, 2p+q). In place of (4.96) we have the following equation, which is a matrix version of (2.93):

$$\mathbf{x}_t^T \mathbf{A} = \varepsilon_t^T \beta. \tag{4.109}$$

where (for p=3 and q=2)

$$\beta = \begin{bmatrix} b_0 & & & & & & \\ b_1 & b_0 & & & \mathbf{0} & & \\ b_2 & b_1 & b_0 & & & & \\ & b_2 & b_1 & b_0 & & & \\ & & b_2 & b_1 & b_0 & & \\ & & & b_2 & b_1 & b_0 & \\ & & & & b_2 & b_1 & b_0 \\ & \mathbf{0} & & & & b_2 & b_1 \\ & & & & & & b_2 \end{bmatrix} \begin{bmatrix} 1 \\ -a_1 \\ -a_2 \\ -a_3 \\ \vdots \\ -a_6 \end{bmatrix} = \mathbf{BA}.$$

Square both sides of (4.109) and carry out the expectation operation. We get

$$\mathbf{A}^T\mathbf{C}_x\mathbf{A} = \sigma_\varepsilon^2\mathbf{A}^T\mathbf{B}^T\mathbf{BA} = \sigma_\varepsilon^2\mathbf{A}^T\mathbf{QA}. \tag{4.110}$$

where $\mathbf{Q} = \mathbf{B}^T\mathbf{B}$. Rewrite (4.110) as $\mathbf{A}^T[\mathbf{C}_x\mathbf{A} - \sigma_\varepsilon^2\mathbf{QA}] = 0$, but since $\mathbf{A}^T \neq 0$ we must have

$$\mathbf{C}_x\mathbf{A} = \sigma_\varepsilon^2\mathbf{QA},$$

or

$$\mathbf{Q}^{-1}\mathbf{C}_x\mathbf{A} = \sigma_\varepsilon^2\mathbf{IA}.$$

Hence **A** must be the eigenvector corresponding to the smallest eigenvalue of $\mathbf{Q}^{-1}\mathbf{C}_x$. Now, given the MA parameters we can evaluate **Q** and its inverse (since **Q** is full rank). Multiplying the covariance matrix with $\mathbf{Q}^{-1}$ is equivalent to prewhitening of the colored noise given the spectral characteristics of the background noise as suggested in Reference 40.

D. Error Analysis

The errors in the estimated covariance matrix will in turn introduce errors in the estimation of eigenvectors and eigenvalues. The frequency estimates based on Pisarenko's decomposion or Schmidt's subspace method naturally will contain errors. Our interest here is to assess the nature of frequency estimation errors caused by the errors in the covariance matrix obtained by using finite data. First, we shall summarize the main results of perturbation theory for eigenvalues and the corresponding eigenvectors of the covariance matrix.[41,42]

Let $\hat{\mathbf{C}}_x = \mathbf{C}_x + \Delta\mathbf{C}_x$, where $\Delta\mathbf{C}_x$ is the estimation error which we model as $\Delta\mathbf{C}_x = \mu\mathbf{B}$, where $|\mu| \ll 1$ and **B** is a symmetric matrix whose elements are uncorrelated zero mean, unit variance Gaussian random variables. Let $\hat{\lambda}_i$ be the ith simple eigenvalue and $\hat{\mathbf{e}}_i$ be the corresponding eigenvector (unnormalized) of $\hat{\mathbf{C}}_x$. We state the following result:[41,42]

$$\hat{\lambda}_i = \lambda_i + \mu k_{i,1} + \mu^2 k_{i,2} + \ldots$$

$$\hat{\mathbf{e}}_i = \mathbf{e}_i + \sum_{\substack{j=1 \\ j\neq i}}^{N} \left(\sum_{k=1}^{\infty} t^i_{k,j} \mu^k\right) \mathbf{e}_j. \tag{4.111}$$

From the definition of eigen decomposition we have

$$(\mathbf{C}_X + \mu \mathbf{B})\hat{\mathbf{e}}_i = \hat{\lambda}_i \hat{\mathbf{e}}_i$$

in which we make use of (4.111) and equate terms of the same power of μ. We shall consider only the first-order term,

$$\sum_{\substack{j=1 \\ j\neq i}}^{N} (\lambda_j - \lambda_i) t^i_{1,j} \mathbf{e}_j + \mathbf{B}\mathbf{e}_i = k_{1,i}\, \mathbf{e}_i. \tag{4.112}$$

Remembering that $\mathbf{e}_j^H \mathbf{e}_i = 0$ and $\mathbf{e}_i^H \mathbf{e}_i = 1$ we obtain from (4.112)

$$t^i_{1,j} = \frac{\mathbf{e}_j^H \mathbf{B}\mathbf{e}_i}{\lambda_i - \lambda_j} \quad \text{and} \quad k_{i,1} = \mathbf{e}_i^H \mathbf{B}\mathbf{e}_i.$$

Finally, the perturbed eigenvector and eigenvalue, to a first-order approximation, are expressed below:

$$\tilde{\mathbf{e}}_i \approx \mathbf{e}_i + \mu \sum_{\substack{j=1 \\ j\neq i}}^{N} \frac{\mathbf{e}_j^H \mathbf{B}\mathbf{e}_i}{\lambda_j - \lambda_i} \mathbf{e}_j,$$

$$\hat{\lambda}_i = \lambda_i + \Delta\lambda_i \approx \lambda_i + \mu \mathbf{e}_i^H \mathbf{B}\mathbf{e}_i. \tag{4.113}$$

To normalize the above perturbed eigenvector divide both sides of (4.113) by the norm of the vector. Retaining terms up to second order we obtain the normalized perturbed eigenvector,[43]

$$\hat{\mathbf{e}}_i \approx \mathbf{e}_i + \sum_{\substack{j=1 \\ j\neq i}}^{N} \beta_{i,j} \mathbf{e}_j = \mathbf{e}_i + \Delta\mathbf{e}_i, \tag{4.114}$$

where

$$\beta_{i,j} = \left\{ -\frac{1}{2}\left| \mu \frac{\mathbf{e}_j^H \mathbf{B}\mathbf{e}_i}{\lambda_j - \lambda_i} \right|^2 + \mu \frac{\mathbf{e}_j^H \mathbf{B}\mathbf{e}_i}{\lambda_j - \lambda_i} \right\}.$$

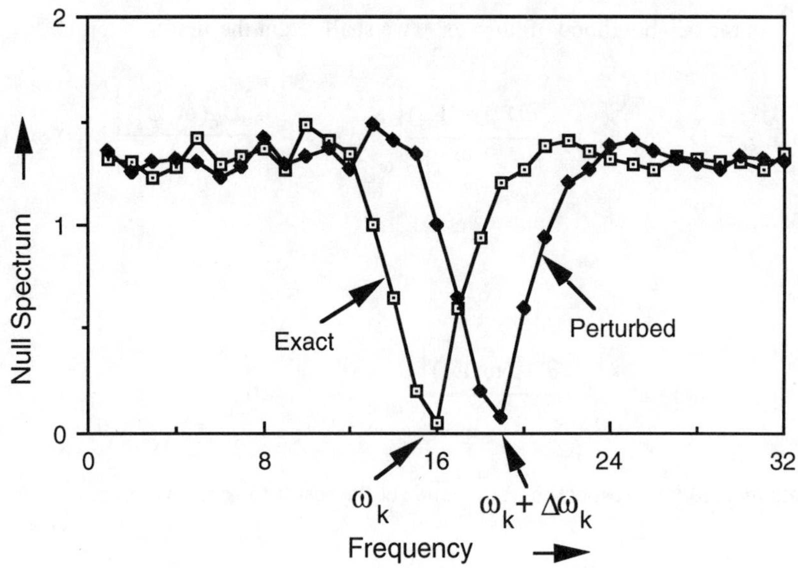

Figure 4.12: The minimum of the null spectrum gets shifted when the exact eigenvector is replaced by its estimate. The shift in the minimum is equal to error in the frequency estimate.

Next, we would like to relate the errors in the estimated eigenvectors to those in the estimated frequencies from the position of peaks of $S_{EV}(\omega)$ or the position of zeros or nulls of $\mathbf{h}^H(\omega)\mathbf{E}_n\mathbf{E}_n^H\mathbf{h}(\omega)$, which we shall call a null spectrum,

$$D(\omega,\mathbf{E}_n) = \mathbf{h}^H(\omega)\mathbf{E}_n\mathbf{E}_n^H\mathbf{h}(\omega),$$

where $\mathbf{E}_n$ is noise subspace spanned by $N-2p$ noise eigenvectors, and ω represents any one of the unknown frequencies. The null spectrum vanishes at true frequencies,

$$D(\omega,\mathbf{E}_n) = 0, \quad \omega = \omega_k, \; k = 1, 2, \ldots, 2p.$$

In deriving (4.113) we had assumed that the eigenvalues are simple; therefore, the above analysis is not applicable to noise eigen-space. To overcome this problem we introduce the complementary null spectrum,[43]

$$D_c(\omega,\mathbf{E}_s) = 1 - \mathbf{h}^H(\omega)\mathbf{E}_s\mathbf{E}_s^H\mathbf{h}(\omega),$$

where $\mathbf{E}_s$ is the signal space spanned by the signal eigenvectors. We shall assume that $D_c(\omega,\mathbf{E}_s)$ is differentiable with respect to unknown frequencies. The zeros of $D_c(\omega,\mathbf{E}_s)$ will be shifted when $\mathbf{E}_s$ is replaced by its estimate, $\hat{\mathbf{E}}_s$. For example, the zero at ω_k is shifted to $\omega_k + \Delta\omega_k$, where $\Delta\omega_k$ is an error in the frequency estimate (see Figure 4.12).

We will assume that the frequency shift is small; hence, in the Taylor's series expansion of $\hat{D}_c(\omega, \hat{\mathbf{E}}_s)$ in the neighborhood of $\omega = \omega_k$, we shall retain the first two terms,

$$\hat{D}_c(\omega, \hat{\mathbf{E}}_s) = \hat{D}_c(\omega_k, \hat{\mathbf{E}}_s) + \left.\frac{\partial \hat{D}_c(\omega, \hat{\mathbf{E}}_s)}{\partial \omega}\right|_{\omega_k} \Delta\omega_k + \left.\frac{\partial^2 \hat{D}_c(\omega, \hat{\mathbf{E}}_s)}{2\partial\omega^2}\right|_{\omega_k} [\Delta\omega_k]^2 + \ldots. \tag{4.115}$$

Note that

$$\left.\frac{\partial \hat{D}_c(\omega, \hat{\mathbf{E}}_s)}{\partial \omega}\right|_{\omega = \omega_k + \Delta\omega_k} = 0.$$

Differentiate (4.115) with respect to $\Delta\omega_k$ and set the result to zero. We obtain

$$\Delta\omega_k = -\frac{\dfrac{\partial \hat{D}_c(\omega_k, \hat{\mathbf{E}}_s)}{\partial \omega_k}}{\dfrac{\partial^2 \hat{D}_c(\omega_k, \hat{\mathbf{E}}_s)}{\partial \omega_k^2}} \qquad k = 1, 2, \ldots, 2p.$$

The differentiation of $\hat{D}_c(\omega, \hat{\mathbf{E}}_s)$ is straightforward though tedious. We shall skip that part and give the final result (details of differentiation are given in Reference 44),

$$\Delta\omega_k = \frac{\operatorname{Re}\{\mathbf{h}^1(\omega_k)^H \{\mathbf{E}_s \Delta\mathbf{E}_s^H + \Delta\mathbf{E}_s \mathbf{E}_s^H\} \mathbf{h}(\omega_k)\}}{\mathbf{h}^1(\omega_k)^H \mathbf{E}_n \mathbf{E}_n^H \mathbf{h}^1(\omega_k)}, \tag{4.116a}$$

where $\mathbf{h}^1(\omega_k)$ stands for the first derivative of $\mathbf{h}(\omega_k)$. Substituting for $\Delta\mathbf{E}_s$ from (4.114),

$$\Delta\mathbf{E}_s = \{\Delta\mathbf{e}_i = \sum_{\substack{j=1 \\ j \neq i}}^{N} \beta_{i,j} \mathbf{e}_j, \quad i = 1, 2, \ldots, 2p\},$$

the numerator of (4.116a) can be expressed in a compact form as

$$\operatorname{Re}\{\mathbf{h}^1(\omega_k)^H \mathbf{T}\, \mathbf{h}(\omega_k)\} = \operatorname{tr}[\mathbf{T} \operatorname{Re}\{\mathbf{h}(\omega_k) \mathbf{h}^1(\omega_k)^H\}],$$

where $\mathbf{T}$ is a real matrix whose (r,s)th element is given by

$$[\mathbf{T}]_{rs} = T_{rs} = \sum_{i=1}^{2p}[\sum_{\substack{j=1\\ j\neq i}}^{2p}\{\beta_{ij}e_j(r)e_i(s) + \beta_{ij}e_i(r)e_j(s)\} + \sum_{j=2p+1}^{N}\beta_{in}e_j(r)e_i(s)]$$,

and tr[ ] stands for the trace of the matrix inside the brackets. Let

$$\mathbf{H}_k = \mathrm{Re}\{\,\mathbf{h}(\omega_k)\mathbf{h}^1(\omega_k)^H\}$$

$$= \begin{bmatrix} 0 & \sin(\omega_k) & 2\sin(2\omega_k) & \cdots & (N-1)\sin((N-1)\omega_k) \\ 0 & 0 & 2\sin(\omega_k) & \cdots & (N-1)\sin((N-2)\omega_k) \\ 0 & \sin(-\omega_k) & 0 & \cdots & (N-1)\sin((N-3)\omega_k) \\ \vdots & & & & \vdots \\ 0 & \sin(-(N-2)\omega_k) & 2\sin(-(N-3)\omega_k) & \cdots & 0 \end{bmatrix}.$$

The numerator of (4.116a) may be expressed in a compact form using $\mathbf{H}_k$.

$$\Delta\omega_k = \frac{\sum_{i=1}^{2p}[\sum_{\substack{j=1\\ j\neq i}}^{2p}\beta_{ij}\mathbf{e}_i^T\mathbf{H}_k\mathbf{e}_j + \beta_{ij}\mathbf{e}_j^T\mathbf{H}_k\mathbf{e}_i + \sum_{j=2p+1}^{N}\beta_{ij}\mathbf{e}_i^T\mathbf{H}_k\mathbf{e}_j]}{\sum_{i=2p+1}^{N}\mathbf{h}^1(\omega_k)^H\mathbf{e}_i\mathbf{e}_i^H\mathbf{h}^1(\omega_k)}. \tag{4.116b}$$

From (4.116b) we observe that the error in the frequency estimate is linearly related to the errors in the estimation of the covariance matrix (if we agree to drop the quadratic term in the second equation of (4.114)). Therefore, $E\{\Delta\omega_k\} \approx 0$; that is, the frequency estimates are asymptotically unbiased (that is, for large data length). The variance of $\Delta\omega_k$ is obtained by squaring and evaluating the expected value on both sides of (4.116b) where we make use of the identity,[42]

$$E\{(\mathbf{A}_1^H\mathbf{B}\mathbf{A}_2)(\mathbf{A}_3^H\mathbf{B}\mathbf{A}_4)\} = \frac{(\mathbf{A}_1^H\mathbf{C}_X\mathbf{A}_4)(\mathbf{A}_3^H\mathbf{C}_X\mathbf{A}_2)}{\mu^2 M},$$

where $\mathbf{A}_i$, $i = 1,2,3,4$ are any four complex vectors and obtain

$$E\{\Delta\omega_k^2\} = \frac{\frac{1}{M}\sum_{j}^{2p}\sum_{\substack{j'=1\\ j\neq j'}}^{2p}\frac{\lambda_j\lambda_{j'}}{(\lambda_j - \lambda_{j'})^2}[\mathbf{e}_{j'}^T\mathbf{H}_k\mathbf{e}_j + \mathbf{e}_j^T\mathbf{H}_k\mathbf{e}_{j'}]^2}{[\mathbf{h}^1(\omega_k)^H\mathbf{E}_n\mathbf{E}_n^H\mathbf{h}^1(\omega_k)]^2}. \tag{4.117a}$$

Using the Schwartz inequality, namely, $\|\mathbf{A}.\mathbf{B}\| \leq \sqrt{\|\mathbf{A}\|\|\mathbf{B}\|}$, where **A** and **B** are vectors, we obtain a lower bound on $E\{\Delta\omega_k^2\}$,

$$E\{\Delta\omega_k^2\} \geq \frac{\frac{1}{M}\sum_{j}^{2p}\sum_{\substack{j'=1\\ j\neq j'}}^{2p}\frac{\lambda_j\lambda_{j'}}{(\lambda_j-\lambda_{j'})^2}[\mathbf{e}_{j'}^T\mathbf{H}_k\mathbf{e}_j+\mathbf{e}_j^T\mathbf{H}_k\mathbf{e}_{j'}]^2}{[(2p-1)\frac{\Delta T(\Delta T-1)(2\Delta T-1)}{6}]^2}. \tag{4.117b}$$

The error analysis outlined above will not be valid when the frequencies of the sinusoids are multiples of $\frac{2\pi}{N}$. Then, as stated in Exercise 4.17, the eigenvalues will become equal to the power of the sinusoids when $\omega_0 = r\frac{2\pi}{\Delta T}$, where r is any integer. If all p sinusoids are of equal power, the signal eigenvalue will be repeated 2p times. In fact, when the frequencies are integer multiples of the fundamental harmonic, the signal eigenvalue will be repeated at least twice. To overcome the problem of closely spaced eigenvalues an alternate approach to the estimation of subspace, perturbation has been suggested.[45] In this approach only the orthogonal component of the perturbation is shown to affect the frequency estimation error.

In (4.116a), we express

$$\Delta\mathbf{E}_s = \Pi_s\Delta\mathbf{E}_s + \Pi_n\Delta\mathbf{E}_s = \Delta\mathbf{E}_s^{\|} + \Delta\mathbf{E}_s^{\perp}.$$

where

$$\Pi_s = \mathbf{E}_s\mathbf{E}_s^H$$

and

$$\Pi_n = \mathbf{E}_n\mathbf{E}_n^H.$$

(note, $\mathbf{I} = \mathbf{E}_s\mathbf{E}_s^H + \mathbf{E}_n\mathbf{E}_n^H$). Since $\Delta\mathbf{E}_s^{\|}$ has no effect on the frequency estimation error,[45] we shall drop it and rewrite (4.116a) as follows:

$$\Delta\omega_k = \frac{\mathrm{Re}\{\mathbf{h}^1(\omega_k)^H\Pi_n\Delta\mathbf{E}_s\mathbf{E}_s^H\mathbf{h}(\omega_k)\}}{\mathbf{h}^1(\omega_k)^H\mathbf{E}_n\mathbf{E}_n^H\mathbf{h}^1(\omega_k)}.$$

From (4.114) it is straightforward to show that $\Pi_n\Delta\mathbf{E}_s = \Pi_n\Delta\mathbf{C}\,\mathbf{E}_s\Lambda_s^{-1}$, and using this result we simplify above equation:

$$\Delta\omega_k = \frac{\mathrm{Re}\{\mathbf{h}^1(\omega_k)^H\Pi_n\Delta\mathbf{C}\,\mathbf{E}_s\Lambda_s^{-1}\mathbf{E}_s^H\mathbf{h}(\omega_k)\}}{\mathbf{h}^1(\omega_k)^H\mathbf{E}_n\mathbf{E}_n^H\mathbf{h}^1(\omega_k)}, \tag{4.118a}$$

where

$$\Lambda_s^{-1} = \text{diag}\ \{\lambda_1^{-1}, \lambda_2^{-1}, 0, \ldots, 0\}.$$

The MSE in the frequency estimate may be computed from (4.118a) by squaring and computing the expected value. The final simplified result taken from Reference 45 is given below:

$$E\{\Delta\omega_k^2\} = \frac{\dfrac{\sigma_\eta^2}{2M}\displaystyle\sum_{i=1}^{2p}\frac{\lambda_i}{(\lambda_i - \sigma_\eta^2)^2}\left|\mathbf{E}_i^H \mathbf{h}(\omega_k)\right|^2}{\mathbf{h}^1(\omega_k)^H \Pi_n \mathbf{h}^1(\omega_k)}. \tag{4.118b}$$

As in (4.117b) a lower bound on MSE may be given:

$$E\{\Delta\omega_k^2\} \geq \frac{\dfrac{\sigma_\eta^2}{2M}\displaystyle\sum_{i=1}^{2p}\frac{\lambda_i}{(\lambda_i - \sigma_\eta^2)^2}\left|\mathbf{E}_i^H \mathbf{h}(\omega_k)\right|^2}{[(2p-1)\dfrac{\Delta T(\Delta T-1)(2\Delta T-1)}{6}]}. \tag{4.118c}$$

Now compare (4.118c) with (4.117b) and observe that in place of the difference between two signal eigenvalues, we have now a difference between the signal eigenvalue and noise eigenvalue which are not likely to be equal unless the sinusoids are strongly correlated.

Example 4.13 Consider a single sinusoid, that is, p=1. The covariance matrix is given by

$$\begin{bmatrix} \frac{\alpha^2}{2} + \sigma_\eta^2 & \frac{\alpha^2}{2}\cos(\omega_1) & \frac{\alpha^2}{2}\cos(2\omega_1) & \cdots & \frac{\alpha^2}{2}\cos((\Delta T - 1)\omega_1) \\ \frac{\alpha^2}{2}\cos(\omega_1) & \frac{\alpha^2}{2} + \sigma_\eta^2 & \frac{\alpha^2}{2}\cos(\omega_1) & \cdots & \frac{\alpha^2}{2}\cos((\Delta T - 2)\omega_1) \\ \vdots & & & & \vdots \\ \frac{\alpha^2}{2}\cos((\Delta T - 1)\omega_1) & & & \cdots & \frac{\alpha^2}{2} + \sigma_\eta^2 \end{bmatrix},$$

where $\frac{\alpha^2}{2}$ stands for power in the sinusoid. The first two nonzero eigenvalues are (see Exercise 4.17)

$$\lambda_1 = \frac{\alpha^2 \Delta T}{4}\left(1 + \frac{\sin(\omega_0 \Delta T)}{\Delta T \sin(\omega_0)}\right) + \sigma_\eta^2,$$

$$\lambda_2 = \frac{\alpha^2 \Delta T}{4}\left(1 - \frac{\sin(\omega_0 \Delta T)}{\Delta T \sin(\omega_0)}\right) + \sigma_\eta^2,$$

and the correponding eigenfunctions are

ΔT	Ratio λ_1/λ_2	Theory (Lower Bound) (4.117b)	Theory (Lower Bound) (4.118c)	Simulation MSE
4	0.3975	1.26×10^{-4}	0.12×10^{-4}	3.74×10^{-4}
8	0.6958	7.62×10^{-5}	0.94×10^{-6}	7.20×10^{-6}
10	0.6842	2.92×10^{-5}	0.46×10^{-6}	4.51×10^{-6}
12	0.9154	2.57×10^{-4}	0.25×10^{-6}	$2.28\times$ 10-6
16	0.8429	1.08×10^{-5}	0.13×10^{-6}	7.14×10^{-7}

Table 4.9: A comparison of theoretically predicted and simulated MSE of frequency estimate by the MUSIC algorithm. A real random sinusoid of frequency 0.5 rad in the presence of white noise (SNR=10 dB) is assumed. 800 snapshots were used for the estimation of the covariance matrix. Fifty independent trials were performed for calculation of MSE.

$$\text{Eigenvectors} \;:\; \begin{cases} \mathbf{e}_1 = a\{\cos\left(\left(\frac{\Delta T-1}{2}-k\right)\omega_0\right), & k=0,1,\ldots,\Delta T-1\}, \\ \mathbf{e}_2 = b\{\sin\left(\left(\frac{\Delta T-1}{2}-k\right)\omega_0\right), & k=0,1,\ldots,\Delta T-1\}. \end{cases}$$

Note that when $\omega_0 = \frac{2\pi}{N}i$, where i is an integer, or when $\Delta T \to \infty$, the eigenvalues become equal, and the first order perturbation theory fails. On substituting for λ_1 and λ_2 in (4.117b) and assuming that $\text{SNR}\,\Delta T >> 1$, we obtain

$$E\{\Delta\omega_1^2\} \geq \frac{\frac{1}{M}\left\{\frac{1-\delta}{4\delta} + \frac{1}{\text{SNR}\;\Delta T\;\delta}\right\}[\mathbf{e}_1^T\mathbf{H}_1\mathbf{e}_2 + \mathbf{e}_2^T\mathbf{H}_1\mathbf{e}_1]^2}{\left[\frac{\Delta T(\Delta T-1)(2\Delta T-1)}{6}\right]^2},$$

where

$$\delta = \frac{\sin^2(\Delta T\omega_1)}{\Delta T^2 \sin^2(\omega_1)}$$

and

$$\text{SNR} = \frac{\alpha^2}{2\sigma_\eta^2}.$$

Further, the expression inside the square brackets in the numerator can be reduced to, after some tedious algebra,

$$[\mathbf{e}_1^T \mathbf{H}_1 \mathbf{e}_2 + \mathbf{e}_2^T \mathbf{H}_1 \mathbf{e}_1] =$$

$$ab\,\Delta T(\Delta T - 1)\left[\frac{\sin(\Delta T\omega_1)\sin((\Delta T - 1)\omega_1)}{4\,\Delta T(\Delta T - 1)\sin^2(\frac{\omega_1}{2})} - \frac{\sin(\Delta T\omega_1)}{4\,\Delta T\sin(\omega_1)} - \frac{\cos((2\Delta T - 1)\frac{\omega_1}{2})\sin((\Delta T - 1)\omega_1)}{2(\Delta T - 1)\sin(\frac{\omega_1}{2})} + 1\right],$$

which may be approximated for $\Delta T\omega_1 > \pi$ (but $\Delta T\omega_1 \neq n\pi$ where n is an integer) as

$$[\mathbf{e}_1^T \mathbf{H}_1 \mathbf{e}_2 + \mathbf{e}_2^T \mathbf{H}_1 \mathbf{e}_1]^2 \approx 4(\Delta T - 1)^2.$$

We use the alternate approach leading to (4.118c). For a single sinusoid this expression reduces to

$$E\{\Delta\omega_1^2\} \geq \frac{\frac{1}{M}\frac{1}{\mathrm{SNR}\,(1 - \delta)}[1 + \frac{2(1 + \delta)}{\mathrm{SNR}\,\Delta T(1 - \delta)}]}{[\frac{\Delta T(\Delta T - 1)(2\Delta T - 1)}{6}]}.$$

A comparison of the relative accuracy of the two approaches for error estimation is shown in Table 4.9.

E. **Damped Sinusoids** (Prony's Algorithm)

It is of interest to consider a time series as an output of a difference equation driven by an impulse function (AR equation in which the random input is replaced by an impulse function, see (2.52)). The solution of a pth-order difference equation may be written as a sum of p damped sinusoids:

$$x(t) = \sum_{k=1}^{p} \alpha_k \gamma_k^t = \sum_{k=1}^{p} \alpha_k e^{\lambda_k t} e^{j\omega_k t}, \tag{4.119}$$

where γ_k, k = 1,..., p are the roots of the coefficient polynomial (2.53), and α_k, k = 1,... ,p are complex constants to be determined from p initial conditions. Let there be p/2 (assume that p is even) pairs of complex poles. Equation (4.119) can be expressed as

$$x(t) = \sum_{k=1}^{\frac{p}{2}} |\alpha_k| e^{\lambda_k t} \cos(\omega_k t + \phi_k). \tag{4.120}$$

Define the following matrices:

$$\mathbf{x}_t = \begin{bmatrix} x(t), & x(t-1), & \cdots & x(t-p) \\ x(t+1), & x(t), & \cdots & x(t+1-p) \\ \vdots & & \cdots & \vdots \\ x(t+T-1) & x(t+T-2), & \cdots & x(t+T-1-p) \end{bmatrix}$$

$$\mathbf{A} = \{1,\ a_1, \ldots, a_p\}^T.$$

Using the above matrices, (2.52) may be expressed in a matrix form,

$$\mathbf{x}_t\mathbf{A} = \mathbf{0} \qquad \text{for } T-1 \geq t \geq p. \tag{4.121}$$

where $\mathbf{0}$ stands for a column of zeros. Premultiply (4.121) with $\mathbf{x}_t^T$:

$$\mathbf{x}_t^T\mathbf{x}_t\mathbf{A} = \rho\ \mathbf{A} = \mathbf{0}, \tag{4.122}$$

where ρ is a $(p+1) \times (p+1)$ autocorrelation matrix whose $(m,n)^{th}$ term is given by

$$\rho_{m,n} = \sum_{k=0}^{T-1} X(t-m+k)X(t-n+k).$$
$$m,n = 0,1, \ldots, p$$

Note that ρ is symmetric but is not necessarily a Toeplitz matrix. The rank of ρ, in the absence of any measurement noise, is equal to p. To show this we rewrite (4.119) as follows:

$$x(t) = \alpha^H \gamma^t 1, \tag{4.123}$$

where

$$\alpha = \{\alpha_1\ \alpha_2\ \ldots\ \alpha_p\}^T,$$
$$\gamma = \text{diag}\,\{\gamma_1\ \gamma_2\ \cdots\ \gamma_p\},$$

and

$$1 = \text{col}\,\{1\ \ 1\ \ldots\ \ 1\}.$$

Now use (4.123) to replace each element of the $\mathbf{x}_t$ matrix. After some simple manipulations we obtain

$$\rho = \mathbf{x}_t^H\mathbf{x}_t = \mathbf{G}^H\mathbf{D}\,\mathbf{G}, \tag{4.124}$$

where

$$\mathbf{x}_t = \begin{bmatrix} \alpha^H \gamma^t \\ \alpha^H \gamma^{t+1} \\ \cdot \\ \cdot \\ \cdot \\ \alpha^H \gamma^{t+T-1} \end{bmatrix} \begin{bmatrix} 1 & \gamma_1^{-1} & \dots & \gamma_1^{-p} \\ 1 & \gamma_2^{-1} & \dots & \gamma_2^{-p} \\ & & \dots & \\ & & \dots & \\ & & \dots & \\ 1 & \gamma_p^{-1} & \dots & \gamma_p^{-p} \end{bmatrix}$$

$$(T \times p) \qquad p \times (p+1)$$

and $\mathbf{D}$ is a $p \times p$ square matrix whose $(m,n)^{th}$ element is given by

$$d_{m,n} = \alpha^*_m \alpha_n \mathbf{f}^H_m \mathbf{f}_n,$$

$$\mathbf{f}_m = \mathrm{col}\{\gamma_m^t,\ \gamma_m^{t+1},\ \dots,\gamma_m^{t+T-1}\}.$$

Hence, $\mathbf{D}$ may be expressed as a product of a matrix and its Hermitian transpose,

$$\mathbf{D} = [\alpha_1 \mathbf{f}_1\ \alpha_2 \mathbf{f}_2\ \dots\ \alpha_p \mathbf{f}_p]^H [\alpha_1 \mathbf{f}_1\ \alpha_2 \mathbf{f}_2\ \dots\ \alpha_p \mathbf{f}_p].$$

It is of size $T \times p$ ($T>p$) and of rank p. (Note that it is a p column Vandermonde matrix.) Thus, the rank of $\mathbf{D}$ is p. Also, the rank of $\mathbf{G}$ given below is p as it is also a p column Vandermonde matrix,

$$\mathbf{G} = \begin{bmatrix} 1 & \gamma_1^{-1} & \dots & \gamma_1^{-p} \\ 1 & \gamma_2^{-1} & \dots & \gamma_2^{-p} \\ & & \dots & \\ & & \dots & \\ & & \dots & \\ 1 & \gamma_p^{-1} & \dots & \gamma_p^{-p} \end{bmatrix}.$$

Finally, since ρ is a product of three p rank matrices, its rank will be equal to p. The AR coefficient vector will therefore be equal to the null eigenvector of ρ. An example of a pole location from the transient solution of a difference equation is given in Example 4.14.

Example 4.14: A signal was generated using the following difference equation:

$$x(t) - 2.7607\ x(t-1) + 3.8106\, x(t-2) - 2.6535\, x(t-3) - 0.9238\ x(t-4) = \delta(t)$$

with initial conditions $x(-1)=x(-2)=x(-3)=x(-4)=0.0$. One hundred samples of the signal were generated as shown in Figure 4.13. The actual pole locations and those computed from the signal with and without background noise are shown in Table 4.10. It may be noted that the presence of background noise can seriously affect the estimates of pole locations (End of Example).

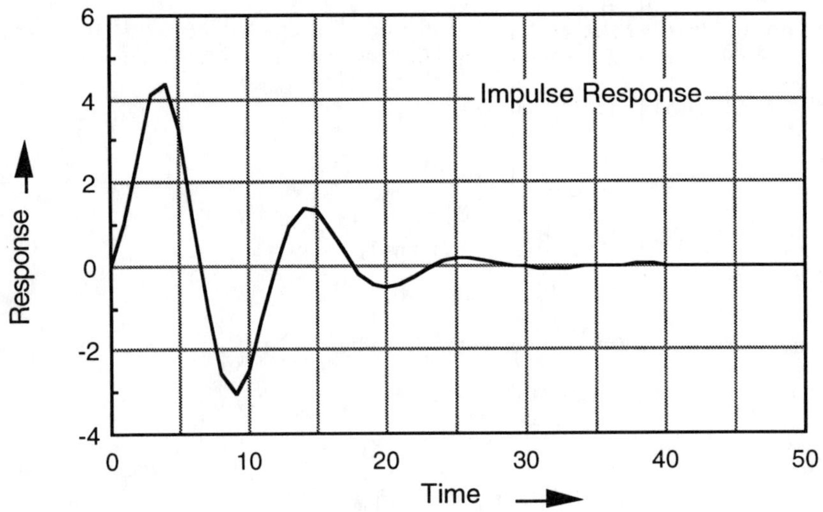

Figure 4.13: Solution of the difference equation with zero initial conditions. Note that this function was called a transient solution of AR model (see page 98).

Pole Position	Computed Pole Position without noise	Computed Pole Position with noise (SNR=8.75 dB)
$0.9802\, e^{\pm j39.58^0}$	$0.9802\, e^{\pm j39.58^0}$	$0.9917\, e^{\pm j43.97^0}$
$0.9805\, e^{\pm j50.41^0}$	$0.9805\, e^{\pm j50.41^0}$	$1.0430\, e^{\pm j86.15^0}$

Table 4.10: The effect of noise on the estimated pole position of damped sinusoids.

F. **Signal Subspace**

We shall now take a look at the signal subspace. In particular, we shall show that the signal subspace is spanned by frequency vectors, and the eigenvalues are related to the signal power. Let $\mathbf{E}_S$ represent the signal subspace spanned by the eigenvectors corresponding to unequal eigenvalues of the covariance matrix. Then

$$\mathbf{E}_s^H(\mathbf{C}_x - \sigma_\varepsilon^2\mathbf{I})\mathbf{E}_s = \mathbf{E}_s^H\mathbf{H}\mathbf{C}_0\mathbf{H}^H\mathbf{E}_s = \begin{bmatrix} \gamma_1, 0, 0, & \dots 0 \\ 0, \gamma_2, 0, & \dots 0 \\ & \\ 0, 0, 0, & \dots \gamma_{2p} \end{bmatrix}, \tag{4.125}$$

where $\gamma_1, \gamma_2, \dots, \gamma_{2p}$ are signal eigenvalues, that is, the eigenvalues of the covariance matrix without background noise. Now we shall make use of polar decomposition[28] of **H** (rank 2p). There exists a matrix, **P** (N × 2p), and a nonsingular (full rank) matrix, **G** (2p × 2p), such that $\mathbf{H} = \mathbf{PG}$ and $\mathbf{P}^H\mathbf{P} = \mathbf{I}$. Equation (4.125) reduces to

$$\mathbf{E}_s^H\mathbf{PGC}_0\mathbf{G}^H\mathbf{P}^H\mathbf{E}_s = \begin{bmatrix} \gamma_1, 0, 0, & \dots 0 \\ 0, \gamma_2, 0, & \dots 0 \\ \\ 0, 0, 0, & \dots \gamma_{2p} \end{bmatrix}. \tag{4.126}$$

Let $\mathbf{P} = \mathbf{E}_s$. This selection of the **P** matrix is consistent with the requirement that $\mathbf{P}^H\mathbf{P} = \mathbf{I}$ as the eigenvectors of a Hermitian matrix satisfy $\mathbf{E}_s^H\mathbf{E}_s = \mathbf{I}$. Hence, (4.126) reduces to

$$\mathbf{GC}_0\mathbf{G}^H = \begin{bmatrix} \gamma_1 & 0 & 0 & \dots & 0 \\ 0 & \gamma_2 & 0 & \dots & 0 \\ \\ 0 & 0 & 0, & \dots & \gamma_{2p} \end{bmatrix} = \Lambda. \tag{4.127}$$

Thus, the polar decomposition of **H** may be expressed as

$$\mathbf{H} = \mathbf{E}_s\mathbf{G}. \tag{4.128a}$$

From (4.128a) we observe that the columns of **H** matrix are indeed the linear combinations of eigenvectors spanning the signal subspace. Naturally, the frequency vectors will also span the same space. Using (4.128a) we can estimate the **G** matrix as

$$\mathbf{G} = \mathbf{E}_s^H\mathbf{H}. \tag{4.128b}$$

Since **G** is a full-rank square matrix, its inverse exists and hence,

$$\mathbf{E}_s = \mathbf{HG}^{-1}. \tag{4.128c}$$

Let $g_{i,k}$ be the (i,k)th element of $\mathbf{G}^{-1}$. Each signal eigenvector can be expressed as a linear combination of frequency vectors,

$$\mathbf{e}_i = \sum_{k=1}^{2p} g_{i,k}\mathbf{h}_k, \quad i = 1, 2, \dots, 2p, \tag{4.128d}$$

where $\mathbf{h}_k$ is the kth column of the **H** matrix or the frequency vector corresponding to the kth frequency. It is indeed a complex sinusoid (see the expansion of the **H** matrix on page 290). Thus, a signal eigenvector has been expressed as a linear combination of 2p complex sinusoids; however, the signal eigenvectors themselves are real. Prony's method described earlier can be used to estimate the unknown frequencies. It is not necessary to know the coefficients, $g_{i,k}$. The above simple approach was suggested by Cadzow,[46] while a more elaborate approach on the same lines was given by Reddi.[47] The chief advantage of estimating unknown frequencies in the signal subspace is that it is no longer necessary to compute the EV spectrum (4.106) over a dense grid of frequencies and then search for the maxima. This is a very time-consuming computation. In the next example (4.15) we shall show how to estimate directly the frequencies of sinusoids in the signal subspace.

Example 4.15: Consider two equal amplitude (ten units) random sinusoids of normalized frequency, 0.10 and 0.15 Hz. The background noise is a zero mean, unit variance Gaussian white noise. A covariance matrix of size 16×16 was computed using 500 independent snapshots (data vectors). The signal eigenvectors are shown in Figure 4.14, and the frequency estimates using the Prony's algorithm described earlier are shown in Table 4.11. The experiment was repeated for two more sets of frequencies, and the estimated frequencies are shown in the same table (End of Example).

The power in each sinusoid can be estimated from the signal eigenvalues. We rewrite (4.127) as follows:

$$\text{diag}\,\{\frac{A_1^2}{4}, \frac{A_1^2}{4}, \ldots \frac{A_{2p}^2}{4}, \frac{A_{2p}^2}{4}\} = \mathbf{G}^{-1}\text{diag}\{\gamma_1, \gamma_2, \ldots \gamma_{2p}\}\mathbf{G}^{-H}, \tag{4.129}$$

where $\mathbf{G}^{-1}$ is given by (4.128b).

Example 4.16: The spread of signal values is closely controlled by the frequency interval between the sinusoids. To show this consider the following numerical example: Consider two equal power sinusoids in unit variance white noise. The frequencies of the sinusoids are a) 0.10 Hz and 0.12 Hz, b) 0.10 Hz and 0.15 Hz, and c) 0.10 Hz and 0.18Hz. A covariance matrix of size 16×16 is formed using 500 data vectors. The computed signal eigenvalues, the signal power estimated from (4.127a) and the actual signal power for different pairs of frequencies, are in Table 4.12 (End of Example).

In (4.128) we have seen that the signal eigenvectors and the frequency vectors (i.e., columns of **H** matrix) are linearly related, but there is a significant coupling among different frequency vectors. We shall now show that it is possible to practically decouple the frequency vectors by increasing the dimension of the covariance matrix. For this, premultiply (4.127) by $\mathbf{HG}^{-1}$ and postmultiply by $(\mathbf{HG}^{-1})^H$. Carry out a similar operation on the right-hand side of the equation but with a different multiplier, $\mathbf{E}_s$. On account of (4.128c) the equality of the right- and left-hand products remains in force. We obtain

$$\mathbf{H}\,\mathbf{C}_0\mathbf{H}^H = \mathbf{E}_s \Lambda\, \mathbf{E}_s^H.$$

Further simplification yields

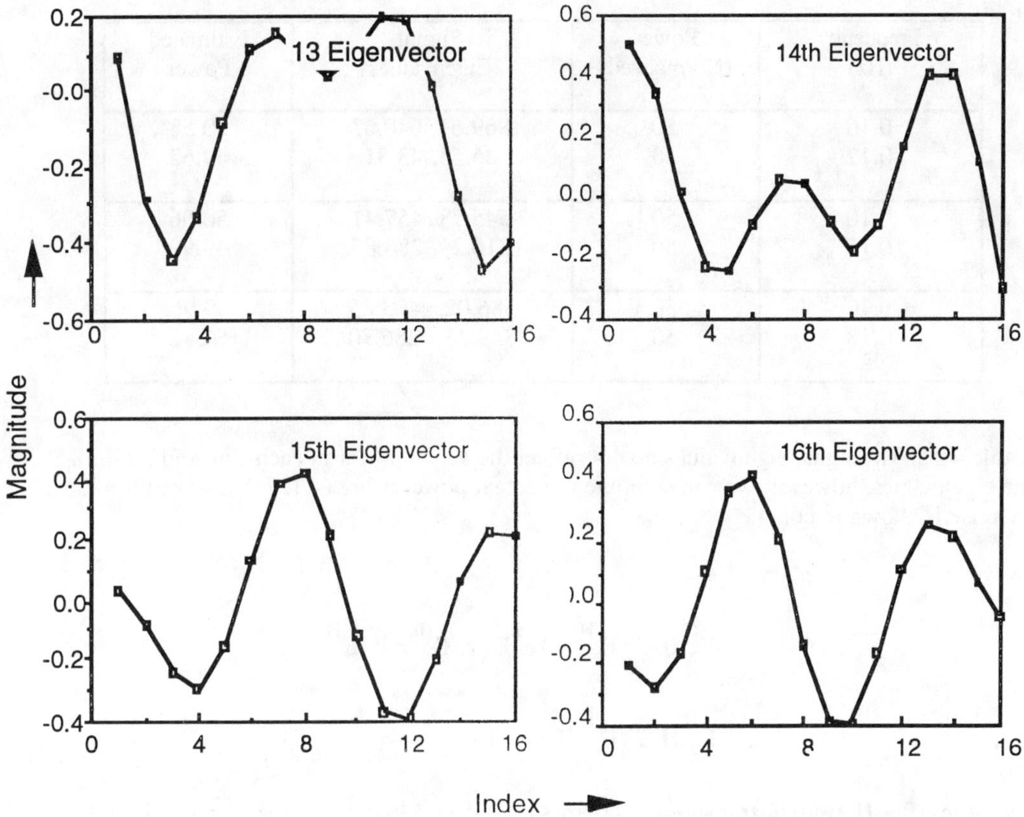

Figure 4.14: The signal eigenvectors 13th to 16th. The frequency estimates using Prony's method are shown in the Table 4.11.

Freq.	13th	14th	15th	16th	Average
0.10	0.0969	0.1082	0.1089	0.1004	0.1036
0.12	0.1256	0.1143	0.1545	0.1109	0.1263
0.10	0.1018	0.0983	0.1028	0.0996	0.1006
0.15	0.1480	0.1493	0.1525	0.1512	0.1502
0.10	0.1004	0.1003	0.1006	0.0987	0.1000
0.18	0.1805	0.1804	0.1794	0.1806	0.1802

Table 4.11: Frequency estimates (three sets) using signal subspace and Prony's algorithm. The signal subspace is spanned by 13th, 14th, 15th, and 16th eigenvectors of the covariance matrix of size 16×16. See Example 4.15 for more details.

Frequency (Hz)	Power (Normalized)	Signal Eigenvalues	Estimated Power
0.10 0.12	50 50	869.61, 640.62 86.53, 43.31	50.58 49.62
0.10 0.15	50 50	545.78, 457.41 316.29, 286.85	50.06 49.90
0.10 0.18	50 50	586.02, 403.85 349.25, 286.70	49.96 50.42

Table 4.12: The signal eigenvalues do not reflect the actual power in each sinusoid (col. 3). From the eigenvalues, however, we can compute the actual power using (4.127a). The result of such an exercise is shown in col. 4.

$$\mathbf{E}_s^H \mathbf{H} \mathbf{C}_0 \mathbf{H}^H = \mathbf{E}_s^H \mathbf{E}_s \Lambda \mathbf{E}_s^H = \Lambda \mathbf{E}_s^H$$

$$\mathbf{E}_s^H \mathbf{H} \mathbf{C}_0 \mathbf{H}^H \mathbf{H} = \Lambda \mathbf{E}_s^H \mathbf{H}. \tag{4.130a}$$

By taking the Hermitian transpose on both sides of (4.130a) we can express it in a standard eigenvalue-eigenvector form,

$$\mathbf{H}^H \mathbf{H} \mathbf{C}_0 \mathbf{H}^H \mathbf{E}_s = \mathbf{H}^H \mathbf{E}_s \Lambda, \tag{4.130b}$$

where $\mathbf{H}^H \mathbf{E}_s$ is the eigenvectror matrix of $\mathbf{H}^H \mathbf{H} \mathbf{C}_0$, and Λ is eigenvalue (diagonal) matrix. Note that the eigenvalues of $\mathbf{H}^H \mathbf{H} \mathbf{C}_0$ are the nonzero eigenvalues of $\mathbf{H} \mathbf{C}_0 \mathbf{H}^H$. Further, on account of (4.128b), $\mathbf{G}^H$ equals the eigenvector matrix of $\mathbf{H}^H \mathbf{H} \mathbf{C}_0$. We shall now show that $\mathbf{H}^H \mathbf{H}$ is close to a diagonal matrix. To show this let us examine any one element of the matrix $\mathbf{H}^H \mathbf{H}$,

$$\left\{\mathbf{H}^H \mathbf{H}\right\}_{k\,l} = \frac{\sin\left(\frac{N}{2}(\omega_k - \omega_l)\right)}{\sin\left(\frac{1}{2}(\omega_k - \omega_l)\right)} e^{j\frac{N-1}{2}(\omega_k - \omega_l)}.$$

We have come across the expression on the right-hand side above in Chapter Three (page 190) where we have shown that the limiting value as $N \to \infty$ is $N\delta_{kl}$, where δ_{kl} is a Kronecker delta function. Hence, $\mathbf{H}^H \mathbf{H}$ as $N \to \infty$ becomes diagonal, that is, $N\mathbf{I}$. In practice a good approximation to the diagonal matrix can be achieved for

$$N >> \frac{2\pi}{(\omega_k - \omega_l)_{min}}.$$

Further, since the sinusoids are uncorrelated, the source covariance matrix, $\mathbf{C}_0$, will be diagonal. As a result $\mathbf{H}^H\mathbf{H}\mathbf{C}_0$ becomes diagonal, and its eigenvectors are simply the euclidean basis vectors. Returning to (4.128a), the matrix **G** connecting $\mathbf{E}_s$ and **H** is a matrix whose columns are euclidean basis vectors scaled by a factor, $\sqrt{N}$. The role of the **G** matrix then is to rearrange the columns of $\mathbf{E}_s$ in the same order as the sinusoids. Further, on account of (4.129), the eigenvalues of the covariance matrix are proportional to the power of the sinusoids,[48]

$$\frac{\gamma_i}{N} \rightarrow \frac{A_i^2}{4} \quad \text{as } N \rightarrow \infty .$$

Hu[48] arrived at a similar result using the more general approach of degenerate Toeplitz matrices.

Example 4.17: We consider a sum of four complex sinusoids and white noise,

$$x(t) = \sum_{k=1}^{4} a_k e^{j(\omega_k t + \phi_k)} + \eta(t) \quad ,$$

where

$$\omega_1 = 0.2\pi,\ \omega_2 = 0.3\pi,\ \omega_3 = 0.6\pi,\ \omega_4 = 0.8\pi \quad ,$$

$$a_1 = 4,\ a_2 = 3,\ a_2 = 2 \quad ,$$

and

$$a_4 = 1 \quad .$$

The overall SNR=14.77 dB. The theoretical covariance matrix was obtained from the above data model. The minimum size of the covariance matrix for the above frequencies is 20. The signal eigenvectors of the covariance matrix were computed and finally $\mathbf{G} = \mathbf{E}_s^H \mathbf{H}$ was evaluated for two different matrix sizes, equal to 16 and 64. The results are shown below:

G Matrix (size: 16×16)

3.791, 0.870	-0.9655, -1.5408	0.1337, 0.031	0.041, -0.786
-0.153, 0.903	**-2.8750, 2.0414**	-0.2033, 1.199	0.212, 0.037
0.068, 0.156	-0.1865, 0.4753	**-1.520, 3.483**	-0.901, 0.393
-0.006, -0.001	0.0076, 0.0111	-0.243, 0.046	**0.721,-3.798**

G Matrix (size: 64×64)

7.972,0.295	0.9645, 0.447	-0.052, -0.1835	0.088, 0.032
0.561, -0.203	**-5.411, 0 5.73**	0.2724, 0.3494	0.0244, -0.001
-0.023, 0.022	-0.0701, 0.1903	**-6.968, -3.898**	-0.4674, 0.215
0.002, 0.006	0.001, 0.001	-0.098, 0.083	**-0.593, 7.960**

Notice that even for N=16 columns of **G** matrix have some resemblance to the euclidean basis vectors. But for N=64 the **G** matrix is much closer to the euclidean basis vectors. The scale factor is, as expected, approximately eight.

G Matrix (size:16×16) for a different power assignment ($a_1 = 1$, $a_2 = 3$, $a_3 = 2$, and $a_4 = 4$) is shown below (End of Example).

0.014, 0.197	-0.044, 0.051	0.075, 1.012	**3.981,-0.294**
-0.917, -0.428	**1.361, 3.743**	0.764, 0.356	-0.037, 0.080
-0.498, 0.249	0.331, 0.111	**-3.377,1.685**	-0.106, -0.213
-3.398, 1.755	-0.113 -0.036	0.105, -0.054	-0.002, -0.005

Refering to (4.128a) it is clear that the role of the **G** matrix is to rearrange the signal eigenvectors, which are otherwise arranged in the decreasing order of eigenvalues, to the same order as the frequency vectors are arranged in the **H** matrix. In this example the first eigenvector corresponding to the largest eigenvalue is placed in the fourth position corresponding to frequency ω_4 where the power is highest, and the fourth eigenvector is placed in the first position corresponding to ω_1 where the power is lowest. It is assumed that the columns of the **H** matrix are arranged in the order of increasing frequency.

G. Subspace Rotation

On page 290 we have shown how to represent a data vector in a matrix form (see (4.101)). We shall use that representation to compute the cross-covariance matrix between two temporally displaced data vectors, each of length equal to 2p.

$$\mathbf{C}_{x_t x_{t+2p+1}} = E\{\mathbf{x}_t \mathbf{x}^H_{t+2p+1}\} = \mathbf{HG}_t \mathbf{C}_0 \mathbf{G}^H_{t+2p+1} \mathbf{H}^H + E\{\eta_t \eta^H_{t+2p+1}\}, \tag{4.131}$$

where $\mathbf{C}_0$ is a diagonal matrix as shown on page 291. Further, note that since $\mathbf{G}_t$, $\mathbf{C}_0$, and $\mathbf{G}^H_{t+2p+1}$ are diagonal matrices, the order of multiplication can be changed, and hence

$$\mathbf{G}_t \mathbf{C}_0 \mathbf{G}^H_{t+2p+1} = \mathbf{C}_0 [\phi^{2p+1}]^H ,$$

where

$$\phi = \begin{bmatrix} e^{j\omega_1} & 0 & \dots & & 0 \\ 0 & e^{-j\omega_1} & \dots & & 0 \\ \vdots & & & & \\ 0 & 0 & \dots & e^{j\omega_p} & 0 \\ 0 & 0 & \dots & & e^{-j\omega_p} \end{bmatrix}.$$

We can further show by actual multiplication that

$$E\{\eta_t \eta^H_{t+2p+1}\} = 0\,.$$

Thus, (4.121) may be reduced to

$$\mathbf{C}_{x_t x_{t+2p+1}} = \mathbf{H}\mathbf{C}_0[\phi^{2p+1}]^H \mathbf{H}^H\,. \tag{4.132}$$

The rank of the cross-covariance matrix is same as that of $\mathbf{C}_0$, namely, 2p.

Example 4.18: The cross-covariance matrix between two temporally displaced data vectors can be expressed in terms of the covariance function. Consider the following example: Let 2p=4 and compute.

$$\mathbf{C}_{x_t x_{t+5}} = E\{\mathbf{x}_t \mathbf{x}_{t+5}\} = \begin{bmatrix} E\{x(t)x(t+5)\} & \dots & E\{x(t)x(t+8)\} \\ E\{x(t+1)x(t+5)\} & \dots & E\{x(t+1)x(t+8)\} \\ E\{x(t+2)x(t+5)\} & \dots & E\{x(t+2)x(t+8)\} \\ E\{x(t+3)x(t+5)\} & \dots & E\{x(t+3)x(t+8)\} \end{bmatrix}$$

$$= \underset{(4\times 4)}{\begin{bmatrix} C_x(5) & C_x(6) & C_x(7) & C_x(8) \\ C_x(4) & C_x(5) & C_x(6) & C_x(7) \\ C_x(3) & C_x(4) & C_x(5) & C_x(6) \\ C_x(2) & C_x(3) & C_x(4) & C_x(5) \end{bmatrix}}.$$

(End of Example).

Now let us consider the following expression:

$$\mathbf{C}_{x_t x_{t+2p}} - \gamma \mathbf{C}_{x_t x_{t+2p+1}} = \mathbf{H}\mathbf{C}_0[\phi^{2p}]^H \mathbf{H}^H - \gamma \mathbf{H}\mathbf{C}_0[\phi^{2p+1}]^H \mathbf{H}^H$$

$$= \mathbf{H}\mathbf{C}_0[\mathbf{I} - \gamma\phi^H][\phi^{2p}]^H \mathbf{H}^H\,. \tag{4.133}$$

The rank of the matrix expression on the left-hand side of (4.133)

Angular Frequencies in radians	*Generalized Eigenvalues*	
	White Noise	MA Process
$\omega_1=1.0$	$0.9981\,e^{\pm j.9985}$	$0.9808\,e^{\pm j.9589}$
$\omega_1=1.0$ $\omega_2=2.0$	$1.0003\,e^{\pm j.9893}$ $1.0044\,e^{\pm j1.9879}$	$1.1416\,e^{\pm j.9299}$ $1.0818\,e^{\pm j2.0689}$
$\omega_1=0.5$ $\omega_2=1.5$ $\omega_3=2.5$	$1.0122\,e^{\pm j.4937}$ $1.0148\,e^{\pm j1.4898}$ $0.9942\,e^{\pm j2.5022}$	$1.0007\,e^{\pm j.5127}$ $1.0321\,e^{\pm j1.4929}$ $1.0492\,e^{\pm j2.5066}$

Table 4.13: Estimation of frequencies using the generalized eigenvalues in presence of white or colored noise (SNR=13 dB).

solely depends upon the rank of $[\mathbf{I} - \gamma\phi^H]$ as all the other matrices on the right-hand side of (4.133) are full-rank matrices. Observe that whenever $\gamma = \exp(\pm j\omega_i)$, one of the columns of $[\mathbf{I} - \gamma\phi^H]$ becomes a column of zeros, and hence it becomes a singular matrix. Consequently,

$$\det[\mathbf{C}_{x_t x_{t+2p}} - \gamma \mathbf{C}_{x_t x_{t+2p+1}}] = 0 \qquad \text{for } \gamma_i = \exp(\pm j\omega_i), \quad i = 1, 2, \ldots p. \tag{4.134}$$

Note that (4.134) is indeed a statement of the generalized eigenvalues of a pair of matrices, namely, $\mathbf{C}_{x_t x_{t+2p}}$ and $\mathbf{C}_{x_t x_{t+2p+1}}$. Interestingly, the unknown frequencies are directly obtained from the generalized eigenvalues of a pair of cross-covariance matrices. This forms the basis of a new method known as ESPRIT proposed by Paulraj et al.[49] and Roy et al.[50] When the background noise is colored, say, as in the qth-order MA process, the temporal displacement of the data vectors should be at least 2p+q.

Example 4.19: A numerical illustration of the above method is described below. A sum of p (p=1, 2, 3) random sinusoids in presence of random noise (in the first instance white noise but later a 3rd-order MA process with parameters $b_0 = 2.5, b_1 = 1.8,$ $b_2 = 4.2, b_3 = 1.2$, and $b_2 = 4.2, b_3 = 1.2$) was generated. Four hundred snapshots were used to compute the cross-covariance matrices. The SNR was set to 13 dB. Table 4.13 summarizes the numerical results (End of Example).

H. Frequency Estimation Using Cumulant

Consider a model of a time series as a sum of random sinusoids and a Gaussian, but colored, noise, for example, a qth-order MA Gaussian process:

$$x(t) = \sum_{i=1}^{p} A_i \cos(\omega_i t + \phi_i) + \eta(t).$$

We have shown in Chapter Two that such a time series may be modeled as an ARMA process,

$$\sum_{k=0}^{2p} a_k x(t-k) = \sum_{l=0}^{2p+q} \beta_l \varepsilon(t-l), \quad \text{where} \quad a_0 = 1, \tag{4.135}$$

where the background noise is modeled as a qth-order MA process,

$$\eta(t) = \sum_{i=0}^{q} b_i \varepsilon(t-i),$$

and $\varepsilon(t)$ is a Gaussian process of zero mean and of variance σ_ε^2. The coefficients b_i and β_i are related to each other (see (2.93)). We shall compute the fourth-order cumulant on the diagonal by multiplying (4.135) by $X^3(t+t)$ and carrying out the expectation operation,

$$\sum_{k=0}^{2p} a_k E\{x(t-k)x^3(t+\tau)\} = \sum_{l=0}^{2p+q} \beta_l E\{\varepsilon(t-l)x^3(t+\tau)\}$$

$$\sum_{k=0}^{2p} a_k T_c(\tau+k) = \sum_{l=0}^{2p+q} \beta_l E\{\varepsilon(t-l)[\sum_{n=0}^{\infty} h_n \varepsilon(t+\tau-n)]^3\}$$

$$= 3\sigma_\varepsilon^4 \sum_{l=0}^{2p+q} \beta_l h_{\tau+l} \sum_{n=0}^{\infty} h_n^2, \tag{4.136}$$

where we assumed an MA representation, $x(t) = \sum_{n=0}^{\infty} h_n \varepsilon(t-n)$. It is easy to show that

$$C_x(0) = \sigma_\varepsilon^2 \sum_{n=0}^{\infty} h_n^2$$

and

$$\sum_{k=0}^{2p} a_k C_x(\tau+k) = \sigma_\varepsilon^2 \sum_{i=0}^{2p+q} \beta_i h_{\tau+i}.$$

Using these results we can simplify the right hand side of (4.136) as follows:

$$\sum_{k=0}^{2p} a_k T_c(\tau+k) = 3C_x(0) \sum_{k=0}^{2p} a_k C_x(\tau+k).$$

Next we shall replace the tricovariance function by the cumulant, obtained by subtracting the Gaussian component, namely, $3C_x(0)C_x(\tau)$, from the tricovariance function. We obtain the following important result:

$$\sum_{k=0}^{2p} a_k T_0(\tau + k) = 0 ,$$

or in the matrix form,

$$\mathbf{T}_0\mathbf{A} = \mathbf{0} , \tag{4.137}$$

where $\mathbf{T}_0$ is the cumulant matrix (of size: (2p+1) × (2p+1)) having properties similar to those of the covariance matrix and $\mathbf{A} = \text{col}\{1, -a_1, -a_2, \ldots, -a_{2p}\}$. The rank of $\mathbf{T}_0$ is equal to 2p, and hence there is one null vector which will be proportional to $\mathbf{A}$. As in Pisarenko's method, which is based on the covariance matrix, the frequencies are estimated from the roots of the coefficient polynomial, $A(z) = [1 - a_1 z^{-1} - a_2 z^{-2}, \ldots, - a_{2p} z^{-2p}]$.[51] A subspace approach based on the fourth-order cumulant has recently been proposed.[52] It may be noted that the background noise need be Gaussian but not necessarily white. Recall that the subspace methods based on the covariance matrix require that the noise be either white or of known covariance structure. In practice the covariance function of the background noise is rarely fully known; on the other hand, the assumption of Gaussian nature is closer to reality as the noise is the result of many independent sources.

I. Number of Sinusoids

We have so far assumed that the number of sinusoids is known *a priori* or can be estimated from the knowledge of the significant and repeated eigenvalues of the covariance matrix. We have shown earlier how the order of an AR process can be estimated from the eigenvalue distribution of the covariance matrix. This straightforward approach is inadequate whenever the sinusoids are closely spaced and/or the data length is too short. We shall describe a test known as sphericity test for estimating the number of repeated eigenvalues.

We consider a model of time series as p real sinusoids in white noise. A covariance matrix of size $\Delta T \times \Delta T$ ($\Delta T > 2p$) is computed from a time series of length T. Let $\hat{\lambda}_1 > \hat{\lambda}_2 > \hat{\lambda}_3 > \ldots > \hat{\lambda}_{\Delta T}$ be the eigenvalues of the estimated covariance matrix. We begin with the hypothesis that there are k sinusoids; then a subset of smaller eigenvalues, $\hat{\lambda}_{2k+1}, \hat{\lambda}_{2k+2}, \ldots, \hat{\lambda}_{\Delta T}$, are equal (or repeated). This null hypothesis is tested starting at k=0 till the test fails. Mauchley[53] suggested a test known as sphericity test to verify the equality of all eigenvalues of an estimated covariance matrix. Define a likelihood ratio statistic,

$$\Gamma(\hat{\lambda}_{2k+1}, \hat{\lambda}_{2k+2}, \ldots, \hat{\lambda}_{\Delta T}) = \ln\left[\frac{\left(\frac{1}{\Delta T - 2k}\sum_{i=2k+1}^{\Delta T}\hat{\lambda}_i\right)^{\Delta T - 2k}}{\prod_{i=2k+1}^{\Delta T}\hat{\lambda}_i}\right],$$

λ_1	94.6997
λ_2	55.1989
λ_3	12.2363
λ_4	4.9362
λ_5	1.7333
λ_6	1.3828
λ_7	1.2389
λ_8	1.1566
λ_9	1.0876
λ_{10}	1.0298
λ_{11}	.8896
λ_{12}	.7590
λ_{13}	.7945
λ_{14}	.6283
λ_{15}	.5863
λ_{16}	.4946

Table 4.14: in decreasing order

which is then compared with a threshold γ. Whenever

$$\Gamma(\hat{\lambda}_{2k+1}, \hat{\lambda}_{2k+2}, \ldots, \hat{\lambda}_{\Delta T}) > g$$

the test is said to have failed. Modified forms of the sphericity tests have been suggested in References 54 and 55.

The choice of γ is subjective. Alternate approaches which do not require a subjective judgement have been proposed.[56] The number of sinusoids is given by that value of k in the range 0 to $\frac{\Delta T}{2} - 1$, where either

$$\frac{T}{\Delta T}\Gamma(\hat{\lambda}_{2k+1}, \hat{\lambda}_{2k+2}, \ldots, \hat{\lambda}_{\Delta T}) + 4k(\Delta T - k),$$

or

$$\frac{T}{\Delta T}\Gamma(\hat{\lambda}_{2k+1}, \hat{\lambda}_{2k+2}, \ldots, \hat{\lambda}_{\Delta T}) + 2k(\Delta T - k)\ln[\frac{T}{\Delta T}], \tag{4.138}$$

is minimum. The first form is known as Akaike information criterion (AIC), and the second form is known as minimum description length (MDL). The expression consists of two parts; the first part is the log likelihood ratio which decreases monotonically, and the second part is a penalty term which increases monotonically as k increases from 0 to $\frac{\Delta T}{2} - 1$. The penalty term is equal to the adjusted free parameters. The log likelihood ratio around k=p has an interesting behavior which is responsible for the occurrence of a minimum at k=p. This is demonstrated in Example 4.20.

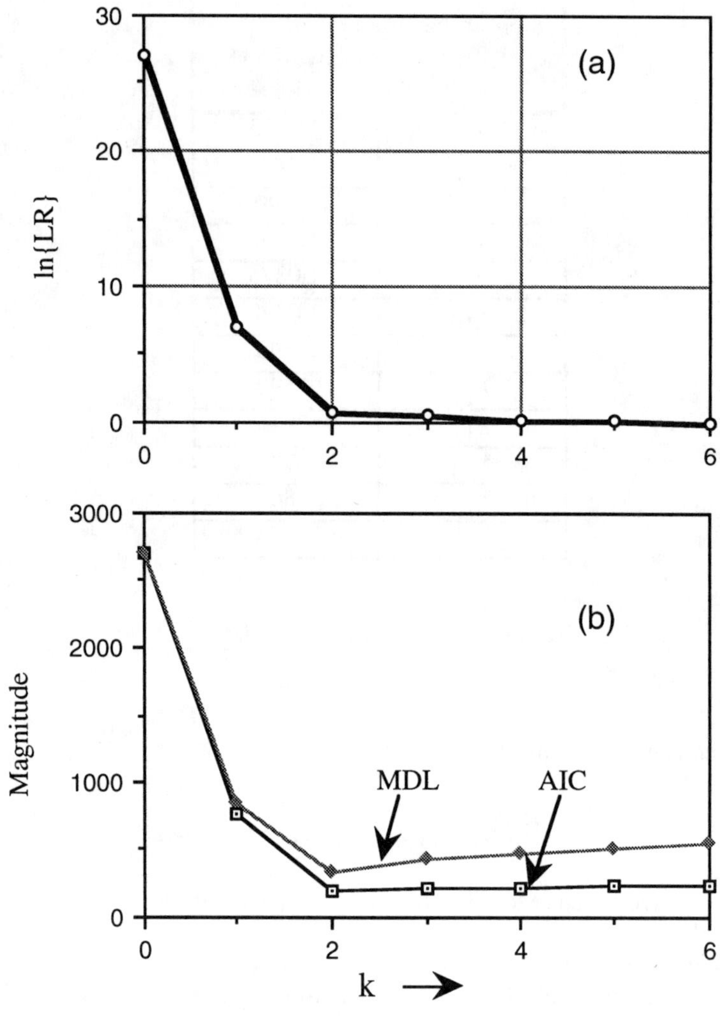

Figure 4.15: (a) Log likelihood ratio as a function of k, number of assumed sinusoids. (b) AIC and MDL as a function of k. Note the minimum at k=2.

Example 4.20: Consider two real random sinusoids with angular frequencies $\omega_1 = 0.18\pi$ and $\omega_2 = 0.24\,\pi$ in the background of white Gaussian noise. A covariance matrix of size 16×16 was computed using 100 independent segments each of length 16. The eigenvalues of the estimated covariance matrix are listed in Table 4.14 in decreasing order. It is possible to say, just by looking at the table, how many sinusoids are present in the time series, but the transition from the signal eigenvalues (shaded) to noise eigenvalues is somewhat blurred. We now apply the information theoretic criteria. First we have plotted in Figure 4.15a the log likelihood ratio (LR) as a function of k (assumed number of sinusoids). Notice at k=2 there is an abrupt change in the slope of the curve. In Figure 4.13b we have plotted both AIC and MDL functions (End of Example).

The performance of AIC and MDL criteria has been studied in the context of detection of plane waves incident on a linear array in Reference 57. They define a probability of error as

$$P_{error} = \text{prob}\,(k_{min} < p|H_p) + \text{prob}\,(k_{min} > p|H_p),$$

where k_{min} stands for the position of the minimum of (4.138), and H_p denotes the hypothesis that the true number of signal sources is p. The probability of error for the MDL criterion monotonically goes to zero for a large number of snapshots or high SNR. In contrast, the probability of error for the AIC criterion tends to a small finite value. However, for a small number of snapshots or low SNR the performance of AIC is better than that of MDL.

4.6 Nonlinear Transformation

The fundamental problem in the estimation of the spectrum lies in the need for evaluation of an infinite sum (1.30) given a finite number of terms as in (3.25). There exist nonlinear transformations known as G_n and ε_n, transformations which are known to accelerate the convergence; in particular, Shanks[58] has shown how by means of ε_n transformation the convergence of the Leibnitz expansion, $\pi = 4 - \frac{4}{3} + \frac{4}{5} - \frac{4}{7} + \ldots$, can be accelerated; for example, to achieve an eight-figure accuracy, straightforward summation would require 4×10^7 terms but using ε_n transformation the same accuracy can be achieved with just ten terms. G_n transformation does to an infinite integral what ε_n transformation does to an infinite sequence. Since our interest is in the infinite sum (1.30), we shall confine ourselves to ε_n transformation.

A. Definition of ε_n Transformation

Let there be an infinite sum,

$$\underset{k\to\infty}{S_k} = \sum_{i=0}^{k} C_i .$$

Define the following transformation:

$$\varepsilon_n[S_k] = \frac{\begin{vmatrix} S_k & S_{k+1} & \cdots & S_{k+n} \\ C_k & C_{k+1} & \cdots & C_{k+n} \\ \vdots & & & \vdots \\ C_{k+n-1} & C_{k+n} & \cdots & C_{k+2n-1} \end{vmatrix}}{\begin{vmatrix} 1 & 1 & \cdots & 1 \\ C_k & C_{k+1} & \cdots & C_{k+n} \\ \vdots & & & \vdots \\ C_{k+n-1} & C_{k+n} & \cdots & C_{k+2n-1} \end{vmatrix}} = \frac{H(S_{k+j}, C_{k+i+j};\ i,j = 0,1,\ldots,n)}{H(1\ \ C_{k+i+j};\ \ i,j = 0,1,\ldots,n)}, \tag{4.139}$$

where H(.) stands for the determinant of the matrix in the numerator or denominator of (4.139). We shall merely state the following important result[59] without proof but use it extensively in the estimation of spectrum. Let C_i for $i \geq 1+q$ $(q \geq 0)$ satisfy a difference equation,

$$C_i + a_1 C_{i-1} + a_2 C_{i-2} + \ldots + a_p C_{i-p} = 0, \tag{4.140}$$

then

$$\varepsilon_n[S_k] \equiv S_\infty$$

for $n \geq p$ and $k \geq 1$. For an AR process the covariance function satisfies a difference equation of order, p, for all lags greater than zero (see (2.54)) which we reproduce here for easy reference:

$$C_x(\tau) + a_1 C_x(\tau - 1) + a_2 C_x(\tau - 2) + \ldots + a_p C_x(\tau - p) = 0 \qquad \tau > 0.$$

A similar relation exists for an ARMA process (see (2.82)), but now $\tau \geq 1 + q$.

Multiply both sides of (2.54) by $\exp(\pm j\omega\tau)$ and rearrange the terms:

$$\begin{aligned} &C_x(\tau)\exp(\pm j\omega\tau) + a_1 \exp(\pm j\omega)C_x(\tau - 1)\exp(\pm j\omega(\tau - 1)) \\ &+ a_2 \exp(\pm j2\omega)C_x(\tau - 2)\exp(\pm j\omega(\tau - 2)) + \ldots \\ &+ a_p \exp(\pm jp\omega)C_x(\tau - p)\exp(\pm j\omega(\tau - p)) = 0 \qquad \tau > 0. \end{aligned} \tag{4.141}$$

From (4.141) we claim that $C_x(\tau)\exp(\pm j\omega\tau)$ also satisfies a difference equation similar to (4.140) but now with $a_k \exp(\pm j\omega k)$ as coefficients. Hence, the ε_n transform of the partial sums, defined in (4.139), will be

$$\varepsilon_m\left[1 + 2\sum_{\tau=1}^{m} C_x(\tau)\cos(\omega\tau)\right] = S_\infty = 1 + 2\sum_{\tau=1}^{\infty} C_x(\tau)\cos(\omega\tau) \tag{4.142}$$

for $p \leq m \leq 2p$.

B. Recursive Algorithm

We now describe a recursive algorithm to compute the ε-spectrum. Define the following quantities:

$$\varepsilon_0^k = 1 + 2\sum_{\tau=1}^{k} C_x(\tau)\cos(\omega\tau),$$

$$C_k = C_x(k)\cos(\omega k),$$

$$H_n^k = \begin{vmatrix} C_k & C_{k+1} & \cdots & C_{k+n-1} \\ C_{k+1} & C_{k+2} & \cdots & C_{k+n} \\ \vdots & & & \\ C_{k+n-1} & C_{k+n} & \cdots & C_{k+2n-2} \end{vmatrix},$$

$$G_n^k = \begin{vmatrix} 1 & 1 & \cdots & 1 \\ C_k & C_{k+1} & \cdots & C_{k+n} \\ C_{k+1} & C_{k+2} & \cdots & C_{k+n+1} \\ \vdots & & & \\ C_{k+n-1} & C_{k+n} & \cdots & C_{k+2n-1} \end{vmatrix}.$$

with $H_0^k = 1$. We define two other quantities, namely,

$$r_n^k = \frac{H_n^k}{G_n^k} \quad \text{and} \quad s_n^k = \frac{G_{n+1}^k}{H_n^k}.$$

In terms of the above quantities, the ε-spectrum is recursively expressed as

$$\varepsilon_{n+1}^k = \varepsilon_n^k + \frac{\varepsilon_n^{k+1} - \varepsilon_n^k}{1 - R_{n+1}^k}. \tag{4.143}$$

where $R_{n+1}^k = \dfrac{r_{n+1}^{k+1}}{r_{n+1}^k}$. Further, we can express r_n^k recursively:[60]

$$r_{n+1}^k = \begin{cases} r_n^{k+1}\left[\dfrac{s_n^{k+1}}{s_n^k} - 1\right], & s_n^k \neq 0 \\ -r_n^{k+1}, & s_n^k = 0 \end{cases}$$

$$s_{n+1}^k = \begin{cases} s_{n-1}^{k+1}\left[\dfrac{r_n^{k+1}}{r_n^k} - 1\right], & r_n^k \neq 0 \\ -s_{n-1}^{k+1}, & r_n^k = 0 \end{cases}$$

for n= 1, 2, ... and k= 0, 1, ... ; and $s_0^k = 1$, $k = 0,1, \ldots$, and $r_1^k = C_k$, $k = 0,1, \ldots$.

By observing the sequences s_n^k and r_{n+1}^k it is possible to ascertain the order of the ARMA(p,q) model governing the time series under study. It is shown in Reference 59 that at $\omega =$ 0 or π,

$$\left.\begin{array}{l} s_p^k = c \quad \text{and} \\ s_p^{q-p} \neq c \\ \\ r_p^k = 0 \quad \text{and} \\ r_p^{q-p} \neq 0 \end{array}\right\} k \geq (q - p + 1)$$

$S_X(0.0)=0.32$

k	ε_0^k	ε_1^k	ε_2^k	ε_3^k
0	0.0	2.30	**0.32**	**0.32**
1	1.39	1.32	**0.32**	
2	1.22	8.87		
3	0.03			

$S_X(0.1)=0.5673$

k	ε_0^k	ε_1^k	ε_2^k	ε_3^k	ε_4^k	ε_5^k
0	0.0	1.936	1.488	1.512	**0.567**	**0.567**
1	1.319	1.412	1.157	1.438	**0.567**	**0.567**
2	1.463	1.476	1.496	1.934	**0.567**	**0.567**
3	1.491	1.616	1.306	-1.3233	**0.567**	**0.567**

$S_X(0.1950)=9.053$

k	ε_0^k	ε_1^k	ε_2^k	ε_3^k	ε_4^k	ε_5^k
0	0.0	1.308	-0.679	**9.053**	**9.045**	**9.045**
1	1.133	0.882	1.352	**9.052**	**9.044**	**9.046**
2	1.701	-1.679	4.774	**9.052**	**9.046**	**9.046**
3	2.698	3.283	-2.135	**9.051**	**9.045**	**9.048**

Table 4.15: ε-matrices at different frequencies. The process of stabling off is demonstrated.

at a frequency corresponding to a peak in the spectrum,

$$\left.\begin{array}{l} s_{2p-1}^{k} = c \text{ and} \\ s_{2p-1}^{q-p} \neq c \\ \\ r_{2p}^{k} = 0 \text{ and} \\ r_{2p}^{q-p} \neq 0 \end{array}\right\} k \geq (q-p+1)$$

and at other frequencies,

S Matrix (at frequency 0.195 Hz)

k	S^k_0	S^k_1	S^k_2	S^k_3	S^k_4	S^k_5
-2	1.0	-0.693	-1.177	-0.206	0.385	-0.340
-1	1.0	6.494	0.499	-0.349	0.336	-0.027
0	1.0	-0.867	-0.484	**-0.335**	0.030	-0.121
1	1.0	2.255	1.605	**-0.335**	0.027	-0.047
2	1.0	0-.295	-0.754	**-0.335**	0.037	-0.395

R Matrix (at frequency 0.195 Hz)

k	R^k_1	R^k_2	R^k_3	R^k_4	R^k_5	R^k_6
-2	0.869	-2.769	3.228	1.31	0.017	0.0
-1	0.267	-2.267	1.893	-0.134	0.0	0.0
0	2.0	-0.961	3.260	**0.0**	0.0	0.0
1	0.267	-0.755	2.579	**0.0**	0.0	0.0
2	0.869	-4.864	1.433	**0.0**	0.0	0.0

Table 4.16: **S** and **R** matrices.

$$\left.\begin{array}{ll} s^k_{2p} = c & \text{and} \\ s^{q-p}_{2p} \neq c & \\ & \\ r^k_{2p+1} = 0 & \text{and} \\ r^{q-p}_{2p+1} \neq 0 & \end{array}\right\} k \geq (q-p+1), \tag{4.144}$$

where c is constant. At ω=0 or π the ε-spectrum is equal to the true spectrum,

$$\varepsilon^k_{p+i} = S_x(\omega) \quad , \tag{4.145a}$$

for $i = 0, 1, 2, \ldots$ and $k \geq \max\{1, q-p+1\}$.

At the frequency corresponding to the peak in the spectrum, that is, at the angular position of the pole,

$$\varepsilon^k_{2p-1+i} = S_x(\omega) \tag{4.145b}$$

for $i = 0, 1, 2, \ldots$ and $k \geq \max\{1, q-p+1\}$. And at other frequencies we have

$$\varepsilon^k_{2p+i} = S_x(\omega) \tag{4.145c}$$

for $i = 0, 1, 2, \ldots$ and $k \geq \max\{1, q-p+1\}$.

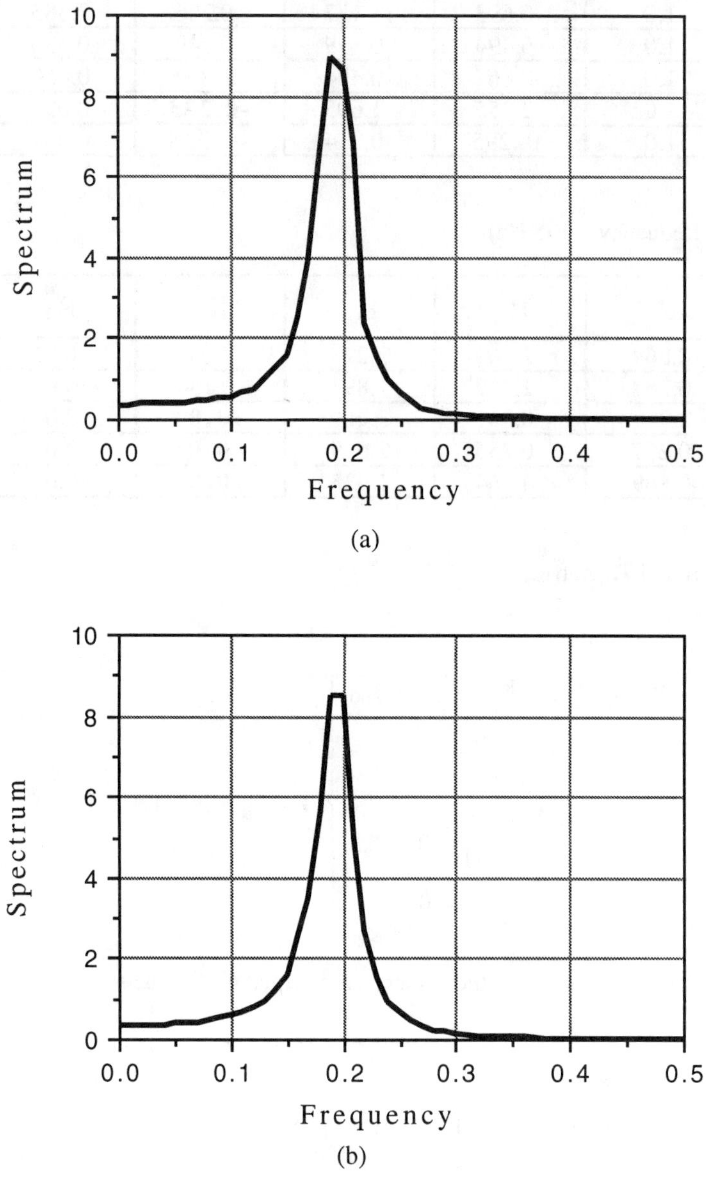

Figure 4.16: (a) A plot of ε-spectrum obtained by using 2000-point ARMA time series. (b) Theoretically computed ARMA spectrum. The computed spectrum and the theoretical spectrum are very close to each other with a mean square error equal to 0.124.

The recursive precedure outlined above is carried on up to some preassigned values of k and n, and the estimates of the ε-spectrum are tabulated. It will be noticed that starting from some value of k and n, the ε-spectrum reaches a stable value. This phenomenon is called stabling off. The true

k	ε_0^k	ε_1^k	ε_2^k	ε_3^k	ε_4^k	ε_5^k	ε_6^k	ε_7^k	ε_8^k
0	0.0	1.54	1.251	5.166	.053	1.906	2.89	2.704	2.894
1	1.213	-2.841	.223	4.505	**2.106**	**2.069**	**2.661**	**2.664**	
2	1.662	1.151	3.473	4.136	**1.857**	**2.451**	**2.732**		
3	2.541	3.11	2.801	3.883	**2.700**	**2.539**			
4	3.145	3.170	3.308	3.721	**2.157**				

Table 4.17: ε-Spectrum at frequency 0.16, on the left flank of the spectral peak in Figure 4.14b.

spectrum is then given by the stable estimate of the ε-spectrum. For estimation of the ε-spectrum we need autocorrelation coefficients ranging from lag equal to k to lag $k+2n-1$.

Example 4.21: We consider a numerical example of an ARMA(2,1) process,

$$X(t) - 0.6X(t-1) + 0.8X(t-2) = \eta(t) + 0.6\eta(t-1).$$

It has a pole at $\omega=0.19556$ and zero at $\omega=-0.6$. The exact covariance function is assumed to be known. e-matrices at different frequencies are as shown in Table 4.15. The figures in bold indicate the stabling off phenomenon in ε-spectrum matrix. The correct order of the ARMA model can be estimated, noting the stabling off phenomenon and applying the rules in (4.145). As noted earlier, the model order information can also be derived from s_n^k and r_{n+1}^k. We form two matrices, **S** and **R,** where $\{\mathbf{S}\}_{k,n} = s_n^k$ and $\{\mathbf{R}\}_{k,n} = r_{n+1}^k$ as shown in Table 4.16.

Next, we generated a 2000-point ARMA time series and estimated the covariance function for the first ten lags. The ε-spectrum was estimated using the previously estimated covariance function. A plot of the estimated ε-spectrum (Figure 4.16a) was compared with the theoretical ARMA spectrum in Figure 4.16b.

The order of the ARMA model was correctly estimated at all frequencies except where the gradient was very high, that is, on the flanks of the spectrum peak. It was noticed that stabilization at these frequencies was poor, leading to incorrect order estimate. However, the spectrum estimate, when averaged over those columns where stabilization was expected but not found, was close to the true spectrum. As an example, consider the ε-spectrum at frequency 0.16 for different (n,k) (Table 4.17). Columns 4 to 8 should have been stabilized, but they were not. The average of all the bold-faced numbers turned out to be 2.393, which is close to the true spectrum, 2.256. Thus, while the order of the model could not be estimated, the true spectrum was close to the average (End of Example).

4.7 **Extrapolation of Band-Limited Time Series**

It was pointed out in Chapter One that a band-limited process is predictable. We shall explore this property of a band-limited process for extrapolation beyond the interval over which the data is given. The basic idea behind extrapolation is repeated imposition of band limitedness under the constraint that the extrapolated time series exactly matches the given data. Let $x_0(t),\ t=0,1,\ldots T_0-1$ be the observed band-limited time series (bandwidth=$\pm\,\omega_0$) and $x^e(t),\ t=0,1,\ldots,T_1-1\ (T_1 \ge T_0)$ be the extrapolated version such that $x^e(t)=x_0(t),\ t=0,1,\ldots,T_0-1$ and that it has the same bandwidth. $x^e(t)$ is obtained by passing $x_0(t)$ through an ideal lowpass filter (bandwidth=$\pm\,\omega_0$) but in the process, since the length of the filtered sequence becomes infinite, it will have to be limited to some finite value, say, $T_1 \geq T_0$, while matching with the given data. These two operations are incompatible but can be approximately achieved through repeated iterations. The above idea was first postulated by Gerchberg[61] and realized through FFT by Papoulis.[62] We shall here give a matrix formulation and bring out some of the issues of existence.

A. Extrapolation Matrix

Define a lowpass filter matrix operator,

$$\mathbf{B} = \mathbf{H}^H \mathbf{A} \mathbf{H} \;,$$

where

$$\{\mathbf{H}\}_{k,t} = \frac{1}{\sqrt{T_1}} \exp\left(-j\frac{2\pi}{T_1}kt\right)$$

is a matrix of size $(T_1 \times T_1)$, and $\mathbf{A}$ is a diagonal matrix of filter weights of size $(T_1 \times T_1)$. For example,

$$\mathbf{A} = \text{diag}\,\{\underbrace{1,1,1,\ldots 1}_{k_0},0,0,0,\ldots 0,0,0,\underbrace{1,1,1,\ldots 1}_{k_0}\},$$

where $\omega_0 = \frac{2\pi k_0}{T_1}$. Define a data vector, $\mathbf{X}_0$,

$$\mathbf{x}_0 = \underset{T_1 \times 1}{\text{col}\,\{x(0)\;\; x(1)\;\; x(2)\;\; \ldots\;\; x(T_0-1)\;\; 0\;\; 0\;\; \ldots 0\;\; 0\}}$$

and a limiting matrix of size $(T_1 \times T_1)$,

$$\mathbf{D} = \text{diag}\,\{\underbrace{1\;\;1\;\;1\;\;\ldots\;\;1}_{T_0}\;\;0\;\;0\;\;0\;\;\ldots\;\;0\}$$

and its complement $\bar{\mathbf{D}} = \mathbf{I} - \mathbf{D}$. The extrapolated data vector, $\mathbf{x}_1^e$, in the first iteration is given by

$$\mathbf{x}_1^e = \mathbf{x}_0^e + \bar{\mathbf{D}}\mathbf{B}\mathbf{x}_0^e,$$

where $\mathbf{x}_0^e = \mathbf{x}_0$, and in the second iteration,

$$\mathbf{x}_2^e = \mathbf{x}_0^e + \bar{\mathbf{D}}\mathbf{B}\mathbf{x}_1^e,$$

and so on. Thus, at the end of nth iteration we have

$$\mathbf{x}_n^e = \mathbf{x}_0^e + \bar{\mathbf{D}}\mathbf{B}\mathbf{x}_{n-1}^e. \tag{4.146}$$

The solution of the above difference equation with the boundary condition, $\mathbf{x}^e_{-1} = \mathbf{0}$, is given by

$$\begin{aligned}\mathbf{x}^e_n &= \mathbf{x}^e_0 + \bar{\mathbf{D}}\mathbf{B}\mathbf{x}^e_0 + (\bar{\mathbf{D}}\mathbf{B})^2\mathbf{x}^e_0 + \ldots + (\bar{\mathbf{D}}\mathbf{B})^n\mathbf{x}^e_0 \\ &= \{\mathbf{I} + \bar{\mathbf{D}}\mathbf{B} + (\bar{\mathbf{D}}\mathbf{B})^2 + \ldots + (\bar{\mathbf{D}}\mathbf{B})^n\}\mathbf{x}^e_0 \\ &= \frac{\mathbf{I} - (\bar{\mathbf{D}}\mathbf{B})^{n+1}}{\mathbf{I} - \bar{\mathbf{D}}\mathbf{B}}\mathbf{x}^e_0 . \end{aligned} \tag{4.147}$$

Define an extrapolation matrix,

$$\mathbf{E}_n = \frac{\mathbf{I} - (\bar{\mathbf{D}}\mathbf{B})^{n+1}}{\mathbf{I} - \bar{\mathbf{D}}\mathbf{B}} . \tag{4.148}$$

To obtain a limiting value of the extrapolation matrix, $\mathbf{E}_{\substack{n \\ n\to\infty}}$, we must ensure that eigenvalues of $\bar{\mathbf{D}}\mathbf{B}$ satisfy the condition that $|\lambda_k| < 1$ for all k, where λ_k is the kth eigenvalue. An alternate approach is to show that since the norm of $\bar{\mathbf{D}}\mathbf{B}$ must satisfy a condition $\|\bar{\mathbf{D}}\mathbf{B}\| \le \lambda_k$ for all k, we must have $\|\bar{\mathbf{D}}\mathbf{B}\| < 1$. The extrapolation matrix becomes singular whenever one of the eigenvalues of $\bar{\mathbf{D}}\mathbf{B}$ approaches unity. It is shown in Reference 63 that the extrapolation matrix becomes singular as $T_1 \to \infty$.

Example 4.22: The ability to extrapolate a band-limited process is seriously impaired in the presence of background noise, finite precision arithmetic, etc. The main difficulty lies in the stability of the extrapolation matrix. which becomes singular when the largest eigenvalue of $\bar{\mathbf{D}}\mathbf{B}$ approaches unity. As pointed out above, the largest eigenvalue approaches unit value as the extrapolation duration becomes large. In practice the limiting value is approached much faster than expected, depending upon the bandwidth. In this example we show the growth of the largest eigenvalue (squared) of $\bar{\mathbf{D}}\mathbf{B}$ as a function of extrapolation duration, T_1 (seconds). The data length is assumed to be 64 seconds and the bandwidth $(= 0.04\text{ Hz})$. The growth of the largest eigenvalue (squared) is shown in Figure 4.17 (End of Example).

There is an important difference between the continuous-time signal and discrete-time signal. A continuous-time signal and its Fourier transform cannot both be identically zero over intervals of any length,[20] but for a discrete-time signal it is possible to find a signal which is zero over a finite interval as well as band limited. In support of the above assertion we provide a constructive proof.[64] Let f(t) be a band-limited signal and consider a sequence g(t) defined as

$$g(t) = f(t) + \sum_{n=1}^{P} a_n f(t-n). \tag{4.149}$$

Since the spectrum of g(t) is related to the spectrum of f(t) through a relation $S_g(\omega) = |A(\omega)|^2 S_f(\omega)$ (see (2.11)), g(t) is also band limited like f(t). Now it is possible to impose a constraint that g(t)=0, t=1, 2, ... , p. This leads to a set of p equations in p unknowns; that is,

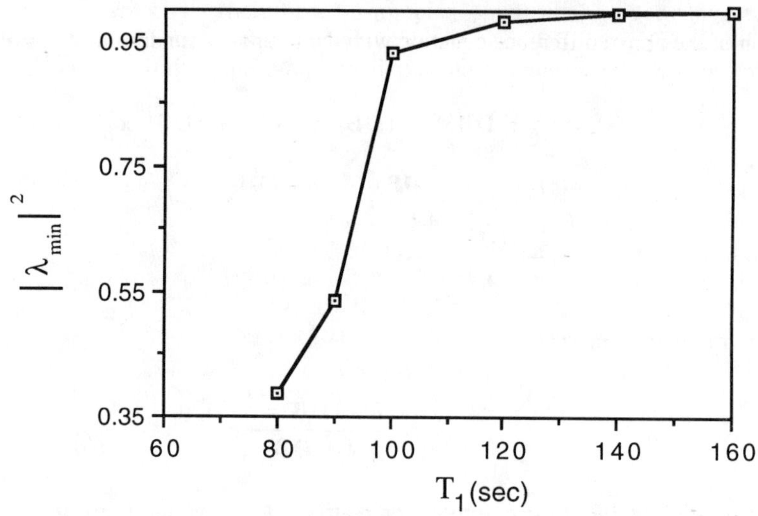

Figure 4.17: Growth of the largest eigenvalue (squared) of $\bar{\mathbf{D}}\mathbf{B}$ with the increasing duration of extrapolation $(T_1 - T_0)$. As the eigenvalue approaches unity the extrapolation matrix becomes singular. This happens at a faster rate as the bandwidth increases.

$$f(t) + \sum_{n=1}^{P} a_n f(t-n) = 0, \quad t = 1, 2, \ldots, p \qquad (4.150)$$

Assuming for the time being that a solution to (4.150) exists, we can generate a time series, g(t), which is zero over the interval $1 \le t \le p$ as well as band limited. Now, since f(t)+g(t) will also be an acceptable extrapolation, there is no unique solution to the discrete band-limited extrapolation problem. However, there are special cases of f(t) where g(t) is identically zero everywhere.

Consider a special case where f(t) is a sum of p random sinusoids:

$$f(t) = \sum_{n=1}^{P} \alpha_n \cos(\omega_n t + \varphi_n),$$

for which (4.150) has a unique solution valid for all t (see page 123). Now, g(t) as defined in (4.149) will be identically zero for all t. Hence a time series, which consists of a sum of random sinusoids, can uniquely be extrapolated. Consider another interesting case where f(t) is a solution of a pth-order difference equation whose solution is expressed as a sum of p damped sinusoids,

$$f(t) = \sum_{i=1}^{p} \alpha_i \exp(\gamma_i t) \quad t > 0,$$

where γ_i, $i = 1, 2, \ldots, p$ are the roots of a polynomial,

$$1 + a_1 z^{-1} + a_2 z^{-2} + \ldots + a_p z^{-p} = 0.$$

Hence, all signals having their poles inside or on the unit circle in the z-plane will possess the property of infinite extrapolation on the positive time axis. Even in the presence of noise, the extrapolation of a sum of random sinusoids is possible, but with the help of a prediction filter determined from the data.[65]

B. Minimum Energy Signal

In general, a band-limited signal which is consistent with the observed segment and has finite energy can be extrapolated to an infinite interval. To show this we shall reformulate the problem of extrapolation in terms of minimum norm least squares (MNLS) solution. Define the data vector, the extrapolated data vector, and the low pass filter as follows:

Data vector : $\mathbf{x}_0 = \text{col}\,\{x(0)\ \ x(1)\ \ x(2)\ \ \ldots\ \ x(T_0 - 1)\}$,

Extrapolated
Data Vector : $\mathbf{x}^e = \text{col}\,\{x(0)\ \ x(1)\ \ x(2)\ \ \ldots\ \ x(T_1 - 1)\}$,

Lowpass filter matrix: $\{\mathbf{B}\}_{k,l} = \dfrac{\sin(k-1)\omega_0}{\pi(k-1)}$,

$k,l = 0, \pm 1, \pm 2, \ldots, \frac{T_1 - 1}{2}, \quad T_1 : \text{odd},$

Selection matrix: $\{\mathbf{D}\}_{k,l} = 1$ for $k = 1, \quad 0 \le k \le T_0 - 1$

$= 0$ otherwise.

($T_0 \times T_1$ matrix)

$\mathbf{x}^e$ is the unknown vector which satisfies the following under-determined linear equations:

$$\mathbf{x}_0 = \mathbf{D}\mathbf{B}\mathbf{x}^e, \tag{4.151a}$$

whose least squares solution is made unique by requiring that the norm of the extrapolated data vector be minimum,[28] that is,

$$\|\mathbf{x}^e\|^2 = \text{min}. \tag{4.151b}$$

The solution of (4.151) is given by[62]

$$\hat{\mathbf{x}}^e = \mathbf{B}\mathbf{D}^T(\mathbf{D}\mathbf{B}\mathbf{B}^T\mathbf{D}^T)^{-1}\mathbf{x}_0, \tag{4.152}$$

where $\mathbf{B}\mathbf{D}^T(\mathbf{D}\mathbf{B}\mathbf{B}^T\mathbf{D}^T)^{-1}$ is the pseudoinverse of $\mathbf{DB}$. We can also define the pseudoinverse via the singular value expansion of $\mathbf{DB}$ (see (4.91)),

$$\mathbf{DB} = \sum_{k=1}^{T_0} \lambda_k \mathbf{U}_k \mathbf{V}_k^H,$$

and

$$\mathbf{B}^+ = \sum_{k=1}^{p} \frac{1}{\lambda_k} \mathbf{U}_k \mathbf{V}_k^H, \tag{4.153}$$

where $p \le T_0$ stands for the number of significant singular values. Since the remaining singular values represent noise power, their elimination improves the performance of the extrapolation algorithm. The solution given by (4.153) is also the minimum norm least squares solution of (4.151). Filtering implied in (4.152) may be realized in two steps as follows:

$$(\mathbf{DBB}^T\mathbf{D}^T)\,\mathbf{r} = \mathbf{x}_0, \tag{4.154a}$$

$$\hat{\mathbf{x}}^e = \mathbf{BD}^T\mathbf{r}. \tag{4.154b}$$

In the first step we solve a system of linear equations. Since $(\mathbf{DBB}^T\mathbf{D}^T)$ is a full-rank matrix, the vector $\mathbf{r}$ is uniquely determined. In the second step we filter the vector $\mathbf{r}$ to obtain the extrapolated signal. The importance of extrapolation in the context of spectrum estimation lies in the possibility of improved spectral resolution. It must be remembered that extrapolation does not bring in any extra information or degrees of freedom (equal to $\frac{2\omega_0}{\pi}T_0$). Consequently, the stability of the spectral estimate remains unchanged. Similarly, the spectral resolution of the MNLS extrapolated time series $\hat{\mathbf{x}}^e$ remains the same as that of $\mathbf{x}_0$, that is, on the order of $\frac{1}{T_0}$, even though the extrapolated time series is of length, T_1, where $T_1 > T_0$. From (4.154a) we note that $\mathbf{x}_0$ is a bandpass filtered version of $\mathbf{r}$; hence both will have similar spectrums. Since $\mathbf{x}_0$ has a smooth spectrum, $\mathbf{r}$ will have a smooth spectrum. The Fourier transform of the extrapolated time series can be shown (see Exercise 4.15) as the Fourier transform of $\mathbf{r}$ padded with $T_1 - T_0$ zeros. Thus, MNLS extrapolation is of little significance from the point of enhancing the spectral resolution.

4.8 Exercises

1. Show that the maximum entropy gain of a minimum phase filter driven by a zero mean and unit variance white noise is given by (see (4.72))

$$\Delta H = \log_2\left(\frac{1}{a_0}\right).$$

2. Let

$$x(t) = \cos\left(\frac{\pi}{4}t + \phi_1\right) + 0.5\cos\left(\frac{2\pi}{3}t + \phi_2\right) + \varepsilon(t) + 0.4\varepsilon(t-1),$$

where ϕ_1 and ϕ_2 are random phases uniformly distributed over $\pm\pi$, and $\varepsilon(t)$ is a zero mean and σ_ε^2 variance white noise process. Develop an equivalent ARMA model for x(t) and compute the ARMA parameters (see (4.109)).

3. Let

$$x(t) = \sum_{i=1}^{p} A_i \cos(\omega_i t + \phi_i).$$

Compute the covariance matrix of x(t) for the following situations:

(a) A_i is a constant for all i, and ϕ_i is a uniformly distributed ($\pm\pi$) random phase for all i.

(b) A_i is constant for all i, $\omega_i = i\omega_0$, and $\phi_i = \phi_0 + \Delta\phi_i$, where $\Delta\phi_i$ is a random variable which is uniformly distributed in the range $\pm\alpha_0$.

Express the covariance matrix in a form $\mathbf{C}_x = \mathbf{H}\mathbf{C}_0\mathbf{H}^H$. What can you conclude about matrix $\mathbf{C}_0$?

4. Now consider a model used in the Example 1.6 for x(t) where it was shown that x(t) is not an ergodic process (its covariance function does not tend to zero for large lags). Hence its autocorrelation function computed from a single realization,

$$R_x(\tau) = \frac{1}{2T}\sum_{t=-T}^{T} x(t)x(t+\tau)$$

will be different from its covariance function. Express the autocorrelation matrix, $\mathbf{R}_x$, in a form, $\mathbf{R}_x = \mathbf{H}\mathbf{C}_0\mathbf{H}^H$. What can you say about the matrix $\mathbf{C}_0$?

5. Given that a covariance function is known only over a finite segment, say, $\tau_0 \le \tau \le \tau_1$, where τ_0 and τ_1 are some positive integers, suggest a method to estimate the maximum entropy spectrum.

6. Note that the difference equation for a sum of p sinusoids (4.112) is similar to the extended Yule-Walker (2.63). On account of this similarity it may be possible to solve the Yule-Walker equations using Prony's algorithm. Explore this possibility by writing the ρ matrix in terms of the estimated covariance function.

7. Show that a covariance matrix, $\mathbf{C}_x$, can be expressed as follows:

$$\mathbf{C}_x = \mathbf{A}\sigma\mathbf{A}^H,$$

where

$$\mathbf{A} = \begin{bmatrix} 1 & & & & \\ a_{1,1} & 1 & & \mathbf{0} & \\ a_{2,2} & a_{1,2} & 1 & & \\ \vdots & & & \ddots & \\ a_{p,p} & a_{p-1,p} & \cdots & a_{1p} & 1 \end{bmatrix} \quad \text{and} \quad \sigma = \begin{bmatrix} \sigma_0 & & & \\ & \sigma_1 & & o \\ & & \ddots & \\ & o & & \sigma_p \end{bmatrix},$$

where $\sigma_0, \sigma_1, \sigma_2, \ldots, \sigma_p$ are prediction error power values using prediction filters of different order, 0, 1, 2, ..., p, respectively. The columns of the matrix **A** are the corresponding filter coefficients. [Hint:Write (4.84) for different orders, 0,1, 2, ..., p, and extend each equation to the same order p by including zero valued elements and then combine them into a form as above. This is known as Cholesky decomposition.]

8. The ML spectrum as defined in (4.12) essentially provides power at the steering frequency, but does not give power density, that is, power/Hz. To obtain the power spectrum density we must divide the ML spectrum given by (4.12) by bandwidth of the filter. Define the effective bandwidth, $\Delta\omega_{eff}$, as the width of a rectangle whose height is equal to the filter response at the steering frequency and area equal to the area under the filter response curve,

$$\Delta\omega_{eff}|A(\omega_0)|^2 = \frac{1}{2\pi}\int_{-\pi}^{\pi}|A(\omega)|^2 d\omega.$$

Show that the ML spectrum density (also known as normalized ML spectrum[66]) is given by

$$S^n_{ML}(\omega) = \frac{\mathbf{E}^H\mathbf{C}_X^{-1}\mathbf{E}}{\mathbf{E}^H\mathbf{C}_X^{-H}\mathbf{C}_X^{-1}\mathbf{E}}.$$

It is shown in Reference 66 that the normalized ML spectrum has a better resolution capability than the unnormalized one. Why?

9. In the maximum likelihood signal power estimation in the presence of background noise, it was shown on page 244 (4.3) that the mean square error in the estimation of signal power is given by

$$\frac{1}{\mathbf{E}_0^H\mathbf{C}_\eta^{-1}\mathbf{E}_0},$$

where $\mathbf{C}_\eta$ is the noise covariance matrix. Given the choice of selecting the signal waveform, how would you select the signal so that the mean square error is the minimum possible? Discuss the result when the background noise is white.

10. Let x(t), t=0, 1, 2, ..., $T\times 1$ be a stationary stochastic process and h(t) be a temporal window. The autocorrelation function of the window is given by

$$\sum_{t=0}^{T-\tau} h(t)h(t+\tau) = \beta_0 + \beta_1\tau^2 + \beta_2\tau^4 + \ldots,$$

where β_0, β_1, and β_2 are constants. Show that the bias in the estimated spectrum is given by

$$\text{bias} = \beta_1 \frac{\partial^2 S_x(\omega)}{\partial\omega^2} + \beta_2 \frac{\partial^4 S_x(\omega)}{\partial\omega^4} + \ldots$$

11. Consider a wideband ML method where the filter preserves the amplitude over a bandwidth, $\Delta\omega$, about some center frequency, ω_c. Show that the ML spectrum at ω_c is given by

$$S_{ML}(\omega_c) = \frac{1}{\mathbf{E}^H_{\Delta\omega}\mathbf{C}^{-1}_X\mathbf{E}_{\Delta\omega}},$$

where

$$\mathbf{E}_{\Delta\omega} = \text{col}\{1, e^{j\omega_c}\sin c(\frac{\Delta\omega}{2}), e^{j2\omega_c}\sin c(\frac{2\Delta\omega}{2}), + \ldots + e^{j(p-1)\omega_c}\sin c(\frac{(p-1)\Delta\omega}{2})\}.$$

Observe that as $\Delta\omega \to 0$ the solution approaches the narrowband case. However, as $\Delta\omega \to 2\pi$ the optimum filter approaches the prediction error filter of the maximum entropy method. Show that the output power of such a filter is equal to the prediction error power.

12. On page 291 it was stated that the **H** matrix must have full column rank. For this to be true show that Δt, the sampling interval, must be less than or equal to half the smallest period. Consider a numerical example. Let $\omega = 0.6(2\pi)$ and $\Delta t = 1$. Where will be the spectral peak corresponding to the above frequency?

13. Consider a minimum phase whitening filter, $\mathbf{A} = \text{col}\{1\ a_1\ a_2\ \ldots\ a_p\}$. Let σ^2_ε be the variance of the whitened output. Show that

$$\sigma^2_\varepsilon = \exp[\frac{1}{2\pi}\int_{-\pi}^{\pi} \ln S_X(\omega)\, d\omega].$$

[Hint: First show that for minimum phase filter $\frac{1}{2\pi}\int_{-\pi}^{\pi} \ln|A(\omega)|^2 d\omega = 0$.]

14. Show that the reflection coefficient defined in (4.80) becomes zero for m>p, [Hint: Use the Yule-Walker equation (2.54) for m=p+1, where p is the true order of the AR process and repeat this step for m=p+2, m=p+3, and so on.]

15. MNLS extrapolation of a band-limited process given by (4.154) does not offer any improved spectral resolution. Show this by computing the Fourier transform of the extrapolated time series, $\hat{\mathbf{x}}^e$. It turns out that the Fourier transform of $\hat{\mathbf{x}}^e$ has about the same resolution as that of $\mathbf{r}$ or $\mathbf{x}_0$.

16. An alternate formulation for Pisarenko's harmonic decomposition may be expressed as a solution to the following constraints minimization:

$$\mathbf{A}^T \mathbf{C}_x \mathbf{A} = \min,$$

$$\mathbf{A}^T \mathbf{A} = 1.$$

Show that the solution of the above is identical to Pisarenko's harmonic decomposition.

17. Show that the nonzero eigenvalues and the corresponding eigenvectors of a covariance matrix of a stochastic process consisting of a single real random sinusoid and white noise are given by

$$\text{Eigenvalues} : \begin{cases} \lambda_1 = \dfrac{\alpha^2 \Delta T}{4}\left(1 + \dfrac{\sin(\Delta T \omega_0)}{\Delta T \sin(\omega_0)}\right) + \sigma_\eta^2, \\ \lambda_2 = \dfrac{\alpha^2 \Delta T}{4}\left(1 - \dfrac{\sin(\Delta T \omega_0)}{\Delta T \sin(\omega_0)}\right) + \sigma_\eta^2, \end{cases}$$

$$\text{Eigenvectors} : \begin{cases} \mathbf{e}_1 = a\{\cos\left(\left(\frac{\Delta T - 1}{2} - k\right)\omega_0\right), \quad k = 0, 1, \ldots \Delta T - 1\}, \\ \mathbf{e}_2 = b\{\sin\left(\left(\frac{\Delta T - 1}{2} - k\right)\omega_0\right), \quad k = 0, 1, \ldots \Delta T - 1\}, \end{cases}$$

where a is amplitude and w_0 is frequency of the random sinusoid. σ_η^2 is noise power and ΔT is the size of the covariance matrix. The normalizing constants, a and b, are given by

$$a = \frac{1}{\sqrt{\dfrac{\Delta T}{2} + \dfrac{\sin \Delta T \omega_0}{2 \sin \omega_0}}},$$

$$b = \frac{1}{\sqrt{\dfrac{\Delta T}{2} - \dfrac{\sin \Delta T \omega_0}{2 \sin \omega_0}}}.$$

[Hint: One easy method is to compute the eigenvalues and the corresponding eigenvectors of a smaller matrix, $\mathbf{H}^H \mathbf{H} \mathbf{C}_0$. Following the result shown in (4.130) we can obtain the desired answer].

18. Consider an alternate measure of entropy which may also be looked upon as a measure of flatness,[67]

$$H(s) = \frac{1}{2\pi} \int_{-\pi}^{\pi} S(\omega) \log \frac{1}{S(\omega)} d\omega.$$

It is interesting to note that H(s) is equal to the entropy of random sinusoid whose frequency is a random variable with distribution function equal to $S(\omega)$ which is normalized to unit area (see Exercise 4.4). The covariance constriants are, as usual,

$$C_x(\tau) = \frac{1}{2\pi} \int_{-\pi}^{\pi} S(\omega) \exp(j\omega\tau) d\omega.$$

The unknown spectrum, $S(\omega)$, is now determined by maximizing the above entropy measure under the covariance constraints. Show that the unknown component has the following form:

$$S_0(\omega) = \exp\left(\sum_{k=-p+1}^{p-1} \lambda_k e^{-j\omega k} \right).$$

where λ_k are unknown coefficients to be determined from the covariance constraints.

References

1. J. Capon, High resolution frequency-wavenumber spectrum analysis, *Proc. IEEE*, 57, 1408-1418, 1969.

2. J. P. Burg, Maximum entropy spectrum analysis, Presented at the International Meeting of Exploration Geophysicists, Orlando, FL, 1967.

3. A. van den Bos, Alternative interpretation of maximum entropy spectral analysis, *IEEE Trans.*, IT-17, 493-494, 1971.

4. S. Ihara, Maximum entropy spectral analysis and ARMA processes, *IEEE Trans.*, IT-30, 377-380, 1984.

5. U. Grenander and G. Szego, *Toeplitz Forms and Their Applications*, University of California Press, Berkeley, 1958.

6. V. F. Pisarenko, The retrieval of harmonics from covariance function, *Geophys. J. R. Astron. Soc.,* 33, 347-366, 1973.

7. R. N. McDonough, Application of the maximum-likelihood method and the maximum-entropy method to array processing, in *Nonlinear Methods of Spectrum Analysis*, S. Haykin, Ed, Springer-Verlag, New York, 1979.

8. S. M. Kay, *Modern Spectrum Analysis*, Prentice-Hall, Englewood Cliffs, 24, 1989.

9. B. Musicus, Fast MLM power spectrum estimation from uniformly spaced correlation, *IEEE Trans.*, ASSP-33, 1333-1335, 1985.

10. R. E. Davies and L. A. Regier, Methods for estimating directional spectra from multi-element arrays, *J. Mar. Res.*, 35, 453-477, 1977.

11. R. T. Lacoss, Data adaptive spectral analysis methods, *Geophysics*, 36, 661-675, 1971.

12. J. Munier and G. R. Delisle, Spatial analysis in passive listening using adaptive techniques, *Proc. IEEE*, 75, 1458-1471,1987.

13. P. S. Naidu and V. V. Krishna, Improved maximum likelihood (IML) spectrum estimation: Performance analysis, *J. Inst. Electron Telecommun. Eng.,* 34, 383-390,1988.

14. M. A. Lagunas, M. E. Santamaria, A. Gasuli, and A. Moreno, Cross spectrum ML estimate, *IEEE Proc. ICASSP-85*, 77-80, 1985.

15. J Capon and N.R. Goodman, Probability distribution for estimators of the frequency-wavenumber spectrum, *Proc. IEEE*, 58, 1785-1786,1970

16. D. F. Gingras and S.L.Hobbs, Asymptotic statistics for a generalized frequency-wavenumber estimator, *IEEE Trans.*, ASSP-38, 804-813,1990.

17. J. G. Ables, Maximum entropy spectral analysis, *Astron. Astrophys. Suppl. Series*, 15, 383-393, 1974.

18. A. Papoulis, Maximum entropy and spectral estimation: A review, *IEEE Trans.*, ASSP-29, 1176-1186, 1981.

19. E. T. Jaynes, On the rationale of maximum-entropy methods, *Proc. IEEE*, 70, 939-952, 1982.

20. A. Papoulis, *Signal Analysis*, McGraw-Hill, New York, 1977.

21. J. A. Edward and M.M. Fitelson, Notes on maximum-entropy processing, IEEE, Trans., IT-19, 232-234, 1973.

22. J. P. Burg, The relationship between maximum entropy spectra and maximum likelihood spectra, *Geophysics*, 37, 375-376, 1972.

23. S. Kullback, *Information Theory and Statistics*, Dover Pub, New York,1969.

24. J. E. Shore, Minimum cross entropy spectral analysis, *IEEE Trans.*, ASSP-29, 230-237, 1981.

25. T. S. Durrani and A. S. Arslanian, Window associated with parametric spectral estimation, *IEEE Proc. ICASSP-82,* 1021-1024, 1982.

26. J. A. Cadzow, Spectral estimation: An over determined rational model equation approach, *Proc. IEEE, Special Issue on Spectral Estimation,* 70, 907-939, 1982.

27. B. Friedlander and B. Porat, A non-iterative method for ARMA spectrum estimation, IEEE Workshop, 331-335, 1983.

28. G. H. Golub and C. F. Van Loan, *Matrix Computations*, The John Hopkins University Press, Baltimore, Maryland, 1983.

29. R. Vaccaro, Ed., *SVD and Signal Processing. II. Algorithms, Analysis and Applications*, Elsevier, New York, 1991

30. J. A. Cadzow, B. Baseghi, and T. Hsu, Singular value decomposition approach to time series modelling, *Proc. IEE*, 130, F, 3, 202-210.

31. K. N. Berk, Consistent autoregressive spectral estimates, *Ann. Statist.*, .2, 489-502, 1974.

32. H. Sakai, Statistical Properties of AR Spectral Analysis, *IEEE Trans.*, ASSP-27, 402-409, 1979.

33. H. Sakai, Statistical analysis of Pisarenko's method for sinusoidal frequency estimation, *IEEE Trans.*, ASSP-32, 95-101,1984.

34. B. Friedlander and B. Porat, A general lower bound for parametric spectrum estimation, *IEEE Trans.*, ASSP-32, 728-733, 1984.

35. P. Whittle, The analysis of multiple stationary time series, *J. R. Stat. Soc.*, 15, 125-139, 1953.

36. R. O. Schmidt, Multiple emitter location and signal parameter estimation, in Proc. of RADC Spectral Estimation Workshop, Griffis AFB, New York, 243-258. Reprinted in *IEEE Trans.*, AP-34, 276-280, 1986.

37. G. Bienvenu and L. Kopp, Principe de la goniometri passive adaptive, Proc. Colloque, CRESTI, Nice (France), 106/1-10, May, 1979.

38. R. O. Schmidt, A signal subspace approach to multiple emitter location and spectral estimation, Ph. D. dissertation, Stanford University, 1981.

39. S. M. Kay and S. L. Marple, Jr., Spectrum analysis — A modern perspective, *Proc. IEEE*, 69, 1380-1419, 1981.

40. G. Bienvenu and L. Kopp, Adaptivity to background noise spatial coherence for high resolution passive methods, *IEEE Proc. ICASSP-79*, 306-309,1979.

41. J. H. Wilkinson, *Algebraic Eigenvalue Problem*, Oxford University Press, New York, 69 1965.

42. D. R. Brillinger, *Time Series: Data Analysis and Theory*, Holt, Rinehart and Winston, New York, 1975.

43. M. Kaveh and A. J. Barabell, The statistical performance of the MUSIC and the minimum norm algorithms in resolving plane waves in noise. *IEEE Trans.*, ASSP-34, 331-341, 1986.

44. H. Srinath and V. U. Reddy, Analysis of MUSIC algorithm with sensor gain and phase perturbations, *Signal Process*, 23, 245-256, 1991,

45. B.D. Rao and K. V. S. Hari, Effect of spatial smoothing on the performance of MUSIC and the minimum-norm method, *Proc. IEE,* 137, pt. F, 449-458, 1990.

46. J. A. Cadzow, Y. S. Kim, D. C. Shiue, Y. Sun, and G. Xu, Resolution of coherent signals using a linear array, *IEEE Proc. ICASSP, 37*.4.1-37.4.4, 1987.

47. S. S. Reddi, Multiple source location — A digital approach, *IEEE Trans.*, AES-15, 95-105, 1979.

48. Yu Hen Hu, Notes on eigenvalue distribution of Toeplitz matrices, *IEEE Proc. ICASSP*, 2480-2483, 1988.

49. A. Paulraj, R. H. Roy, and T. Kailath, Estimation of signal parameters via rotation invariance techniques-ESPRIT, in Proc of the 19th Asilomar Conference on Circuits, Systems and Computers, Asilomar, CA, 83-89, 1985.

50. R. Roy, T. Kailath, ESPRIT — estimation of signal parameters via rotational invariance techniques, *IEEE Trans.*, ASSP-37, 984-995, 1989.

51. A Swami and J. M. Mendel, Cumulant-based approach to the harmonic retrieval problem, *IEEE Proc. ICASSP-88*, 2264-2267, 1988.

52. A. Ferrari and G. Alengrin, Estimation of the frequencies of a complex sinuoidal noisy signal using fourth order statistics, *IEEE Proc. ICASSP-91*, 3457-3459, 1991.

53. J. W. Mauchley, Significance test for sphericity of a normal n-variate distribution, *Ann. Math. Stat.*, 11. 204-209, 1940.

54. W. S. Ligget, Jr., Passive sonar: fitting models to multiple time series, in Signal Processing: Proc. NATO ASI Signal Processing, J. W. R. Griffiths and P.L. Stocklin, Eds., Academic, New York, 327-345, 1973.

55. D. B. Williams and D. H. Johnson, Using the sphericity test for source detection with narrow band passive array, *IEEE Trans.*, ASSP-38, 2008-2014, 1990.

56. M. Wax and T. Kailath, Detection of signals by information theoretic criteria, *IEEE Trans.*, ASSP-33, 387-392, 1985.

57. Qi-Tu Zhang, Kon Max Wong, P. C. Yip, and J. P. Reilly, Statistical analysis of the performance of information theoretic criteria in the detection of the number of signals in array processing, *IEEE Trans.*, ASSP-39, 1557-1567,1989.

58. D. Shanks, Nonlinear transformation of divergent and slowly convergent sequences, *J. Math. Phys.*, 34, 1-42, 1955.

59. H. L. Gray and A. G. Houston, On G-spectral estimation, in *Applied Time Series*, Academic Press, New York, 39-138, 1978.

60. W. C. Pye and T. A. Atchison, An algorithm for the computation of the higher order G-transformation, *SIAM J. Numer. Anal.*, 10, 1-7, 1973.

61. R. W. Gerchberg, Super resolution through error energy reduction, *Opt. Acta*, 21, 709-720, 1974.

62. A. Papoulis, A new algorithm in spectral analysis and bandlimited signal extrapolation, *IEEE Trans.*, CAS-22, 735-742, 1975.

63. A. K. Jain and S. Ranganath, Extrapolation algorithms for discrete signals with application in spectral estimation, *IEEE Trans.*, ASSP-29, 830-845, 1981.

64. M. H. Hayes and R. W. Schafer, On the bandlimited extrapolation of discrete signals, *IEEE Proc. ICASSP-83,* 1450-1453, 1983.

65. R. D. Hippensteil, Spectral estimation of short duration signals using time domain extrapolation, *IEEE Proc. ICASSP-91*, 3133-3136, 1991.

66. M. A. Lagunas, A. Gasuli, An improved maximum likelihood method for power spectral density estimation, *IEEE Trans.*, ASSP-32, 170-173 1984.

67. C Nadeu, Maximum flatness spectral modeling, *IEEE Trans.*, ASSP-38, 2006-2008, 1990.

Chapter Five

Spectrum Estimation: *Data Adaptive Approach*

We shall explore different means of improving the spectral estimates by adapting to the true nature of data actually available for spectrum estimation. The role of spectral window has been documented in Section 2.6, where it was pointed out that the bias due to a window is minimum when the true spectrum is relatively flat. A simple filter whose transfer function is approximately equal to the inverse square root of the spectrum may be used for spectrum flattening or prewhitening; alternatively, an iterative procedure is suggested to devise an improved inverse filter which is appropriate to the available data, and thus it is data adaptive. The Burg spectrum is based on the reflection coefficients which are estimated directly from the data. The concept of forward and backward prediction error plays a significant role. The Burg spectrum, however, is not without problems. In particular, for a pure sinusoidal input with arbitrary phase, a peak in the Burg spectrum may be shifted and split into two or more closely spaced peaks. Several remedies have been suggested, of which we shall look into the popular method of least square estimation of all prediction coefficients. Singular value decomposition (SVD) plays an important role in the removal of noise in the matrix having a structure induced by the signal which can be expressed as a sum of sinusoids or damped sinusoids; for example, an impulse response function of a time invariant linear system can be expressed as a sum of damped sinusoids and background noise. The singular vectors of the data matrix are used to get an estimate of the frequencies and damping coefficients. The eigen decomposition of a covariance matrix computed from a finite data and updating the same as more data arrives at the sensors is the primary goal of the adaptive approach in subspace-based methods for frequency estimation.

5.1 Data Adaptive Approach

The central theme in this chapter is the estimation of spectrum directly from the data. We shall avoid the usual approach to estimate the covariance function first and then proceed to estimate the spectrum. Instead we fit a model (MA/AR/ARMA or sum of sinusoids plus noise) to the raw data in the least squares sense (or in the maximum likelihood sense) and then compute the spectrum. There are at least three reasons why we should take recourse to such a data-adaptive approach. In many real-time applications we cannot afford to wait till all the required data has reached the processor. After the required amount of data is collected, the processor would take some time to compute the spectrum of the stored data. Now, consider an alternate strategy wherein, after collecting enough data to achieve the desired resolution, we simply update the computed spectrum as new data arrives. This reduces the waiting time. As new data arrives it is not necessary to recompute the entire spectrum but only provide a correction. Such a scheme is indeed very attractive for real-time implementation. There is another angle to the above scheme of adaptive approach. It is possible to monitor the changes in the spectrum after the arrival of each data or a block of data. Under nonstationary environments such a facility is ideal for tracking some feature, for example, a time-varying spectral peak. Real-time spectrum estimation is outside the scope of this book.

We note that the statistical efficiency of parameter estimation, based on the observed data, is higher than that based on the reduced data, such as the estimated covariance function.[1-3] The statistical efficiency refers to the Cramer-Rao lower bound ([CRLB] which was introduced in Chapter Four, page 285) on the variance of the estimated parameter using the reduced data as against that using the raw data via the maximum likelihood method. The statistical efficiency is defined as a ratio of the determinant of Fisher's information matrix of the reduced data to that of raw data; it lies between zero and one. The reduced data is said to be a sufficient statistic whenever the statistical efficiency is one; for example, the estimated covariance coefficients for lags zero to p are asymptotically sufficient for the parameters of an AR(p) process, but this is no longer true when there is background white noise, because even a small amount of noise can lower the statistical efficiency; for example, it is shown in Reference 3 that for AR(2) process (poles at $0.95e^{\pm j0.25\pi}$) in white noise with SNR=20 dB, the statistical efficiency reduces to 0.84.

The real-life data is often far from the ideal, assumed in most algorithms. Apart from the usual background noise, there are other factors such as data outliers, missing data, gaps in the data, non-uniform spacing, etc. In the usual spectrum analysis literature not much attention is paid to the problems arising from the above-mentioned degradation of data, other than the background noise. In this section we plan to focus on some of these problems, in particular, their role in spectrum analysis. In this context the readers would benefit by referring to a recent book entitled *Randomized Signal Processing* by Bilinskis and Mikelsons,[4] which addresses processing of data which has been intentionally or accidentally randomly quantized or sampled.

A. Robust Filtering

Often, a real-life time series may contain a small fraction of unusual observations, generally caused by an abnormal experimental environment. Naturally, they do not belong to the same class to which the bulk of observations do. Such observations are called outliers in the statistical literature. A time series characterized by outliers would seem to possess a long and heavy tail. Assuming that the bulk of time series is derived from a Gaussian population, the probability of observing a heavy tail is quite small when the data is finite. Hence, one should eliminate the outliers which are known to seriously affect the spectral estimate as demonstrated in Reference 5. Elimination of outliers, also known as robustification, can be achieved through filtering and thresholding[6] of each bit of data as it arrives at the processor before it is integrated into the spectral estimation algorithm.

Let us model the observed time series as a pth-order AR process contaminated with outliers. It is assumed that an estimate of the AR coefficients based on the past data is available; also it is assumed that the past data was free of the outliers, or that it has been cleaned following a scheme to be described here. Let us predict the new arrival using the past p samples,

$$\hat{x}(t) = -\sum_{i=1}^{p} a_i x(t-i) \quad , \tag{5.1}$$

and then compute the difference between the new data and the predicted one; that is, $\hat{\varepsilon}(t) = x(t) - \hat{x}(t)$. The difference is now compared with the standard deviation of the noise process, σ_ε, driving the AR model. We shall assume that along with AR coefficients, the standard deviation of the driving noise process is also available. The new data is accepted if $|\hat{\varepsilon}(t)| < \gamma\sigma_\varepsilon$, where γ is a preassigned constant; otherwise it is rejected as an outlier and is replaced by the predicted value. An alternate strategy is to replace each value by a linear combination of the predicted and the observed. The coefficient of linear combination is a nonlinear function of the residual, $w(\frac{\hat{\varepsilon}(t)}{\hat{\sigma}_\varepsilon})$, where $\hat{\sigma}_\varepsilon$ is the standard deviation of the prediction errors, and w(.) is a nonlinear function to be defined later (see (5.4)):

$$\tilde{X}(t) = w(\frac{\hat{\varepsilon}(t)}{\sigma_\varepsilon})X(t) + (1 - w(\frac{\hat{\varepsilon}(t)}{\sigma_\varepsilon}))\hat{X}(t) \quad . \tag{5.2}$$

The effect of the above scheme, known as robust filtering, is to leave the data unmodified where the prediction error is small and to replace the data with its prediction at points where the error is large.[7] The AR coefficients were assumed to be known, but in practice one would need to compute these from the available but contaminated data. An iterative approach in which the data is subjected

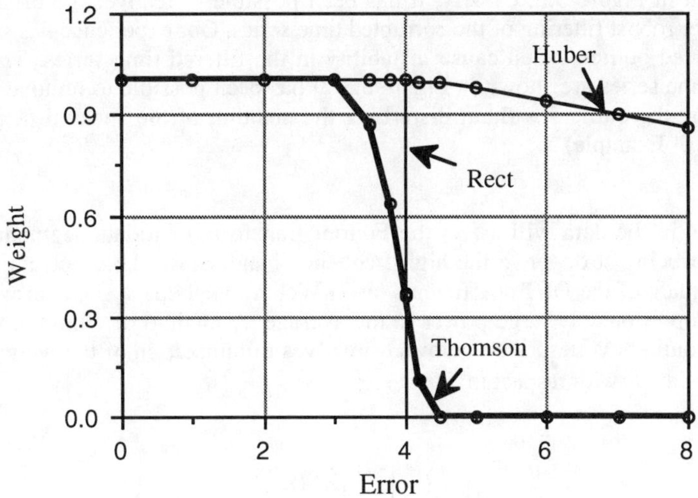

Figure 5.1: Commonly used weight functions for robust filtering (5.2). For Thomson's weight function, $\beta = 4.0$, and for Huber's weight function, $\gamma = 4.0$.

to repeated robust filtering and at the end of each filtering step the AR coefficients are refreshed was suggested by Thomson.[7]

There are different types of weighting coefficients; for example, the one introduced by Thomson[7] is

$$w(x) = \exp\left(-e^{\beta(|x|-\beta)}\right), \tag{5.3a}$$

where parameter β determines the scale at which downweighting begins. Thomson's weight coefficients are an improved version of Huber's weight coefficients,[8]

$$w(x) = 1 \qquad |x| \le \gamma$$

$$= \sqrt{\frac{2\gamma}{|x|} - \frac{\gamma^2}{x^2}} \qquad |x| > \gamma \quad . \tag{5.3b}$$

The simplest type of weight function is a rectangular function, where the moment the prediction error exceeds a certain preassigned number, it is concluded that an outlier is present and it needs to be replaced by the predicted value. All three types of the weight functions are illustrated in Figure 5.1.

Example 5.1: This example shows the effect of outliers on the spectrum of a synthetically generated AR(4) process which was introduced earlier in Example 2.9. A time series of length 1024 point was generated using a zero mean and unit variance pseudorandom Gaussian noise. One hundred outliers whose amplitudes are uniformly distributed in the range +13 to +20 and – 13 to – 20 were randomly introduced into the time series. To remove the outliers, robust filtering using

Thomson's weight coefficients (5.3a) was used. The question of optimum β in (5.3a) was settled by trial and error. The samples of pure time series, corrupted time series, and the filtered time series are shown in Figure 5.2. Clearly, it has been possible to recover the original time series shown in (a) after robust filtering of the corrupted time series. Our experience has shown, however, that closely spaced outliers often cause instability in the filtered time series. The spectra of all three types of time series are shown in Figure 5.3. It has been possible to remove the background clutter caused by the outliers without disturbing the position of the spectral peaks as shown in Figure 5.3 (End of Example).

The outliers in the data will affect the Fourier transform of a data segment in which they occur, the effect being stronger in the high-frequency band. Since the spectrum is estimated by averaging the square of the DFT coefficients (as in Welch's method), a segment with outliers will contribute disproportionately large power to the average. A method of eliminating the segments affected by the outliers is described below. It involves minimization of the weighted sum of the squares of the residuals with respect to $S_x(k)$,

$$\min \sum_{i=1}^{p} w^2\left(\frac{|X_i(k)|^2}{S_x(k)}\right)\left\{\frac{|X_i(k)|^2}{S_x(k)} - 2\log 2\right\}^2 , \tag{5.4}$$

where 2log2 is a median of a random variable, $\frac{|X_i(k)|^2}{S_x(k)}$, which has χ_2^2 distribution with two degrees of freedom for $k \neq 0$ or $\frac{\Delta T}{2} + 1$, where ΔT is the length of each segment. In Thomson's weight function (5.3a), the parameter b, equal to the Nth quintile of the χ_2^2 distribution, that is, b=2log2N, gives good results.[8] We minimize (5.4) at a fixed frequency with respect to $S_x(k)$. To start the minimization process we need an initial estimate of $S_x(k)$, for which the sample median is a good candidate. Using the sample median the weighting coefficients are first calculated and then used for the minimization of (5.4). The newly estimated $S_x(k)$ is used in the second step for computation of the weighting coefficients, which are then used to get a fresh estimate of $S_x(k)$. This iterative procedure is continued until there is no substantial change in the estimate of the spectrum.

B. Incomplete Time Series

There are occasions when the time series is incomplete, either because data is missing or there is a gross error in the observations. The time series with known outliers can be considered as incomplete time series. The missing data points may be filled with the interpolated ones when the interpolation is carried out under an additional assumed property; namely, the time series is a band-limited process. Linear prediction method[9] and Lagrange interpolation of oversampled time series[10] were suggested for filling in the missing data points. In the linear prediction method the forward and backward errors are computed by running the prediction filter over the data such that the known bad data points are excluded. The rest of the procedure is the same as in the Burg spectrum described in Section 5.3. The Gerchberg-Papoulis algorithm[11] for extrapolation of a band-limited process makes use of FFT and is found to be fast and yields satisfactory results, provided the background (white) noise power is moderate. An outline of the algorithm is given below.

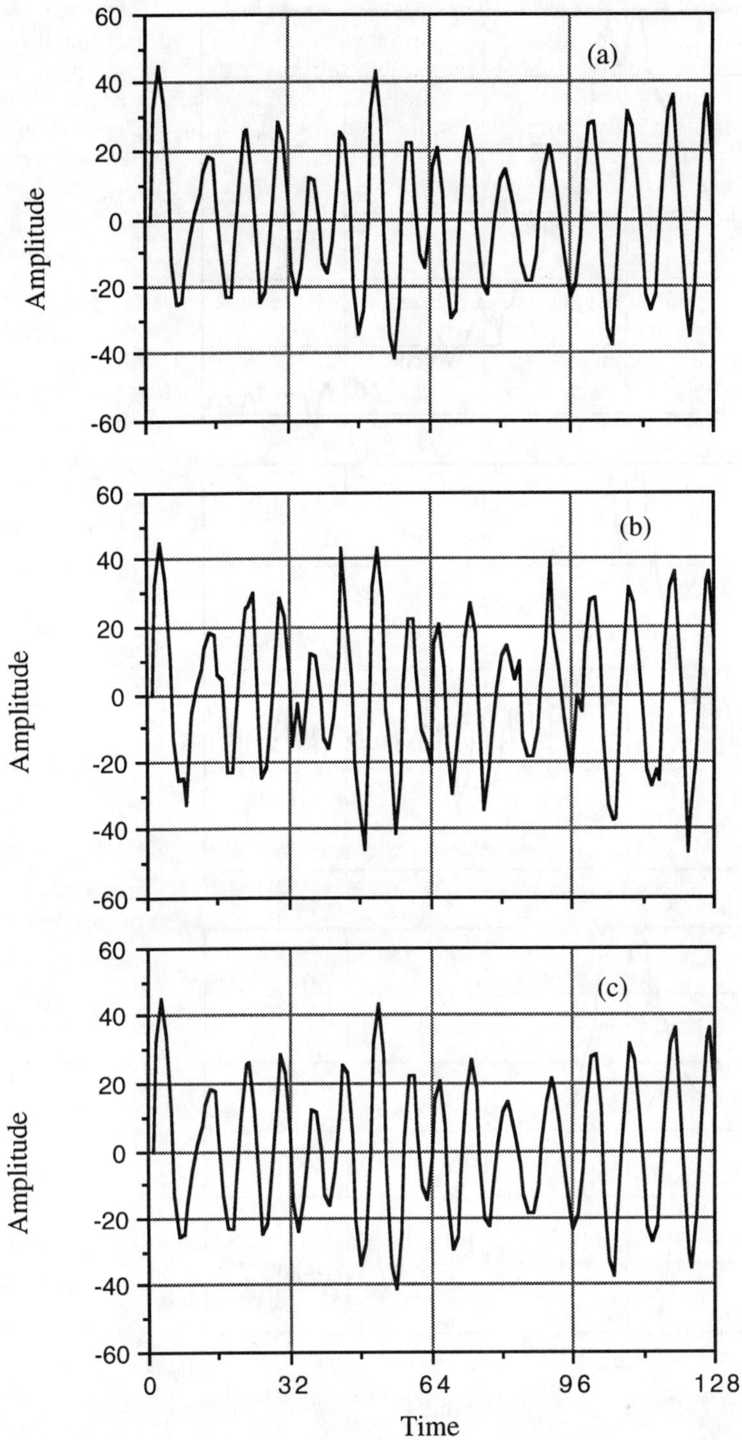

Figure 5.2: (a) Error-free time series (AR(4) process), (b) corrupted with outliers in the range 13 to 20, and (c) filtered time series. Thomson's weight function (5.3a) with b=11.0 was used.

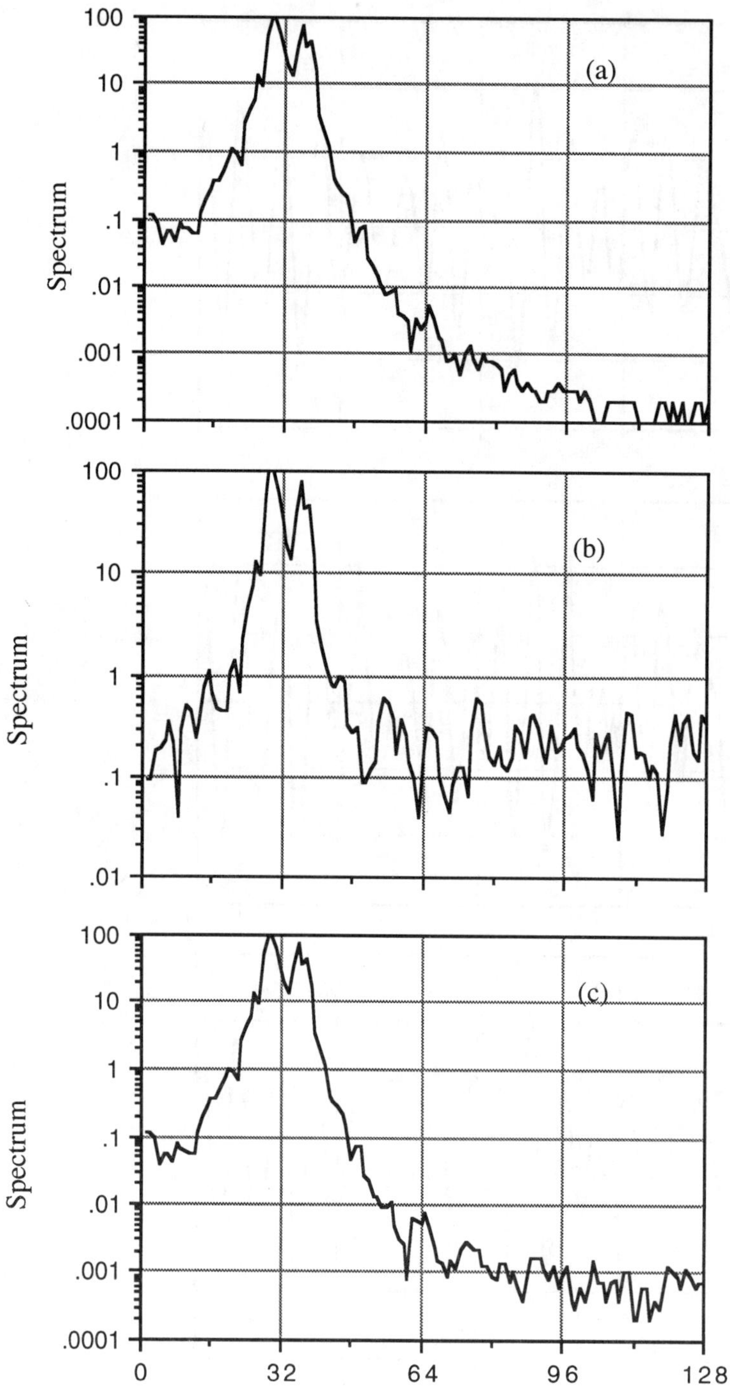

Figure 5.3: Spectrum of (a) pure time series, (b) corrupted time series, and (c) filtered time series.

Consider a time series, x(t), with Fourier transform, $X(\omega)$, such that $X(\omega) = 0 \quad |\omega| > \omega_0$. Define a finite sequence, D(t), t=0, 1, ... , T − 1, such that

$$D(t) = 0, \quad \text{at} \quad t = t_i \le T - 1, \quad i = 1, 2, \dots, r,$$
$$= 1 \quad \text{otherwise} \quad ,$$

and form an incomplete time series as $x_0(t) = D(t)x(t)$. The Gerchberg-Papoulis algorithm has the following steps:

(1) Compute DFT of $x_0(t)$.

(2) Band limit the DFT of $x_0(t)$ to 0 to k_0 and $T - 1 - k_0$ to $T - 1$,

where $\omega_0 = \frac{2\pi k_0}{T}$. Let $X_b(k)$ represent the band-limited DFT of $x_0(t)$.

(3) Compute the inverse DFT of $X_b(k)$.

(4) Form a new time series combining $x_b(t)$ and $x_0(t)$, where $x_b(t)$ is the inverse DFT of $X_b(k)$,

$$x_1(t) = (1 - D(t))x_b(t) + x_0(t) \, .$$

$x_1(t)$ is the first estimate of the interpolated time series. To get the next improved estimate we go back to step one with $x_1(t)$ as input.The iterative process is continued until the relative mean square difference is less than a prescribed number,

$$\sum_t \left| x_n(t) - x_{n-1}(t) \right|^2 \le \delta .$$

Example 5.2: Consider an ideal example of a band-limited process consisting of two sinusoids,[12]

$$x(t) = 1.25 \cos\left(20\pi t + \frac{\pi}{2}\right) + 1.5 \cos\left(24\pi t + \frac{\pi}{3}\right) .$$

Twenty-six samples of the signal were randomly picked up over a duration of one second by first sampling the signal uniformly at a rate of 256 samples/sec followed by random picking. Large gaps of missing data were filled by the iterative method described above. In Figure 5.4 the reconstructed waveform after 30 iterations along with the known data points are shown. To the above signal a small amount of white noise was added (SNR=22.8 dB), and the reconstruction algorithm was applied. The mean square error as a function of number of iterations is shown in Figure 5.5 where we also show the convergence in the case of an error-free signal. It is noticed that the mean square error flattens out after a sufficiently large number of iterations, and the minimum error that can be attained depends upon the background noise.

C. Unevenly Sampled Data

There are instances when the sampling is necessarily at uneven time intervals as the observer cannot completely control the time of observation as is common in observational sciences like astronomy, geophysics, etc. This situation is different from the one discussed in the previous section where there were occasional missing or bad data points in otherwise evenly sampled time

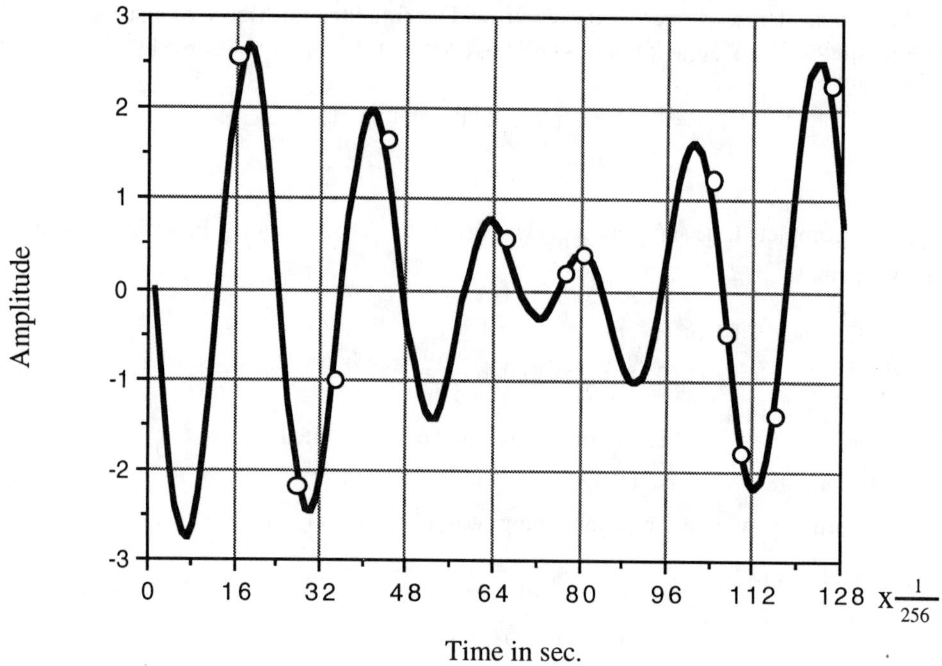

Figure 5.4: Interpolated time series after 30 iterations. The actual data points are also shown in the figure by means of empty circles. (Taken from Reference 12. © 1984 IEEE)

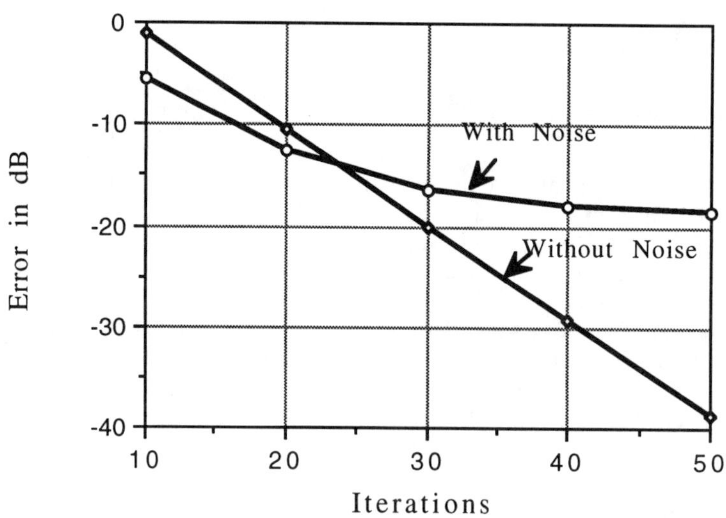

Figure 5.5: Mean square interpolation error as a function of the number of iterations. In the presence of noise the interpolation error appears to get saturated; that is, the error does not decrease any further even if we increase the number of iterations.(Taken from Reference 12. © 1984 IEEE)

series. One benefit derived from the uneven or random sampling is absence of aliasing; indeed, if the sampling were to be a completely random, Poisson process, it is known that the spectrum would be completely alias free.[4]

Let a continuous signal x(t) be sampled at time instants,

$$t = t_i, \quad i = 0, 1,3, \ldots, T-1.$$

We would like to look for a harmonic component,

$$h(t) = a\cos(\omega(t-\tau)) + b\sin(\omega(t-\tau)),$$

of frequency ω, where a, b, and τ are constants. The delay parameter τ enables us to select any arbitrary origin of time. To estimate a and b we minimize the mean square difference,

$$\sum_{i=0}^{T-1}\{x(t_i) - a\cos(\omega(t_i-\tau)) - b\sin(\omega(t_i-\tau))\}^2$$

with respect to the unknowns. We obtain the following normal equations:

$$\sum_{i=0}^{T-1} x(t_i)\cos(\omega(t_i-\tau)) = a\sum_{i=0}^{T-1}\cos^2(\omega(t_i-\tau)) + b\sum_{i=0}^{T-1}\cos(\omega(t_i-\tau))\sin\omega((t_i-\tau))$$

$$\sum_{i=0}^{T-1} x(t_i)\sin(\omega(t_i-\tau)) = a\sum_{i=0}^{T-1}\cos(\omega(t_i-\tau))\sin(\omega(t_i-\tau)) + b\sum_{i=0}^{T-1}\sin^2(\omega(t_i-\tau)).$$

Since τ is a free parameter, we shall select it so as to simplify the normal equations; that is,

$$\sum_{i=0}^{T-1}\cos(\omega(t_i-\tau))\sin(\omega(t_i-\tau)) = 0$$

or

$$\sum_{i=0}^{T-1}\sin(2\omega(t_i-\tau)) = 0$$

$$\sum_{i=0}^{T-1}\sin(2\omega t_i)\cos(2\omega\tau) - \cos(2\omega t_i)\sin(2\omega\tau) = 0,$$

which yields a simple equation in τ,

$$\tan(2\omega\tau) = \frac{\sum_{i=0}^{T-1}\sin(2\omega t_i)}{\sum_{i=0}^{T-1}\cos(2\omega t_i)}. \tag{5.5}$$

It is easy to show that if the zero time is shifted by an amount, δ, the delay parameter is also shifted by the same amount; that is,

$$\tan(2\omega(\tau+\delta)) = \frac{\sum_{i=0}^{T-1} \sin(2\omega(t_i+\delta))}{\sum_{i=0}^{T-1} \cos(2\omega(t_i+\delta))} .$$

The resulting simplified normal equations are easily solved:

$$a = \frac{\sum_{i=0}^{T-1} x(t_i)\cos(\omega(t_i-\tau))}{\sum_{i=0}^{T-1} \cos^2(\omega(t_i-\tau))}$$

$$b = \frac{\sum_{i=0}^{T-1} x(t_i)\sin(\omega(t_i-\tau))}{\sum_{i=0}^{T-1} \sin^2(\omega(t_i-\tau))} . \tag{5.6}$$

The power in the harmonic component at frequency ω is given by

$$\begin{aligned} P(\omega) &= \frac{1}{2}\sum_{i=0}^{T-1} h^2(t_i) \\ &= \frac{a^2}{2}\sum_{i=0}^{T-1} \cos^2(\omega(t_i-\tau)) + \frac{b^2}{2}\sum_{i=0}^{T-1} \sin^2(\omega(t_i-\tau)) . \end{aligned} \tag{5.7}$$

The factor $\frac{1}{2}$ is included to account for the fact that power is equally distributed between positive and negative frequencies. Substituting for a and b from (5.6), we obtain the power as a function of ω:

$$P(\omega) = \frac{\frac{1}{2}\left(\sum_{i=0}^{T-1} x(t_i)\cos(\omega(t_i-\tau))\right)^2}{\sum_{i=0}^{T-1} \cos^2(\omega(t_i-\tau))} + \frac{\frac{1}{2}\left(\sum_{i=0}^{T-1} x(t_i)\sin(\omega(t_i-\tau))\right)^2}{\sum_{i=0}^{T-1} \sin^2(\omega(t_i-\tau))} . \tag{5.8}$$

The above estimate of the spectrum from unevenly sampled data was first given by Lomb[13] and implemented by Press,[14] requiring only $\gamma T \log_2 T$ (where $\gamma \approx 100$) arithmetic operations. The straightforward implementation of (5.8) would require γT^2 operations.

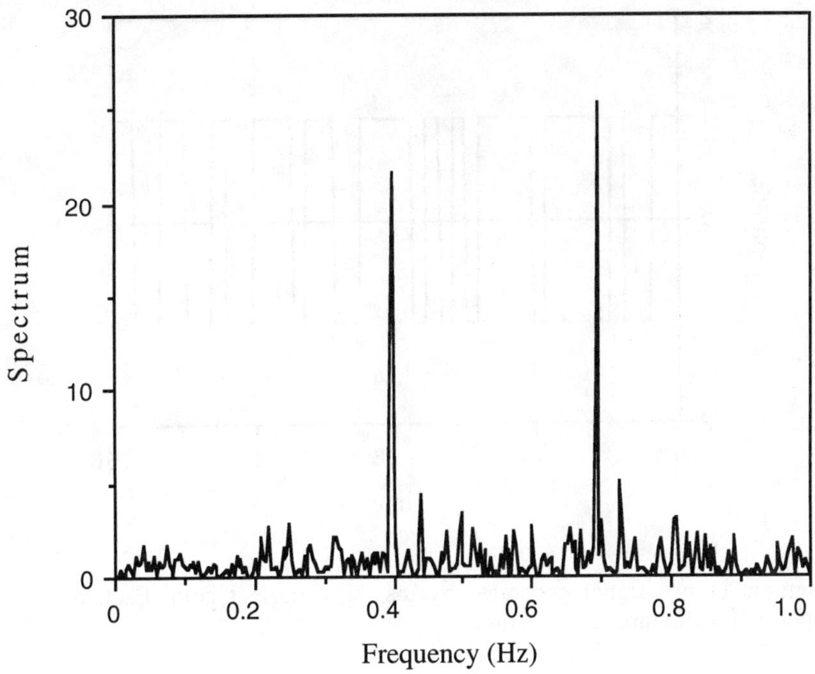

Figure 5.6: Spectrum of unevenly spaced time series obtained by sampling a sum of two sinusoids at random intervals.

Example 5.3: In this example we shall apply the above method for estimation of spectrum of a time series consisting of two equal amplitude sinusoids of frequency

$$\omega_1 = 0.8\pi \text{ and } \omega_2 = 1.4\pi$$

plus white noise sampled at random intervals having a uniform distribution in the range 0 to 2. The mean sample interval is one. Unlike in the periodically sampled time series, the spectrum is *not* periodic *nor* is there any aliasing phenomenon as demonstrated in Figure 5.6. There is a slight error in the spectral peak position (End of Example).

D. Choice of Data Models

Perhaps the most difficult question in modern spectrum analysis is, *how does one determine the right model of a time series whose finite duration sample is available as the observed data?* As of now, we do not have any clear-cut answer to this question; we have to depend upon the physics of the problem to surmise on the nature of signal and noise sources. In the absence of such *a priori* knowledge the safe passage is to use the low resolution methods which require very minimal *a priori* knowledge. However, we have an example to show how incorrect modeling can lead to disastrous results. Consider a binary random signal shown in Figure 5.7. We have used the Burg spectral estimation method where the assumed data model is an AR model. The result of application of the Burg method is shown in Figure 5.8a and 5.8b for two different model orders. The theoretical spectrum is shown in Figure 5.8c where we have also shown the Welch spectrum, where the assumed data model is simply a stationary stochastic process.

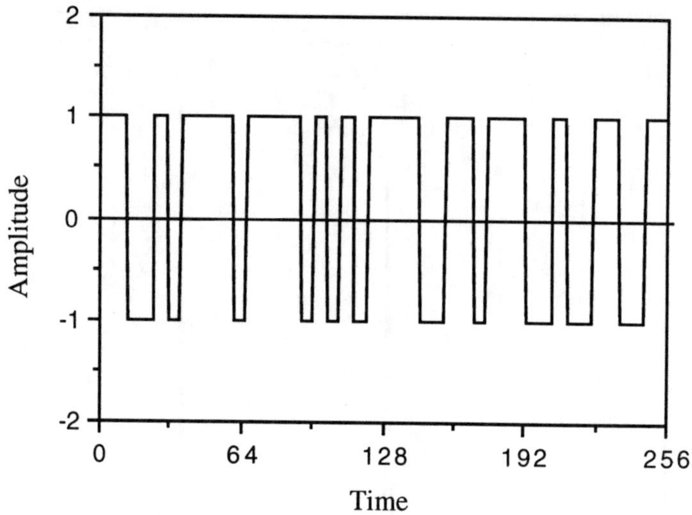

Figure 5.7: A random binary signal generated by tossing a perfect coin. Each outcome is represented by a pulse of width three units of time.

E. **Bootstrap Methods**

In Chapter Three we noticed many instances of estimation error, that is, errors caused by finite data in mean, variance, covariance, spectrum, cross-spectrum, etc., where the error depends upon the quantity being estimated or other unknown quantities. Naturally, these error estimates are of limited practical utility. As an example consider (3.3) for variance of the estimated mean. For white noise case let us recall the result (3.3b),

$$\mathrm{var}\{\hat{\mu}_x\} = \frac{\mathrm{Var}_x}{T}.$$

Evidently the error in $\hat{\mu}_x$ cannot be found unless Var_x is known. An ingenious (known as "jackknife") method was introduced by John Tukey in 1958[15] to overcome the above impasse. Tukey's method avoided formula (3.3b) entirely. Instead, the method estimates a series of $\hat{\mu}_i$, one from each sequence generated by deleting one data sample. Let

$$x^i = \{x(0), x(1), \ldots x(i-1), x(i+1), \ldots, x(T-1)\},$$

where the ith sample, $x(i)$, has been deleted and let the corresponding estimate of the mean be $\hat{\mu}_i$. It is assumed that all data samples are independent and identically distributed (iid). The jackknife estimate of error is given by

$$\mathrm{var}\{\hat{\mu}_x\}_{jack} = \frac{T-1}{T}\sum_{i=1}^{T}(\hat{\mu}_i - \hat{\mu}_x)^2 \qquad (5.9)$$

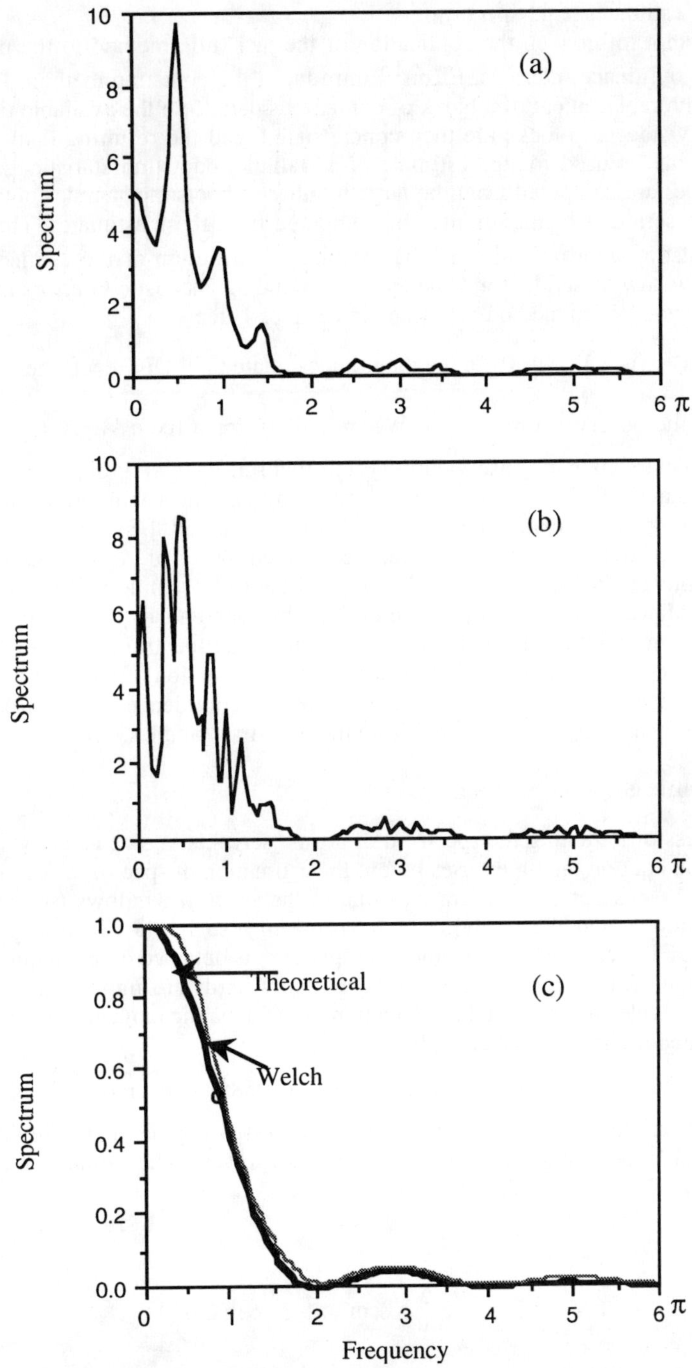

Figure 5.8: Spectrum of random binary signal shown in Figure 5.7, which was sampled at the rate 6 samples/unit time. The Burg spectrums for two different orders are shown above: (a) order=20, (b) order=80. The theoretical spectrum and Welch spectrum are shown in (c).

The peculiarity of Tukey's jackknife is that it automatically produces an error estimate for even other types of estimates, e.g., spectrum.

In an attempt to broaden the application of the jackknife method to the problem of finding approximate confidence intervals, Efron[15] introduced the bootstrap method. It involves random resampling with replacement of a block of contiguous data from the available time series. Several such randomly selected blocks are then concatenated, and the resulting data sample, known as bootstrap sample, is used for the estimate of a statistic, bootstrap statistic. A large number of bootstrap values are computed from the large number of bootstrap samples obtained by sampling the given time series. A histogram may be computed from these estimates. This histogram is the proposed bootstrap approximation of the sampling distribution of the estimation errors.[16] We shall now show how these ideas may be applied to finding a statistic of errors in the estimation of spectrum using the Welch method, described in Chapter Three.

Let $|X_i(k)|^2$, $i = 1, \ldots, P$ be the magnitude squared k^{th} DFT coefficients from P segments derived from the observed time series. We may treat, for a fixed k, $|X_i(k)|^2$, $i = 1, \ldots, P$ as weakly dependent stationary observations. From this set, $P - q + 1$ groups, each of q contiguous magnitude squared DFT coefficients, are formed by sliding a window of width q. From this collection, r groups are picked randomly with replacement and then concatenated to form a long sequence called a bootstrap sample. This has been called blocks of blocks resampling in Reference 17. For example, let P=16, q=4, and r=4; a typical bootsrap sample will appear as {2345, 6789, 2345, 7891}. Many bootstrap samples are derived by random sampling from the available DFT coefficients. From each bootstrap sample an estimate of the spectrum is made. All these estimates of the spectrum constitute a population drawn from a distribution of which the Welch spectrum is also a member. It is now easy to compute variance and confidence bounds from the distribution function of the spectral estimates derived from the bootstrap samples.

5.2 **Prewhitening**

The process of flattening the spectrum of a time series is known as prewhitening, a filtering operation carried out on the time series before the estimation of spectrum. A flat spectrum, as that of a white noise, ensures a minimum bias due to the spectral windows (see page 203 for more details on windows and their role). A flat spectrum also implies lower dynamic range and, therefore, the DFT coefficients from which the spectrum is obtained do not require too large a word for their storage. This is a definite advantage in any hardware implementation of a spectrum analyzer. There is also a corresponding reduction in the dynamic range of the whitened time series. This is illustrated in the example given below.

Example 5.4: Consider a simple second-order AR process with poles at $0.95e^{\pm j30^0}$. The AR filter is driven by a zero mean and unit variance white noise process. The variance of the output is given by (see Example 2.6 on page 97)

$$\frac{\sigma_\varepsilon^2}{1 - 2r_0^2\cos(\omega_0) + r_0^4\cos(2\omega_0)},$$

which for the above example reduces to 18.6915. For this time series the prewhitening filter is simply the inverse of the AR filter; namely,

$$H^{Pre}(z^{-1}) = 1 - 0.95\,z^{-1} + 0.9025\,z^{-2}.$$

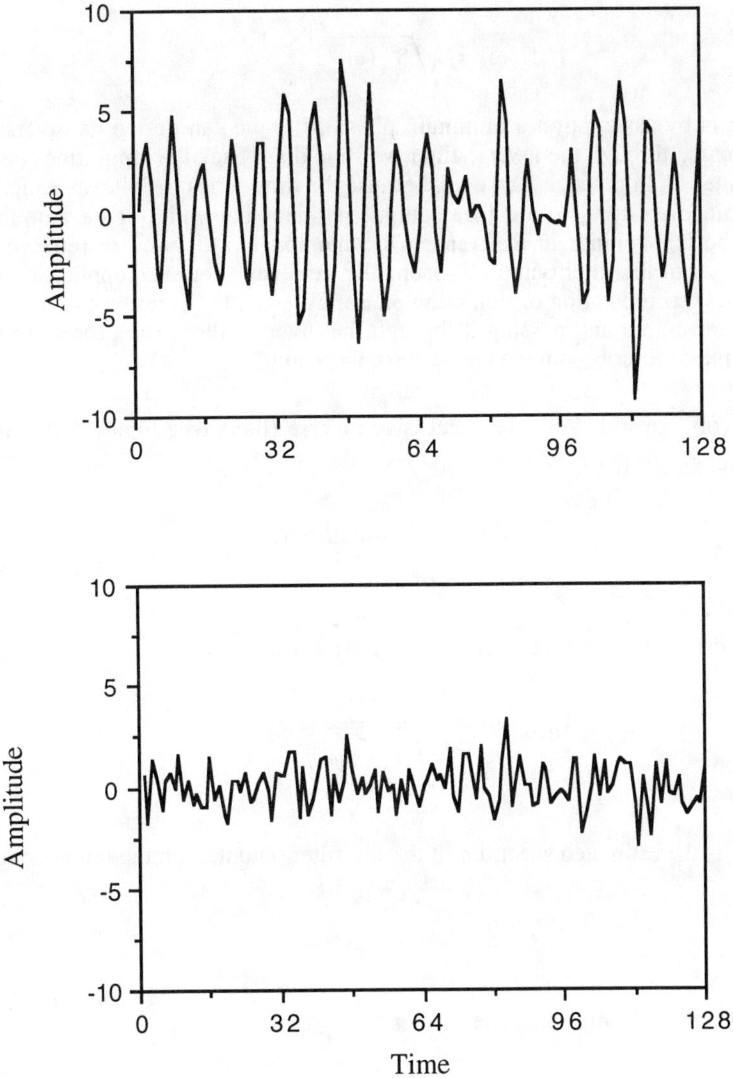

Figure 5.9: A sample of AR(2) process (Top) and the input white noise of zero mean and unit variance (Bottom).

The example of the above time series and its prewhitened version are shown in Figure 5.9 with the same vertical axis. From the figure it is evident that the dynamic range of the prewhitened time series has been dramatically reduced. In the spectral domain the dynamic range of the spectrum of the original time series is also reduced from 20 to 0 dB after prewhitening (End of Example).

A. Iterative Reduction of Bias

In Section (2.2) we introduced a filter model for a stationary time series where the time series was modeled as an output of a minimum phase filter driven by white noise. The filter impulse response can be determined from the spectrum alone by first computing its corresponding minimum phase (see (2.13)) and then computing the inverse Fourier transform of the transfer function,

$$H(\omega) = \sqrt{S_x(\omega)}\, e^{j\phi_{min}(\omega)} .$$

Since the filter is by assumption a minimum phase filter we can define its inverse. If the time series is now passed through the inverse filter we should get back the input time series which is a white noise sequence. In practice, however, because the time series is finite, an exact inverse filter cannot be obtained. Naturally, the filter output will not be white; some amount of colored character will be left behind. In Reference 18 a method is a devised to retrieve this leftover nonwhite character of the filter output. It essentially involves a repeated application of a two-step procedure; namely, computation of minimum phase inverse filter from the estimated spectrum of the previous filter output and passing it through the inverse filter. The procedure is terminated whenever the inverse filter becomes a constant (or close to it).

Let $H^n_{INV}(\omega)$, $n = 1, 2, \ldots$ be successive inverse filters (see Figure 5.10). It is shown in Reference 18 that for a fixed ω

$$H^n_{INV}(\omega) \underset{n \to \infty}{\to} \text{constant} .$$

In the above algorithm the nth inverse filter is defined as follows:

$$H^n_{INV}(\omega) = \frac{1}{\sqrt{\hat{S}_{\varepsilon_{n-1}}(\omega)}}\, e^{j\phi_{min}(\omega)} .$$

where $\hat{S}_{\varepsilon_n}(\omega)$ is the estimated spectrum of the nth filter output, with the initial condition,

$$\hat{S}_{\varepsilon_0}(\omega) = \hat{S}_x(\omega).$$

We now introduce a composite inverse filter,

$$\tilde{H}^n_{INV}(\omega) = \prod_{i=1}^{n} H^i_{INV}(\omega). \tag{5.10}$$

On account of the reported property of H^n_{INV} it is reasonable to assume that the composite filter will converge to a finite value for every ω as $n \to \infty$. The composite filter for large n will render the given time series totally white (zero mean and unit variance) when it is passed through the inverse filter. But for finite n the filter output may not be totally white; then its spectrum can be expressed as follows:

$$\begin{aligned} E\{\hat{S}_{\varepsilon_n}(\omega) \big| \tilde{H}^n_{INV}(\omega)\} &= [\left|\tilde{H}^n_{INV}(\omega)\right|^2 S_X(\omega)] * S_w(\omega) \\ &\approx [\left|\tilde{H}^n_{INV}(\omega)\right|^2 S_X(\omega)], \end{aligned} \tag{5.11}$$

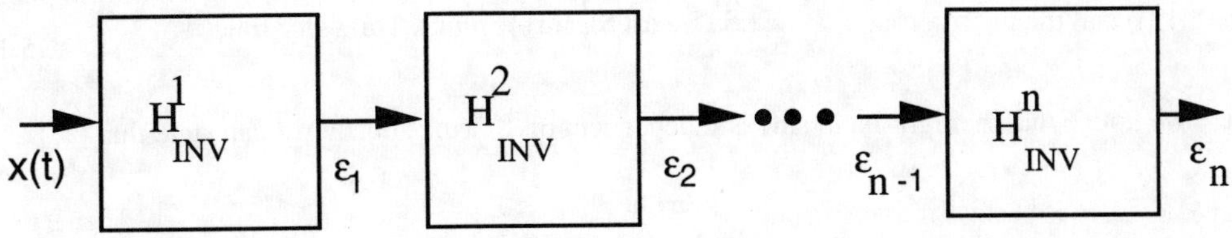

$$H^n_{INV}(\omega) = \prod_{i=1}^{n} H^i_{INV}(\omega)$$

Figure 5.10: Iterative inverse filtering to whiten a time series. Each whitening filter is defined in terms of the spectrum of the output of the previous stage.

where we have assumed that the spectrum of the filter output is at least uniform over the bandwidth of the window spectrum.

Next, we define the nth-order spectrum,

$$\hat{S}^n_X(\omega) = \frac{1}{\left|\tilde{H}^n_{INV}(\omega)\right|^2}, \tag{5.12}$$

with the initial value, $\hat{S}^0_X(\omega) = 1$. From (5.12) we can get a recursive relation between the n+1st and the nth-order spectrum,

$$\hat{S}^{n+1}_X(\omega) = \frac{\hat{S}^n_X(\omega)}{\left|H^{n+1}_{INV}(\omega)\right|^2},$$

or

$$\hat{S}^n_\varepsilon(\omega) = \frac{\hat{S}^{n+1}_X(\omega)}{\hat{S}^n_X(\omega)}. \qquad 5.13$$

Using (5.12) in (5.11) we obtain

$$\hat{S}^n_x(\omega)E\{\hat{S}_{\varepsilon_n}(\omega)\big|S^n_x(\omega)\} \approx S_x(\omega). \qquad 5.14$$

Next, using (5.13) in (5.14) we obtain

$$E\{\hat{S}_x^{n+1}(\omega)|S_x^n(\omega)\} \approx S_x(\omega), \tag{5.15}$$

where we notice that the right-hand side is independent of $S_x^n(\omega)$; hence we can state that

$$E\{\hat{S}_x^{n+1}(\omega)\} \approx S_x(\omega). \tag{5.16}$$

The approximation in (5.16) will improve as n tends to infinity. Thus, $\hat{S}_x^n(\omega)$ $n \to \infty$ becomes an unbiased estimate of the actual spectrum. The variance of $\hat{S}_x^n(\omega)$ has been shown[18] to be of the same order as that of the BT spectrum estimate (see (3.34)).

Example 5.5: This example demonstrates the effectiveness of the iterative bias reduction. An AR(4) process (from Example 2.11 with poles at $0.9802\, e^{\pm j39.58^0}$ and $0.9805\, e^{\pm j50.41^0}$) of length 256 points was used for computation. An autocorrelation function of the process up to a lag of 64 was computed. The successive inverse filters were of length 64, and their frequency response beyond $\frac{15}{32}\pi$ was set to zero so that the high frequency noise is attenuated. The raw spectrum (BT spectrum) is shown in Figure 5.11(a). The spectral peaks are resolved, but they are shifted. The next two figures represent the 3rd and 6th iterations where there is considerable improvement in the peak position. The correct peak positions are shown by long arrows in (c). Through iterative bias reduction it has been possible to correctly locate the spectrun peaks. Further examples of iterative bias reduction are given in Reference 18 (End of Example).

5.3 **Burg Spectrum**

In solving for AR coefficients using the Durbin-Levinson procedure, the only instance of use of autocorrelation function is in the computation of so-called reflection coefficients (4.88). Burg[19] suggested the use of raw data for estimation of the reflection coefficients by minimizing the prediction error. He defined two types of prediction errors, namely, forward and backward prediction errors, and minimized a linear combination of the two types of prediction errors. It may be pointed out that the maximum entropy spectrum (see Section 4.3) invented by Burg is quite different from what is known as Burg spectrum.

A. **Forward and Backward Prediction Errors**

Let $\mathbf{x}_t^{-p}$ stand for the data vector using the past p data points,

$$\mathbf{x}_t^{-p} = \text{col}\,\{x(t)\;\; x(t-1)\;\; x(t-2)\;\; \ldots\;\; x(t-p)\},$$

and $\mathbf{x}_t^{+p}$ for the data vector using future data,

$$\mathbf{x}_t^{+p} = \text{col}\,\{x(t)\;\; x(t+1)\;\; x(t+2)\;\; \ldots\;\; x(t+p)\}.$$

Now we define the forward and backward prediction errors as follows:

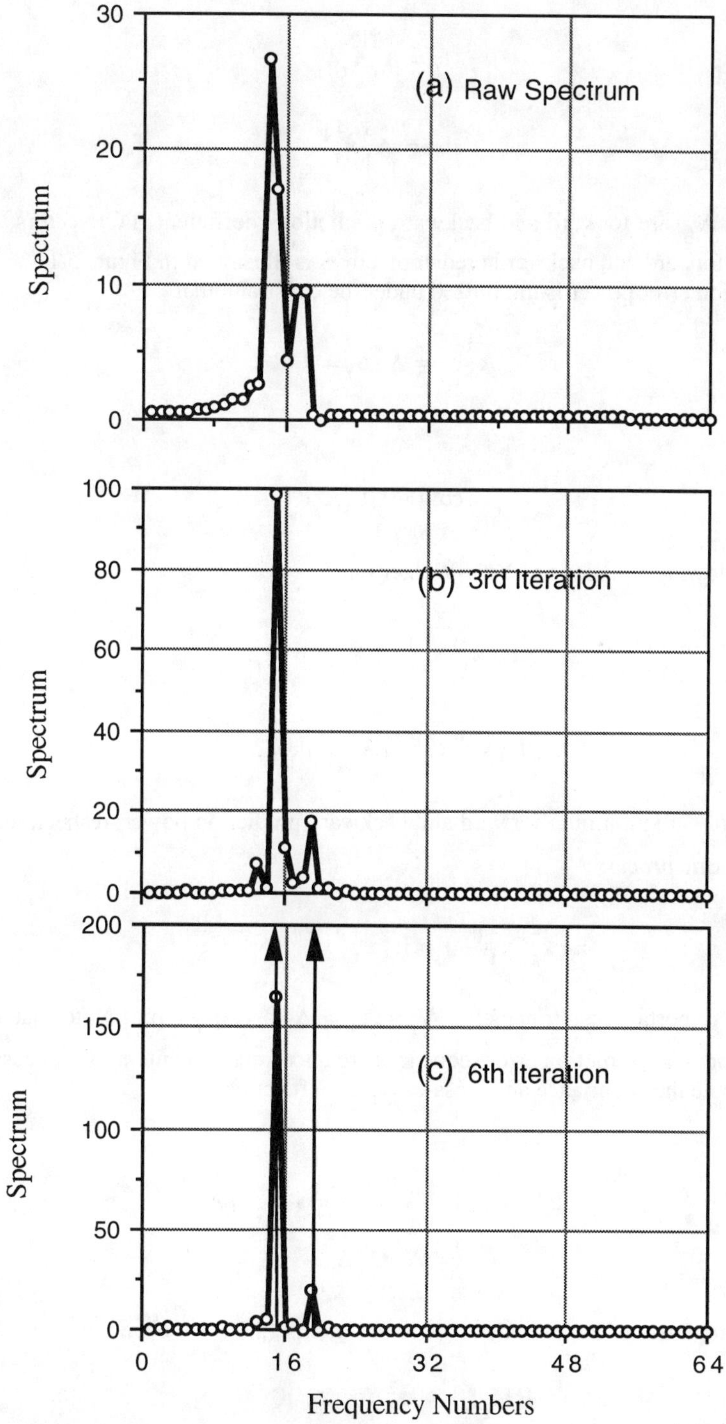

Figure 5.11: An example of iterative bias reduction. A 256-point AR(4) process with poles at $0.9802\, e^{\pm j39.58^0}$ and $.9805\, e^{\pm j50.41^0}$ is used to demonstrate the effectiveness of iterative bias reduction. (a) BT spectrum of the time series, (b) after 3rd iteration, and (c) after 6th iteration. Arrows represent true positions $0.9802\, e^{\pm j39.58^0}$ and $0.9805\, e^{\pm j50.41^0}$ of spectral peaks.

Forward: $$f_t = \mathbf{A}_f^T \mathbf{x}_t^{-p}, \tag{5.17a}$$

Backward: $$b_t = \mathbf{A}_b^T \mathbf{x}_t^{+p}, \tag{5.17b}$$

where $\mathbf{A}_f$ and $\mathbf{A}_b$ are forward and backward prediction coefficient (PC) vectors, respectively. Computation of forward and backward prediction errors is illustrated in Figure 5.12.

The prediction error power is minimized under the constraint that

$$\mathbf{A}_b^T \delta_t = \mathbf{A}_f^T \delta_t = 1 ,$$

where

$$\delta_t = \text{col}\,\{1\ 0\ 0\ \ldots\ 0\}.$$

This leads to following equations of Yule-Walker type:

$$E\{\mathbf{x}_t^{-p} \mathbf{x}_t^{-p^T}\} \mathbf{A}_f = p_f \delta_t,$$

$$E\{\mathbf{x}_t^{+p} \mathbf{x}_t^{+p^T}\} \mathbf{A}_b = p_b \delta_t, \tag{5.18}$$

where p_f and p_b are minimum forward and backward prediction power, respectively. For a *real stationary stochastic process* ,

$$E\{\mathbf{x}_t^{-p} \mathbf{x}_t^{-p^T}\} = E\{\mathbf{x}_t^{+p} \mathbf{x}_t^{+p^T}\} = \mathbf{C}_x .$$

This leads to an important result; namely, $\mathbf{A}_f = \mathbf{A}_b = \mathbf{A}$, and $p_f = p_b$. Note that $\mathbf{A}$ is the AR coefficient vector. The prediction error sequences are uncorrelated white noise processes. To show this, let us compute the covariance and cross-covariance functions:

$$\begin{aligned} E\{f_t f_{t+\tau}\} &= \mathbf{A}^T E\{\mathbf{x}_t^{-p} \mathbf{x}_{t+\tau}^{-p^T}\} \mathbf{A} \\ &= \mathbf{A}^T \mathbf{C}_x^e \mathbf{A}.. \end{aligned} \tag{5.19}$$

Following the properties of the extended covariance matrix (2.64) we obtain

$$E\{f_t f_t\} = \sigma_\varepsilon^2 \text{ for } \tau = 0$$

and

$$E\{f_t f_{t+\tau}\} = 0 \text{ for } \tau > 0 .$$

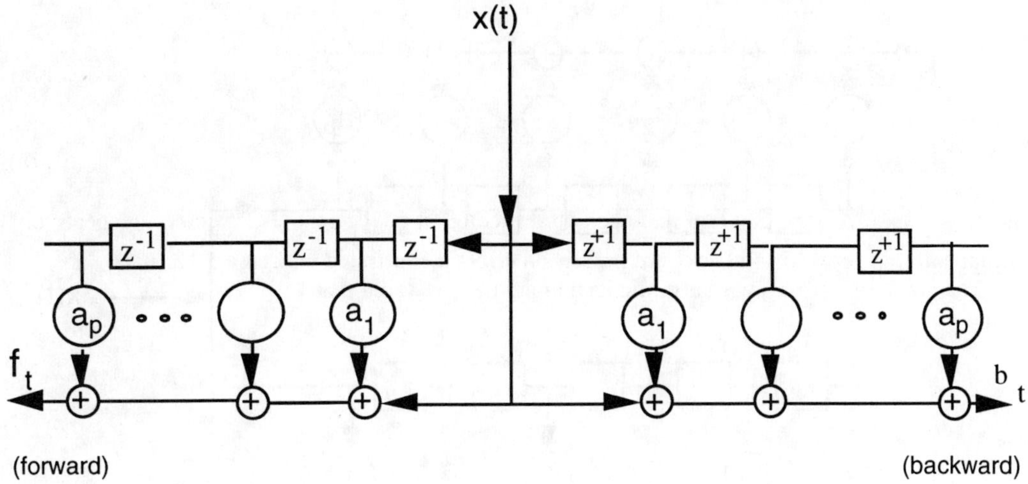

Figure 5.12: A schematic representation of forward and backward prediction error computation.

The backward error sequence also behaves in the same form. Next, we compute the cross-covariance function,

$$
\begin{aligned}
E\{f_t b_{t-\tau}\} &= \mathbf{A}^T E\{\mathbf{x}_t^{-p} \mathbf{x}_{t-\tau}^{+p^T}\} \mathbf{A} \\
&= \mathbf{A}^T \mathbf{C}_x^e \mathbf{A} = 0 \qquad \tau > p, \\
&\neq 0 \qquad \tau \leq p.
\end{aligned}
\tag{5.20}
$$

For $\tau \leq p$ we note that

$$
E\{\mathbf{x}_t^{-p} \mathbf{x}_{t-\tau}^{+p^T}\} \neq \mathbf{C}_x^e,
$$

which may be verified by actual multiplication of forward and backward vectors with lag less than or equal to P. Consider the following example.

Example 5.6: Let p=2 and τ =2. Define the following two data vectors,

$$
\mathbf{x}_t^{-2} = \text{col}\{x(t)\ \ x(t-1)\ \ x(t-2)\},
$$

and

$$
\mathbf{x}_{t-2}^{+2} = \text{col}\{x(t-2)\ \ x(t-1)\ \ x(t)\}.
$$

The outer product of the above two vectors is given by

$$
E\{\mathbf{x}_t^{-2} \mathbf{x}_{t-2}^{+2^T}\} = \begin{bmatrix}
E\{x(t)x(t-2)\}, & E\{x(t)x(t-1)\}, & E\{x(t)x(t)\} \\
E\{x(t-1)x(t-2)\}, & E\{x(t-1)x(t-1)\}, & E\{x(t-1)x(t)\} \\
E\{x(t-2)x(t-2)\}, & E\{x(t-2)x(t-1)\}, & E\{x(t-2)x(t)\}
\end{bmatrix}
$$

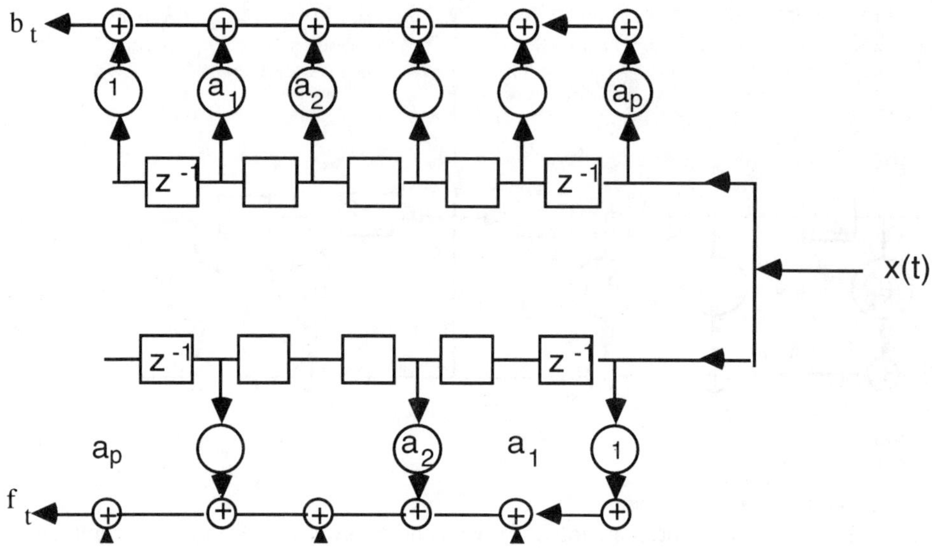

Figure 5.13: A practical scheme of computing forward and backward prediction errors.

$$= \begin{bmatrix} C_x(-2), & C_x(-1), & C_x(0) \\ C_x(-1), & C_x(0), & C_x(1) \\ C_x(0), & C_x(1), & C_x(2) \end{bmatrix}.$$

The resulting matrix is different from the extended covariance matrix as defined in (2.62). Now consider $\tau = 3$. By repeating the above calculation we obtain

$$\tau = 3E\{\mathbf{x}_t^{-2}\mathbf{x}_{t-3}^{+2}{}^T\} = \begin{bmatrix} C_x(-3) & C_x(-2) & C_x(-1) \\ C_x(-2) & C_x(-1) & C_x(-0) \\ C_x(-1) & C_x(0) & C_x(1) \end{bmatrix}.$$

The above matrix is of the form of an extended covariance matrix. Thus, the forward and backward prediction error sequences are indeed uncorrelated only for lag t >p.

B. Real Time

In the above procedure for computation of backward prediction errors, we require future samples, that is, for t+1, t+2, t+3, ..., etc. This, naturally, is not possible in any real-time processing. To overcome this difficulty we define a second type of backward prediction errors as follows:

$$b_t^{II} = a_p x(t) + a_{p-1} x(t-1) + \ldots + a_1 x(t-p+1) + x(t-p)$$

$$= \mathbf{A}^{R^T} \mathbf{x}_t, \tag{5.21}$$

where the prediction coefficients are applied in reverse order. The modified scheme of computing the backward prediction error is shown in Figure 5.13. The modified scheme is also recommended when dealing with finite data. The prediction filter should not go outside the range of data, that is, at the two ends of the finite data record. The two types of backward prediction coefficients defined in (5.17b) and (5.21), respectively, differ by a time lag of p units; that is, $b_t^{II} = b_{t-p}^{I}$. We shall hereafter consider only the second type of backward prediction errors and denote them simply as b_t.

The prediction coefficients may be estimated in the least squares sense by minimizing the error power. The resulting normal equations are similar to Yule-Walker equations for an AR process of order m. The AR coefficients of order m are related to those of order m – 1. For convenience the recursive equations (4.87) are reproduced here:

$$\begin{aligned} a_{0,m} &= 1.0 \\ a_{1,m} &= a_{1,m-1} + \alpha_m a_{m-1,m-1} \\ a_{2,m} &= a_{2,m-1} + \alpha_m a_{m-2,m-1} \\ &\vdots \\ a_{m,m} &= \alpha_m . \end{aligned}$$

In matrix notation (4.87) may be written as follows:

$$\begin{bmatrix} 1 \\ \mathbf{a}_m \end{bmatrix} = \begin{bmatrix} 1 \\ \mathbf{a}_{m-1} + \alpha_m \mathbf{a}^R_{m-1} \\ \alpha_m \end{bmatrix},$$

where

$$\mathbf{a}_{m-1} = \text{col}\{a_1 \ a_2 \ \dots \ a_{m-1}\}$$

and

$$\mathbf{a}^R_{m-1} = \text{col}\{a_{m-1} \ a_{m-2} \ \dots \ a_1\}.$$

The forward and backward prediction errors of order m, that is, the ones obtained by using the mth-order prediction coefficients may be written in terms of those of order m – 1. We write the forward prediction error as

$$\begin{aligned} f_t^m &= \mathbf{A}_m^T \mathbf{x}_t \\ &= [1 \ \vdots \ \mathbf{a}^T_{m-1} + \alpha_m \mathbf{a}^{R^T}_{m-1} \ \vdots \ \alpha_m] \begin{bmatrix} x(t) \\ \dots \\ \mathbf{x}_{t-1} \\ \dots \\ x(t-m) \end{bmatrix} \\ &= x(t) + \mathbf{a}^T_{m-1}\mathbf{x}_{t-1} + \alpha_m \mathbf{a}^{R^T}_{m-1}\mathbf{x}_{t-1} + \alpha_m x(t-m). \end{aligned}$$

We note that the first two terms on the right-hand side are equal to forward error of order m – 1. Therefore,

$$f_t^m = f_t^{m-1} + \alpha_m [\mathbf{a}_{m-1}^{R^T} \mathbf{x}_{t-1} + x(t-m)], \tag{5.22a}$$

and similarly for the backward prediction errors (using.(5.21)),

$$b_t^m = b_{t-1}^{m-1} + \alpha_m [\mathbf{a}_{m-1}^T \mathbf{x}_{t-1} + x(t)]. \tag{5.22b}$$

The quantity inside the square brackets in (5.22a) and (5.22b) may be shown to be equal to the backward and forward prediction errors of order m-1, respectively. Writing out in full form,

$$[\mathbf{a}_{m-1}^{R^T} \mathbf{x}_{t-1} + x(t-m)] = a_{m-1,m-1} x(t-1) + a_{m-2,m-1} x(t-2) + \dots + a_{1,m-1} x(t-m+1) + x(t-m)$$

$$= x(t-m) + a_{1,m-1} x(t-m+1) + \dots \dots + a_{m-2,m-1} x(t-2) + a_{m-1,m-1} x(t-1)$$

$$= b_{t-1}^{m-1}$$

Similarly, the quantity inside the square brackets of (5.22b) can be shown to be forward prediction error of order m – 1 but evaluated at t. Equations (5.22a and 5.22b) may be expressed in a recursive form as follows:

$$\begin{aligned} f_t^m &= f_t^{m-1} + \alpha_m b_{t-1}^{m-1} \\ b_t^m &= b_{t-1}^{m-1} + \alpha_m f_t^{m-1}. \end{aligned} \tag{5.23}$$

Thus, the forward and backward prediction errors of order m may be obtained from those of order m – 1 through a linear combination. To start the recursion in (5.21) we need initial conditions; namely, $f_t^0 = b_t^0 = x(t),\ t = 0, \dots, T-1$. The solution may be shown to be

m=2

$$f_t^2 = x(t) + (\alpha_1 + \alpha_1\alpha_2) x(t-1) + \alpha_2 x(t-2)$$

$$b_t^2 = x(t-2) + (\alpha_1 + \alpha_1\alpha_2) x(t-1) + \alpha_2 x(t)$$

m=3

$$f_t^3 = x(t) + (\alpha_1 + \alpha_1\alpha_2 + \alpha_2\alpha_3) x(t-1) + (\alpha_2 + \alpha_1\alpha_3 + \alpha_1\alpha_2\alpha_3) x(t-2) + \alpha_3 x(t-3)$$

$$b_t^3 = x(t-3) + (\alpha_1 + \alpha_1\alpha_2 + \alpha_2\alpha_3) x(t-2) + (\alpha_2 + \alpha_1\alpha_3 + \alpha_1\alpha_2\alpha_3) x(t-1) + \alpha_3 x(t)$$

and so forth. Notice the reversal of the order of the time series in the solution for f_t^m and b_t^m while the coefficients are identical.

The coefficient of linear combination, often known as reflection coefficient, is still a free parameter. We shall choose this so as to minimize the power in the mth-order forward or backward prediction errors or sum of forward and backward prediction errors. Straightforward least square minimization of the prediction power leads to

$$\alpha_m = -\frac{E\{f_t^{m-1} b_{t-1}^{m-1}\}}{E\{|b_{t-1}^{m-1}|^2\}},$$

or

$$= -\frac{E\{b_{t-1}^{m-1} f_t^{m-1}\}}{E\{|f_t^{m-1}|^2\}}, \tag{5.24}$$

or when the sum of the prediction error powers is minimized,

$$\alpha_m = \frac{-2E\{f_t^{m-1} b_{t-1}^{m-1}\}}{E\{|f_t^{m-1}|^2 + |b_{t-1}^{m-1}|^2\}}. \tag{5.25}$$

The minimum prediction error power of order m is related to that of order $m-1$ through a recursive relation,

$$\sigma_m^2 = \sigma_{m-1}^2(1 - \alpha_m^2)$$
$$\sigma_0^2 = \sigma_x^2. \tag{5.26}$$

The least mean square estimation of the reflection coefficient implies orthogonality of the prediction errors of successive orders. We shall here briefly derive the properties of the prediction errors, but their full account may be found in Reference 20.

The prediction errors given by (5.17) and (5.21) are indeed least squares estimates of the difference between x(t) and predicted value given by (5.17a) for forward prediction error and between $x(t-m)$ and backward predicted value given by (5.21) for backward prediction error. The least squares criterion also implies orthogonality between the error and the past values used for prediction.[20] Naturally, it is also orthogonal to the predicted component, which is a linear combination of the past values.

$$E\{f_t^m x(t-k)\} = 0 \quad \text{for} \quad 1 \le k \le m \tag{5.27a}$$

and

$$E\{b_t^m x(t-k)\} = 0 \quad \text{for} \quad 0 \le k \le m-1. \tag{5.27b}$$

From (5.27) we can show that

$$E\{f_t^m f_{t-1}^{m-1}\} = E\{b_t^m b_t^{m-1}\} = 0 \ ,$$

and further, multiplying the first equation in (5.23) by b_{t-1}^{m-1} and the second equation by f_t^{m-1} and using the least squares estimate of α_m given by (5.24), we obtain

$$E\{[b_t^m f_t^{m-1}]\} = E\{[f_t^m b_{t-1}^{m-1}]\} = 0$$

for all m. Further,

$$E\{\left|f_t^m\right|^2\} = E\{\left|b_{t-1}^m\right|^2\} = \sigma_m^2 \ ,$$

and the prediction errors become a white noise sequence whenever $m \geq p$, assuming that the time series is an AR process of order p (see (5.19)).

C. Lattice Representation

We shall express (5.23) in a matrix form as follows:

$$\begin{bmatrix} f_t^m \\ b_t^m \end{bmatrix} = \begin{bmatrix} 1 & \alpha_m \\ \alpha_m & 1 \end{bmatrix} \begin{bmatrix} f_t^{m-1} \\ b_{t-1}^{m-1} \end{bmatrix}, \tag{5.28}$$

whose graphical representation, known as a lattice element, is shown in Figure 5.14a. Several such lattice elements connected in series (see Figure 5.14b) constitute a lattice network. There will be p lattice elements to represent a pth-order AR process. In the last lattice element the reflection coefficient is equal to zero and the output; namely, the forward and backward prediction errors, will be totally uncorrelated. Note that (5.20), when expressed in terms of the backward prediction errors as defined in (5.21), reduces to

$$E\{f_t^p b_{t-1}^p\} = 0 \ . \tag{5.29}$$

Further, the prediction error sequence becomes a white noise sequence with power equal to σ_ε^2. Taking the z-transform on both sides of (5.23) we obtain

$$\begin{bmatrix} F^m(z) \\ B^m(z) \end{bmatrix} = \begin{bmatrix} 1 & \alpha_m z^{-1} \\ \alpha_m & z^{-1} \end{bmatrix} \begin{bmatrix} F^{m-1}(z) \\ B^{m-1}(z) \end{bmatrix}. \tag{5.30a}$$

The input to the first element of the lattice network is the time series itself. From this we can write down the input/output relation of a lattice network shown in Figure 5.14b,

$$\begin{bmatrix} F^p(z) \\ B^p(z) \end{bmatrix} = \prod_{m=1}^{p} \begin{bmatrix} 1 & \alpha_m z^{-1} \\ \alpha_m & z^{-1} \end{bmatrix} \begin{bmatrix} X(z) \\ X(z) \end{bmatrix}. \tag{5.30b}$$

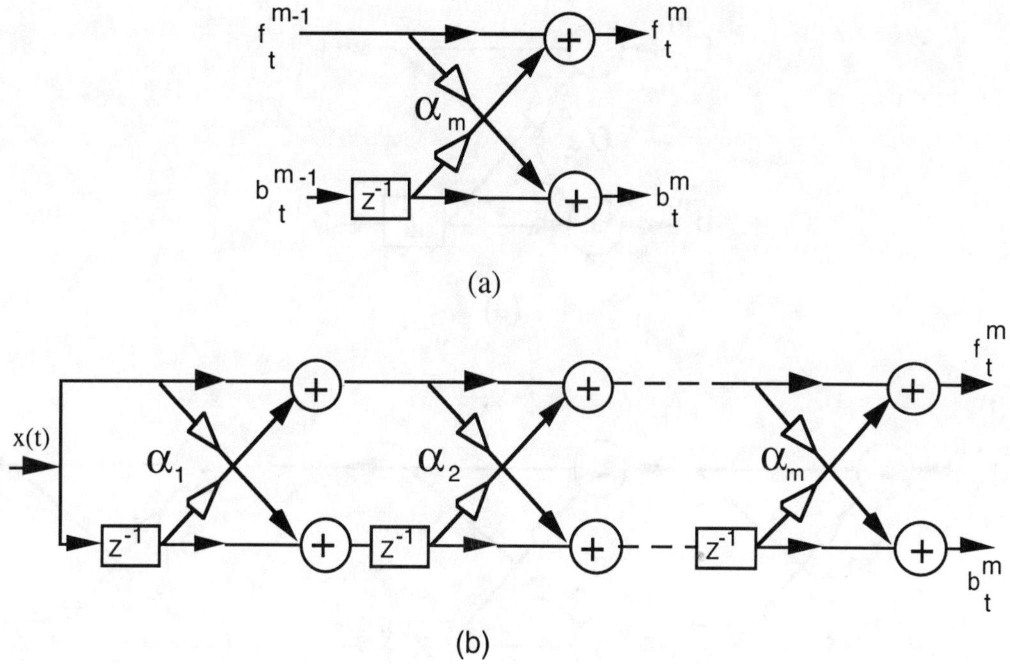

Figure 5.14: Lattice network for generating forward/backward error sequence. (a) Single element of lattice and (b) a network of lattice elements.

The recursive relation (5.23) connecting the forward and backward prediction errors may be written in a different form,

$$f_t^{m-1} = f_t^m - \alpha_m b_{t-1}^{m-1}$$

$$b_t^m = b_{t-1}^{m-1} + \alpha_m f_t^{m-1},$$

and then solved with the boundary conditions,

$$f_t^0 = b_t^0 \text{ and } f_t^m = x(t).$$

The solutions for the first few orders are

m=2

$$f_t^2 = f_t^0 + (\alpha_1 + \alpha_1\alpha_2) f_{t-1}^0 + \alpha_2 f_{t-2}^0$$

$$b_t^2 = f_{t-2}^0 + (\alpha_1 + \alpha_1\alpha_2) f_{t-1}^0 + \alpha_2 f_t^0,$$

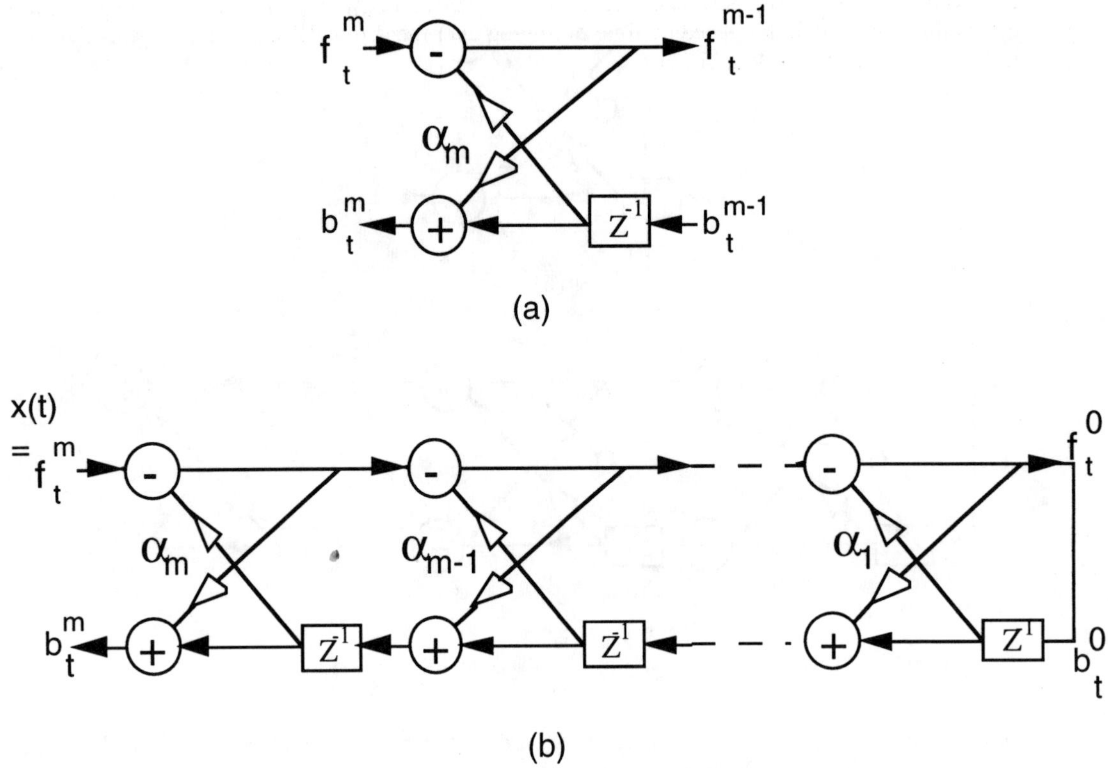

Figure 5.15: Reverse lattice network. a) single element of the lattice and (b) a network of lattice elements.

m=3

$$f_t^3 = f_t^0 + (\alpha_1 + \alpha_1\alpha_2 + \alpha_2\alpha_3) f_{t-1}^0 + (\alpha_2 + \alpha_1\alpha_3 + \alpha_1\alpha_2\alpha_3) f_{t-2}^0 + \alpha_3 f_{t-3}^0$$

$$b_t^3 = f_{t-3}^0 + (\alpha_1 + \alpha_1\alpha_2 + \alpha_2\alpha_3) f_{t-2}^0 + (\alpha_2 + \alpha_1\alpha_3 + \alpha_1\alpha_2\alpha_3) f_{t-1}^0 + \alpha_3 f_t^0,$$

and so forth. Since $x(t) = f_t^m$, x(t) may be expressed as a linear combination of m+1 outputs of the reverse lattice (see Figure 5.15).

D. Estimation of Reflection Coefficients

From the above discussion it is apparent that the reflection coefficients play a central role in the modeling of time series. Indeed, they constitute a lattice filter for whitening a time series. From the transfer function of the whitening filter we can obtain the spectrum of the time series. Hence, the question of estimation of the reflection coefficients directly from the observed data is of great significance to spectrum estimation. The reflection coefficients are related to the forward and backward prediction error sequences as shown in (5.24) and (5.25). The expectation operation in these equations may be replaced by time averaging over the actual duration of the error sequences. Let us in particular consider (5.25):

$$\alpha_m = \frac{-2\sum_{t=1}^{T-m+1} f_t^{m-1} b_{t-1}^{m-1}}{\sum_{t=1}^{T-m+1} \{|f_t^{m-1}|^2 + |b_{t-1}^{m-1}|^2\}}. \tag{5.31}$$

The denominator in (5.31) may be computed recursively (order recursion). Let N_{m-1} and D_{m-1} represent the numerator and the denominator in (5.31). A recursive relation for the denominator can be easily shown as

$$D_m = (1+\alpha_m^2)D_{m-1} - \alpha_m^2 N_{m-1}.$$

However, the same is not possible for the numerator.

It may be observed from (5.26) that $1+\alpha_m \geq 0$ and $1-\alpha_m \geq 0$, which is possible only when $|\alpha_m| \leq 1$, m=1, ..., p. This ensures that the poles of the AR filter are inside the unit circle, and hence its stability is assured. To show this we take the z-transform on both sides of (4.87):

$$\sum_{k=0}^{m} a_{k,m} z^{-k} = \sum_{k=0}^{m} a_{k,m-1} z^{-k} + \sum_{k=0}^{m} \alpha_m a_{m-k,m-1} z^{-k}.$$

After some simplification we obtain

$$A^m(z) = A^{m-1}(z) + \alpha_m z^{-m} A^{m-1}(z^{-1}), \tag{5.32}$$

which may be factored as

$$A^m(z) = A^{m-1}(z)\left[1 + \alpha_m z^{-m}\frac{A^{m-1}(z^{-1})}{A^{m-1}(z)}\right]. \tag{5.33}$$

Assuming that $A^{m-1}(z)$ is a minimum phase filter, it may be seen that on the unit circle

$$|G(z)| < 1,$$

where

$$G(z) = \alpha_m z^{-m}\frac{A^{m-1}(z^{-1})}{A^{m-1}(z)}$$

is analytic in the region bounded by the unit circle and the infinite circle. Following Rouche's theorem[21] it may be shown that the quantity inside the square brackets of (5.33) has no zeros outside the unit circle. Since the recursion in (5.33) begins with an initial value, $A^0(z) = 1$, which is a minimum phase filter, the successive filters obtained from (5.33) will be minimum phase. As the forward and backward prediction error sequences become uncorrelated (5.29), the reflection coefficient vanishes for $m \geq p$. Whenever $\alpha_m = 1$, the prediction error power becomes zero (4.87) and, naturally, the prediction coefficients are terminated.

Example 5.7: The forward prediction filter coefficients can be recursively generated using (5.33), given the reflection coefficients. Consider a three-stage lattice with reflection coefficients,[22] $\alpha_1 = \frac{1}{4}$, $\alpha_2 = \frac{1}{2}$, and $\alpha_3 = \frac{1}{3}$. After the first recursion we obtain

$$A^1(z) = 1 + \frac{1}{4}z^{-1}.$$

After the second recursion we get

$$A^2(z) = (1 + \frac{1}{4}z^{-1})[1 + \frac{1}{2}z^{-2}\frac{1 + \frac{1}{4}z^{+1}}{1 + \frac{1}{4}z^{-1}}]$$

$$= 1 + \frac{3}{8}z^{-1} + \frac{1}{2}z^{-2},$$

and finally at the end of the third recursion we get

$$A^3(z) = (1 + \frac{3}{8}z^{-1} + \frac{1}{2}z^{-2})[1 + \frac{1}{3}z^{-3}\frac{1 + \frac{3}{8}z^{+1} + \frac{1}{2}z^{+2}}{1 + \frac{3}{8}z^{-1} + \frac{1}{2}z^{-2}}]$$

$$= 1 + \frac{13}{24}z^{-1} + \frac{5}{8}z^{-2} + \frac{1}{3}z^{-3}.$$

Note that the zeros of the prediction filter polynomial are always inside the unit circle. In the above example the zeros are located as enumerated below:

$A^1(z)$: zero is at $\frac{1}{4}$

$A^2(z)$: zeros are at $-0.375 \pm j0.5995$

$A^3(z)$: zeros are at $-0.5359, -0.0028 \pm j0.7886$.

(End of Example)

E. **Burg Spectrum**

Finally, the Burg spectrum of order m is simply given by

$$S^m_{Burg}(\omega) = \frac{1}{\left|A^m(e^{-j\omega})\right|^2}. \tag{5.34}$$

As the order is increased the Burg spectrum becomes sharper, yielding high resolution. There is an optimum order beyond which the Burg spectrum becomes unreliable because of problems such as splitting of a single peak (at high SNR) into two or more peaks, appearance of spurious peaks, error in the peak position, etc.

It has been suggested[23,24] that by using a window function on the error sequence, the spectral estimate can be improved in the sense that line splitting can be avoided and an improved frequency estimate can be obtained. Frequency splitting and large frequency shift error have been reported when the data length is equal to an odd number of quarter cycles, and the initial phase of a sinusoid is an odd multiple of 45 degrees (for more details see References 25 and 26). The initial phase of the sinusoid plays an important role in controlling the shift in the frequency as well as the prediction error power.[28,29] Consider a highly idealized model, a single sinusoid with arbitrary phase but without any noise.

$$x(t) = \cos(\omega_0 t + \phi), \qquad t = 0, 1, \ldots, T-1$$

The exact linear prediction coefficients ($T \rightarrow \infty$) are $\{1, -2\cos(\omega_0), 1\}$ and the prediction error power is equal to zero. For finite data length ($T>>1$) the prediction filter coefficients are approximately given by $\{1, -2\cos(\omega_0 - \Delta\omega), 1\}$,[28] and the prediction error power is approximately given by $2\sin^2(\omega_0)\Delta\omega^2$,[29] where

$$\Delta\omega \approx \frac{1}{T}\cos(2\phi + \omega_0 T)\sin \omega_0 T.$$

The maximum frequency shift is on the order of $\frac{1}{T}$. The prediction error is maximum whenever the frequency shift error is maximum. Since the reflection coefficients are constrained to lie very close to but inside the unit circle, they may be expressed as $\alpha_m = r \sin(\theta_m)$ where r is close to, but less than, 1.0, say, r=0.99999. The prediction error power is then minimized with respect to θ_m, m=1,2,..., p, independently. Absence of frequency splitting and shift of the frequency have been attributed to the increased freedom in the reduction of error power.[30] The line splitting problem was overcome by minimizing the forward and backward error power with respect to all reflection coefficients independently, thus gaining more degrees of freedom.[9,30]

F. **Model Order Estimation**

Determination of the order of the time series model is the most difficult task on account of the fact that the same finite time series is used both for the estimation of reflection coefficients by minimizing the prediction error, and for the estimation of order of the model, also by minimizing the prediction error. As a result, higher model orders with more parameters will always give a smaller prediction error power. To overcome this drawback the prediction error should be computed on another independent realization of the same stochastic process.[31] Thus, we would require two independent realizations of the stochastic process whose model order is being estimated. Many different order determination rules based on the prediction error power have been suggested. We shall review a few of the most commonly used rules. First, we shall examine the statistical properties of the reflection coefficients. The reflection coefficients, α_m, $m > p$, are zero mean Gaussian random variables with variance equal to $\frac{1}{T}$.[32] This means that with 95% probability the reflection coefficients of order greater than p will lie within $\pm\frac{1.96}{\sqrt{T}}$ limits. This property of higher order reflection coefficients can be used to estimate the order of the model. The order of the model is m for which all higher order reflection coefficients lie within $\pm\frac{1.96}{\sqrt{T}}$ limits. Next, let us examine the prediction error power, σ_m^2, $m > p$. From (5.26) we obtain

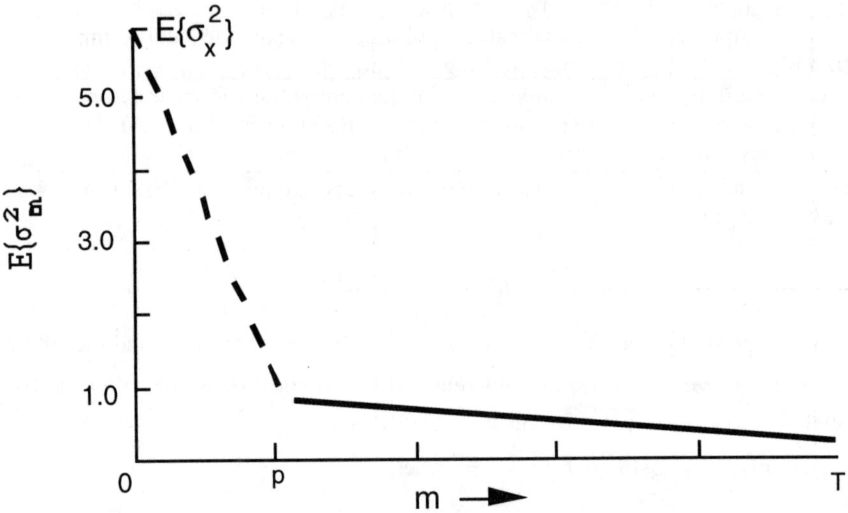

Figure 5.16: A schematic showing the variation of $E\{\hat{\sigma}_m^2\}$ as a function of m. The dashed line in the range m=0 to p is purely schematic, but the line in the range m>p to m=T is given by (5.33). Note the knee joining the two line segments is indicative of the true model order.

$$\hat{\sigma}_m^2 = \sigma_p^2 \prod_{i=p+1}^{m} (1 - \alpha_i^2), \tag{5.35}$$

where

$$\sigma_p^2 = \prod_{i=1}^{p} (1 - \alpha_i^2).$$

For large T the reflection coefficients α_i for $i > p$ will be small (of the order $\frac{1}{\sqrt{T}}$); hence (5.35) may be approximated as follows:

$$\hat{\sigma}_m^2 \approx \sigma_p^2 [1 - \sum_{i=p+1}^{m} \alpha_i^2]. \tag{5.36}$$

It is easy to show that the mean prediction error power is given by

$$E\{\hat{\sigma}_m^2\} \approx \sigma_p^2 [1 - \frac{m-p}{T}]. \tag{5.37}$$

It is interesting to note that the mean prediction error power decreases linearly and

$$E\{\hat{\sigma}_m^2\} \to \sigma_p^2 \frac{p}{T} \qquad m \to T.$$

Sl. No.	Type	Formula	Remarks[38,39]
1	FPE	$\frac{T+m}{T-m}\hat{\sigma}_m^2$	Limited to AR process. Underestimates the model order.
2	AIC	$\ln(\hat{\sigma}_m^2)+\frac{2m}{T}$	Useful for AR/MA/ARMA model determination. Tends to overestimate the order. Good for short data records.
3	MDL	$\ln(\hat{\sigma}_m^2)+m\frac{\ln(T)}{T}$	Asymptotically consistent. Known to yield good results even for finite data.
4	CAT	$\frac{1}{T}\sum_{k=1}^{m}\frac{T-k}{T\hat{\sigma}_k^2}-\frac{T-m}{T\hat{\sigma}_m^2}$	Accounts for the presence of noise in the AR model.

Table 5.1: Relative perfomance of different order estimation methods.

If we were to monitor the prediction error power as a function of order, m, we will encounter a typical variation as sketched in Figure 5.16. First, in the range m=0 to p there is a rapid decrease starting at $E\{\hat{\sigma}_0^2\} = \sigma_x^2$; thereupon there is a slow decrease. The first segment shown by a dashed line is purely schematic. The actual variation will depend upon the AR coefficients and other factors. The slope of the second segment is, however, inversely proportional to the data length (see (5.33)). The true model order lies somewhere near the knee joining the two segments. Various model estimation techniques based on the prediction error power are designed to enhance the knee effect.

Following are some of the important model order estimators based on the prediction error power. The time series is assumed to be Gaussian. There are other methods of order estimation, for example, those based on the rank estimates of a covariance matrix or a data matrix (see Chapter Two). A comparative study of many of these order estimators may be found in Reference 33 (see Table 5.1).

Final prediction error (FPE):[34]

$$FPE(m) = \frac{T+m}{T-m}\hat{\sigma}_m^2.$$

Akaike information criterion (AIC):[35]

$$AIC(m) = \ln(\hat{\sigma}_m^2) + \frac{2m}{T}.$$

Minimum description length (MDL):[36, 27]

$$MDL(m) = \ln(\hat{\sigma}_m^2) + m\frac{\ln(T)}{T},$$

Criterion autoregressive transfer (CAT):[37]

$$\mathrm{CAT}(m) = \frac{1}{T}\sum_{k=1}^{m} \frac{T-k}{T\hat{\sigma}_k^2} - \frac{T-m}{T\hat{\sigma}_m^2}.$$

The problem of order estimation is by no means resolved. As made out in Reference 33 there is no single method which will work under all conditions. Most methods seem to yield good results for low-order AR process with poles deep inside (i.e., wideband process) and high SNR.

G. Adaptive AR Spectrum

The reflection coefficients may be mapped into AR coefficients using (4.87). The AR coefficients are then used to estimate the spectrum as explained above. We can avoid the step of mapping the reflection coefficients into AR coefficients by directly using the reflection coefficients themselves to compute the transfer function of the prediction filter using (5.30a). Thus, as the new observation arrives and the reflection coefficients are updated, we will be able to compute the up-to-date spectrum. An expression for the transfer function, in terms of the reflection coefficients, may be obtained by rewriting (5.32) on the unit circle:

$$A^{m}(\omega) = A^{m-1}(\omega) + \alpha_m e^{-jm\omega} A^{m-1^*}(\omega) \tag{5.38}$$

with $A^{0}(\omega) = 1$. Let us write down the first few prediction filter transfer functions:

$$A^{1}(\omega) = 1 + \alpha_1 e^{-j\omega},$$

$$A^{2}(\omega) = (1 + \alpha_1 e^{-j\omega}) + \alpha_2 e^{-j2\omega}(1 + \alpha_1 e^{j\omega}),$$

$$A^{3}(\omega) = (1 + \alpha_1 e^{-j\omega}) + \alpha_2 e^{-j2\omega}(1 + \alpha_1 e^{j\omega}) + \alpha_3 e^{-j3\omega}((1 + \alpha_1 e^{j\omega}) + \alpha_2 e^{j2\omega}(1 + \alpha_1 e^{j\omega})).$$

The reflection coefficients need to be updated as new observations arrive. First, the forward and backward prediction errors will have to be computed. As soon as a new observation arrives at the input of the lattice (Figure 5.14), the forward and backward prediction errors of all orders are computed. These prediction errors are now used in (5.23) to compute current reflection coefficients. Only a finite number of past prediction errors along with the current estimate are used; that is, the estimate is done over a finite time window. Equation (5.31) now assumes a slightly different form,

$$\alpha_m(t) = \frac{-2\sum_{i=t-\Delta T}^{t-m+1} f_i^{m-1} b_{i-1}^{m-1}}{\sum_{i=t-\Delta T}^{t-m+1} \{|f_i^{m-1}|^2 + |b_{i-1}^{m-1}|^2\}}, \tag{5.39}$$

where DT is the time window used for the estimation of reflection coefficients. It is also possible to recursively compute the reflection coefficients.[40]

H. Least Squares Estimation

In the Burg method the reflection coefficients alone were estimated by minimizing the forward and backward prediction error power. The AR coefficients were then computed by using Levinson's recursive approach (4.87). We can go a step further and estimate all AR coefficients in the least squares sense without invoking Levinson's recursion.[41,9] It is claimed that such an estimate overcomes the problems of line splitting and line shift encountered in Burg's method. However, this view is not shared by others, for example, by Fougere.[42] The AR model for finite data (2.51) may be written in a matrix form as follows:

$$\begin{bmatrix} x_0 & & & \\ x_1 & x_0 & & \mathbf{0} \\ \vdots & & \ddots & \\ x_p & x_{p-1} & \cdots & x_0 \\ \vdots & & & \\ x_{T-1} & x_{T-2} & \cdots & x_{T-p-1} \\ & \mathbf{0} & \ddots & \vdots \\ & & & x_{T-1} \end{bmatrix} \begin{bmatrix} 1 \\ a_1 \\ a_2 \\ \vdots \\ a_p \end{bmatrix} = \begin{bmatrix} \varepsilon_0 \\ \varepsilon_1 \\ \vdots \\ \varepsilon_p \\ \vdots \\ \varepsilon_{T-1} \\ \vdots \\ \varepsilon_{T-1+p} \end{bmatrix}, \tag{5.40}$$

where we have, for the sake of compactness, dropped the parentheses, e.g., $x_{0}, x_{1}, x_{2}, \ldots, x_{T-1}$. It is assumed that the time series is zero outside the given interval of time, that is, 0 to $T-1$. The above assumption is unreasonable as the finiteness of the data is often due to not having observed the time series rather than it being zero. We must redefine the AR equation without imposing the above condition. For this it is necessary to ensure that the AR filter does not cross the boundaries of the finite data. This results in four different forms of the data matrices depending upon whether the filter is allowed or not allowed to cross one or both boundaries.

$$\mathbf{x}_1 = \underset{(T+p)\times(p+1)}{\begin{bmatrix} x_0 & & & \\ x_1 & x_0 & & \mathbf{0} \\ & & \ddots & \\ x_p & x_{p-1} & \cdots & x_0 \\ \vdots & & & \vdots \\ x_{T-1} & x_{T-2} & \cdots & x_{T-p-1} \\ & x_{T-1} & \cdots & x_{T-p-2} \\ \mathbf{0} & & \ddots & \\ & & & x_{T-1} \end{bmatrix}}, \qquad \mathbf{x}_2 = \underset{(T-p)\times(p+1)}{\begin{bmatrix} x_p & x_{p-1} & \cdots & x_0 \\ x_{p+1} & x_p & \cdots & x_1 \\ \vdots & & & \vdots \\ x_{T-1} & x_{T-2} & \cdots & x_{T-p-1} \end{bmatrix}},$$

$$\mathbf{x}_3 = \begin{bmatrix} x_0 & & & \\ x_1 & x_0 & \mathbf{0} & \\ & & \ddots & \\ x_p & x_{p-1} & \cdots & x_0 \\ \vdots & & & \vdots \\ x_{T-1} & x_{T-2} & \cdots & x_{T-p-1} \end{bmatrix}, \qquad \mathbf{x}_4 = \begin{bmatrix} x_p & x_{p-1} & \cdots & x_0 \\ x_{p+1} & x_p & \cdots & x_1 \\ \vdots & & & \vdots \\ x_{T-1} & x_{T-2} & \cdots & x_{T-p-1} \\ & x_{T-1} & \cdots & x_{T-p-2} \\ & \mathbf{0} & \ddots & \vdots \\ & & & x_{T-1} \end{bmatrix},$$

$$(T) \times (p+1) \qquad\qquad (T) \times (p+1)$$

where $\mathbf{x}_1$: autocorrelation form, $\mathbf{x}_2$: covariance form, $\mathbf{x}_3$: prewindowed form, and $\mathbf{x}_4$: postwindowed form. Observe that all four forms of the data matrices possess Toeplitz property. The AR coefficients are determined by minimizing the prediction error energy,

$$\varepsilon^H \varepsilon = \mathbf{A}^H \mathbf{x}_m^H \mathbf{x}_m \mathbf{A} = \min, \quad m = 1,2,3,4$$
$$\mathbf{A}^H \delta = 1, \tag{5.41}$$

where $\mathbf{A}$ stands for AR coefficient vector, ε stands for prediction error vector, and $\delta=\{1, 0, 0, ..., 0\}$. Minimization of (5.41) leads to the following normal equations:

$$\mathbf{R}_m \mathbf{A} = \sigma_\varepsilon^2 \delta, \qquad m = 1, 2, 3, 4 \,, \tag{5.42}$$

where

$$\mathbf{R}_m = \frac{1}{T_m} \mathbf{x}_m^H \mathbf{x}_m$$

is a correlation matrix and

$$\sigma_\varepsilon^2 = \frac{1}{T_m} \sum_i \varepsilon_i^2 .$$

Further,

$$T_1 = T + p,$$
$$T_2 = T - p,$$
$$T_3 = T_4 = T.$$

The elements of the correlation matrices are given by

$$\{\mathbf{R}_1\}_{k,l} = \frac{1}{T_1} \sum_{i=1}^{T+p} x_{i-k} x_{i-l},$$

$$\{\mathbf{R}_2\}_{k,l} = \frac{1}{T_2}\sum_{i=p+1}^{T} x_{i-k}x_{i-l}.$$

$$\{\mathbf{R}_3\}_{k,l} = \frac{1}{T_3}\sum_{i=1}^{T} x_{i-k}x_{i-l},$$

$$\{\mathbf{R}_4\}_{k,l} = \frac{1}{T_4}\sum_{i=p}^{T+p} x_{i-k}x_{i-l}. \tag{5.43}$$

Out of the four correlation matrices only $\mathbf{R}_1$ has the Toeplitz symmetry which permits the use of Levinson's recursive algorithm (4.87).

Example 5.8 In this example we shall demonstrate through actual multiplication of the data matrices that the correlation matrices are non-Toeplitz except in the first case. Let p=2 and T=8 and the observed data be $x_0, x_1, x_2, x_3, x_4, x_5, x_6, x_7$. The four data matrices are given below. Observe the Toeplitz property of the data matrices.

$$\mathbf{x}_1 = \begin{bmatrix} x_0 & 0 & 0 \\ x_1 & x_0 & 0 \\ x_2 & x_1 & x_0 \\ x_3 & x_2 & x_1 \\ x_4 & x_3 & x_2 \\ x_5 & x_4 & x_3 \\ x_6 & x_5 & x_4 \\ x_7 & x_6 & x_5 \\ 0 & x_7 & x_6 \\ 0 & 0 & x_7 \end{bmatrix}, \quad \mathbf{x}_2 = \begin{bmatrix} x_2 & x_1 & x_0 \\ x_3 & x_2 & x_1 \\ x_4 & x_3 & x_2 \\ x_5 & x_4 & x_3 \\ x_6 & x_5 & x_4 \\ x_7 & x_6 & x_5 \end{bmatrix}, \quad \mathbf{x}_3 = \begin{bmatrix} x_0 & 0 & 0 \\ x_1 & x_0 & 0 \\ x_2 & x_1 & x_0 \\ x_3 & x_2 & x_1 \\ x_4 & x_3 & x_2 \\ x_5 & x_4 & x_3 \\ x_6 & x_5 & x_4 \\ x_7 & x_6 & x_5 \end{bmatrix}, \quad \mathbf{x}_4 = \begin{bmatrix} x_2 & x_1 & x_0 \\ x_3 & x_2 & x_1 \\ x_4 & x_3 & x_2 \\ x_5 & x_4 & x_3 \\ x_6 & x_5 & x_4 \\ x_7 & x_6 & x_5 \\ 0 & x_7 & x_6 \\ 0 & 0 & x_7 \end{bmatrix}.$$

(10× 3) (6 × 3) (8 × 3) 8 × 3)

The correlation matrices were computed using (5.43).

$$\mathbf{R}_1 = \frac{1}{10}\begin{bmatrix} \sum_{i=0}^{7} x_i^2 & \sum_{i=0}^{6} x_i x_{i+1} & \sum_{i=0}^{5} x_i x_{i+2} \\ \sum_{i=0}^{6} x_i x_{i+1} & \sum_{i=0}^{7} x_i^2 & \sum_{i=0}^{6} x_i x_{i+1} \\ \sum_{i=0}^{5} x_i x_{i+2} & \sum_{i=0}^{6} x_i x_{i+1} & \sum_{i=0}^{7} x_i^2 \end{bmatrix} \qquad \mathbf{R}_2 = \frac{1}{6}\begin{bmatrix} \sum_{i=2}^{7} x_i^2 & \sum_{i=1}^{6} x_i x_{i+1} & \sum_{i=0}^{5} x_i x_{i+2} \\ \sum_{i=1}^{6} x_i x_{i+1} & \sum_{i=1}^{6} x_i^2 & \sum_{i=0}^{5} x_i x_{i+1} \\ \sum_{i=0}^{5} x_i x_{i+2} & \sum_{i=0}^{5} x_i x_{i+1} & \sum_{i=0}^{5} x_i^2 \end{bmatrix}$$

$$\mathbf{R}_3 = \frac{1}{8}\begin{bmatrix} \sum_{i=0}^{7} x_i^2 & \sum_{i=0}^{6} x_i x_{i+1} & \sum_{i=0}^{5} x_i x_{i+2} \\ \sum_{i=0}^{6} x_i x_{i+1} & \sum_{i=0}^{6} x_i^2 & \sum_{i=0}^{5} x_i x_{i+1} \\ \sum_{i=0}^{5} x_i x_{i+2} & \sum_{i=0}^{5} x_i x_{i+1} & \sum_{i=0}^{5} x_i^2 \end{bmatrix} \qquad \mathbf{R}_4 = \frac{1}{8}\begin{bmatrix} \sum_{i=2}^{7} x_i^2 & \sum_{i=1}^{6} x_i x_{i+1} & \sum_{i=0}^{5} x_i x_{i+2} \\ \sum_{i=1}^{6} x_i x_{i+1} & \sum_{i=1}^{7} x_i^2 & \sum_{i=0}^{6} x_i x_{i+1} \\ \sum_{i=0}^{5} x_i x_{i+2} & \sum_{i=0}^{6} x_i x_{i+1} & \sum_{i=0}^{7} x_i^2 \end{bmatrix}$$

Notice that the entries differ only with respect to the limits on the summation. It turns out that only $\mathbf{R}_1$ is Toeplitz, but they are all symmetric matrices. It is interesting to note that $\mathbf{R}_2$, $\mathbf{R}_3$, and $\mathbf{R}_4$ may be expressed as a sum of a Toeplitz matrix and outer product of a lower triangular matrix (or matrices). For example, $\mathbf{R}_3$ can be expressed as

$$\mathbf{R}_3 = \frac{1}{8}\begin{bmatrix} \sum_{i=0}^{5} x_i^2 & \sum_{i=0}^{5} x_i x_{i+1} & \sum_{i=0}^{5} x_i x_{i+2} \\ \sum_{i=0}^{5} x_i x_{i+1} & \sum_{i=0}^{5} x_i^2 & \sum_{i=0}^{5} x_i x_{i+1} \\ \sum_{i=0}^{5} x_i x_{i+2} & \sum_{i=0}^{5} x_i x_{i+1} & \sum_{i=0}^{5} x_i^2 \end{bmatrix} + \frac{1}{8}\begin{bmatrix} x_6^2 + x_7^2 & x_6 x_7 & 0 \\ x_6 x_7 & x_6^2 & \\ 0 & 0 & 0 \end{bmatrix}$$

$$= \text{Toeplitz matrix} \quad + \frac{1}{8}\begin{bmatrix} 0 & x_6 & x_7 \\ 0 & 0 & x_6 \\ 0 & 0 & 0 \end{bmatrix}\begin{bmatrix} 0 & 0 & 0 \\ x_6 & 0 & 0 \\ x_7 & x_6 & 0 \end{bmatrix}.$$

(End of Example).

The correlation matrices, $\mathbf{R}_2$, $\mathbf{R}_3$, and $\mathbf{R}_4$, belong to the class of so-called α-stationary matrices for which efficient techniques for solving a system of linear equations are available.[43] For stationary time series all four forms of the linear equation (5.42) asymptotically yield the same result; but for finite data they will lead to different results. It is, however, not known which one of these will give the most satisfactory result. Some discussion on the relative performance of the above least squares techniques and comparison with Burg's maximum entropy method may be found in Reference 44.

It is observed that the least squares estimate using the forward prediction error only is much inferior to the one where a sum of the forward and backward prediction error power is minimized.[41] This observation is supported by theoretical analysis (first order approximation) and numerical simulation.[45] The normal equations obtained after minimizing the sum of forward and backward error power are given by

$$[\bar{\mathbf{R}}_m]\mathbf{A} = \sigma_0^2 \delta, \qquad m = 1, 2, 3, 4 \quad , \tag{5.44}$$

where

$$\bar{\mathbf{R}}_m = \mathbf{R}_m + \mathbf{R}^R_m,$$

$$\mathbf{R}^R_m = \frac{1}{T_m}\mathbf{x}^{R\,H}_m \mathbf{x}^R_m ,$$

and

$$\mathbf{x}^R_m$$

is a data matrix where the data is arranged in the reverse order, for example,

$$\mathbf{x}^R_2 = \begin{bmatrix} x_{T-p-1} & x_{T-p} & \cdots & x_{T-1} \\ x_{T-p} & x_{T-p+1} & \cdots & x_{T-2} \\ \vdots & & & \vdots \\ x_1 & x_2 & \cdots & x_{p+1} \\ x_0 & x_1 & \cdots & x_p \end{bmatrix}.$$

σ^2_0 is minimum error power. $\bar{\mathbf{R}}_m$ has a special symmetry; that is, it is symmetric ($r_{k,l} = r_{l,k}$) and persymmetric ($r_{k,l} = r_{p-k,p-l}$) but not Toeplitz. By exploiting the special structure of $\bar{\mathbf{R}}_m$ Marple[46] developed a fast algorithm, on par with Burg's algorithm.

The performance of the least squares algorithm in comparison with the Burg algorithm is found to be superior when the time series consists of a sum of sinusoids with random phase. There is no shift in the peak position, and there is no splitting of the peak. In contrast, the performance of the least squares algorithm, when the time series is truly stochastic, for example, an AR(4) process, is at best as good as that of Burg's algorithm.[9,46] The important difference between the Burg and the least squares algorithms lies in the fact that in the Burg method the poles are necessarily within the unit circle, but this is not so in the least squares method. The poles are free to drift to even outside the unit circle. Further, the Burg method and the least squares method, which directly use the data, perform better than the AR coefficients derived by solving the Yule-Walker equations with the estimated autocorrelation coefficients. The example described below clarifies the relative performance of the three methods.

Example 5.9: This example is taken from Reference 9 to illustrate the relative performance of the least squares algorithm for the AR(4) process introduced in Example 2.9. One hundred independent sequences of the AR(4) process in steady state, each of 40 samples, were generated. The Burg spectrum and the least squares AR spectrum of each sequence were computed and averaged over one hundred samples. The results are shown in Figure 5.17 below. The autocorrelation coefficients were also computed (3.1) and used in the Yule-Walker equations (2.60) in place the covariance coefficients to estimate the AR coefficients. The AR spectrum from these coefficients is shown in Figure 5.16a. Both Burg (b) and least squares (c) AR spectra are close to the true spectrum, but the AR spectrum from the estimated autocorrelation function is much inferior in every respect. The least squares AR spectrum is more spiky than the Burg spectrum, a consequence of the poles drifting closer to the unit circle (End of Example).

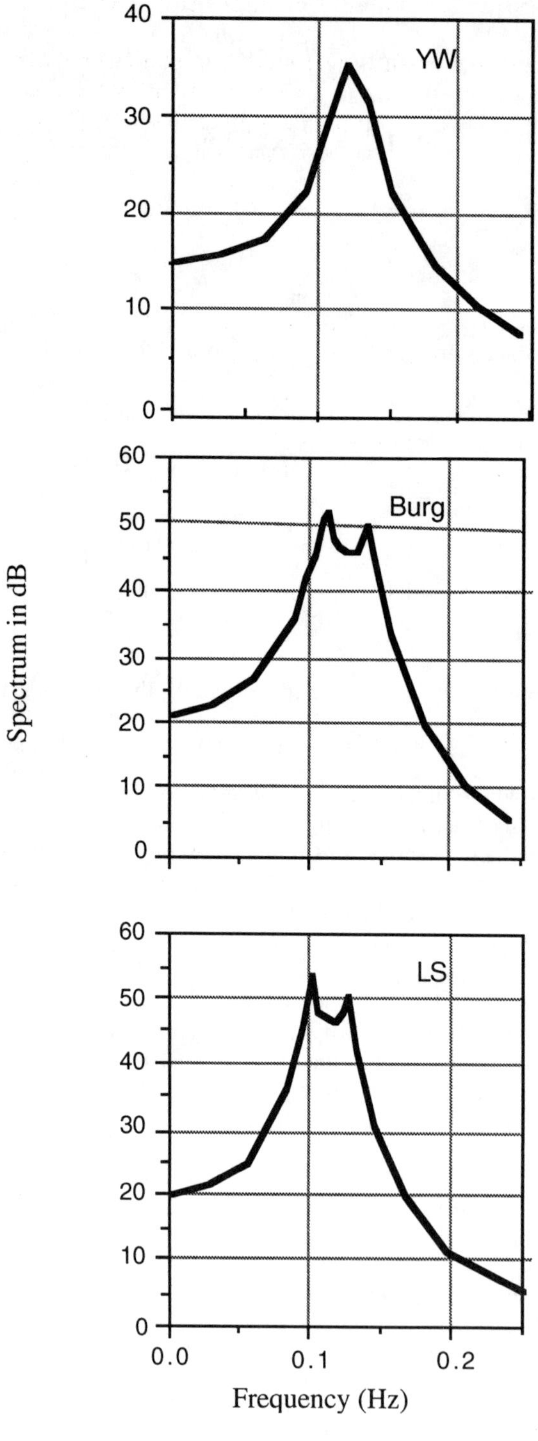

Figure 5.17: The AR spectra obtained from finite data using three different methods are compared. The Burg and least squares methods appear to be superior to the YW method. (YW) Yule-Walker, (BURG) Burg, and (LS) Least squares.

5.4 Data Matrix and Singular Value Decomposition (SVD)

A sum of random sinusoids is a useful signal model, in particular, in array signal processing. Similarly, an impulse response function of a time invariant system, covariance function of a stationary stochastic function, or any decaying signal can be modeled as a sum of damped sinusoids. All these signals possess a certain structure which forms a basis for their spectrum estimation.

A. Data Matrix

A time series may be arranged in a matrix form having Hankel symmetry where all elements on a right subdiagonal are the same. Let us consider a time series, x(t), t=0, 1, ..., T − 1, arranged in a matrix form as follows:

$$\mathbf{x}_f = \begin{bmatrix} x(0) & x(1) & x(2) & \cdots & x(T-\Delta T) \\ x(1) & x(2) & x(3) & \cdots & x(T-\Delta T+1) \\ x(2) & x(3) & x(4) & \cdots & x(T-\Delta T+2) \\ \vdots & & & & \vdots \\ x(\Delta T-1) & x(\Delta T) & & \cdots & x(T-1) \end{bmatrix}.$$

$$\Delta T \times (T - \Delta T + 1)$$

We can also generate another matrix by first reversing the order of the time series and then arranging it in a matrix form,

$$\mathbf{x}_b = \begin{bmatrix} x^*(T-1) & x^*(T-2) & x^*(T-3) & \cdots & x^*(\Delta T-1) \\ x^*(T-2) & x^*(T-3) & x^*(T-4), & \cdots & x^*(\Delta T-2) \\ x^*(T-3), & x^*(T-4), & x^*(T-5), & \cdots & x^*(\Delta T-3) \\ \vdots & & & & \vdots \\ x^*(T-\Delta T) & x^*(T-\Delta T-1), & x^*(T-\Delta T-2) & \cdots & x^*(0) \end{bmatrix},$$

$$\Delta T \times (T - \Delta T + 1)$$

where * stands for complex conjugation. The forward and backward data matrices may be combined into a single data matrix,

$$\mathbf{x} = [\mathbf{x}_f \,\vdots\, \mathbf{x}_b]$$

$$\Delta T \times 2(T - \Delta T + 1) \; .$$

The data matrix and the correlation matrix are related; in fact, the $(m,n)^{th}$ element of the matrix product $\mathbf{x}\mathbf{x}^H$ can be expressed in terms of the correlation function.

$$\{\mathbf{x}\mathbf{x}^H\}_{m,n} = \sum_{t=0}^{T-\Delta T} x(t+m)\,x^*(t+n) + \sum_{t=0}^{T-\Delta T} x^*(T-t-1-m)\,x(T-t-1-n)$$

$$0 \le m,n \le \Delta T - 1 \tag{5.45}$$

The two terms on the right-hand side of (5.45) are the forward and backward correlation functions, respectively, of a deterministic time series. Notice that in evaluating the summations we do not go outside the range of the data.

We shall call $\mathbf{x}\mathbf{x}^H$ a correlation matrix, which is different from the covariance matrix of a stochastic process. The advantage of the correlation function is that it is defined for finite data, and it is not an approximation to the true covariance function of an ideal stationary stochastic process. However, when the time series consists of a sum of sinusoids,[47] the correlation matrix possesses an interesting structure similar to that of a covariance matrix.

B. Sinusoids

To demonstrate the above property of the correlation matrix let us consider a sum of p sinusoids,

$$x(t) = \sum_{i=0}^{p-1} a_i e^{j\omega_i t},$$

where a_i is the complex amplitude of the ith sinusoid, and substitute the same in (5.45). We obtain

$$R(m,n) = \{\mathbf{x}\mathbf{x}^H\}_{m,n} = \sum_{k=0}^{p-1}\sum_{l=0}^{p-1} S_{k,l}\, e^{j(\omega_k m - \omega_l n)},$$
$$0 \le m,n \le \Delta T - 1 \tag{5.46}$$

where

$$S_{kl} = \sum_{t=0}^{T-\Delta T} [a_k a_l^* e^{j(\omega_k - \omega_l)t} + a_k^* a_l e^{-j(\omega_k - \omega_l)(T-t-1)}]. \tag{5.47}$$

Define a matrix **S** whose $(k,l)^{th}$ element is S_{kl} and a matrix **H** whose columns are p frequency vectors; for example, the kth vector is $\text{col}\{1 \; e^{j\omega_k} \; e^{j2\omega_k} \; \ldots \; e^{j(\Delta T-1)\omega_k}\}$. The correlation matrix can now be expressed in terms of **S** and **H**.

$$\mathbf{R} = \mathbf{H}\mathbf{S}\mathbf{H}^H. \tag{5.48}$$

The matrix **S** ($p \times p$) is full rank as each element of the matrix can be expressed as an inner product of two linearly independent vectors.[47] The structure of **R** as revealed by (5.48) is identical to that of the covariance matrix as shown in (4.104). Following the explanation tendered to show the property that the eigenvectors corresponding to significant eigenvalues span a subspace, known as signal subspace, which is spanned by the columns of the **H** matrix, we can here also show the same property for **R**.

We shall now show that the singular value decomposition of the data matrix achieves the same result as eigen decomposition of the correlation matrix. Let $\mathbf{x} = \mathbf{U}\boldsymbol{\Sigma}\mathbf{V}^H$ be the SVD of the data matrix. Σ is a diagonal matrix of singular values. There are only p significant singular values, and the rest are either zero or very small, depending upon the noise. It is now easy to show that

$$\mathbf{x}\mathbf{x}^H = \mathbf{U}\boldsymbol{\Sigma}\mathbf{V}^H\mathbf{V}\boldsymbol{\Sigma}\mathbf{U}^H$$
$$= \mathbf{U}\boldsymbol{\Gamma}\mathbf{U}^H, \tag{5.49}$$

where Γ is a diagonal matrix made up of P significant eigenvalues of the correlation matrix and (DT-P) zero eigenvalues. Let $\mathbf{U}_s$ represent eigenvectors corresponding to the significant singular values and $\mathbf{U}_\eta$ represent eigen vectors corresponding to the null or insignificant singular values. Comparing (5.48) and (5.49) we can show

$$\mathbf{U}_s = \mathbf{HQ},$$
$$\mathbf{U}_\eta \perp \mathbf{H}, \tag{5.50}$$

where $\mathbf{Q}$ satisfies a relation, $\mathbf{S} = \mathbf{Q}\Gamma\mathbf{Q}^H$, and it is nonsingular.

The columns of the **H** matrix possess an interesting property; namely, each column of the matrix can be expressed in terms a similar column but of a lower dimension; for example,

$$\text{col}\left\{1, e^{j\omega_k}, e^{j2\omega_k}, e^{j3\omega_k}, \ldots, e^{j(\Delta T-1)\omega_k}\right\}$$

$$= \text{col}\left\{1,\ e^{j\omega_k}\{1,\ e^{j\omega_k}, e^{j2\omega_k}, \ldots, e^{j(\Delta T-2)\omega_k}\}\right\}.$$

Using the above property that the **H** matrix of dimension $\Delta T \times p$ can be expressed in terms of the similar matrix but of a lower dimension, we have

$$\mathbf{U}_s = \begin{bmatrix} 1 \quad 1 \quad 1 \quad \ldots \quad 1 \\ \\ \mathbf{H}_{\Delta T-1xP}\Gamma \\ \\ \end{bmatrix}\mathbf{Q}, \tag{5.51}$$

where

$$\Gamma = \text{diag}\left\{e^{j\omega_1}\ e^{j\omega_2}\ e^{j\omega_3}\ \ldots.\ e^{j\omega_P}\right\}.$$

Using (5.51) in (5.50) we obtain

$$\mathbf{U}_s^1 = \{1\ 1\ 1\ \ldots\ 1\}\,\mathbf{Q}$$

and

$$\begin{bmatrix} \mathbf{U}_s^2 \\ \mathbf{U}_s^3 \\ \vdots \\ \mathbf{U}_s^{\Delta T} \end{bmatrix} = \mathbf{H}_{\Delta T-1xP}\Gamma\mathbf{Q}, \tag{5.52}$$

where $\mathbf{U}_s^1, \mathbf{U}_s^2, \mathbf{U}_s^3, \ldots \mathbf{U}_s^{\Delta T}$ are rows of $\mathbf{U}_s$. Next, taking the first $\Delta T - 1$ rows of $\mathbf{U}_s$, that is, after dropping the last row, we can similarly show that

$$\begin{bmatrix} \mathbf{U}_s^1 \\ \mathbf{U}_s^2 \\ \vdots \\ \mathbf{U}_s^{\Delta T-1} \end{bmatrix} = \mathbf{H}_{\Delta T-1xP}\mathbf{Q}. \tag{5.53}$$

Let $\mathbf{U}'_s$ and $\mathbf{U}''_s$ represent the left-hand side of (5.52) and (5.53), respectively. Note that $\mathbf{U}'_s$ and $\mathbf{U}''_s$ are obtained from $\mathbf{U}_s$ by dropping the top and bottom row, respectively.

$$\mathbf{U}'_s = \mathbf{H}_{\Delta T-1xP}\Gamma\mathbf{Q} \tag{5.54a}$$

and

$$\mathbf{U}''_s = \mathbf{H}_{\Delta T-1xP}\mathbf{Q}. \tag{5.54b}$$

From equations (5.54a) and (5.54b) we eliminate $\mathbf{H}_{\Delta T-1xP}$, giving us

$$\mathbf{U}''_s = \mathbf{U}'_s\mathbf{Q}^{-1}\Gamma^H\mathbf{Q}. \tag{5.55}$$

Now, to get a least squares solution for $\mathbf{Q}^{-1}\Gamma^H\mathbf{Q}$ we multiply on both sides of (5.55) by the pseudoinverse of $\mathbf{U}'_s$ and obtain

$$\mathbf{U}'^{+}_s\mathbf{U}''_s = \mathbf{Q}^{-1}\Gamma^H\mathbf{Q}, \tag{5.56}$$

where the pseudoinverse of $\mathbf{U}'_s$ is defined as

$$\mathbf{U}'^{+}_s = (\mathbf{U}'^{H}_s\mathbf{U}'_s)^{-1}\mathbf{U}'^{H}_s.$$

The eigenvalues of $\mathbf{U}'^{+}_s\mathbf{U}''_s$ are same as those of Γ^H. Since

$$\Gamma = \text{diag}\{\exp(j\omega_1), \exp(j\omega_2), \ldots, \exp(j\omega_p)\},$$

the eigenvalues of $\mathbf{U}'^{+}_s\mathbf{U}''_s$ are simply $\exp(-j\omega_1), \exp(-j\omega_2), \ldots, \exp(-j\omega_p)$. Thus, one can estimate the unknown frequencies from the eigenvalues of $\mathbf{U}'^{+}_s\mathbf{U}''_s$, a quantity defined in terms of the signal eigenvectors. The method described above was first suggested in Reference 48, and it is similar to ESPRIT (see page 312).

C. Damped Sinusoids

A sum of damped sinusoids is a useful model to characterize an impulse response function of a system. We shall demonstrate the use of the data matrix for estimation of the frequency and the damping factor. We can use either $\mathbf{x}_f$ or $\mathbf{x}_b$ but not both. Here we shall use $\mathbf{x}_f$ and leave the use of $\mathbf{x}_b$ as an exercise. A sum of damped sinusoids is expressed as

$$x(t) = \sum_{i=0}^{p-1} a_i e^{(\alpha_i + j\omega_i)t} \quad , \tag{5.57}$$

where $\alpha_i < 0$ is the damping coefficient. We shall now evaluate the correlation matrix, $\mathbf{x}_f\mathbf{x}_f^H$. By straightforward substitution of (5.57) for each element in the data matrix and evaluation of the matrix product, we can express any one element of the correlation matrix, say, the (m,n)th element as

$$\begin{aligned} R(m,n) &= \left\{\mathbf{x}_f\mathbf{x}_f^H\right\}_{m,n} \\ &= \sum_{k=0}^{p-1}\sum_{l=0}^{p-1} s_{k,l}\, e^{(\alpha_k + j\omega_k)m + (\alpha_l - j\omega_l)n}, \\ & 0 \le m,n \le \Delta T - 1 \end{aligned} \tag{5.58}$$

where

$$S_{kl} = \sum_{t=0}^{T-\Delta T} [a_k a_l^* e^{(\alpha_k + \alpha_l)t + j(\omega_k - \omega_l)t}] \quad .$$

The correlation matrix may be expressed as in (5.48):

$$\mathbf{R} = \mathbf{HSH}^H ,$$

where the columns of $\mathbf{H}$ depend both on frequency and damping coefficient. A typical column is given by

$$\text{col}\left\{1 \quad e^{(\alpha_k + j\omega_k)} \quad e^{2(\alpha_k + j\omega_k)} \quad e^{3(\alpha_k + j\omega_k)} \ldots e^{(\Delta T - 1)(\alpha_k + j\omega_k)}\right\}.$$

Notice that the columns of $\mathbf{H}$ possess the same property as in the case of pure sinusoids (without damping coefficient). Naturally, going through all steps as in the previous subsection we arrive at a result identical to (5.55),

$$\mathbf{U}''_s = \mathbf{U}'_s\mathbf{Q}^{-1}\Gamma^H\mathbf{Q} ,$$

where

$$\Gamma = \text{diag}\left\{e^{\alpha_1 + j\omega_1} \quad e^{\alpha_2 + j\omega_2} \quad e^{\alpha_3 + j\omega_3} \ldots . \; e^{\alpha_p + j\omega_p}\right\}.$$

The eigenvalues of $\mathbf{Q}^{-1}\Gamma^H\mathbf{Q}$ are the same as those of Γ. Thus, from the complex eigenvalues of $\mathbf{U}'^{+}_s\mathbf{U}''_s$ (see (5.56)) we can obtain both the frequencies and the damping coefficients of the damped sinusoids.

T=16		32		64	
Mean	Std	Mean	Std	Mean	Std
-0.0217	0.0458	-0.0331	0.0241	-0.0272	0.0136
0.4708	0.0853	0.4579	0.0281	0.4692	0.0181
-0.0467	0.0229	-0.0500	0.0125	-0.0518	0.0136
0.8825	0.0327	0.8816	0.0111	0.8750	0.0110

Table 5.2: Mean and standard deviation of the estimates of the damping coefficients and frequencies for different data lengths. Ten independent trials were carried out.

Example 5.10: We shall consider a case of two damped sinusoids in the presence of zero mean unit variance white noise $\eta(t)$ (complex) in the background,

$$x(t) = a_1 e^{(\gamma_1 + j\omega_1)t} + a_2 e^{(\gamma_2 + j\omega_2)t + \frac{\pi}{4}} + \eta(t) .$$

The parameters of the damped sinusoids are as follows:

$$\frac{a_1^2}{\sigma_\eta^2} = \frac{a_2^2}{\sigma_\eta^2} = 10 \text{ dB SNR}$$

$$\gamma_1 = -0.03, \quad \omega_1 = 0.4712$$
$$\gamma_2 = -0.05 \quad \omega_2 = 0.8706 .$$

Three time series of three different lengths, namely, 16, 32, and 64 points, were generated. A data matrix, $\mathbf{x}_f$, of size $8 \times (T-8)$ was formed and eigenvectors corresponding to the significant eigenvalues (two) of $\mathbf{x}_f \mathbf{x}_f^H$ were computed. From (5.56) the parameters of the damped sinusoids are obtained from the eigenvalues (complex) of $\mathbf{U}_s^{\prime +} \mathbf{U}_s''$. The experiment was repeated ten times with a new random phase ϕ each time. The statistics (mean and standard deviation) of the estimates are listed in Table. 5.2 (End of Example).

D. **Total Least Squares** (TLS)

In the least squares solution of (5.55) for $\mathbf{Q}^{-1}\Gamma^H\mathbf{Q}$ it is assumed that estimation errors are present only in $\mathbf{U}_s''$ but that they are absent in $\mathbf{U}_s'$. Since both quantities are derived from the actual data, they are likely to be equally contaminated with the estimation errors. Naturally, they will not strictly satisfy (5.55). In the least squares method we had minimally perturbed $\mathbf{U}_s''$ alone till it satisfied (5.55). We must now minimally perturb both $\mathbf{U}_s''$ and $\mathbf{U}_s'$ till they satisfy (5.55). Mathematically, the TLS approach solves the following problem:

$$\left\| \delta \mathbf{U}_s' \vdots \delta \mathbf{U}_s'' \right\|_F = \text{min}. \tag{5.59a}$$

and

$$(\hat{\mathbf{U}}''_s - \delta\mathbf{U}''_s) = (\hat{\mathbf{U}}'_s - \delta\mathbf{U}'_s)\mathbf{Q}^{-1}\Gamma^H\mathbf{Q},$$

(5.59b)

where $\hat{\mathbf{U}}'_s$ and $\hat{\mathbf{U}}''_s$ are signal eigenvectors with estimation errors, and $\delta\mathbf{U}'_s$ and $\delta\mathbf{U}''_s$ are perturbations introduced in $\mathbf{U}'_s$ and $\mathbf{U}''_s$, respectively. Equation (5.59b) can be rewritten as

$$\left([\hat{\mathbf{U}}'_s \vdots \hat{\mathbf{U}}''_s] - [\delta\mathbf{U}'_s \vdots \delta\mathbf{U}''_s]\right)\begin{bmatrix}\mathbf{Q}^{-1}\Gamma^H\mathbf{Q}\\ \cdots\cdots\\ -\mathbf{I}\end{bmatrix} = \mathbf{0}$$

or

$$[\hat{\mathbf{U}}'_s \vdots \hat{\mathbf{U}}''_s]\begin{bmatrix}\mathbf{Q}^{-1}\Gamma^H\mathbf{Q}\\ \cdots\cdots\\ -\mathbf{I}\end{bmatrix} = [\delta\mathbf{U}'_s \vdots \delta\mathbf{U}''_s]\begin{bmatrix}\mathbf{Q}^{-1}\Gamma^H\mathbf{Q}\\ \cdots\cdots\\ -\mathbf{I}\end{bmatrix}. \tag{5.60}$$

Premultipling on both sides of (5.60) by their respective Hermitian transpose forms and rearranging the product we obtain

$$\begin{bmatrix}\mathbf{Q}^{-1}\Gamma^H\mathbf{Q}\\ \cdots\cdots\\ -\mathbf{I}\end{bmatrix}^H\left\{[\hat{\mathbf{U}}'_s \vdots \hat{\mathbf{U}}''_s]^H[\hat{\mathbf{U}}'_s \vdots \hat{\mathbf{U}}''_s] - [\delta\mathbf{U}'_s \vdots \delta\mathbf{U}''_s]^H[\delta\mathbf{U}'_s \vdots \delta\mathbf{U}''_s]\right\}\begin{bmatrix}\mathbf{Q}^{-1}\Gamma^H\mathbf{Q}\\ \cdots\cdots\\ -\mathbf{I}\end{bmatrix} = \mathbf{0}. \tag{5.61}$$

The rank of the matrix expression inside the curly brackets is atmost equal to p (the number of sinusoids); however, the matrices inside the square brackets are full rank, namely, DT, as they contain error terms which presumably span the entire space. This rank reduction is possible only when the SVD of $[\delta\mathbf{U}'_s \vdots \delta\mathbf{U}''_s]$ is a subset of the SVD of $[\hat{\mathbf{U}}'_s \vdots \hat{\mathbf{U}}''_s]$. Further, on account of (5.59a) we must choose the least significant singular values of $[\hat{\mathbf{U}}'_s \vdots \hat{\mathbf{U}}''_s]$ to make up $[\delta\mathbf{U}'_s \vdots \delta\mathbf{U}''_s]$. Hence, from (5.60) the total least squares solution matrix is given by the right singular vectors of $[\hat{\mathbf{U}}'_s \vdots \hat{\mathbf{U}}''_s]$ corresponding to insignificant singular values. Let $\mathbf{v}$ be eigenvectors of $[\hat{\mathbf{U}}'_s \vdots \hat{\mathbf{U}}''_s]^H[\hat{\mathbf{U}}'_s \vdots \hat{\mathbf{U}}''_s]$. We rearrange $\mathbf{v}$ as follows:

$$\mathbf{v} = \begin{bmatrix}\mathbf{v}_{11} & \mathbf{v}_{12}\\ \mathbf{v}_{21} & \mathbf{v}_{22}\end{bmatrix},$$

where $\mathbf{v}_{11}$, $\mathbf{v}_{12}$, $\mathbf{v}_{21}$ and $\mathbf{v}_{22}$ are square matrices of size pxp. The TLS solution will be given by $[\mathbf{v}_{12} \;\; \mathbf{v}_{22}]^T$ or

True values	SNR=5 dB		SNR=10 dB	
	Mean	Std	Mean	Std
$\gamma_1 = -0.03$,	-0.0471	0.0498	-0.0344	0.0236
$\omega_1 = 0.4712$	0.3923	0.1756	0.4582	0.0283
$\gamma_2 = -0.05$,	-0.0525	0.0226	-0.0504	0.0126
$\omega_2 = 0.8796$	0.8858	0.0278	0.8814	0.0114

Table 5.3: Total least squares (TLS) approach. Average of ten independent trials.

True values	SNR=5 dB		SNR=10 dB	
	Mean	Std	Mean	Std
$\gamma_1 = -0.03$,	-0.0156	0.0800	-0.0331	0.0241
$\omega_1 = 0.4712$	0.3923	0.1746	0.4579	0.0281
$\gamma_2 = -0.05$	-0.0498	0.0208	-0.0500	0.0125
$\omega_2 = 0.8796$	0.8854	0.0202	0.8816	0.0111

Table 5.4: Least squares (LS) approach. Average of ten independent trials.

$$\begin{bmatrix} \mathbf{Q}^{-1}\Gamma^H\mathbf{Q} \\ \cdots\cdots\cdots \\ -\mathbf{I} \end{bmatrix} = \begin{bmatrix} -\mathbf{v}_{12}\mathbf{v}_{22}^{-1} \\ \cdots\ \cdots\ \cdots \\ -\mathbf{I} \end{bmatrix}. \tag{5.62}$$

Finally, we can estimate the unknown parameters of the damped sinusoids from the eigenvalues of $-\mathbf{v}_{12}\mathbf{v}_{22}^{-1}$.[49]

Example 5.11: We shall now use the total least squares (TLS) method for estimation of the parameters of damped sinusoids from example 5.10 and compare the results with those obtained by using the least squares (LS) method. As in Example 5.10, we have two damped sinusoids with the same set of parameters in the presence of white noise. A time series consisting of 32 samples is used to estimate the parameters. Ten independent trials were conducted, and the results were averaged to yield the mean and standard deviation of the estimates. The results are summarized in Tables 5.3 and 5.4 (End of Example).

From the above example it may be seen that the TLS does not seem to provide any significant gain over the LS, except perhaps at low SNR. Since the above comment is based on a limited number of trials, the comment must be taken as tentative.

5.5 Adaptive Subspace

We have noted that the estimation of frequencies of sinusoids in the background of white noise is intimately tied up with the estimation of the eigenvectors of a covariance matrix; some of the examples are Pisarenko's method based on eigenvector corresponding to the smallest eigenvalue, MUSIC using the entire noise subspace, and Prony's method using the signal subspace. Thus, eigen decomposition of a covariance matrix computed from finite data and updating the same as more data arrives at the sensors is the primary goal of the adaptive approach, which we would like to pursue in this section. The first step is to compute the covariance matrix and update the same after the arrival of a new data vector consisting of DT (>p) consecutive observations. Alternatively, we define a data matrix by rearranging the data column into several smaller columns, each of length DT. The next step is to compute the rank of the estimated covariance matrix or data matrix. This would enable us to separate the noise subspace from the signal subspace. The final step is to carry out the eigen decomposition, usually in a recursive manner. All three steps need to be repeated after every update of the covariance matrix.

Adaptive methods for the estimation of the eigenvector corresponding to the minimum (or maximum) eigenvalue have been described by many researchers.[50-52] An attempt to extend these methods to estimation of the entire subspace (signal or noise subspace) has also been reported.[53] All the above methods require computation of the gradient of the cost function which is minimized. The gradient computation is slow and, further, in the presence of noise it is unstable. Additionally, the rate of convergence of a gradient method is known to be very slow.[54] The alternate approach is to use the well-known power method for estimation of an eigenvector corresponding to the minimum (or maximum) eigenvalue.[55] The power method is easy to understand and easy to implement on a computer. We shall briefly describe the power method for estimation of the entire signal or noise subspace.[54]

A. Power Method for Estimation of Subspaces

Assume that we have a time series of length T, $x(t)$, $t = 0,1,\dots, T-1$. A data vector is formed as follows:

$$\mathbf{x}_m = \text{col}\,\{x(m),\ x(m+1),\ \dots,\ x(m+\Delta T)\}, \qquad m = 0, 1, \dots, T-\Delta T-1,$$

where $\Delta T >$ rank of the covariance matrix. We form a correlation matrix using these data vectors:

$$\hat{\mathbf{C}}_x = \frac{1}{T-\Delta T} \sum_{m=0}^{T-\Delta T-1} \mathbf{x}_m \mathbf{x}_m^H. \tag{5.63a}$$

When a new sample arrives, making the data of length T+1, we can update the correlation matrix,

$$\hat{\mathbf{C}}_x^{T} = \left(1 - \frac{1}{T-\Delta T+1}\right)\hat{\mathbf{C}}_x^{T-1} + \frac{1}{T-\Delta T+1}\mathbf{x}_{T-\Delta T+1}\mathbf{x}_{T-\Delta T+1}^H, \tag{5.63b}$$

where T in $\hat{\mathbf{C}}_x^{T}$ denotes that the covariance function is estimated up to time instant, T. The power method for estimation of P eigenvectors corresponding to p largest eigenvalues (it is assumed that p is known or it can be estimated, see page 367) is outlined here.

Let $\hat{\mathbf{e}}_i^{T-1}$, $i = 1,2, \dots, p$ be the eigenvectors based on the data till time $T-1$, that is, the eigenvectors of the covariance matrix, $\hat{\mathbf{C}}_x^{T-1}$. When a new sample arrives the eigenvectors are updated using the following recursive equation followed by Gram-Schmidt orthonormalization (GSO),

$$\mathbf{g}_i^{T} = \hat{\mathbf{C}}_X^{T}\,\hat{\mathbf{e}}_i^{T-1},$$

$$\hat{\mathbf{e}}_i^{T} = \mathrm{GSO}\,(\mathbf{g}_i^{T}), \quad i = 1, 2, \ldots, p. \tag{5.64}$$

The choice of initial eigenvectors to start the recursion is important for rapid convergence. A suggested choice in Reference 54 is to orthonormalize P initial data vectors and use them as the initial eigenvectors. For this, the minimum length of the time series required is DT+p. The starting value of the correlation matrix is chosen as

$$\hat{\mathbf{C}}_x^{\Delta T+p} = (\mathbf{I} + \mathbf{x}_0\mathbf{x}_0^H).$$

The Gram-Schmidt orthonormalization process involves finding p orthonormal vectors which are linearly equivalent to data vectors.[56] Let $\mathbf{x}_1, \mathbf{x}_2, \ldots, \mathbf{x}_p$ be a set of data vectors and $\mathbf{I}_1, \mathbf{I}_2, \ldots, \mathbf{I}_p$ be a set of linearly equivalent orthonormal vectors which satisfy the following conditions:

$$\mathbf{I}_k \perp \mathbf{X}_r \text{ for } r < k$$

$$\left|\mathbf{I}_k\right|^2 = 1 \text{ for all } k, \tag{5.65}$$

The Gram-Schmidt method determines the orthonormal vectors by solving a system of linear equations,

$$\begin{aligned}
\mathbf{I}_1 &= \gamma_1^1 \mathbf{X}_1\\
\mathbf{I}_2 &= \gamma_1^2 \mathbf{X}_1 + \gamma_2^2 \mathbf{X}_2\\
&\vdots\\
\mathbf{I}_p &= \gamma_1^p \mathbf{X}_1 + \gamma_2^p \mathbf{X}_2 + \ldots + \gamma_p^p \mathbf{X}_p,
\end{aligned} \tag{5.66a}$$

which may be expressed in a matrix form

$$\mathbf{I} = \Gamma\mathbf{X}. \tag{5.66b}$$

The entries in **G** matrix are determined by requiring $\mathbf{I}_1, \mathbf{I}_2, \ldots, \mathbf{I}_p$ to satisfy (5.65).

Estimation of noise subspace of dimension $\Delta T - p$ follows the same procedure except that now an inverse of the correlation matrix is used. Fortunately, the inverse correlation matrix can also be updated when a new sample becomes available.

$$\hat{\mathbf{C}}_x^{-1\,T} = \frac{T}{T-1}\left\{\hat{\mathbf{C}}_x^{-1\,T-1} - \frac{\hat{\mathbf{C}}_x^{-1\,T-1}\,\mathbf{x}_{T-\Delta T+1}\mathbf{x}^H_{T-\Delta T+1}\,\hat{\mathbf{C}}_x^{-1\,T-1}}{T-1+\mathbf{x}^H_{T-\Delta T+1}\,\hat{\mathbf{C}}_x^{-1\,T-1}\,\mathbf{x}_{T-\Delta T+1}}\right\}$$

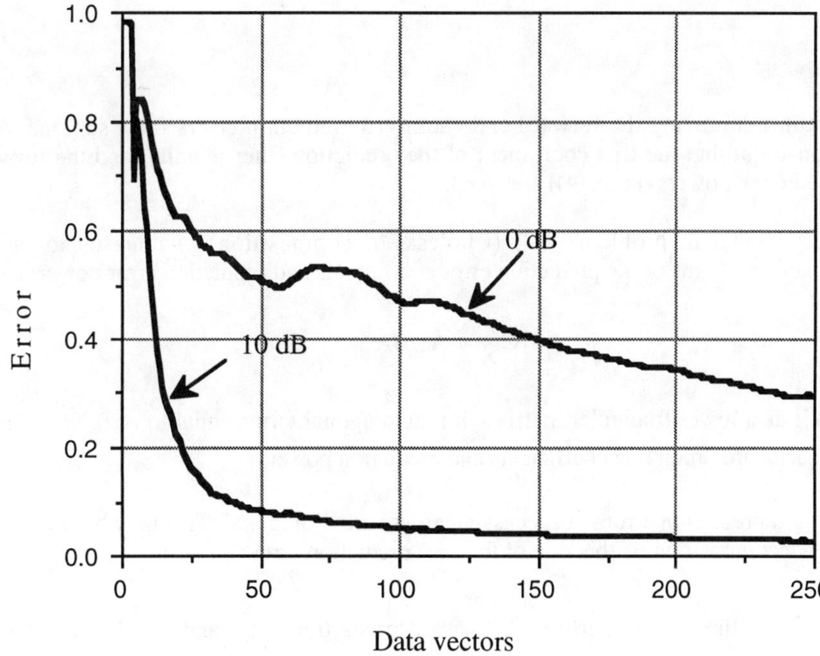

Figure 5.18: Convergence property of the power method for adaptive estimation of noise subspace. A model of two random sinusoids in the presence of white noise was considered here.

The convergence property of the power method is claimed to be superior to that of the gradient method.[54]

Example 5.12: In this example we demonstrate the convergence rate for two different noise levels. Two equal amplitude real random sinusoids, $f_1 = 0.2$ and $f_2 = 0.24$ Hz, in the presence of white noise were considered. A set of 250 data vectors, each of dimension seven, were generated. The noise subspace is now spanned by three eigenvectors. The power method was used to recursively estimate the noise subspace. A measure of the quality of the subspace used is

$$\text{Error} = \left\| (\mathbf{I} - \sum_{i=5}^{7} \mathbf{e}_i \mathbf{e}_i^H) \hat{\mathbf{e}}_n^T \right\|,$$

where

$$\hat{\mathbf{e}}_n^T = \{\hat{\mathbf{e}}_5^T, \hat{\mathbf{e}}_6^T, \hat{\mathbf{e}}_7^T\},$$

and $\mathbf{e}_i$, i=5, 6, 7 are the eigenvectors spanning the noise subspace of the error-free covariance matrix presumably obtained from infinite data. We have plotted the above error measure as a function of the number of data vectors used in the estimation of the correlation matrix. The result is shown in Figure 5.18 (End of Example).

5.6 Exercises

1. Show that in minimizing the forward and backward prediction errors for a stationary process under the constraint that the first coefficient of the prediction filter is unity, and the forward and backward prediction powers (in (5.14)) are equal.

2. The result shown in problem (4.7) (Cholesky decomposition) enables us to obtain the covariance matrix in terms of the prediction error coefficients and prediction error power; namely,

$$\mathbf{C}_X = \mathbf{A}_m \sigma \mathbf{A}_m^H .$$

Note that $\mathbf{A}_m$ is a lower triangular matrix with the diagonal terms equal to one. Now derive this result using backward prediction coefficients and prediction power.

3. The backward prediction errors of successive orders are orthogonal. This may be shown[11] using the Cholesky decomposition. Is this true of forward prediction errors?

4. Let $A_m(z)$ be the z-transform of the forward prediction filter and $A_m^R(z)$ be that of the backward prediction filter. Show that $A_m^R(z) = z^{-m} A_m(z^{-1})$ and the backward prediction filter is a maximum phase filter.

5. Derive the recursive (order) relation for forward and backward prediction filter transfer functions using (5.30). [Hint: Divide both sides of the equation by X(z).] The result should be consistent with (5.34).

6. Show that each element of matrix **S** in (5.48) is an inner product of linearly independent vectors. Further, show that **S** becomes diagonal whenever the frequencies satisfy the following relation:

$$\omega_k - \omega_l = \frac{2\pi}{T - \Delta T} r, \qquad 0 \le k, l \le p - 1,$$

where r is an integer. For the above special case the correlation matrix becomes Toeplitz.

7. Analogous to the covariance function of the forward prediction errors (5.19) derive the covariance function of the backward prediction errors and the cross-covariance function between the forward and backward prediction errors. In particular verify the following results:

$$C_{f_t^m f_{t+\tau}^m} = \sum_{k=0}^{m} \sum_{l=0}^{m} a_k^m a_l^m C_x(\tau + k - l),$$

$$C_{b_t^m b_{t+\tau}^m} = \sum_{k=0}^{m} \sum_{l=0}^{m} a_k^m a_l^m C_x(-\tau + k - l),$$

and

$$C_{f_t^m b_{t+\tau}^m} = \sum_{k=0}^{m} \sum_{l=0}^{m} a_k^m a_l^m C_x(-\tau + m + k - l).$$

8. Using the above defined quantities, show that the recursive relation (5.23) reduces to[57]

$$C_{f^{m+1}}(i-j) = C_{f^m}(i-j) + \alpha_{m+1}^2 C_{f^m}(i-j) + \alpha_{m+1}\{C_{f^m b^m}(i-j) + C_{f^m b^m}(j-i)\},$$

$$C_{b^{m+1}}(i-j) = C_{b^m}(i-j) + \alpha_{m+1}^2 C_{f^m}(i-j) + \alpha_{m+1}\{C_{f^m b^m}(i-j) + C_{f^m b^m}(j-i)\},$$

and

$$C_{f^{m+1} b^{m+1}}(i-j) = C_{f^m b^m}(i-j-1) + \alpha_{m+1}^2 C_{f^m b^m}(j-i+1)$$
$$+ \alpha_{m+1}\{C_{f^m}(i-j-1) + C_{b^m}(i-j-1)\}.$$

9. Show that the correlation matrix $\mathbf{R}_m$ in (5.42) or $\bar{\mathbf{R}}_m$ in (5.44) is positive semidefinite for all m= 1, 2, 3, 4.

10. A running Fourier transform is defined as in Reference 58:

$$X(t,\omega) = \sum_{\tau=0}^{T-1} x(t-\tau)\exp(-j\omega\tau),$$

where it is assumed that the data is indexed backwards, that is, x(t), x(t − 1), x(t − 2), ..., x(t − T+1). As the new sample becomes available, the running Fourier transform is given by

$$x(t+1,\omega) = \sum_{\tau=0}^{T-1} x(t+1-\tau)\exp(-j\omega\tau).$$

Show the following recursive relation:

$$x(t+1,\omega) = e^{-j\omega} x(t,\omega) + x(t+1) - e^{-j\omega T} x(t-T+1).$$

References

1. M. Arato, On the sufficient statistics for stationary Gaussian random processes, *Theor. Prob. Appl.*, 6, 199-201, 1961.

2. S. P. Bruzzone and M. Kaveh, A criterion for selecting information preserving data reduction for use in the design of multiple parameter estimators, *IEEE Trans.*, IT-29, 466-470, 1983.

3. S. P. Bruzzone and M. Kaveh, Information tradeoffs in using the sample autocorrelation function in ARMA parameter estimation, *IEEE Trans.*, ASSP-32, 701-715, 1984.

4. I. Bilinskis and A. Mikelsons, *Randomized Signal Processing,* Prentice-Hall, Hertfordshire (UK), 1992.

5. A. D. Chave, D. J. Thomson, and M. E. Ander, On the robust estimation of power spectra, coherences and transfer functions, *J. Geophys. Res.*, 92, 633-647, 1987.

6. B. Kleiner, R. D. Martin, and D. J. Thomson, Robust estimation of power spectra, *J. R. Stat. Soc, Ser. B*, 41, 313-351, 1979.

7. D. J. Thomson, Spectrum estimation techniques for characterization and development of WT4 waveguide-I, *Bell Syst. Tech. J.*, 56, 1769-1851, 1977.

8. P. J. Huber, Robust estimation of a location parameter, *Ann. Math. Stat.*, 35, 73-101, 1964.

9. A. H. Nuttal, Spectral analysis of univariate process with bad points via maximum entropy and linear prediction techniques, NUSC Tech. Rep. 5303, Naval Underwater Systems Center, New London, CT 06320, 1976.

10. R. Radzyner and P. T. Bason, An error bound for lagrange interpolation of a low pass function, *IEEE Trans.*, IT-18, 669-671, 1972.

11. A. Papoulis, A new algorithm in spectral analysis and band limited extrapolation, *IEEE Trans.*, CAS-22, 735-742, 1975.

12. P. S. Naidu and B. Paramasivaiah, Estimation of sinusoids from incomplete time series. *IEEE Trans.*, ASSP-32, 559-562, 1984.

13. N. R. Lomb, Astrophysics and Space Science, 39, 447-451, 1976.

14. W. H. Press and G. B. Rybicki, Fast algorithm for spectral analysis of unevenly sampled data, *Astrophys. J.*,.338, 277-280, 1989.

15. B. Efron, Bootstrap method: Another look at the jackknife. *Ann. of Stat.*, 7, 1-26, 1979.

16. R. Y. Liu and K. Singh, Moving blocks jackknife and bootstrap capture of weak dependance, in *Exploring the Limits of Bootstrap*, Raoul LePage and Lynne Billard, (Eds), John Wiley & Sons, Inc., New York, 225-248, 1992.

17. D. N. Politis, J. P. Romano, and Tze-Leung Lai, Bootrstrap confidence bands for spectra and cross-spectra. *IEEE Trans.*, SP-40, 1206-1215, 1992.

18. I. S. Konvalinka, Iterative nonparametric spectrum estimation, *IEEE Trans.*, ASSP-32, 59-69, 1984.

19. J. P. Burg, A new analysis technique for time series data, NATO Advanced Study Institute on Signal Processing, Enschede, Netherlands,.1, 1968.

20. S. Haykin, *Adaptive Filter Theory,* Prentice-Hall, Englewood Cliffs, NJ., 1986.

21. K. Kopp, *Theory of Functions,* Part II, Dover Publications, New York, 1947.

22. J. P. Proakis, C. M. Rader, F. Ling, and C. L. Nikias, *Advanced Digital Signal Processing*, Maxwell Macmillan International Edition, 1992.

23. D. N. Swingler, A modified Burg algorithm for maximum entropy spectral analysis, *Proc. IEEE*, 67, 1368-1369, 1979.

24. D. N. Swingler, A comparison between Burg's maximum entropy method and a nonrecursive technique for the spectral analysis of deterministic signals, *J. Geophys. Res.*, 84, 679-685, 1979.

25. P. F. Fougere, E. J. Zawalick, and H.R. Radoski, Spontaneous line splitting in maximum entropy power spectrum analysis, *Phys. Earth Planet. Inter.*, 12, 201-207, 1976.

26. W. Y. Chen and G. R. Stegen, Experiments with maximum entropy power spectra of sinusoids, *J. Geophs. Res.*, 79, 3019-3022, 1974.

27. G. Schwartz, Estimation of the dimension of a model, *Anns. Stat.*, 6, 461-464, 1978.

28. D. N. Swingler, Frequency errors in MEM processing, IEEE Trans., ASSP-28, 257-259, 1980.

29. S. Pignario and F. G. Canavero, Amplitude errors in the Burg spectrum estimation of sinusoidal signal, *Signal Process.*, 22, 107-112, 1991.

30. P. F. Fougere, A solution to the problem of spontaneous line splitting in maximum entropy power spectrum analysis, *J. Geophys. Res.*, 82, 1051-1054, 1991.

31. P. M. T. Broersen, The prediction error of autoregressive small sample models, *IEEE Trans.*, ASSP-38, 858-860, 1990.

32. G. E. P. Box and G. M. Jenkins, *Time series Analysis: Forecasting and Control*, Holden-Day, San Francisco, 1970.

33. J. R. Dickie and A. K. Nandi, A comparative study of AR order selection methods, *Signal Process.*, 40, 239-255, 1994.

34. H. Akaike, Statistical prediction identification, *Ann. Inst. Stat. Math.*, 22, 203-217, 1970.

35. H. Akaike, A new look at the statistical model identification, IEEE Trans., AC-19, 716-723, 1974.

36. J. Rissanen, Modeling by shortest data description, *Automatica*, 14, 465-471, 1978.

37. E. Parzen, Some recent advances in time series modeling, *IEEE Trans.*, AC-19, 723-730, 1974.

38. S. M. Kay, Modern Spectrum Estimation, Prentice Hall, Englewood Cliffs, 235, 1988.

39. M. Wax and T. Kailath, Detection of signals by information theoretic criterion, *IEEE Trans.*, ASSP-33, 387-392, 1985.

40. B. Friedlander, Lattice methods for spectral Estimation, *Proc. IEEE*, 70, 990-1017, 1982.

41. T. J. Ulrych and R. W. Clayton, Time series modelling and maximum entropy, *Phys. Earth and Planet. Inter.*, 12, 188-200, 1976.

42. P. F. Fougere, A Review of the problem of spontaneous line splitting in maximum entropy power spectral analysis, in Maximum Entropy and Bayesian Mthods in *Inverse Problems*, 303-315, D. Riedel Publishing Co., Boston, 1985.

43. B. Friedlander, M. Morf, T. Kailath, and L. Ljung, New inversion formulas for matrices classified in terms of their distance from Toeplitz matrices, *Linear Algebra Appl.*, 27, 31-60, 1979.

44. T. Thorvaldsen, A comparison of the least squares method and the Burg method for autoregressive spectral analysis, *IEEE Trans*. AP-29, 675-679, 1981.

45. S. W. Lang and J. H. McClellan, Frequency estimation with maximum entropy spectral estimators, *IEEE Trans*., ASSP-28, 716-724, 1980.

46. L. Marple, A new autoregressive spectrum analysis algorithm, *IEEE Trans*., ASSP-28, 441-454,1980.

47. D. W. Tufts, and R. Kumaresan, Estimation of frequencies of multiple sinusoids: Making linear prediction perform like maximum likelihood, *Proc. IEEE*,.70, 975-989, 1982.

48. S. Y. Fung, K. S. Arun, and D. V. Bhaskar Rao, State space singular value decomposition based method for the harmonic retrieval problem, *J Opt. Soc of Am*., 73, 1799-1811, 1983.

49. B. D. Bhaskar Rao, and K. S. Arun, Model based processing of signals: A state space approach, *Proc. IEEE*, 80, 283-309, 1992.

50. P. Thompson, An adaptive spectral analysis technique for unbiased frequency estimation in the presence of white noise, Proc. Asilomar Conf., 529-533, 1979.

51. U. V. Reddy, B. Egardt, and T. Kailath, Least squares type algorithm for adaptive implementation of Pisarenko's harmonic retrieval method, *IEEE Trans*., ASSP-30, 339-405, 1982.

52. H. Chen, T. K. Sarkar, J. Brule, and S. A. Dianat, Adaptive spectral estimation by the conjugate gradient method, *IEEE. Trans*., ASSP-34, 272-284, 1986.

53. J. P. Yang and M. Kaveh, Adaptive eigensubspace algorithm for direction or frequency estimation and tracking, *IEEE Trans*., ASSP-36, 241-251, 1988.

54. J. Karhunen, Adaptive algorithms for estimating eigenvectors of correlation type matrices, I*EEE Proc. ICASSP-84*, 14.6.1-4, 1984.

55. G. Stewart, *Introduction to matrix computations*, Academic Press, New York, 1973.

56. A. Papoulis, *Probability, Random variables and Stochastic Processes*, Third Edition, McGraw-Hill Inc., New York, 1991.

57. A. Cumani, On a covariance-lattice algorithm for linear prediction, *IEEE Proc. ICASSP*, 851-854, 1982.

58. A. Papoulis, *Signal Analysis*, McGraw Hill Inc., New York, 1977.

Index

D

N

O/P/Q

R

U-Z